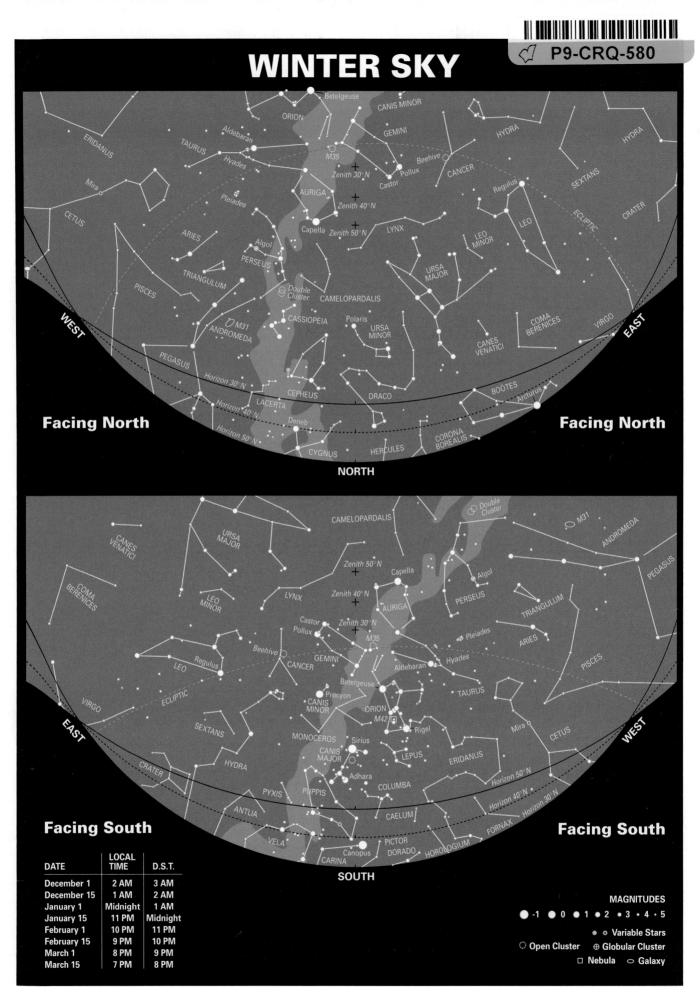

WINTER SKY

P9-CRQ-580

Facing North

Facing North

NORTH

Facing South

Facing South

SOUTH

DATE	LOCAL TIME	D.S.T.
December 1	2 AM	3 AM
December 15	1 AM	2 AM
January 1	Midnight	1 AM
January 15	11 PM	Midnight
February 1	10 PM	11 PM
February 15	9 PM	10 PM
March 1	8 PM	9 PM
March 15	7 PM	8 PM

MAGNITUDES

● -1 ● 0 ● 1 ● 2 • 3 • 4 · 5

◌ ○ Variable Stars

◌ Open Cluster ⊕ Globular Cluster

□ Nebula ○ Galaxy

MAP BY WIL TIRION; FOR JAY M. PASACHOFF

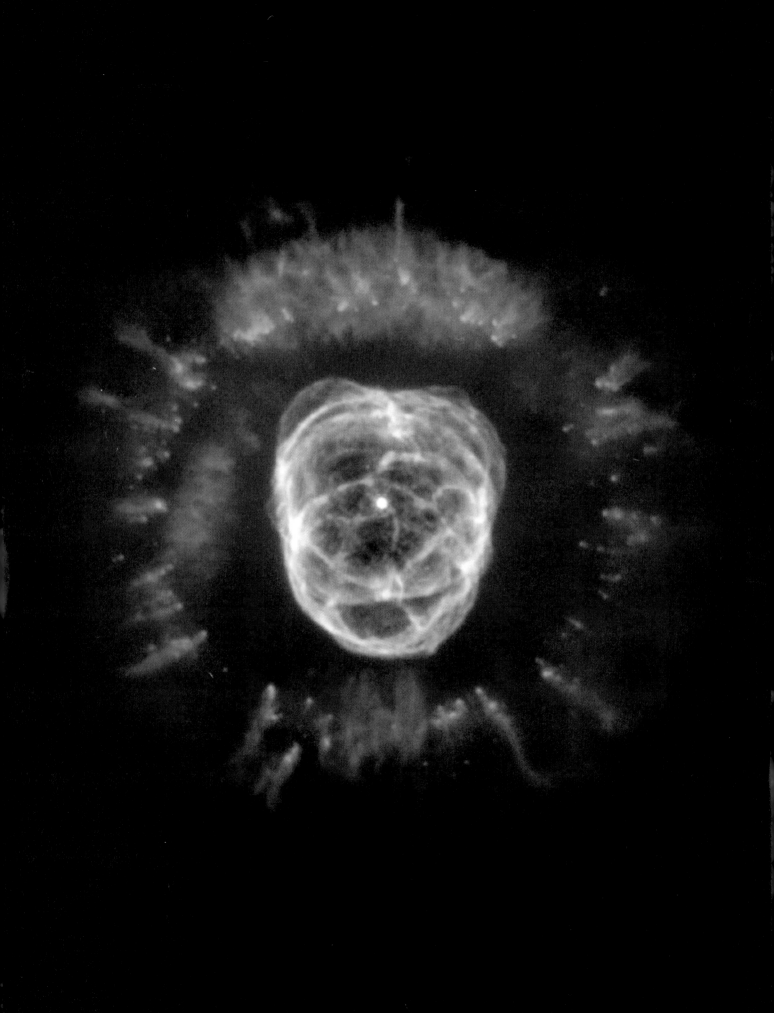

THE Cosmos:

ASTRONOMY IN THE NEW MILLENNIUM

Jay M. Pasachoff
Williams College

Alex Filippenko
*University of California
at Berkeley*

Harcourt College Publishers

Fort Worth   Philadelphia   San Diego   New York   Orlando   Austin   San Antonio
Toronto   Montreal   London   Sydney   Tokyo

Publisher: Emily Barrosse
Acquisitions Editor: Kelley Tyner
Marketing Strategist: Kathleen Sharp
Developmental Editor: Jennifer Pine
Project Editor: Robin C. Bonner
Production Manager: Alicia Jackson
Art Director: Caroline McGowan

Frontispiece: The Eskinco Nebula (NGC-2392), a dying star known as a planetary nebula, the first object imaged with the Hubble Space Telescope after its 1999 repair. [*NASA, A. Fruchter and the ERO Team (STScI, ST-EFC)*]

The Cosmos: Astronomy in the New Millennium, First Edition
ISBN: 0-03-005218-1
Library of Congress Catalog Card Number: 99-69166

Front Cover Credit: Through this portal on "spaceship Earth" we see a view of gravitational lensing, one of the first photos taken after the 1999 repair of the Hubble Space Telescope. The gravity from a cluster of galaxies bends light from even more distant galaxies into circular arcs. [*NASA, A. Fruchter and the ERO team (STScI, ST-EFC)*]

Back Cover Credit: Bright nebulae and hot stars in the constellation Chamaeleon, a false-color view made from yellow, red, and infrared images taken with one of the 8-m-diameter component telescopes of the European Southern Observatory's very large telescope in Chile. (*European Southern Observatory*)

Address for domestic orders:
Harcourt College Publishers, 6277 Sea Harbor Drive, Orlando, FL 32887-6777
1-800-782-4479
e-mail collegesales@harcourt.com

Address for international orders:
International Customer Service, Harcourt, Inc.
6277 Sea Harbor Drive, Orlando FL 32887-6777
(407) 345-3800
Fax (407) 345-4060
e-mail hbintl@harcourt.com

Address for editorial correspondence:
Harcourt College Publishers,
Public Ledger Building, Suite 1250, 150 S. Independence Mall West
Philadelphia, PA 19106-3412

Web Site Address
http://**www.harcourtcollege.com**

Printed in the United States of America
0123456789   032   10 987654321

Contents Overview

Mars, centered on the Syrtis Major dark region, imaged with the Hubble Space Telescope.

◀ The spiral galaxy NGC4414, imaged with the Hubble Space Telescope.

Contents

A group of galaxies, Hickson Compact Group 87, linked by gravity to each other. Foreground stars in our own galaxy are also visible in this Hubble Space Telescope view.

The domes of the twin Keck Telescopes in Hawaii.

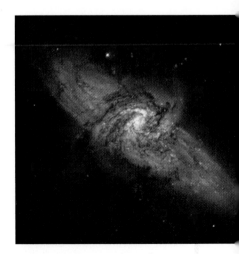

One spiral galaxy silhouetted in front of another, a rare circumstance, imaged with the Hubble Space Telescope.

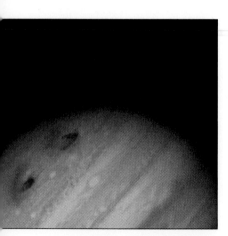

Jupiter, imaged with the Hubble
Space Telescope, showing the results of collisions of two fragments
of Comet Shoemaker-Levy 9.

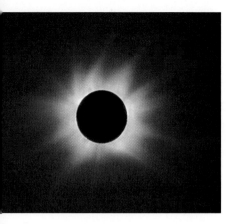

The solar corona at the total solar eclipse of 1999.

The central region of the globular star cluster M4, imaged with one of the Unit Telescopes of the European Southern Observatory's Very Large Telescope in Chile.

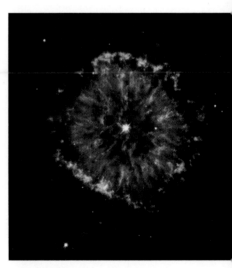

The planetary nebula NGC 6751 imaged with the Hubble Space Telescope. We see gas thrown out by a dying star that once resembled our Sun.

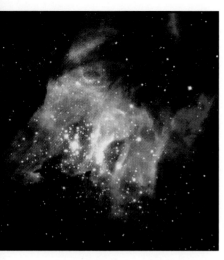

An infrared composite showing a region of our Milky Way Galaxy, RCW38, in which stars are forming out of gas and dust. It was imaged with the Very Large Telescope.

A Hubble Space Telescope image that shows an elliptical galaxy acting as a gravitational lens, providing four images of a farther object. Such lenses image distant galaxies and quasars.

A Hubble Space Telescope view of some of the most distant objects ever studied in a combination of visible and infrared images. It is a small patch of sky in the constellation Tucana, part of the Hubble Deep Field South.

Astronomy continues to flourish. The pair of Keck 10-meter-diameter telescopes in Hawaii and the quartet of 8-meter-diameter telescopes in Chile called the Very Large Telescope are among the new instruments used to make important discoveries from mountaintop observatories. The Hubble Space Telescope sends down exciting data all the time. The latest major space observatory, the Chandra X-ray Observatory, sends Hubble-quality images made with x-rays. Further, new electronic instruments and computer capabilities; new space missions to the comets, asteroids, and the outer planets; and advances in computational astronomy and in theory should continue to bring forth exciting results.

In *The Cosmos: Astronomy in the New Millennium*, we describe the current state of astronomy, both the fundamentals of astronomical knowledge that have been built up over decades and the incredible advances that are now taking place. We want simply to share with you the excitement and magnificence of the Universe. We try to cover all the branches of astronomy without slighting any of them; each teacher and each student may well find special interests that may be different from our own. One of our aims in writing this book is to educate voters and prospective voters in the hope that they will understand the value and methods of scientific research in general and astronomical research in particular.

The Trapezium, the star cluster in the middle of the Orion Nebula, imaged with the Hubble Space Telescope.

In writing this book we share the goals of a commission of the Association of American Colleges, whose report on the college curriculum stated, "A person who understands what science is recognizes that scientific concepts are created by acts of human intelligence and imagination; comprehends the distinction between observation and inference and between the occasional role of accidental discovery in scientific investigation and the deliberate strategy of forming and testing hypotheses; understands how theories are formed, tested, validated, and accorded provisional acceptance; and discriminates between conclusions that rest on unverified assertion and those that are developed from the application of scientific reasoning." The scientific method permeates the book.

What is science? The following statement was originally drafted by the Panel on Public Affairs of the American Physical Society, in an attempt to meet the perceived need for a very short statement that would differentiate science from pseudo-science. This statement has been endorsed as a proposal to other scientific societies by the Council of the American Physical Society, and was endorsed by the Executive Board of the American Association of Physics Teachers in 1999.

Science is the systematic enterprise of gathering knowledge about the world and organizing and condensing that knowledge into testable laws and theories.

The success and credibility of science is anchored in the willingness of scientists to:

1. *Expose their ideas and results to independent testing and replication by other scientists; this requires the complete and open exchange of data, procedures, and materials;*

2. *Abandon or modify accepted conclusions when confronted with more complete or reliable experimental evidence.*

Adherence to these principles provides a mechanism for self-correction that is the foundation of the credibility of science.

An artist's conception of the Next Generation Space Telescope.

Our book, through the methods it describes, should reveal this systematic enterprise of science to the readers.

This book was conceived and written specifically as a *short* book for the one-semester or one-quarter introductory astronomy course. It is not a condensation of Pasachoff's *Astronomy: From the Earth to the Universe*, though obviously some of the narrative and visual content do derive from that book. Some instructors professed a desire to teach from a shorter text than the usual "standard" introductory book, and now *The Cosmos* is the result of those requests. Because one cannot adequately cover the whole Universe in a few months, we have therefore had to pick and choose topics within the various branches of astronomy, while trying to describe a wide range, to convey the spirit of contemporary astronomy and of the scientists working in it. Our mix includes much basic astronomy and many of the exciting topics now at the forefront.

ORGANIZATION

The Cosmos: Astronomy in the New Millennium is organized in an Earth-outward approach. Chapter 1 gives an overview of the Universe. Chapters 2, 3, and 4 present, respectively, fundamental astronomical concepts about light, matter, and energy; the various types of telescopes used to explore the electromagnetic spectrum; and a discussion of easily observed astronomical phenomena and the celestial sphere.

Chapter 5 examines the early history of the study of astronomy and discusses theories of Solar System formation. Chapters 6 through 8 cover the Solar System and its occupants, although an in-depth discussion of the Sun is reserved for later, in Chapter 9. Chapter 6 compares Earth to the Moon and Earth's nearest planetary neighbors, Venus and Mars. Based largely on the Voyager data, Chapter 7 compares and contrasts the jovian gas/liquid giants—Jupiter, Saturn, Uranus, and Neptune. Chapter 8 looks at the outermost part of our Solar System, including Pluto and its moon, Charon. Chapter 8 also spotlights comets and meteoroids, the Solar System's vagabonds.

Chapter 9 discusses the Sun, our nearest star. Moving outward, Chapters 10 through 14 discuss all aspects of stars. Chapter 10 begins by presenting observational traits of stars—their colors and types—and goes on to show how we measure their distances, brightnesses, and motions. Chapter 11 discusses binary stars, variable stars, and star clusters, in the process showing how we derive stellar masses and ages. Chapter 12 answers the question of how stars shine and reveals that all stars have life cycles. Chapter 13 tells what happens when stars die and describes some of the peculiar objects that violent stellar death can create, including neutron stars and pulsars. Black holes, the most bizarre objects to result from star death, are the focus of Chapter 14.

Mars, from Mars Pathfinder on its surface.

As we explore further, Chapter 15 describes the parts of the Milky Way and our place in it. Chapter 16 pushes beyond the Milky Way to discuss galaxies in general, the fundamental units of the Universe, and evidence that they consist largely of dark matter. Ways in which we are studying the evolution of galaxies are also described. Chapter 17 looks at quasars, distant and powerful objects that are probably gigantic black holes swallowing gas in the central regions of galaxies. Chapters 18 and 19 consider the ultimate questions of cosmological creation by analyzing recent findings and current theories. Evidence for an accelerating expansion of the Universe, possibilities for the overall geometry and fate of the Universe, ripples in the cosmic background radiation, the origin and phenomenally rapid early growth of the Universe, and the idea of multiple universes are among the fascinating topics explored. Lastly, Chapter 20 discusses the always intriguing search for extraterrestrial intelligence.

FEATURES

The Cosmos: Astronomy in the New Millennium offers instructors a short text with concise coverage over a wide range of astronomical topics. An early discussion of the scientific method stresses its importance in the verification of observations. The text presents up-to-date coverage of many important findings and theories, such as results from the Galileo orbiter around Jupiter and the Near Earth Asteroid Rendezvous (NEAR-Shoemaker) spacecraft in orbit around Eros, as well as a sampling of the many significant findings from the Hubble Space Telescope and the Chandra X-ray Observatory. Of particular interest is the recent advance in our understanding of the age and expansion of the Universe, both through better measurements of the current expansion rate and through discussion of the new conclusion that the Universe's expansion may be speeding up with time.

The Veil Nebula, part of a supernova remnant.

ORIGINS

The study of our origins, whether it be us as humans, our Earth as a planet, our Sun as a star, or our galaxy as a whole, is as interesting to many of us as it is to look at our own baby pictures. Most of us have gazed at the stars and wondered how they came to be and what their relationship to us is. NASA has chosen *Origins* as one of its major themes for the organization of its missions and has several spacecraft planned in the Origins program discussed later in Chapter 1. We emphasize the study of origins in this text, first by singling out specifics in the headings of each chapter and then by dealing with a variety of relevant material in the text itself.

PEDAGOGY

Mixed in the book we have four kinds of features:

1. *People in Astronomy (interviews with contemporary astronomers)* Each of the interviews presents a notable researcher engaged in conversation about a variety of topics: current and future work in astronomy, what led that person to study and pursue astronomy as a career, why learning about astronomy is an important scientific *and* human endeavor. We hope that you enjoy reading their comments as much as we enjoyed speaking with them and learning about their varied interests and backgrounds.

2. *Starparties* An occasional feature that shows students how to find things in the sky. These include observing exercises and links to the star maps that appear on the inside covers of the book.

3. *Figure It Out* In some astronomy courses it may be appropriate to elaborate on equations. Because we wrote *The Cosmos: Astronomy in the New Millennium* to be a descriptive presentation of modern astronomy for liberal arts students, the use of mathematics has been kept to a minimum. However, we recognize that some instructors wish to introduce their students to more of the mathematics associated with astronomical phenomena.

4. *Lives in Science* These boxes provide biographies of important historical figures like Copernicus and Galileo.

5. *A Closer Look* Using these boxes, students can further explore interesting topics, such as size scales in the Universe, observing with large telescopes, various celestial phenomena, mythology, and naming systems.

We have provided aids to make the book easy to read and to study from. New vocabulary is bold-faced in the text, and in the summaries that are given in context at the end of each chapter, and defined in the glossary. The index

The first diagram showing the Sun as the center of what we now call the Solar-System, from Copernicus's 1543 book.

The Hubble Space Telescope.

provides further aid in finding explanations. End-of-chapter questions cover a range of material and include some that are straightforward to answer from the text and others that require more thought. Appendices provide some standard and recent information on planets, stars, constellations, and nonstellar objects. The exceptionally beautiful Sky Maps by Wil Tirion (inside the cover of the book) will help you find your way around the sky when you go outside to observe the stars. Be sure you do.

Recent educational research has shown that students often need to unlearn incorrect ideas in order to understand the correct ones. We have thus placed a list of such "Misconceptions," including common incorrect ideas and the correct alternatives, on the book's Web site. Topics where these misconceptions commonly arise are marked in the book itself with an icon.

As we go through the millennium—January 1, 2001, for sticklers—access to CD-ROMs and to the World Wide Web has become so commonplace that we are glad to integrate their use into our text. Specific exercises for both the Red-Shift College Edition (RSCE) CD-ROM and the Web are also listed in this book's Web site and are called out with icons. Further, this book's Web page, **http://www.harcourtcollege.com/astro/cosmos**, is updated frequently with links to information and photographs from a wide variety of sources and also is denoted with an icon. Each professor and each student should look at these updates and links before each class.

ANCILLARY MATERIALS

Available to all those who adopt this book are the following ancillaries:

Instructor's Manual and Test Bank.   This manual contains course outlines, answers to all the questions in the text, lists of audiovisual aids and organizations, laboratory experiments and exercises arranged for easy duplication, and sample tests with answers (quizzes and exams). The test bank portion contains approximately 1250 questions in multiple-choice and other formats.

Overhead Transparencies.   This set contains 100 color overhead transparencies of many of the text's images.

Slide Collection.   Available as an alternative to overhead transparencies.

Computerized Test Bank.   All the questions from the printed test bank are presented in a computerized format that allows instructors to edit, add questions, and print assorted versions of the same test. Available for Windows™ and Macintosh® users.

Instructor's Photo Resource Catalog.   Many of the text's images are provided in an electronic file format for use in classroom presentations. This catalog is available both on a CD-ROM and as a downloadable Web file.

RedShift College Edition.   This CD-ROM (available shrink-wrapped with the text for students) is a special version of the powerful RedShift software that allows you to duplicate the look of the celestial sphere on your computer, to watch animated essays on various astronomical topics, to examine several astronomical phenomena, or to see the special features provided to accompany this text.

For information about these ancillaries, visit **www.harcourtcollege.com/ astro/cosmos**, contact your local Harcourt sales representative, or call Harcourt College Publishers at (800) 237–2665.

Harcourt College Publishers may provide complimentary instructional aids and supplements or supplement packages to those adopters qualified under our adoption policy. Please contact your sales representative for more information. If as an adopter or potential user you receive supplements you do not need, please return them to your sales representative or send them to

Attn: Returns Department
Troy Warehouse
465 South Lincoln Drive
Troy, MO 63379

ACKNOWLEDGMENTS

The publishers join us in placing a heavy premium on accuracy, and we have made certain that the manuscript and proof have been read not only by students for clarity and style but also by other professional astronomers for scientific comments. As a result, you will find that the statements in this book, brief as they are, are authoritative. We would particularly like to thank John Dykla (Loyola University), Hollis R. Johnson (Indiana University), Steven L. Kipp (Mankato State University), John P. Oliver (University of Florida), James G. Peters (San Francisco State University), Harley A. Thronson (University of Wyoming), and Louis Winkler (Pennsylvania State University), who commented on *Journey Through the Universe*, which served as a base on which this book was rewritten. We also thank John Martin (San Jose City College) and David Tholen and Richard Canfield, then both at the University of Hawaii, and their students, for their post-publication comments.

We thank the following reviewers: Scott Payson (Wayne State University), Dan Wilkins (University of Nebraska, Omaha), Steinn Sigurdsson (Pennsylvania State University), Robert Egler (North Carolina State University), Thomas Hockey (University of Northern Iowa), Phillip Appleton (Iowa State University), and Kelly Beatty (*Sky and Telescope* magazine). We also found useful the extensive comments on a questionnaire filled out by A.F.'s students at UC–Berkeley and past comments from J.M.P.'s students at Williams.

We are grateful to Nancy Kutner for her excellent work on the index.

At Harcourt College Publishers, we thank Kelley Tyner (Acquisitions Editor), Jennifer Pine (Developmental Editor), Robin Bonner (Senior Project Editor), Caroline McGowan (Senior Art Director), Alicia Jackson (Production Manager), Kim Paschen (Editorial Assistant), Emily Barrosse (Publisher), and Kathleen Sharp (Senior Marketing Strategist).

In Williamstown, we thank Barbara Swanson for her general assistance and Stephan Martin for his computer work on illustrations and this book's Web pages. J.M.P. is especially grateful to Susie Kaufman for her expert editorial work throughout this project.

J.M.P. thanks various members of his family, who have provided vital and valuable editorial services, in addition to their general support. His wife, Naomi, has always been very helpful both personally and editorially. Their children, Eloise and Deborah, now out of college, continue to provide comments and analysis and to read proof. A.F. thanks his wife, Diana Lee, and his young children, Zoe and Simon Filippenko, for their incredible patience and support during the many weekends and long nights when this book was being written and revised.

We are extremely grateful to all the individuals named above for their assistance. Of course, it is we who have put this all together, and we alone are responsible for it. We would appreciate hearing from readers with suggestions for improved presentation of topics, with comments about specific points that need

Jupiter, imaged with the Hubble Space Telescope. The arrow marks the spot where the Galileo Mission's probe entered.

clarification, with typographical or other errors, or just to tell us how you like your astronomy course. We invite readers to write to us respectively at Williams College, Hopkins Observatory, Williamstown, Massachusetts 01267–2630, or Astronomy Department, University of California, Berkeley, CA 94720–3411, or through our web site. We promise a personal response to each writer.

Jay M. Pasachoff
Williamstown, Massachusetts
jay.m.pasachoff@williams.edu

Alex Filippenko
Berkeley, California
alex@astro.berkeley.edu

Astronomy has much variety, and you are sure to find some parts of it that interest you more than others. Do try to get an overview from this text and then go on to do additional reading. (We give some suggestions at the end of the book.) Socrates said, "Education is the kindling of a flame, not the filling of a vessel"; it is in this spirit that we wrote this text. We hope you will continue to learn more about our origins and place in the Cosmos.

To get the most out of this text, you must read each chapter more than once. After you read each chapter the first time, you should carefully go through the summary, trying to understand each of the key concepts and the necessary vocabulary. You can look up boldfaced words in the Glossary (or in the dictionary of the RedShift College Edition CD-ROM) and also find the definition that appears with the word the first time it is used in the chapter; the Index will help you find these references. Next, read through the chapter again especially carefully. Studying the illustrations and captions will also provide you with further explanations of much of the material presented. Finally, answer the questions at the end of the chapter. Make sure that the common misconceptions listed on the Web site aren't ones that you share. Also be sure to examine the animations and explanations in RedShift.

We hope these steps help you understand the Universe we live in, the things in it, and how astronomers find out about them.

NGC 1850, a 40-million-year-old star cluster with an adjacent 4-million-year-old smaller star cluster, both in the Large Magellanic Cloud, a neighboring galaxy to our own. The clusters formed from surrounding gas, some of which remains and which is visible in the red light typical of hydrogen; sharp divisions in the gas may have resulted from supernova explosions. The image is from the Very Large Telescope.

Jay M. Pasachoff

Jay M. Pasachoff is Field Memorial Professor of Astronomy at Williams College, where he teaches the astronomy survey course and works with undergraduate students on a variety of astronomical projects. He is also Director of the Hopkins Observatory there and Chair of the Astronomy Department. His research has brought him to use a wide variety of ground-based telescopes and spacecraft, and he has taken advantage of his broad experience in writing his texts. He recently completed an expedition, with a dozen students, to the 1999 total solar eclipse; it was the 29th solar eclipse he has viewed. His research at the eclipse was sponsored by the National Science Foundation, NASA, and the National Geographic Society. Pasachoff is Chair of the Working Group on Eclipses of the International Astronomical Union. He is a collaborator with a group of scientists observing the atmospheres of moons of outer planets, especially Neptune's moon, Triton. He also works in radio astronomy of the interstellar medium. He has recently been collaborating with a professor of art history on images of comets and eclipses in the art of Renaissance Italy and in the art and science of Great Britain.

Pasachoff has been very active in educational and curriculum matters. He is U. S. National Representative to the Teaching of Astronomy of the International Astronomical Union, has twice been Chair of the Astronomy Division of the American Association for the Advancement of Science, and has been on the astronomy committees of the American Astronomical Society, the American Physical Society, and the American Association of Physics Teachers. In addition to his college astronomy texts, Pasachoff has gone on to write the *Field Guide to the Stars and Planets*, and to be either author or coauthor of textbooks in calculus and in physics, as well as several junior-high-school textbooks. Pasachoff's continued dedication to teaching astronomy through interesting and accurate textbooks is a hallmark of his work.

Pasachoff received his undergraduate and graduate degrees from Harvard and was at the California Institute of Technology before going to Williams. His sabbaticals and other leaves have been taken at the University of Hawaii's Institute for Astronomy, the Institut d'Astrophysique in Paris, the Institute for Advanced Study in Princeton, and the Harvard-Smithsonian Center for Astrophysics.

Alex Filippenko is a Professor of Astronomy at the University of California at Berkeley. After receiving his bachelor's degree in Physics from the University of California at Santa Barbara (1979) and his doctorate in Astronomy from the California Institute of Technology (1984), he went to the University of California of Berkeley as a postdoctoral fellow, and in 1986 he joined the Berkeley faculty. He is an active member of the International Astronomical Union, has served as Councilor of the American Astronomical Society, and is currently Vice President (and President-elect) of the Astronomical Society of the Pacific.

Filippenko's primary areas of research are supernovae (exploding stars), active galaxies, black holes, and cosmology; he has also spearheaded efforts to develop completely robotic telescopes. His research accomplishments have been recognized with several major awards, including the Newton Lacy Pierce Prize of the American Astronomical Society (1992) and the Robert M. Petrie Prize of the Canadian Astronomical Society (1997). In 1999 he was elected to the California Academy of Sciences, and in 2000 he received a Guggenheim Fellowship. He and his collaborators recognized a new class of exploding star, obtained some of the best evidence for the existence of black holes in our Milky Way Galaxy, and found that other galaxies commonly show vigorous activity in their centers that suggests the presence of supermassive black holes. He also made major contributions to the discovery that the expansion rate of the Universe might be speeding up with time—the "Science Breakthrough of 1998," according to the editors of *Science* magazine.

Filippenko has always been strongly motivated to communicate science to the general public in a clear, enthusiastic, and engaging manner. In 1991 he won the two most coveted teaching awards at Berkeley, each of which is generally given only once per career, and in 1995 he was voted "Best Professor" on campus in an informal student poll. He frequently gives public lectures on astronomy, and he has played a prominent role in science newscasts and television documentaries such as "Mysteries of Deep Space" and "Stephen Hawking's Universe." In 1998, working with The Teaching Company, he produced a 40-lecture video course on basic college-level astronomy ("Understanding the Universe: An Introduction to Astronomy").

Alex Filippenko

A Grand Tour of the Heavens

ORIGINS *We comment on "Origins" and "Structure and Evolution" as organizing themes.*

AIMS

To survey the Universe and the methods astronomers use to study it.
•
To see the measuring units used by astronomers.
•
To assess the scientific method and show how pseudoscience fails scientific tests.

Astronomers have deduced that the Universe began about 14 billion years ago. Let us consider that the time between the origin of the Universe and the year 2000 is compressed into one day. If the Universe began at midnight, then it wasn't until slightly after 4 p.m. that the Earth formed; the first fossils date from 6 p.m. The first humans appeared only 2 seconds ago, and it is only 1/300 second since Columbus landed in America. Still, the Sun should shine for another 9 hours; an astronomical time scale is much greater than the time scale of our daily lives (Fig. 1–1; see next page).

One fundamental fact allows astronomers to observe what happened in the Universe long ago: Light travels at a finite speed. As a result, signals don't reach us immediately. Light from the Moon takes about a second to reach us, so we see the Moon as it was a second ago. Light from the Sun takes about eight minutes to reach us. Light from the nearest of the other stars takes over four years to reach us. Once we look beyond the nearest stars, we are seeing much farther back in time.

The Universe is so vast that when we receive light or radio waves from objects across our galaxy, the Milky Way Galaxy, we are seeing back tens of thousands of years. Even for the nearby galaxies, light has taken hundreds of thousands or millions of years to reach us. And for the farthest galaxies and the quasars, the light has been travelling to us for billions of years. New telescopes on high mountains and in orbit around the Earth enable us to study these distant objects much better than we could previously. When we observe these farthest objects, we are seeing them as they were billions of years ago. How have they changed in the billions of years since? Are they still there? What have they evolved into? The observations of distant objects that we make in the present show us how the Universe was long, long ago.

Building on such observations, astronomers use a wide range of technology to gather information and construct theories to find out about the Universe, what is in it, and what its future will be. This book surveys what we have found, how we look, and how we interpret and evaluate the results. The light we see from the past in our present allows us to explore and understand the history of the Universe.

A galaxy, a collection of billions of stars along with gas and dust, viewed edge-on.

◄ The Hubble Space Telescope returns to its observing orbit after it was serviced by Space-Shuttle astronauts in December 1999.

Figure 1-1 A sense of time.

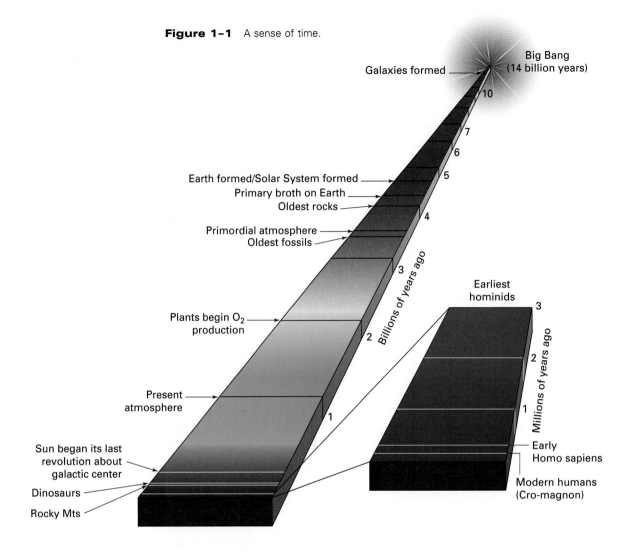

⬤ OBSERVING THE UNIVERSE

The Universe is a place of great variety—after all, it has everything in it! Astronomers study some things of a size and scale that humans might be able to comprehend: the planets, for instance. Most astronomical objects, however, are so large and so far away that we have trouble grasping their sizes and distances. Some of these distant objects are fascinating and bizarre—ultra-dense pulsars that spin on their axes hundreds of times per second, exploding stars that light up the sky and incinerate any planets around them, giant black holes with a billion times the Sun's mass.

In addition to taking photographs of celestial objects, astronomers break down an object's light into its component colors to make a **spectrum** (see Chapter 2), like a rainbow (Fig. 1–2). Today's astronomers study not only the visible part of the spectrum, but also its gamma rays, x-rays, ultraviolet, infrared, and radio waves. We use telescopes on the ground and in space to observe astro-

Figure 1-2 The visible spectrum, light from the Sun spread out in a band. The dark lines represent missing colors, which tell us about specific elements in space absorbing those colors.

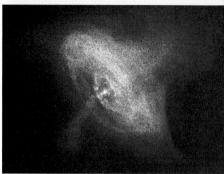

Figure It Out

Scientific Notation

In astronomy, we often find ourselves writing numbers that have strings of zeros attached, so we use what is called either scientific notation or exponential notation to simplify our writing chores. Scientific notation helps prevent making mistakes when copying long strings of numbers, and so aids astronomers (and students) in making calculations.

In scientific notation, which we use in *A Closer Look: A Sense of Scale*, included in this chapter, we merely count the number of zeros, and write the result as a superscript to the number 10. Thus the number 100,000,000, a 1 followed by 8 zeros, is written 10^8. The superscript is called the exponent. (In spreadsheets, like Microsoft Excel, the exponent is written after a caret, ^, as in 10^8.) We also say that "ten is raised to the eighth power."

When a number is not an integer power of 10, we divide it into two parts: a number between 1 and 10, and an integer power of 10. Thus the number 3645 is written as 3.645×10^3. The exponent shows how many places the decimal point was moved to the left.

We can represent numbers between zero and one by using negative exponents. A minus sign in the exponent of a number means that the number is actually one divided by what the quantity would be if the exponent were positive. Thus $10^{-2} = 1/10^2 = 1/100 = 0.01$. Instead of working with 0.00256, for a further example, we would move the decimal point three places to the right and write 2.56×10^{-3}.

Figure 1–3 Objects often look very different in different parts of the spectrum. At top, we see how the center of the Crab Nebula, the remnant of a star that exploded almost 1000 years ago, looks in visible light. This highly detailed image was taken with the new Very Large Telescope in Chile. Below it, we see an x-ray image of the heart of the Crab Nebula taken with the Chandra X-ray Observatory. The image revealed tilted 1-light-year rings of high-energy material thrown out in the explosion, plus jets of gas extending perpendicularly.

nomical objects in almost all parts of the spectrum. We combine views in the visible part of the spectrum with studies of invisible radiation to give us a more complete idea of the astronomical object we are studying than we could otherwise have (Fig. 1–3). On the whole, whether we are looking at nearby or very distant objects, the techniques of studying in various parts of the spectrum are the same.

The tools that astronomers use are bigger and better than ever. Giant telescopes on mountaintops collect visible light with mirrors as big as 10 meters across (Fig. 1–4). Images from these new telescopes can be spectacular

Figure 1–4 The 8.2-meter-diameter Gemini-North telescope on Mauna Kea, a high volcano in Hawaii. The dome is of a new design very open to the air, to improve the quality of the images.

Figure 1–5 Interacting galaxies imaged with one of the 8.2-m telescopes of the Very Large Telescope, in Chile.

Accurate values of constants, including the exact value of the speed of light, appear in Appendix .2.

Figure It Out

Keeping Track of Space and Time

Throughout this book, we shall generally use the metric system, which is commonly used by scientists. The basic unit of length, for example, is the meter, which is equivalent to 39.37 inches, slightly more than a yard. Prefixes are used (Appendix 1) in conjunction with the word "meter," whose symbol is "m," to define new units. The most frequently used prefixes are "milli-," meaning $\frac{1}{1000}$, "centi-," meaning $\frac{1}{100}$, "kilo-," meaning 1000 times, and "mega-," meaning one million times." Thus 1 millimeter is $\frac{1}{1000}$ of a meter, or about 0.04 inch, and a kilometer is 1000 meters, or about $\frac{5}{8}$ of a mile. As we describe in *Figure It Out: Scientific Notation*, we keep track of the powers of 10 by which we multiply 1 m by writing the number of tens we multiply together as an exponent; 1000 m (1000 meters), for example, is 10^3 m, since 1000 has 3 zeros following the 1 and thus represents three tens multiplying each other. The standard symbol for "second" is "s," so km/s is kilometers per second. (We will generally use "sec" in this book for added clarity.) Astronomers measure mass in kilograms (kg), where each kilogram is 10^3 grams (1000 g).

We can keep track of distance not only in the metric system but also in units that are based on the length of time that it takes light to travel. The speed of light is, according to Einstein's special theory of relativity, the greatest speed that is physically attainable for objects travelling through space. Light travels at approximately 300,000 km/sec (186,000 miles/sec), fast enough that if we could bend it enough, it would circle the Earth 7 times in a single second. (Such strong bending takes place only near a black hole, though; for all intents and purposes, light goes straight near the Earth.) Even at that fantastic speed, we shall see that it would take years for us to reach the stars. Similarly, it has taken years for the light we see from stars to reach us, so we are really seeing the stars as they were years ago. Thus, we are looking backward in time as we view the Universe, with distant objects being viewed farther in the past than nearby objects. The distance that light travels in a year is called a **light-year;** note that the light-year is a unit of *length* rather than a unit of time even though the term "year" appears in it. It is equal to 9.5×10^{12} km (nearly 10 trillion km)—an extremely large distance by human standards.

A "month" is an astronomical time unit based on the Moon's orbit, and a "year" is an astronomical time unit based on the Earth's orbit around the Sun. The measurement of time itself is now usually based on processes in atoms, which are used to define the second. So the second from atomic timekeeping is slightly different from the second that we think of as a sixtieth of a minute, which is a sixtieth of an hour, which is a twenty-fourth of a day, which is based on the rotation of the Earth on its axis. For some purposes, weird stars called pulsars keep the most accurate time in the Universe. In this book, when we are talking about objects billions of years old, it won't matter precisely how we define the second or the year.

2 light-years

Figure 1–6 An infrared image of a young star cluster near the center of our galaxy, made with the Hubble Space Telescope. The infrared wavelengths have been transformed into visible colors for printing.

(Fig. 1–5). Up in space, above Earth's atmosphere, the Hubble Space Telescope sends back very clear images (Fig. 1–6). Many things pretty far away are seen as clearly with Hubble as other objects closer to us appear with most ground-based telescopes. That accomplishment gives us many more objects in space that we can study in detail. (The ground-based astronomers are developing methods of seeing more clearly, too.) The Chandra X-ray Observatory gives clear images of a wide variety of objects using the x-rays they emit. Toward the end of this chapter, we give a survey of the kinds of objects that astronomers study.

DISTANCES IN THE UNIVERSE

It is easy to see the direction to an object in the sky but is much harder to find its distance. Astronomers are always seeking new and better ways to measure distances to objects that are too far away to touch. Our direct ability to reach out to astronomical objects is limited to our Solar System. We can send people to the Moon and spacecraft to the other planets. We can even bounce radio waves off the Moon, most of the other planets, and the Sun, and measure how long the radio waves' round trip takes to find out how far they have travelled. The distance d travelled by light or by an object is equal to the constant rate at which it is travelling (its speed v) times the time t spent travelling ($d = vt$).

Once we go farther afield, we must be more ingenious. Repeatedly in this book, we will discuss ways of measuring distance. For the nearest hundreds of thousands of stars, we have recent results from a spacecraft that showed how much their apparent position shifts when we look at them from slightly different angles. For more distant stars, we find out how far away they are by comparing how bright they actually (intrinsically) are and how bright they appear to be. We often tell their intrinsic brightness from looking at their spectra (Chapter 10), though we will also see other methods.

Once we get to galaxies other than our own, we will see that our methods are even less precise. For the nearest galaxies, we search for stars whose specific properties we recognize. Some of these stars are thought to be identical in type to the same kinds of stars in our own galaxy whose brightnesses we know. Again, we can then compare intrinsic brightness with apparent brightness to give distance. For the farthest galaxies, as we shall discuss in Chapter 16, we find distances using the 1920s discovery that shifts in color of the spectrum of a galaxy reveal how far away it is. Of course, we continue to test these methods as best we can, and some of the most exciting investigations of modern astronomy are related to this study. One of the Hubble Space Telescope's Key Projects has been devoted to finding distances in this way, and has made good progress toward resolving a long-term controversy on the size and age of the Universe.

> If we were carrying on a conversation by radio with someone at the distance of the Moon, there would be pauses of noticeable length after we finished speaking before we heard an answer. This is because radio waves, even at the speed of light, take over a second to travel each way. Astronauts on the Moon have to get used to these pauses when their messages travel by radio to people on Earth.

⬤ THE VALUE OF ASTRONOMY

Throughout history, observations of the heavens have led to discoveries that have had a major impact on people's ideas about themselves and the world around them. Even the dawn of mathematics may have stemmed from ancient observations of the sky, made in order to keep track of seasons and seasonal floods in the fertile areas of the Earth (Fig. 1–7). Observations of the motions of the Moon and the planets, which are free of such complicating terrestrial forces as friction and which are massive enough so that gravity dominates their motions, led to an understanding of gravity and of the forces that govern all motion.

We can consider the regions of space studied by astronomers as a cosmic laboratory where we can study matter or radiation, often under conditions that we cannot duplicate on Earth. These studies allow us to extend our understanding of the laws of physics.

A new importance has been given to astronomy by the realization that asteroids and comets have hit the Earth every few tens of millions of years with enough power to devastate our planet. We shall see later on how the dinosaurs and many other species were extinguished 65 million years ago by an asteroid or comet, according to a theory that has become widely favored. The movies that have been made describing how we might search for and react to the discovery of an asteroid en route to the Earth aren't all science fiction.

Many of the discoveries of tomorrow—perhaps the control of nuclear fusion or the discovery of new sources of energy, or perhaps something so revolutionary that it cannot now be predicted—will undoubtedly be based on advances made through such basic research as the study of astronomical systems. Considered in this sense, astronomy is an investment in our future.

Figure 1–7 This long-exposure view of the Pleiades, a star cluster, shows dust around the stars reflecting the bluish starlight. When the Pleiades and the Hyades, another star cluster, rose just before dawn, some ancient peoples knew that the rainy season was about to begin.

A Sense of Scale: Measuring Distances

Let us try to get a sense of scale of the Universe, starting with sizes that are part of our experience and then expanding toward the enormously large. Imagine a series of cubes, of which we see one face (a square).

1 mm = 0.1 cm

We begin our journey through space with a view of something 1 millimeter across, an electron-microscope view of an ant. Every step we take will show a region 100 times larger in diameter than that in the previous picture.

10 cm = 100 mm

A square 100 times larger on each side is 10 centimeters × 10 centimeters. (Since the area of a square is the length of a side squared, the area of a 10-cm square is 10,000 times the area of a 1-mm square.) The area encloses a flower.

10 m = 1000 cm

Here we move far enough away to see an area 10 meters on a side, with Muhammad Ali triumphant.

$1 \text{ km} = 10^3 \text{ m}$

A square 100 times larger on each side is now 1 kilometer square, about 250 acres. An aerial view of Boston shows how big this is.

$100 \text{ km} = 10^5 \text{ m}$

The next square, 100 km on a side, encloses the cities of Boston and Providence. Note that though we are still bound to the limited area of the Earth, the area we can see is increasing rapidly.

$10,000 \text{ km} = 10^7 \text{ m}$

A square 10,000 km on a side covers nearly the entire Earth. We see the southwestern United States, through the clouds, and northwestern Mexico, including the Baja California peninsula.

1,000,000 km = 10^9 m = 3 lt sec

When we have receded 100 times farther, we see a square 100 times larger in diameter: 1 million kilometers across. It encloses the orbit of the Moon around the Earth. We can measure with our wristwatches the amount of time that it takes light to travel this distance. The images, artificially superimposed to give a false perspective, were taken by the Galileo spacecraft in 1990 when it passed by the Earth en route to Jupiter.

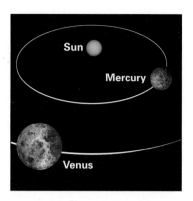

10^{11} m = 5 lt min

When we look on from 100 times farther away still, we see an area 100 million kilometers across, ⅔ the distance from the Earth to the Sun. We can now see the Sun and the two innermost planets in our field of view.

10^{13} m = 8 lt hr

An area 10 billion kilometers across shows us the entire Solar System in good perspective. It takes light about 10 hours to travel across the Solar System. The outer planets have become visible and are receding into the distance as our journey outward continues. Our spacecraft have now visited or passed the Moon, Mercury, Venus, Mars, Jupiter, Saturn, Uranus, Neptune, and their moons.

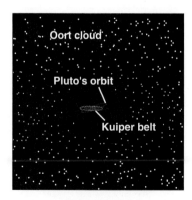

10^{15} m = 38 lt days

From 100 times farther away, we see little that is new. The Solar System seems smaller, and we see the vastness of the empty space around us. We have not yet reached the scale at which another star besides the Sun is in a cube of this size.

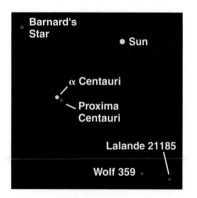

10^{17} m = 10 ly

As we continue to recede from the Solar System, the nearest stars finally come into view. We are seeing an area 10 light-years across, which contains only a few stars, most of whose names are unfamiliar (Appendix 7). The brightest, most familiar stars tend to be intrinsically very powerful, but more distant.

10^{19} m = 10^3 ly

By the time we are 100 times farther away, we can see a fragment of our galaxy, the Milky Way Galaxy. We see not only many individual stars but also many clusters of stars and many "nebulae"—regions of glowing, reflecting, or opaque gas or dust. Between the stars, there is a lot of material (most of which is invisible to our eyes) that can be studied with radio telescopes on Earth, or in ultraviolet or x-rays with telescopes in space.

(box continued next page)

A Sense of Scale: Measuring Distances *(continued)*

10^{21} m = 10^{5} ly

In a field of view 100 times larger in diameter, we would be able to view our entire Milky Way Galaxy, with its spiral arms, in one go. (The picture shown is not actually our galaxy, but we show one similar to ours for the purpose of this illustration—our own Milky Way home probably looks very similar to this Hubble Space Telescope image if we could rise above the plane of the Milky Way and view ourselves from above.)

Next we move sufficiently far away so that we can see an area 10 million light-years across. There are 10^{25} centimeters in 10 million light-years, about as many centimeters as there are grains of sand in all the beaches of the Earth. Our galaxy is in a cluster of galaxies, called the Local Group, that would take up only ⅓ of our angle of vision. In this group are all types of galaxies. The photograph shows part of a cluster of galaxies.

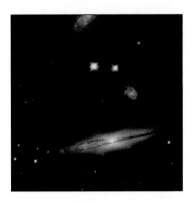

10^{23} m = 10^{7} ly

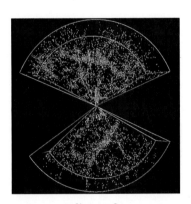

10^{24} m = 10^{9} ly

If we could see a field of view 100 times larger, 1 billion light-years across, our Local Group of galaxies would appear as but one of many clusters, part of a supercluster—a cluster of clusters. If we were to enlarge our field of view only another 10 times, we would be seeing the most distant known galaxies, 10 billion light-years away, whose light took 10 billion years to reach us on Earth. We are thus looking back to times billions of years ago. Since we think that the Universe began about 14 billion years ago, we are looking back almost to the beginning of time.

We have even detected radiation from the Universe's earliest years. This map of the whole sky shows a slight difference in the Universe's temperature in opposite directions (Chapter 19), resulting from our Sun's motion. In more detailed maps, we are seeing the seeds from which today's clusters of galaxies grew. A combination of radio, ultraviolet, x-ray, and optical studies, together with theoretical work and experiments with giant atom smashers on Earth, is allowing us to explore the past and predict the future of the Universe.

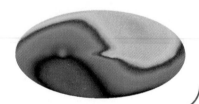

The impact of astronomy on our conception of the Universe has been strong through the years. Discoveries that the Earth is not in the center of the Universe, or that the Universe has been expanding for billions of years, affect our philosophical conceptions of ourselves and our relations to space and time.

Yet most of us study astronomy not for its technological and philosophical benefits but for its beauty and inherent interest. We must stretch our minds to understand the strange objects and events that take place in the far reaches of space. The effort broadens us and continually fascinates us all. Ultimately, we study astronomy because of its fascination and mystery.

Origins

It is always fun to look at our own childhood pictures. Similarly, astronomers are increasingly looking at the baby pictures of the Universe and of various types of objects in it. NASA has made its Origins program a centerpoint of its strategy for investigation. Under the aegis of this program, many scientists will use spacecraft, telescopes on Earth, computers, and just plain clear thought to study all kinds of origins: the origin of the Universe, the origin of our galaxy, the origin of stars, the origin of our Sun, the origin of our Earth and the elements in it, and the origin of life—maybe someday not only life on Earth but also on one of the planets around other stars that are being discovered at an increasing rate. After all, we now know of more planets outside our Solar System than we do inside, a big change in our thinking that occurred in the 1990s.

NASA's Origins program has three official major scientific goals: (1) To understand how galaxies formed in the early Universe and to determine the role of galaxies in the appearance of planetary systems and life. (2) To understand how stars and planetary systems form and to determine whether life-sustaining planets exist around other stars. (3) To understand how life originated on Earth and to determine whether it began and may still exist elsewhere as well. Existing or planned missions relevant to the Origins program include the Hubble Space Telescope, the Space Infrared Telescope Facility, the Keck Interferometer on Mauna Kea (interferometers are ways of linking several telescopes together, as we shall see, making them equivalent for some purposes to a single, huge telescope), the Mars Surveyor Program (which is being restudied after 1999's double failure of Mars Climate Orbiter and Mars Polar Lander), and the Stratospheric Observatory for Infrared Astronomy (SOFIA). Specific near-future space missions that are part of the Origins program are the Space Interferometry Mission and the Next Generation Space Telescope. (www)   In the longer term, perhaps two decades away, we hope for the Terrestrial Planet Finder, an interferometer with a baseline about 100 meters long for operation in the infrared. The hope for the Terrestrial Planet Finder is to detect directly small, rocky planetary companions to other stars as well as to study spectra rich in lines from molecules that might indicate the habitability of any planets that are discovered.

Another major theme of NASA's scientific investigations is Structure and Evolution. In this book, you will learn about the structure of the Solar System, the structure of our galaxy, and the structure of space. You will learn the life stories of stars, and see how they evolve. And you will learn about the evolution of the Universe as a whole.

In this book, we will make these themes explicit as the central organizing structure. In particular, we will try to point out the links to Origins—to our Universe's childhood pictures—on a chapter-by-chapter basis.

◖● THE SCIENTIFIC METHOD

Science is not just a body of facts; it is also a process of investigation. The standards that scientists use to assess their ideas and to decide which to accept—in some sense, which are "true"—are the basis of much of our technological world. One of the major precepts of science is that results should be reproducible; that is, other scientists should be able to get the same result by repeating the same experiment or observation. Science is thus a self-checking way of carrying out investigations.

Though acceptable scientific investigations are actually carried out in many ways, there is a standard model for the scientific method. In this standard model,

Note that "data" is a plural word; the singular form is "datum," a distinction not always followed these days.

one first looks at a body of data and makes educated guesses as to what might explain them. The educated guess is a **hypothesis.** Then one thinks of consequences that would follow if the hypothesis is true and tries to carry out experiments or make observations that test the hypothesis. If, at any time, the results are contrary to the hypothesis, then the hypothesis is discarded or modified. (There may, though, be other assumptions that could be modified or discarded instead, because they are inappropriate. Also, one needs to check for experimental or observational errors.) If the hypothesis passes its tests and is established in some basic framework or set of equations, it can be called a **theory.**

If the hypothesis or theory survives test after test, it is accepted as "true." Still, at any time a new experiment or observation could show it was false after all. This process of "falsification," being able to find out if a hypothesis or theory is false, is basic to the definition of science formulated by a leading set of philosophers of science.

Sometimes, you see something described as a "law" or a "principle" or a "fact." "Laws of nature"—like Kepler's laws of planetary motion, Newton's laws of motion, or Newton's law of gravitation—are really *descriptions* of how nature behaves. Newton used the word "law" (in Latin) over three hundred years ago. The words "law" and "principle" are historical usages associated with certain basic theories. Scientists consider something a "fact" when it is so well established that it would be extremely unlikely that it is incorrect.

What is an example of the scientific method? Albert Einstein advanced a theory in 1916 to explain gravity. (It was worked out so completely that it was far beyond a hypothesis.) His "general theory of relativity" was based on mathematical equations he worked out, based on some theoretical ideas. His result turned out to explain some observations of the orbit of Mercury that had been puzzling up to then (though that hadn't been Einstein's motivation). Still, its basic tests lay ahead. Einstein's theory predicted that starlight would appear to be bent if it could be observed to pass very near the Sun. Such bending would be visible only at a total solar eclipse. The theory seemed so important that expeditions were made to solar eclipses to test it. When the first telegram came back that the strange prediction of Einstein's theory had been verified, the theory was quickly accepted. Successful predictions are usually given more weight than mere explanations of already known facts (Fig. 1–8). Einstein immediately gained a worldwide reputation as a great scientist. Nearly complete acceptance by the scientific community was established as alternatives could not be found.

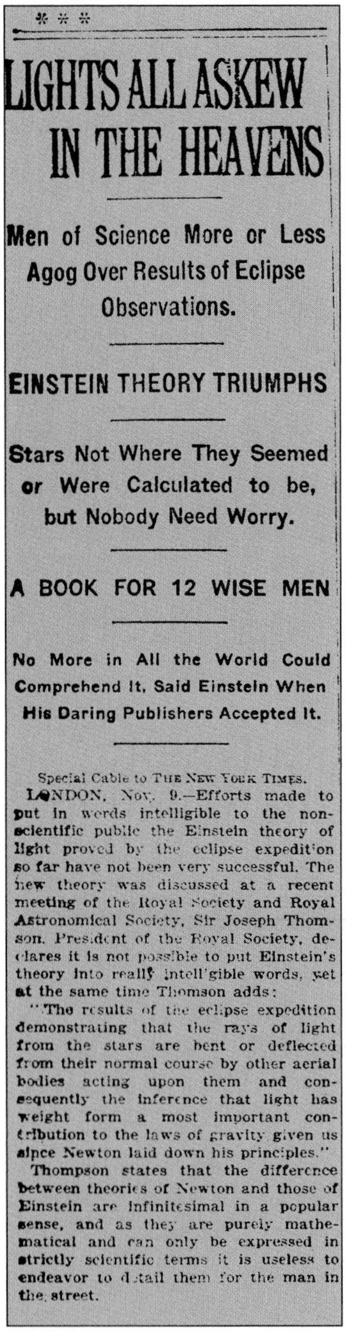

LIGHTS ALL ASKEW IN THE HEAVENS

Men of Science More or Less Agog Over Results of Eclipse Observations.

EINSTEIN THEORY TRIUMPHS

Stars Not Where They Seemed or Were Calculated to be, but Nobody Need Worry.

A BOOK FOR 12 WISE MEN

No More in All the World Could Comprehend It, Said Einstein When His Daring Publishers Accepted It.

Special Cable to THE NEW YORK TIMES.

LONDON, Nov. 9.—Efforts made to put in words intelligible to the non-scientific public the Einstein theory of light proved by the eclipse expedition so far have not been very successful. The new theory was discussed at a recent meeting of the Royal Society and Royal Astronomical Society, Sir Joseph Thomson, President of the Royal Society, declares it is not possible to put Einstein's theory into really intelligible words, yet at the same time Thomson adds:

"The results of the eclipse expedition demonstrating that the rays of light from the stars are bent or deflected from their normal course by other aerial bodies acting upon them and consequently the inference that light has weight form a most important contribution to the laws of gravity given us since Newton laid down his principles."

Thompson states that the difference between theories of Newton and those of Einstein are infinitesimal in a popular sense, and as they are purely mathematical and can only be expressed in strictly scientific terms it is useless to endeavor to detail them for the man in the street.

Figure 1–8 This report was one of those early descriptions of general relativity's success that captured the public's fancy. From the *New York Times* of November 10, 1919.

Most of the time, though, the scientific method does not work so straightforwardly. We can consider, for example, our understanding of how the stars shine. Until the 1930s, it was thought that the Sun got its energy from its contracting gas, but this idea seemed wrong, because it could not explain how the Sun could be as old as rocks on Earth whose ages we measured. (Presumably the Sun and Earth formed nearly simultaneously.) Then scientists suggested that nuclear fusion—the merging of 4 hydrogen nuclei to make a single helium nucleus, in particular—could provide the energy for the Sun to live 10 billion years, and even worked out the detailed ways in which the hydrogen could fuse. Thus a theory of nuclear energy as a source of power for the Sun and stars exists. But how could it be tested? The observed properties of a wide variety of stars around the Universe fit, in several ways, with the deductions of the theory. Still, a direct test had to wait. Over the last thirty years, measurements have finally been made of individual particles called "neutrinos" that should be given off as the fusion process takes place. Neutrinos from fusion in the Sun were indeed measured, but only at one-third to one-half the rate expected. Had our main theory of stellar energy failed? No, because explanations were discovered for why some of the neutrinos may not reach us. Nobody doubts that the Sun and stars get their energy from nuclear fusion, so our theory of the properties of neutrinos themselves is changing. Thus the astronomical study of neutrinos from the Sun may lead to important changes in the most basic ideas of physics.

So the "scientific method" isn't cut and dried. But it does demand a rigor and honesty in scientific testing. The standards of science are high, and we hope that this book will give enough examples to enable you to form an accurate impression of how the process actually works.

SCIENCE AND PSEUDOSCIENCE

Though science is itself fascinating, all too many people have beliefs that may seem related to science but either have no verification or are false. Such beliefs, such as astrology or the idea that UFOs—unidentified flying objects—are now bringing aliens from other planets, are pseudoscience rather than science.

Astrology is not at all connected with astronomy, except in a historical context (they had similar origins), so does not really deserve a place in a text on astronomy. But since so many people associate astrology with astronomy, and since astrologers claim to be using astronomical objects to make their predictions, let us use our knowledge of astronomy and of the scientific method to assess astrology's validity. Millions of Americans believe in astrology—a number that shows no signs of decreasing—so the topic is too widespread to ignore.

Astrology is an attempt to predict or explain our actions and personalities on the basis of the positions of the stars and planets now and at the instants of our births. Astrology has been around for a long time, but it has never definitively been shown to work. Believers may cite incidents that reinforce their faith in astrology, but statistical measurements do not definitively show a real effect. If something happens to you that you had expected because of an astrological prediction, you would more certainly notice that this event occurred than you would notice the thousands of unpredicted things that happened to you that day. Yet we do enough things, have sufficiently varied thoughts, and interact with enough people that if we make many predictions in the morning, some of them are likely to be at least partially fulfilled during the day. We simply forget that the rest ever existed.

In fact, even the traditional astrological alignments are not accurately calculated, for the Earth's pole points in different directions in space as time passes over millennia. In truth, stars are overhead at different times of year from millennia ago, when astrological tables that are often still in current use were computed. At a given time of year, the Sun is usually in a different sign of the zodiac from its traditional astrological one (Fig. 1–9). Further, we know that the constellations are illusions; they don't even exist as physical objects. They are merely projections of the positions of stars that are generally at very different distances from us.

Studies have shown that superstition actively constricts the progress of science and technology in various countries around the world and is therefore not merely an innocent force. It is not just that some people harmlessly believe in astrology. Their lack of understanding of scientific structure may actually impede the training or work of people needed to solve the problems of our age, such as AIDS, pollution, shortages of food, and the energy crisis. Published articles have reported, for example, that widespread superstitious beliefs have even impeded smallpox-prevention programs. Thus many scientists are not content to ignore astrology, and actively oppose its dissemination. Further, if large numbers of citizens do not understand the scientific method and the difference between science and pseudoscience, how can they intelligently vote on or respond to scientific questions that have societal implications?

Figure 1–9 Twelve constellations through which the Sun, Moon, and planets pass make up the zodiac. *(Drawing by Handelsman; © 1978 The New Yorker Magazine, Inc.)*

Figure 1–10 The symbols for the constellations of the zodiac shown on the historic 15th-century astronomical clock in Prague.

In addition to the lack of evidence corroborating astrological predictions, one minor reason why scientists in general and astronomers in particular doubt astrology is that they cannot conceive of a way in which it would work. The human brain is so complex that it seems most improbable that any celestial alignment can affect people, including newborns, in an overall way. The celestial forces that are known are not sufficient to set personalities or influence day-to-day events. Any unknown force must have truly remarkable properties to exert any noticeable effect on humans.

Let us mention a specific study showing that astrology doesn't work, one in which a psychologist tested certain values. Do Libras and Aquarians rank "Equality" highly? Do Sagittarians especially value "Honesty"? Do Virgos, Geminis, and Capricorns treasure the value "Intellectual"? Several astrology books agreed that these and other similar examples are values typical of those signs. Although believers often criticize the objections of skeptics on the grounds that these group horoscopes are not as valuable or accurate as individualized charts, surely some general assumptions and rules hold in common. (Signs of the zodiac appear in Fig. 1–10.)

The subjects, 1600 psychology graduate students, did not know in advance what was being tested. They gave their birthdates, and the questioners determined their astrological signs. The results: no special correlation with the values they were supposed to hold was apparent for any of the signs. Also, when asked to what extent they shared the qualities of each given sign, as many subjects ranked themselves above average as below, regardless of their astrological signs.

But if astrology is so meaningless, why does it still have so many adherents? Well, it could be the bandwagon effect. The same psychology professor took twelve personality descriptions from astrology books, one for each astrological sign, and displayed them to two groups of individuals. The first group was told to which astrological signs the descriptions pertained, and the individuals were asked to write their own signs on the covers of the questionnaires. More than half the members of this group thought that the descriptions listed under their own signs were, for each individual, among the four best descriptions of themselves out of the twelve choices.

Yet, when the second group was given the same twelve descriptions but were told that they came from a book entitled *Twelve Ways of Life* instead of being told that they were astrological, the choices were random. Only 30 per cent chose their own sign's description as being in the group most closely describing them, a result consistent with pure guesswork. So the idea that astrology can predict personality types seems to be the result of self-delusion. When people know what they are expected to be like, they tend to identify themselves with the description. But that doesn't mean that they actually satisfy the description that astrology predicts for them better than any other description.

A team of psychologists from the California State University at Long Beach arranged for a magician to perform three psychic-like stunts in front of psychology classes. Even when they emphasized to the students that the performer was a magician performing tricks, 50 per cent of the class still believed the magician to be psychic. Evidently, a lot of self-deception is involved with believing in pseudoscience.

From an astronomer's view, astrology is meaningless, unnecessary, and impossible to explain without contradicting the broad set of physical laws that we have developed and tested over the years and that very well explains what happens on the Earth and in the sky. Astrology snipes at the roots of all pure science. Astronomers would like nothing better than to discover fundamental new scientific laws, but there is no evidence that astrology provides the way. Let's all learn from the stars, but let's learn the truth.

Similarly, as much fun as it is to watch movies about aliens visiting the Earth or about space travel far in the future, there is no compelling evidence that any of the things seen in the sky at night and called UFOs are evidence of such visits in our time (see Chapter 20).

Science is more than just a set of facts, since a methodology of investigation and standards of proof are involved, but science is more than just a methodology, since many facts have been well established. In this course, you should not only learn certain facts about the Universe but also appreciate the way that theories and facts come to be accepted.

We will be discussing such fascinating things as quasars, pulsars, and black holes. We will consider complex molecules that have spontaneously formed in interstellar space, and try to decide whether the Universe will expand forever. We have sent a rocket into interstellar space bearing a portrait of humans, and have beamed a radio message toward a group of stars 24,000 light-years away. These topics and actions are part of modern astronomy: what contemporary, often conservative, scientists are doing and thinking about the Universe. So, though it would be tremendously exciting if celestial alignments could affect individuals, the evidence is so strongly against this idea that it seems fruitless to spend time on astrology instead of the exciting aspects of the real astronomical Universe.

Concept Review

We gain most of our information about the Universe by studying radiation: gamma rays, x-rays, ultraviolet, visible light, infrared, and radio waves. We break up visible light into its component colors, a **spectrum.**

Astronomers often use a version of the metric system, in which prefixes like "kilo-" for one thousand and "mega-" for one million go with units like meter for distance and second for time, or are found in kilogram for mass. Astronomers also use special units, like the **light-year** for the distance light travels in one year.

Astronomers and other scientists follow the scientific method, which is difficult to define but provides a standard about which scientists agree. In the basic form of the scientific method, a **hypothesis** passes observational tests to become a **theory.** Astrology passes no scientific tests and so is not a science. Further, it has been shown not to work.

Questions

1. Why do we say that our senses have been expanded in recent years?

†**2.** The speed of light is 3×10^5 km/sec. Express this number in m/sec and in cm/sec.

3. Geologists study different strata to determine conditions on Earth long in the past. For example, dinosaur bones are found in strata dating from about 245 million to 65 million years ago. How is it that astronomers are able to study parts of the Universe as they were in the past?

4. Distinguish between a hypothesis and a theory. Give a non-astronomical example of each.

5. Discuss an example of science and an example of pseudo-science, and distinguish between their value.

†**6.** **(a)** Write the following in scientific notation: 4642; 70,000; 34.7. **(b)** Write the following in scientific notation: 0.254; 0.0046; 0.10243. **(c)** Write out the following in an ordinary string of digits: 2.54×10^6; 2.004×10^2; 7.67×10^{-4}.

†This question requires a numerical solution.

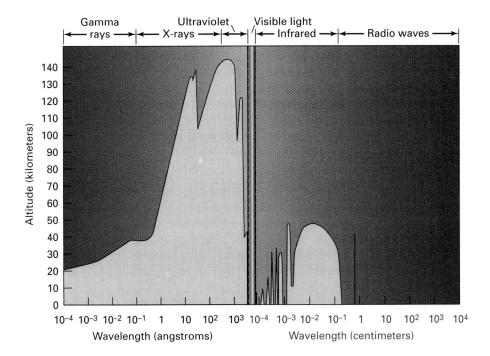

Figure 2-2 In a window of transparency in the terrestrial atmosphere, radiation can penetrate to the Earth's surface. The curve specifies the altitude at which the intensity of arriving radiation is reduced to half its original value. When this level of reduction is reached high in the atmosphere, little or no radiation of that wavelength reaches the ground.

Orange Yellow Green Blue Indigo Violet, going from longer to shorter wavelengths. (The name Indigo is no longer commonly used.)

Astronomers still tend to use angstroms rather than the metric unit nanometers. 10 Å = 1 nm, so divide all the numbers above by 10 to show their values in nanometers: violet light is about 400 nm, for example.

Only certain parts of the electromagnetic spectrum can penetrate the Earth's atmosphere. ☼ We say that our atmosphere has "windows" for the parts of the spectrum that can pass through it. The atmosphere is transparent at these windows and opaque at other parts of the spectrum. One window passes what we generally call "light," and what astronomers technically call visible light, or "the visible." In this book, we will use the term "light" to refer to any form of electromagnetic radiation. Another window falls in the radio part of the spectrum, and modern astronomers can thus base their "radio telescopes" on Earth and still detect that radiation (Fig. 2–2).

☼ **This icon symbolizes that a common misconception on this subject, and the correct reasoning, are given on this book's Web site, chapter by chapter.**

But we of the Earth are no longer bound to our planet's surface. Balloons, rockets, and satellites carry telescopes high up or above our atmosphere. They can observe in parts of the electromagnetic spectrum that do not reach the Earth's surface. In recent years, we have been able to make observations all across the spectrum. It may seem strange, in view of the long-time identification of astronomy with visible-light observations, to realize that optical studies no longer dominate astronomy.

⬤ BLACK BODIES AND THEIR RADIATION

Many things in astronomy seem simple to study. Stars, for example, are balls of gas, and they give off visible light and other electromagnetic radiation that on the whole follows an especially simple rule. If you have two stars of the same size and one is hotter than the other, it is also intrinsically brighter. Furthermore, the hotter star gives off more of its energy at shorter wavelengths.

Let us consider the simplest possible object that gives off radiation. It is called a **black body,** which indicates that it is a simple thing—it isn't like a polka-dotted body, for example, that gives off radiation with different properties from

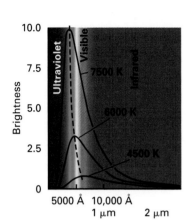

The brightness of radiation at different wavelengths for different temperatures. Black bodies—ideally radiating matter—give off radiation that follows these curves. Radiation from stars follows these curves fairly closely.

Figure It Out

Black-Body Radiation

A "black body" is a perfect absorber: It absorbs all incident radiation, reflecting and transmitting none. It has a certain temperature, which is a measure of the average speed with which the particles within it jiggle around: the higher the temperature, the greater the average speed. The randomly moving particles, some of which are charged (that is, have electric charge), emit electromagnetic radiation. This emission is called thermal radiation, or black-body radiation.

A black body is also a perfect emitter: The shape of the object's emitted spectrum depends only on its temperature, not on its chemical composition or other properties. This spectral shape is called the *Planck curve*, in honor of the physicist Max Planck; its derivation was a fundamental problem in quantum physics.

As can be seen in the figure at left, at all wavelengths the spectrum (Planck curve) of a hot black body is higher (that is, brighter) than that of a colder black body having the same surface area.

An important property is that the spectrum of a hot black body peaks at a shorter wavelength than that of a colder black body. The product of the temperature (T) and the peak wavelength (λ_{max}) is a constant: $\lambda_{max}T = 2.9 \times 10^7$ Å K = 0.29 cm K. This relation is known as *Wien's law*. It is mathematically derivable from the formula for the Planck curve. For example, the spectrum of the Sun, whose surface temperature is about 5800 K, peaks at a wavelength of $\lambda_{max} = (2.9 \times 10^7$ Å K)/ (5800 K) = 5000 Å. A star that is twice as hot as the Sun has a spectrum that peaks at half this wavelength, or about 2500 Å; the product of temperature and peak wavelength must remain constant.

Another useful property is that per unit of surface area, a hot black body emits much more energy per second than a cold black body. In fact, the energy emitted is proportional to the fourth power of the temperature: $T \times T \times T \times T$. This relation is known as the *Stefan-Boltzmann law*, which (like Wien's law) is derivable from the Planck curve. It can be expressed as $E = \sigma T^4$, where E is the energy emitted per unit area (for example, cm^2) per second, T is the surface temperature in kelvins (it is important that the temperature scale start at absolute zero), and σ (sigma) is a constant (known as the Stefan-Boltzmann constant).

For example, if two stars (labeled with subscripts 1 and 2) have the same surface area, but one is twice as hot as the other, the hotter star emits $2^4 = 16$ times as much energy (per second) as the colder star: $E_1/E_2 = (\sigma T_1^4)/(\sigma T_2^4) = T_1^4/T_2^4 = (T_1/T_2)^4 = 2^4 = 16$.

However, in the above example, if the hotter star's radius (R) is ¼ that of the colder star, then its surface area is $(¼)^2 = \frac{1}{16}$ as large as that of the colder star, because the surface area of a sphere is given by $4\pi R^2$. This lesser area exactly balances the greater emission per unit area, making the two stars equally luminous.

In general, the luminosity (L) (that is, the energy emitted per second, also known as power or intrinsic brightness) of a black body is given by its surface area (S) multiplied by the energy emitted per unit area per second (E): $L = SE$. For a sphere of radius R and temperature T, we have $L = 4\pi R^2 \sigma T^4$. Thus, if we know the luminosity and surface temperature of a star, we can derive its radius R.

different areas. A black body is an object that absorbs all radiation that is incident upon it. None is reflected or transmitted. Atomic motions within that object then cause it to emit (radiate) energy in a manner that depends only on its temperature, not on chemical composition or anything else. This result is called **black-body radiation** (or **thermal radiation**). Spectra of hotter black bodies peak (that is, have their highest brightness) at shorter (bluer) wavelengths than spectra of colder black bodies (see *Figure It Out: Black-Body Radiation*). Also, a hot black body emits much more energy per second than a cold black body of equal surface area.

Planets are much cooler than stars, so they give off most of their radiation at long wavelengths, in the infrared. The Sun, on the other hand, gives off most of its radiation in visible light. When we look at the spectrum of a planet, we see that some of the radiation is strongest in the visible, which means it is reflected sunlight, while other radiation is strongest in the infrared and therefore emitted by the planet. Thus a graph of the energy given off by a planet shows two peaks—one in the visible and one in the infrared (Fig. 2–3). Planets, therefore, are not perfect black bodies, but the approximation is nonetheless often useful. Similarly, humans reflect visible light from the Sun or from room lamps, and so are not perfect black bodies. But we do emit our own thermal (black-body) radiation, which is most intense at infrared wavelengths that are visible to certain infrared cameras but not to our eyes.

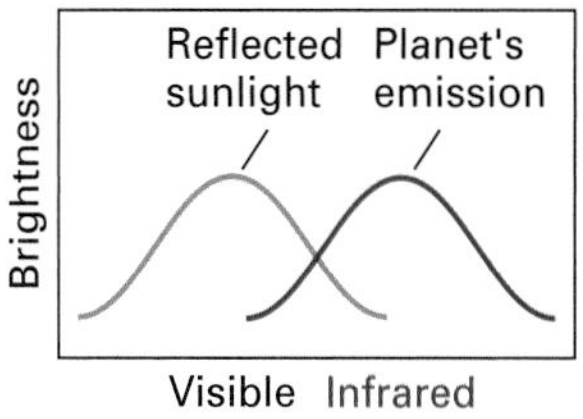

Figure 2–3 The spectrum of a planet includes one peak in the visible reflected from ordinary sunlight and another peak in the infrared. The infrared peak is the thermal emission from the planet itself, based on its surface temperature.

◖◗ ATOMS AND SPECTRAL LINES

In the early 1800s, when Joseph Fraunhofer looked in detail at the spectrum of the Sun, he noticed that the continuous range of colors in the Sun's light was crossed by dark gaps (Fig. 2–4). He saw a dozen or so of these "Fraunhofer lines"; astronomers have since mapped millions.

The dark Fraunhofer lines turn out to be from relatively cool gas absorbing radiation from behind it. We thus say that they are **absorption lines**. Atoms of each of the chemical elements in the gas absorb light at a certain set of wavelengths. By seeing what wavelengths are absorbed, we can tell what elements are in the gas, and the proportion of them present. We can also measure the temperature of the gas.

How does the absorption take place? To understand it, we have to study processes inside the atoms themselves. **Atoms** are the smallest particles of a given chemical element. For example, all hydrogen atoms are alike, all iron atoms are alike, and all uranium atoms are alike. As was discovered by Ernest Rutherford in 1911, atoms contain relatively light particles, which we call **electrons**, distributed around relatively massive central objects, which we call **nuclei** (Fig. 2–5). The nucleus contains **protons**, which have positive electric

Note that "absorption" is spelled with a "p," not with a second "b."

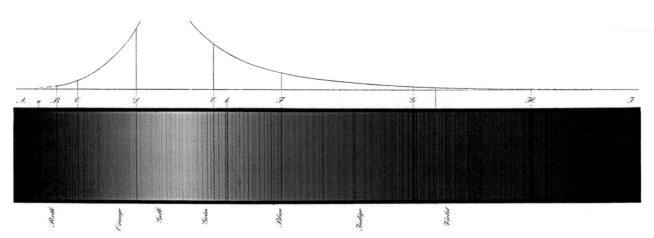

Figure 2–4 The dark lines from top to bottom, marking the absence of color at a specific wavelength, are the Fraunhofer lines. The drawing is Fraunhofer's original. We still use the capital-letter notation today for the D line (sodium) and the H line (from ionized calcium, not hydrogen). We now know that Fraunhofer's C line is a basic line of hydrogen.

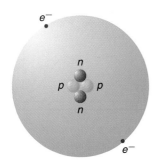

Figure 2-5 A helium atom contains two protons and two neutrons in its nucleus and two "orbiting" electrons. The nucleus is drawn much larger than its real scale with respect to the overall atom.

Figure 2-6 Niels Bohr and his wife Margrethe.

charge, and **neutrons**, which have no electric charge and so are neutral. Thus the nucleus—protons and neutrons together—has a positive electric charge. Electrons, on the other hand, have negative electric charge. The formation of spectral lines depends chiefly on the electrons.

Why aren't the electrons pulled into the nucleus? In 1913, Niels Bohr (Fig. 2–6) made a suggestion that some rule (at that time the rule was arbitrary) kept the electrons at fixed distances, in what are often (inaccurately) thought of as orbits. His suggestion was an early step in the development of the **quantum theory**, which was elaborated in the following two decades. The theory incorporates the idea that light consists of individual packets—quanta of energy—which had been worked out a decade earlier. We can think of each quantum of energy as a particle of light, which is called a **photon**. For some purposes, it is best to consider light as waves, while for others it is best to consider light as particles (that is, as photons). Though we will not explore quantum theory here, we should mention that it is one of the major intellectual advances of the 20th century.

Let us return to absorption lines. We mentioned that they are formed as light passes through a gas (Fig. 2–7). The atoms in the gas take up some of the light at specific wavelengths. The energy of this light goes into giving the electrons in those atoms more energy. But energy can't pile up in the atoms. If we look at the atoms from the side, so that they are no longer seen in silhouette against a background source of light, we would see the atoms giving off just as much energy as they take up (Fig. 2–8). Essentially, they give off this energy at the same set of wavelengths. These wavelengths are the **emission lines**, wavelengths at which there is an abrupt spike in the brightness of light.

All stars have absorption lines. They are formed, basically, as light from layers just inside the star's visible surface passes through atoms right at the surface. The surface is cooler, so the atoms there absorb energy, making absorption lines. (Detailed models are much more complicated.) Emission lines occur only in special cases for stars. We see absorption lines on the Sun's surface, but we can see emission lines by looking just outside the Sun's edge, where only dark sky is the background. Extended regions of gas called "nebulae" (Fig. 2–9) give off emission lines, as we will explain later, because they are not silhouetted against background sources of light.

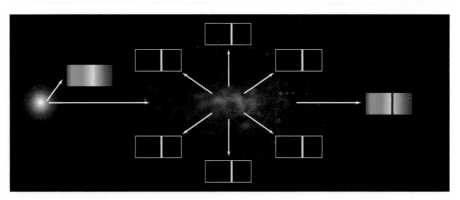

Figure 2-7 When we view a hotter source that emits continuous radiation through a cooler gas, we may see absorption lines. These absorption lines appear at the same wavelengths at which the gas gives off emission lines when viewed with no background or against a cooler, darker background. Each of the spectra in the little boxes is the spectrum you would see looking back along the arrow. Note that the view from the right shows an absorption line. Only when you look through one source silhouetted against a hotter source do you see the absorption lines.

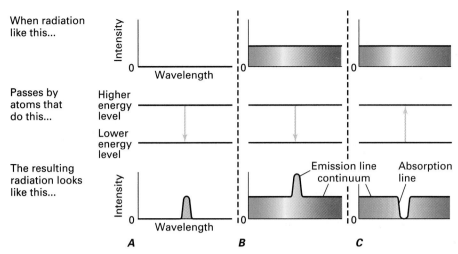

When radiation like this...

Passes by atoms that do this...

The resulting radiation looks like this...

Higher energy level

Lower energy level

Emission line continuum

Absorption line

A   *B*   *C*

Figure 2-8   When photons are emitted, we see an emission line. Continuous radiation may or may not be present (Cases *A* and *B*). An absorption line must absorb radiation from something. Hence an absorption line only appears when there is also continuous radiation (Case *C*). The top and bottom horizontal rows are graphs of the spectrum before and after the radiation passes by an atom. A schematic diagram of the atom's energy levels is shown in the center row (to simplify the situation, only two levels are given).

Figure 2-9   The Rosette Neb-ula, which has a reddish color be-cause of its bright emission lines from hydrogen gas.

Each of the chemical elements has its own distinctive set of spectral lines. So if you take the spectrum of a distant star, you may see several overlapping pat-terns of spectral lines. These patterns allow you to tell what chemical elements are in the star. We know what stars consist of from their spectra!

THE BOHR ATOM

Bohr, in 1913, presented a model of the simplest atom—hydrogen—and showed how it produces emission and absorption lines. Stars often show the spectrum of hydrogen. Laboratories on Earth can also reproduce the spectrum of hydrogen; the same set of spectral lines appears both in emission and in absorption.

Hydrogen consists of a single central particle (a proton) in its nucleus with a single electron surrounding it (Fig. 2–10). The model for the hydrogen atom

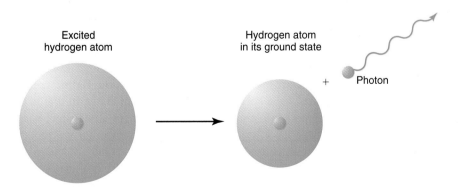

Excited hydrogen atom

Hydrogen atom in its ground state

+   Photon

Figure 2-10   Since the hydrogen atom has only a single electron, it is a particularly sim-ple case to study. The lowest possible energy state of an atom is called its ground state. All other energy states are called excited states. When an atom in an excited state gives off a photon, it drops back to a lower energy state, perhaps even to the ground state. We see the photons for a given transition as an emission line.

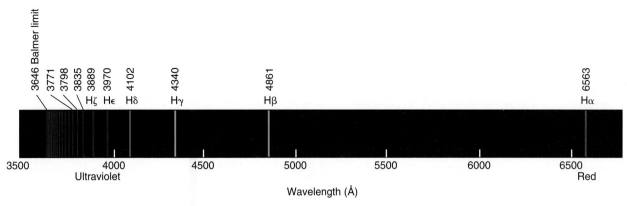

Figure 2-11 The Balmer series, representing transitions down to or up from the second energy state of hydrogen. The strongest line in this series, Hα (H-alpha), is in the red.

explained why hydrogen has only a few spectral lines (Fig. 2–11), rather than continuous bands of color. It postulated that a hydrogen atom could give off or take up energy only in one of a fixed set of amounts, just as you can climb up stairs from step to step but cannot float in between. The position of the electron relative to the nucleus determines its **energy level**, the amount of energy in a hydrogen atom. So the electron can only jump from energy level to energy level, and not hover in between values of the fixed set of energies allowable. When the electron jumps from a higher energy level to a lower one, a photon is given off. Many photons together make an emission line. When light hits an electron and makes it jump to a higher energy level, photons are absorbed and we see an absorption line. Each change in energy corresponds to a fixed wavelength; the greater the energy radiated as a photon, the shorter the wavelength. Photons of blue light have higher energy than photons of red light.

The **Bohr atom** is Bohr's model that explains the hydrogen spectrum. In it, electrons can have orbits of different sizes. Each orbit corresponds to an energy level (Fig. 2–12). Only certain orbits are allowable. (Note, however, that according to modern quantum theory, the electron orbits are not like planetary orbits. Instead, the electron behaves as though it were distributed throughout its orbit.)

We use the letter n to label the energy levels. We call the energy level for $n = 1$ the **ground level**, as it is the lowest possible energy state. **Excited levels** have $n = 2$ and higher. The hydrogen atom's series of transitions from or to the ground level is called the Lyman series, after the American physicist Theodore Lyman. The lines fall in the ultraviolet, at wavelengths far too short to pass through the Earth's atmosphere. Telescopes above the atmosphere now enable us to observe Lyman lines from the stars.

The series of transitions with $n = 2$ as the lowest level is the **Balmer series**, which is in the visible part of the spectrum. The bright red emission line in the visible part of the hydrogen spectrum, known as the Hα (H-alpha) line, corresponds to the transition from level 3 to level 2. The transition from $n = 4$ to $n = 2$ causes the Hβ (H-beta) line, and so on. Because the series falls in the visible where it is so well observed, we usually call the lines simply H-alpha, etc., instead of Balmer alpha, etc. So the Lyman series is transitions to or from lower level 1, and the Balmer series is transitions to or from lower level 2. Other series of hydrogen lines correspond to transitions with still different lower levels.

Since the higher energy levels have greater energy (they are higher above the ground state), a spectral line caused solely by transitions from a higher level to a

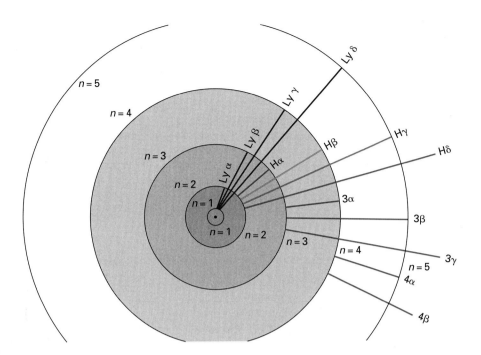

Figure 2–12 The representation of hydrogen energy levels known as the Bohr atom. Each of the circles shows the lone electron in a different energy level.

lower level is an emission line. When, on the other hand, continuous radiation falls on cool hydrogen gas, some of the atoms in the gas can be raised to higher energy levels and absorption lines result.

Each of the chemical elements has its own sets of energy levels. The state of the electrons is all important. Some atoms have lost one or more electrons. They are said to be **ionized** (Fig. 2–13). The spectrum of an ionized element is different from the spectrum of the same element when it is not ionized.

⬤ THE DOPPLER EFFECT

Even though we can't reach out and touch a star, we can examine its light in great detail. Above, you saw how studying starlight reveals the temperature of a star (from its color) and what chemical elements it has near its surface (from its spectral lines). Studying starlight also tells us how fast the star is moving toward or away from us. The technique applies to all kinds of objects, including the planets.

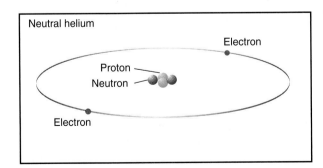

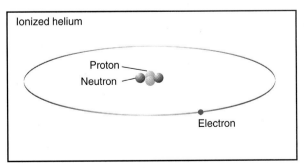

Figure 2–13 An atom missing one or more electrons is ionized. Neutral helium is shown at left; it has two electrons to balance the charge of the two protons. Ionized helium is shown at right. It has only one electron, so its net charge is +1.

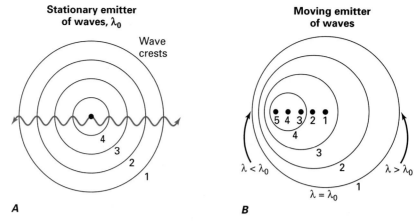

Stationary emitter
of waves, λ_0

Moving emitter
of waves

Wave
crests

$\lambda < \lambda_0$ $\lambda > \lambda_0$

$\lambda = \lambda_0$

A   *B*

Figure 2–14   *(A)* A stationary source is emitting waves of wavelength λ_0, in a circular pattern around it. Four wave crests, labeled 1 through 4, are shown. The observed wavelength is independent of the observer's position relative to the source. *(B)* The source is moving to the left with a constant speed. Along the direction of motion, the source partially keeps up with its most recently emitted wave crest before it emits another wave crest, so the crests get bunched closer together. From the left, the observed wavelength λ is shorter than λ_0; this is a "blueshift." Conversely, an observer on the right would see $\lambda > \lambda_0$; this is a "redshift." Perpendicular to the direction of motion, the spacing of the wave crests is unaffected, and an observer would therefore measure a wavelength $\lambda = \lambda_0$.

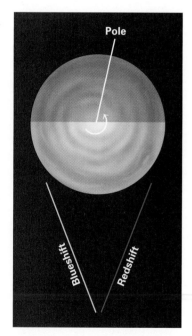

Pole

Blueshift   Redshift

Figure 2–15   If a planet is rotating, one side will be approaching us, blueshifting its spectral lines, while the other side will be receding from us, redshifting its spectral lines.

The effect that motion has on waves is known as the **Doppler effect,** and was studied by Christian Doppler about 150 years ago (see Fig. 2–14). It works for any waves—water waves, sound waves, or light waves. We will give more details about it when we study the motion of stars (Chapter 10). For the moment, we need only realize that when an object is coming toward you, the waves you are getting from it are compressed. The wavelength has gotten shorter. When an object is going away from you, the waves you are getting from it are spread out. The wavelength has gotten longer. Since the visible part of the spectrum has blue at short wavelengths and red at long wavelengths, the wavelengths are shifted toward the blue when objects are approaching and toward the red when objects are receding. Astronomers use the term **blueshift** to describe the shifts in wavelength of objects that are approaching and the term **redshift** to describe the shifts in wavelength of objects that are receding. Note that the light still travels at *c,* the speed of light in a vacuum; wavelengths and frequencies change, but the measured speed is independent of the motion of the source or of the observer.

One example of redshifts and blueshifts deals with planets in our Solar System. If you take spectra of several parts of a planet, and see that one side is redshifted and the other side is blueshifted, then you know that the first side is approaching you and the other side is receding (Fig. 2–15). You have found out that the planet is rotating! By measuring how much the light is shifted, you can even tell how fast the planet is rotating. The Doppler effect is a powerful tool for studying distant objects.

Perhaps the most exciting use of the Doppler effect has been the discovery, in 1999, of the first relatively normal solar system other than our own. Since 1995, over two dozen planets had been discovered around other stars, but only one per star. The discovery of several planets around the same star confirms that these objects are indeed planets rather than some kind of failed companion star; we would not expect there to be so many failed stars in a system.

Figure It Out

Temperature Conversions

Though most Americans use the Fahrenheit temperature scale, in which water freezes at 32°F and boils at 212°F, most of the rest of the world uses the Celsius scale, in which water freezes at 0°C and boils at 100°C. Note that the difference between freezing and boiling points for each scale is 180°F and 100°C, respectively, so a change of 180°F equals a change of 100°C, or 9°F for every 5°C.

There is little, if any, water on the stars or on most planets, so astronomers use a more fundamental scale. Their scale, the kelvin scale (whose symbol is K), begins at absolute zero, the coldest temperature that can ever be approached. Since absolute zero is about −273.16°C, and a rise of one kelvin (1 K) is the same as a rise of 1°C, the freezing point of water (0°C) is about 273 K (see figure below).

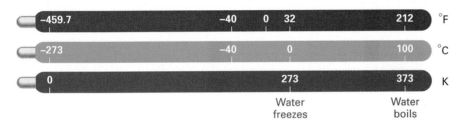

Temperature scales in the range useful for terrestrial and planetary studies.

To convert from kelvins to °C, simply subtract 273. To change from °C to °F, we must first multiply by ⅘ and then add 32. You may (or may not) find it easy to remember: times two, minus point two, plus thirty-two. That is, multiply by 2, subtract two-tenths of the original (which gives you ⅘), and then add 32, to get °F. For stellar temperatures, the 32° is too small to notice, and a sufficient level of approximation is often simply to multiply °C by 2 (see figure below).

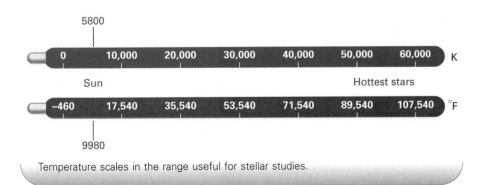

Temperature scales in the range useful for stellar studies.

The technique used to find these planets relies entirely upon the Doppler effect, since the planets are too faint and too close to their parent stars to be seen, even with our best telescopes on Earth or in space. The technique depends on the fact that the star isn't entirely steady in space, but rather moves to and fro as the planet around it moves fro and to. Consider, for example, a pair of dancers waltzing: They usually rotate around a point between them. If one of the dancers were invisible, we could still see the other dancer moving around. Similarly, we can see the visible object, the star, moving slightly toward and away from us, and infer that there must be an invisible object, the planet moving in step but in the

Figure 2-16 The technique used by Geoff Marcy and Paul Butler to detect planets around sun-like stars by searching for Doppler effects in the stars' spectra.

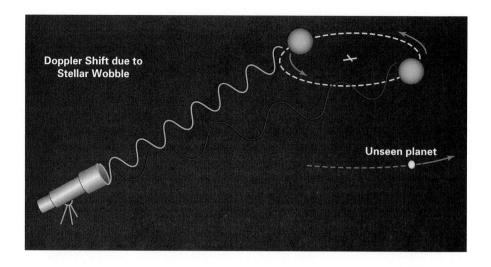

opposite direction at every time. Since the star is so much more massive than the planet, it moves less than the planet (Fig. 2–16).

The breakthrough in 1995 that enabled astronomers to detect these planets came when they developed some extremely precise ways of measuring Doppler shifts, more precise than had been possible before. More planets are now known outside our Solar System than inside it. We will describe those "extrasolar" planets themselves in Chapter 7.

Concept Review

Electromagnetic radiation is a combination of electricity and magnetism changing together in space; it is also known as **electromagnetic waves**. The **wavelength** is the distance between two consecutive crests (or troughs) of the wave, and different wavelengths correspond to different colors. The rainbow produced by passing light through a prism, as well as the graph of color (wavelength) versus the brightness at each color, is called a **spectrum**. Such **spectra** are extremely valuable to astronomers; they can be used to determine the temperature, chemical composition, and other properties of the object being studied. From the Earth's surface, we view through atmospheric windows that pass only certain types of electromagnetic radiation, chiefly visible and radio.

A very dense gas or solid gives off a **continuous spectrum** that changes smoothly in brightness from one color to the next. In the case of a **black body**, an object that absorbs all radiation that is incident upon it (none is reflected or transmitted), the shape of the emitted continuous spectrum depends only on the object's surface temperature, not its chemical composition or other properties. This result is called **black-body radiation** or **thermal radiation**. Spectra of hotter black bodies are brightest at shorter (bluer) wavelengths than spectra of cooler black bodies. Also, a hot black body is intrinsically brighter than a cold black body of equal surface area.

Atoms are composed of positively charged **nuclei** "orbited" by negatively charged **electrons**. The nuclei are made of

positively charged **protons** and neutral **neutrons**. **Quantum theory** explains how to describe atoms, and the rules that prevent electrons from spiralling into nuclei. Light sometimes acts as though it were made of particles, which are called **photons**. **Absorption lines** are dark gaps in a spectrum caused by atoms absorbing some of the photons that are passing through them; the electrons jump from lower to higher energy levels. Conversely, **emission lines** are locations in a spectrum where extra photons are present. They are caused by electrons jumping from higher to lower energy levels, thereby emitting light.

The **Bohr atom**, with electrons at different radii, explains the spectrum of hydrogen in detail. Transitions to the lowest **energy level**, called the **ground level**, do not cause lines in the visible. Transitions between the second energy level and higher levels (called **excited levels**) lead to photons being added or subtracted in the visible part of the spectrum. The resulting spectral lines are known as the **Balmer series**. When an atom loses one or more electrons, it is **ionized**.

The **Doppler effect** is a change in wavelength of light caused by an object's motion along a line toward or away from you. If you and the object are receding from each other, the object's spectrum appears **redshifted**, while **blueshifts** are seen if you and the object are approaching each other. Larger relative speeds produce bigger Doppler shifts.

Questions

1. What is the difference between a gamma ray and a radio wave?

2. Why is electromagnetic radiation so important to astronomers?

3. Discuss the difference between the general concept of the spectrum and the particular spectrum we see as a rainbow.

4. Does white light contain red light? Green light? Discuss.

5. After Newton separated sunlight into its component colors with a prism, he reassembled the colors. What should he have found and why?

6. Identify Roy G Biv. Identify A. J. Ångström. What relation do they have to each other?

7. How can our atmosphere have a "window"? Can our air leak out through it? Explain.

8. Describe the relation of Fraunhofer lines to emission lines and to absorption lines.

9. Describe how the same atoms can sometimes cause emission lines and at other times cause absorption lines.

10. Sketch an atom, showing its nucleus and its electrons.

11. Sketch the energy levels of a hydrogen atom, showing how the Balmer series arises.

12. On a sketch of the energy levels of a hydrogen atom, show with an arrow the transition that matches Hβ in absorption. Also show the transition that matches Hβ in emission.

13. If you were to take an emission nebula and put it in front of a very hot, bright source of continuous radiation, what change would you see in the nebula's spectrum? Explain.

14. If the surface of a star were increasing in temperature as you went outward, would you see absorption lines or emission lines in its spectrum? Explain how this case differs from what we see in normal stars, which decrease in temperature as you go outward in the region from which they give off most of their light.

15. Does the Bohr atom explain iron atoms? Describe from this chapter what type of atom the Bohr atom explains.

16. If someone were to say that we cannot know the composition of distant stars, since there is no way to perform experiments on them in terrestrial laboratories, how would you respond?

17. A spectral line from the left side of Saturn's rings, as you see them, is at a wavelength slightly shorter than the same spectral line measured from the right side of Saturn's rings. Which way is Saturn rotating, and why?

†18. Give the temperature in degrees Celsius for a 60°F day.

†19. What Fahrenheit temperature corresponds to 30°C?

†20. The Earth's average temperature is about 27°C. What is its average temperature in kelvins?

†21. The Sun's surface is about 5800 K. What is its temperature in °F?

22. An iron rod is heated by a welder's torch. Initially it glows a dull red, then a brighter orange, and finally very bright white. Discuss this sequence in terms of black-body radiation.

23. Does the Doppler effect depend on the distance between the source of light and the observer? Explain.

†24. The Hβ line of hydrogen has a wavelength of 4861 Å. (a) What is its frequency? (b) What is the frequency of a line with twice the wavelength of Hα?

†25. Announcers at a certain radio station say that they are at "95 FM on your dial," meaning that they transmit at a frequency of 95 MHz (95 million cycles per second). What is the wavelength of the radio waves from this station?

†26. (a) If one photon has 10 times the frequency of another photon, which photon is the more energetic, and by what factor? (b) Answer the same question for the case where the first photon has twice the wavelength of the second photon.

†27. Consider a black body whose temperature is 3 K. At what wavelength does its spectrum peak? (We will see the importance of this radiation in the discussion of cosmology in Chapter 19.)

†28. Suppose the peak of a particular star's spectrum occurs at about 6000 Å. (a) Use Wien's law to calculate the star's surface temperature. (b) If this star were a factor of four hotter, at what wavelength would its spectrum peak? In what part of the electromagnetic spectrum is this peak?

†29. (a) Compare the luminosity (amount of energy given off per second) of the Sun with that of a star the same size but three times hotter at its surface. (b) Answer the same question, but now assume the star also has twice the Sun's radius.

†30. How many times hotter than the Sun's surface is the surface of a star the same size, but that gives off twice the Sun's energy per second (that is, is twice as luminous)?

---

†These questions require a numerical solution.

Light and Telescopes: Extending Our Senses

ORIGINS *We discuss the telescopes used by astronomers to see the farthest and faintest objects, including a new generation of huge telescopes on the ground and the Hubble Space Telescope and Chandra X-ray Observatory aloft. These instruments allow us to learn about the earliest epochs of the Universe, to study how stars form, and to search for other planetary systems, among other things.*

AIMS

To learn how telescopes work, and about specific types of telescopes on the ground and in space for studying radiation inside and outside the visible spectrum.

Everybody knows that astronomers use telescopes, but not everybody realizes that the telescopes astronomers use are of very different types. Further, very few modern telescopes are used directly with the eye. In this chapter, we will first discuss the telescopes that astronomers use to collect visible light, as they have for hundreds of years. Then we will see how astronomers now also use telescopes to study gamma rays, x-rays, ultraviolet, infrared, and radio waves.

⬤ LIGHT AND TELESCOPES

Almost four hundred years ago, a Dutch optician put two eyeglass lenses together, and noticed that distant objects appeared closer. The next year, in 1609, the English scientist Thomas Harriot built one of these devices and looked at the Moon. But all he saw was a blotchy surface, and he didn't make anything of it. Credit for first using a telescope to make astronomical studies goes to Galileo Galilei. In 1609, Galileo heard that a telescope had been made in Holland, so in Venice he made one of his own and used it to look at the Moon. Perhaps as a result of his training in interpreting light and shadow in drawings—he was surrounded by the Renaissance and its developments in visual perspective—Galileo realized that the light and dark patterns on the Moon meant that there were craters there (Fig. 3–1). With the tiny telescopes he made—only 20 or 30 power, not much more powerful than a modern pair of binoculars and showing a smaller part of the sky—he went on to revolutionize our view of the cosmos. (WWW)

    Whenever he looked at Jupiter through his telescope, he saw that it was not just a point of light, but showed a small disk. He also spotted four points of light

REDSHIFT
http://www.harcourtcollege.com/astro/cosmos/rsce

◀ The Chandra X-ray Observatory in orbit when released from a space shuttle in 1999.

Figure 3–1 An engraving of Galileo's observations of the Moon from his book *Sidereus Nuncius (The Starry Messenger)*, published in 1610. A modern photo appears for comparison. Galileo was the first to report that the Moon has craters. It seems reasonable that Galileo had been sensitized to interpreting surfaces and shadows by the Italian Renaissance and by his related training in drawing. Aristotle and Ptolemy had held that the Earth was imperfect but that everything above it was perfect, so Galileo's observation contradicted them.

that moved from one side of Jupiter to another (Fig. 3–2). He eventually realized that the points of light were moons orbiting Jupiter, the first proof that not all bodies in the Solar System orbited the Earth. The existence of Jupiter's moons contradicted the ancient Greek philosophers—chiefly Aristotle and Ptolemy—who had held that the Earth is at the center of all orbits. Further, the ancient ideas that the Earth could not be in motion because the Moon (and other objects) would be "left behind" was also wrong. Galileo's discovery of the moons thus backed the newer theory of Copernicus, who had said in 1543 that the Sun and not the Earth is at the center of the Universe. And Galileo's lunar discovery—that the Moon's surface had craters—had also endorsed Copernicus's ideas, since the Greek philosophers had held that celestial bodies were all "perfect." Galileo published these discoveries in 1610 in his book *Sidereus Nuncius (The Starry Messenger)*.

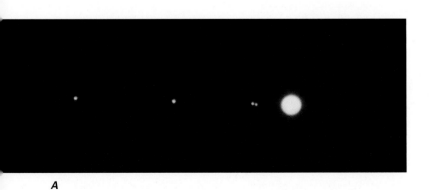

A

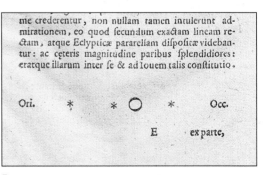

B

Figure 3–2 (A) Jupiter and its four Galilean satellites, observed from the Earth. (B) Some of Galileo's notes about his first observations of Jupiter's moons. Simon Marius also observed the moons at tebout the same time; we now use the names Marius proposed for them: Io, Europa, Ganymede, and Callisto, though we call them the Galilean satellites.

Figure 3-3 Venus goes through a full set of phases, from new to full (crescent to nearly full shown here).

Seldom has a book been as influential as Galileo's slim volume. He also reported in it that his telescope revealed that the Milky Way was made up of myriad individual stars. He drew many individual stars in the Pleiades, which we now know to be a star cluster. And he reported some stars in the middle of what we now know is the Orion Nebula. But one attempt made by Galileo in his *Sidereus Nuncius* didn't stick: he proposed to use the name "Medicean stars," after his financial backers, for the moons of Jupiter. Nowadays, recognizing Galileo's intellectual contribution rather than the Medici's financial contributions, we call them the "Galilean moons."

Galileo went on to discover that Venus went through a complete set of phases, from crescent to nearly full (Fig. 3–3), as it changed dramatically in size (Fig. 3–4). These variations were contrary to the prediction of the Earth-centered theory of Ptolemy and Aristotle that only a crescent phase would be seen (see Chapter 5). The Venus observations were thus the fatal blow to the geocentric hypothesis. He also found that the Sun had spots on it (which we now call "sunspots"), among many other exciting things.

But Galileo's telescopes had deficiencies, among them that white-light images were tinged with color, caused by the way that light is bent as it passes through lenses. Toward the end of the 17th century, Isaac Newton, in England, had the idea of using mirrors instead of lenses to make a telescope. Mirrors do not give the color effect, which is known as "chromatic aberration."

Certain curved lenses and mirrors can bring starlight to a single point, called the **focus** (Fig. 3–5). In astronomy we normally deal with light rays that are parallel to each other, which is the case for light from the stars and planets, since they are far away (Fig. 3–6). When your focusing mirror is only a few centimeters across, your head would block the incoming light if you tried to put your eye to this "prime" focus. Newton had the bright idea of putting a small, flat mirror just

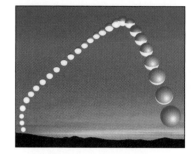

Figure 3-4 The position in the sky of Venus at different phases, in an artist's rendition of a computer plot. Venus is a crescent only when it is in a part of its orbit that is relatively close to the Earth, and so looks larger at those times.

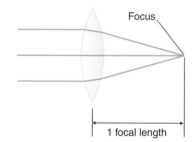

Figure 3-5 The focal length is the distance behind a lens to the point at which objects at infinity are focused. The focal length of the human eye is about 2.5 cm (1 inch).

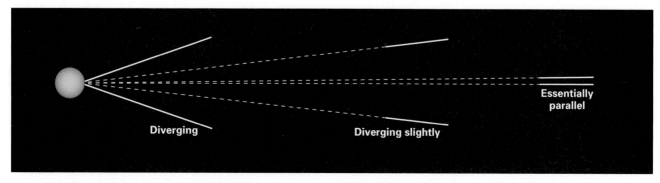

Figure 3-6 Parallel light is diverging imperceptibly, since the stars are so far away.

Figure 3-7 The path of light in a Newtonian telescope; note the diagonal mirror that brings the focus out to the side. Many amateur telescopes are of this type.

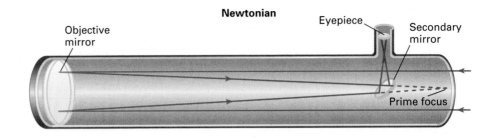

Newtonian

Figure 3-8 In the movie *Roxanne*, Daryl Hannah played an astronomy student. Here we see her with Steve Martin and her telescope, which we can tell is Newtonian by the eyepiece at its top. (© 1986 Columbia Pictures Industries, Inc. All rights reserved.)

in front of the focus to reflect the light out to the side, bringing the focus point outside the telescope tube. This **Newtonian telescope** (Figs. 3–7 and 3–8) is a design still in use by many amateur astronomers. However, many telescopes instead use the **Cassegrain** design, in which a secondary mirror bounces the light back through a small hole in the middle of the primary mirror (Fig. 3–9). Note that the hole, or an obstruction of part of the incoming light, only decreases the apparent brightness of the object; it does not alter its shape. After all, every part of the mirror forms a complete image of the object.

Spherical mirrors reflect light from their centers back onto the same point, but do not bring parallel light to a good focus (Fig. 3–10). This effect is called **spherical aberration.** We now often use mirrors that are in the shape of a **paraboloid,** since only paraboloids bring parallel light to a focus (Fig. 3–11).

Through the 19th century, telescopes using lenses (**refracting telescopes,** or refractors) and telescopes using mirrors (**reflecting telescopes,** or reflectors) were made larger and larger. The pinnacle of refracting telescopes was reached in the 1890s with the construction of a telescope with a lens 40 inches (1 m) across for the Yerkes Observatory in Wisconsin, now part of the University of Chicago (Fig. 3–12). It was difficult to make a lens of clear glass thick enough to support its large diameter; moreover, such a thick lens may sag from the weight, absorbs light, and also suffers from chromatic aberration. And the telescope tube had to be tremendously long. Because of these difficulties, no larger telescope lens has ever been put into service, though the Yerkes telescope is still in research use today, even after more than 100 years.

The size of a telescope's main lens or mirror is particularly important because the primary job of most telescopes is to collect light—to act as a "light bucket." All the light is brought to a common focus, where it is viewed or recorded. (See *Figure It Out: Light-Gathering Power of a Telescope.*) The larger the telescope's lens or mirror, the fainter the objects that can be viewed or the more quickly observations can be made. A larger telescope would also provide better **resolution**—the ability to detect fine detail—if it weren't for the shimmering of the Earth's atmosphere, which limits all large telescopes to about the same resolution (technically, angular resolution). Only if you can improve the resolution is it worthwhile magnifying images. For the most part, then, the fact that telescopes magnify is secondary to their ability to gather light. ✧

Figure 3-9 A Cassegrain telescope, in which a curved secondary mirror bounces light through a hole in the center of the primary mirror.

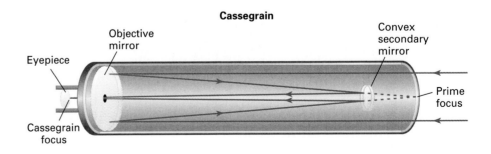

Cassegrain

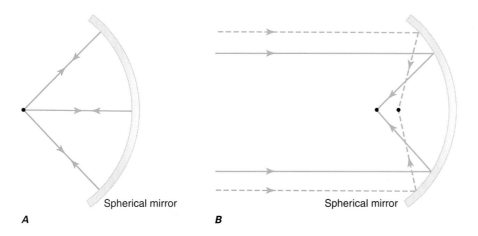

Spherical mirror   Spherical mirror

A   *B*

Figure 3–10 *(A)* A spherical mirror focuses light that originates at its center of curvature back on itself. *(B)* A spherical mirror suffers from spherical aberration in that it does not perfectly focus light from very large distances, for which the incoming rays are essentially parallel. Lenses that don't have the proper shape can also show spherical aberration.

From the mid-19th century onward, larger and larger reflecting telescopes were constructed. But the mirrors, then made of shiny metal, tended to tarnish. This problem was avoided by evaporating a thin coat of silver onto a mirror made of glass. More recently, a thin coating of aluminum turned out to be longer lasting, though silver with a thin transparent overcoat of tough material is now coming back into style. The 100-inch (2.5-m) reflector at the Mt. Wilson Observatory in California became the largest telescope in the world in 1917. Its use led to discoveries about distant galaxies that transformed our view of what the Universe is like and what will happen to it and us in the far future.

In 1948, the 200-inch (5-m) reflecting telescope opened at the Palomar Observatory, also in California, and was for many years the largest in the world. Current electronic imaging devices have made this and other large telescopes many times more powerful than they were when they recorded images on film.

Some of the most interesting astronomical objects are in the southern sky, so astronomers need telescopes at sites more southerly than the continental United States. For example, the nearest galaxies to our own—known as the Magellanic Clouds—are not observable from the continental United States. The

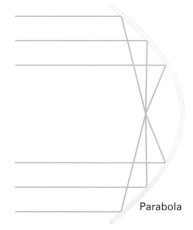

Parabola

Figure 3–11 A paraboloid is a three-dimensional curve created by spinning a parabola on its axis. As shown here, a parabola focuses parallel light to a single point, as does a paraboloid.

Figure It Out

Light-Gathering Power of a Telescope

A telescope's main purpose is to collect light quickly and in large quantities. The mirror or lens of a telescope acts as a gigantic eye pupil that intercepts the light rays from a distant object. The larger its area, the more light it will collect in a given time, allowing fainter objects to be seen. (The length of the telescope tube is irrelevant to this "light-gathering power.")

The area of a circle is proportional to the square of its radius r (or diameter $D = 2r$): $A = \pi r^2 = \pi (D/2)^2 = \pi D^2/4$. Therefore, the ratio of the areas of two telescopes with circular mirrors of diameter D_1 and D_2 is given by $A_2/A_1 = D_2{}^2/D_1{}^2 = (D_2/D_1)^2$.

Suppose $D_2 = 6$ m (a typical size for a large optical telescope), and $D_1 = 6$ mm = 0.006 m (as for the pupil opening of a dilated eye). The ratio of areas is $(D_2/D_1)^2 = (6$ m$/0.006$ m$)^2 = 1000^2 = 10^6$. Thus, looking through the eyepiece of such a telescope, one could see stars a million times fainter than with the unaided eye!

By attaching a detector to the telescope, the exposure time can be made very long, making even fainter stars visible. Some detectors, such as charge-coupled devices (CCDs), are far more sensitive than eyes, and detect most of the photons that hit them. With large telescopes, long exposures, and high-quality CCDs, objects over 10^9 (a billion) times fainter than the limit of the unaided eye have been detected.

Figure 3–12 The Yerkes refractor, still the largest in the world, has a 1-m diameter lens.

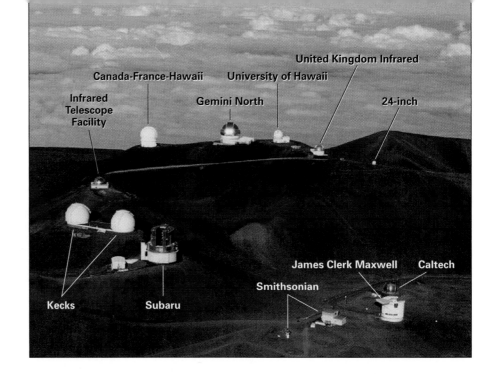

Figure 3–13 Mauna Kea, with its many huge telescopes. The Keck twin domes with the Subaru dome near them are at the mid-left; the NASA Infrared Telescope Facility is above them; and the Canada-France-Hawaii, Gemini North, University of Hawaii 88-inch, United Kingdom Infrared, and University of Hawaii 24-inch telescopes are on a ridge at top. In the "millimeter valley" at lower right, we see the Caltech 10-m submillimeter and James Clerk Maxwell 15-m telescopes (Maxwell, perhaps the third greatest physicist of all time—after Newton and Einstein—unified electricity and magnetism theoretically), as well as the first to be installed of eight 6-m telescopes and the assembly building of the Smithsonian Astrophysical Observatory's submillimeter array.

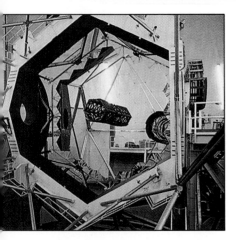

Figure 3–14 The domes of the Keck 10-m telescopes.

Figure 3–15 Each 10-m mirror of the Keck telescopes is made of 36 contiguous hexagonal segments, which are continuously adjustable.

National Optical Astronomy Observatories, supported by the National Science Foundation, have a half-share in two telescopes, each with 8-m mirrors, the northern-hemisphere one in Hawaii and the southern-hemisphere one in Chile. The project is called Gemini, since Gemini are the twins in Greek mythology (and the name of a constellation) and these are twin telescopes. The other half-share in the project is divided among the United Kingdom, Canada, Chile, Australia, Argentina, and Brazil. By sharing, the United States has not only half the time on a telescope in the northern hemisphere, but also half the time on a telescope in the southern hemisphere, a better case than having a full telescope in the north and nothing in the south.

The observatory with the most large telescopes is now on top of the dormant volcano Mauna Kea in Hawaii, partly because its latitude is as far south as +20°, allowing much of the southern sky to be seen, and partly because the site is so high that it is above 40 per cent of the Earth's atmosphere. To detect the infrared part of the spectrum, telescopes must be above as much of the water vapor in the Earth's atmosphere as possible, and Mauna Kea is above 90 per cent of it. In addition, the peak is above the atmospheric inversion layer that keeps the clouds from rising, usually giving about 300 nights each year of clear skies with steady images. Consequently, several of the world's dozen largest telescopes are there (Fig. 3–13). (www)

In particular, the California Institute of Technology and the University of California have built the Keck 10-m telescopes (Fig. 3–14), whose mirrors are each twice the diameter and four times the surface area of Palomar's largest reflector. Hence, each one is able to gather light four times faster. When it was built, a single 10-m mirror would have been prohibitively expensive, so University of California scientists worked out a plan to use a mirror made of 36 smaller hexagons (Fig. 3–15). The first telescope worked so well that a twin (Keck II) was quickly built beside it. Not only the Gemini-North 8-m telescope but also a Japanese 8-m telescope have been completed on Mauna Kea. (www)

The largest project of all is the European Southern Observatory's Very Large Telescope, an array of four 8-m telescopes being opened one at a time and scheduled for completion in 2001 (Fig. 3–16). The images from the first three telescopes are superb (Fig. 3–17).

As we mentioned, the angular size of the finest details you can see (the resolution) is basically limited by the Earth's atmosphere, but a new technique called

Figure 3-16 Domes of the Very Large Telescope, in Chile. The entire project actually consists of four individual 8-m telescopes plus several smaller telescopes.

adaptive optics is improving the resolution of some telescopes. ☼ The light from the main mirror is reflected off a secondary mirror whose shape can be slightly distorted many times a second to compensate for the atmosphere's distortions. So after a long hiatus, the resolution from ground-based telescopes has been improving recently.

⬤ WIDE-FIELD TELESCOPES

Ordinary optical telescopes see a fairly narrow field of view—that is, a small part of the sky. Even the most modern show images of less than about 1° × 1°, which means it would take decades to make images of the entire sky (over 40,000 square degrees). The German optician Bernhard Schmidt, in the 1930s, invented a way of using a thin lens ground into a complicated shape together with a spherical mirror to image a wide field of sky (Fig. 3–18).

The largest **Schmidt telescopes,** except for one of interchangeable design, are at the Palomar Observatory in California and at the United Kingdom Schmidt site in Australia (Fig. 3–19); these telescopes have front lenses 1.25 m (49 inches) in diameter and mirrors half again as large. (The front lens is larger than the back mirror in order to allow study of objects off to the side.) They can observe a field of view some 7° × 7°—almost the size of your fist held at the end of your out-

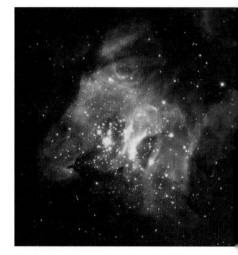

Figure 3-17 The Butterfly Nebula, imaged with one of the telescopes of the Very Large Telescope in Chile.

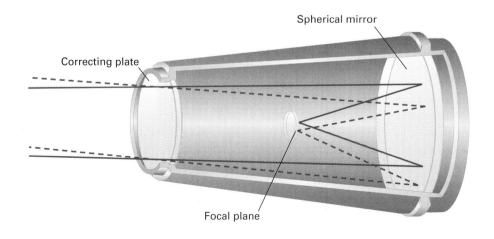

Spherical mirror

Correcting plate

Focal plane

Figure 3-18 By having a nonspherical thin lens called a correcting plate, a Schmidt camera is able to focus a wide angle of sky onto a curved piece of film. Since the image falls at a location where you cannot put your eye, the image is always recorded on film. Accordingly, this device is often called a Schmidt camera rather than a Schmidt telescope.

35

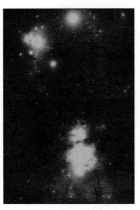
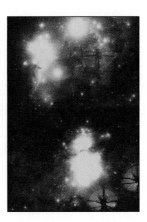

A

Figure 3–19 *(A)* Three individual images, each taken with the U.K. Schmidt telescope (which is in Australia) through a filter in one color, are printed together to make the full-color image. *(B)* This color view of the Horsehead Nebula in Orion is from the top left quarter of the three monochromatic (single-color) images.

B

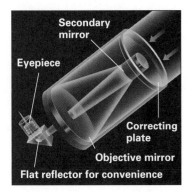

Figure 3–20 An amateur astronomer's Newtonian telescope. Note that the eyepiece is perpendicular to the tube near the tube's top.

Figure 3–21 A cutaway drawing of a compound Schmidt–Cassegrain design, now widely used by amateurs because of its light-gathering power and portability.

stretched arm, compared with only the size of a grain of sand at that distance for the Hubble Space Telescope!

The Palomar Schmidt telescope was used in the 1950s to survey the whole sky visible from southern California with film and filters that made pairs of images in red and blue light. This Palomar Observatory Sky Survey is a basic reference for astronomers. Hundreds of thousands of galaxies, quasars, nebulae, and other objects have been discovered on them. The Schmidt telescopes in Australia and Chile have extended this survey to the southern hemisphere. Moreover, Palomar recently completed a new survey with improved films and more overlap between adjacent regions. Among other things, it is being compared with the first survey to see whether objects have changed or moved. Both surveys have been digitized.

An interesting new method of surveying wide regions of sky has been worked out by the Sloan Digital Sky Survey, using a telescope at Apache Point, New Mexico. **www** It uses electronic detectors to survey the sky as the Earth turns, and observes in several different colors. So many data are being collected that developing new methods of data handling is an important part of the project. The sky mapping is now well under way. (See some examples in Figure 16–6.)

◖◗ AMATEUR TELESCOPES

It is fortunate for astronomy as a science that so many people are interested in looking at the sky. Many are just casual observers, who may look through a telescope occasionally as part of a course or on an "open night," when people are invited to view through telescopes at a professional observatory, but others are quite devoted "amateur astronomers." Some amateur astronomers make their own equipment, ranging up to quite large telescopes perhaps 60 cm in diameter. But most amateur astronomers use one of several commercial brands of telescopes.

Many of the amateur telescopes are Newtonian reflectors, with mirrors 15 cm in diameter being the most popular size (Fig. 3–20). It is quite possible to shape your own mirror for such a telescope. The Dobsonian telescope is a variant of this type, made with very inexpensive mirrors and construction methods.

Increasingly popular are compound telescopes that combine some features of reflectors with some of the Schmidt telescopes. This Schmidt–Cassegrain de-

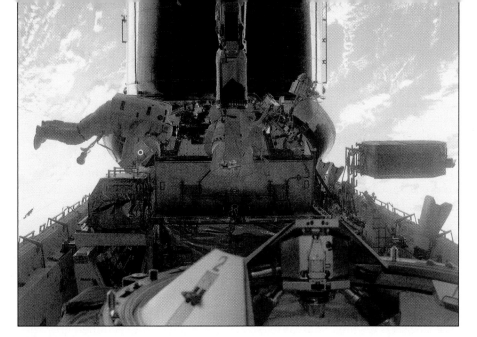

Figure 3-22 An astronaut works on the Hubble Space Telescope during its 1999 servicing mission. The gyroscopes that are used to stabilize and point the telescope were replaced. The telescope is named after Edwin Hubble, who discovered the true nature of galaxies and the expansion of the Universe (see Chapter 16).

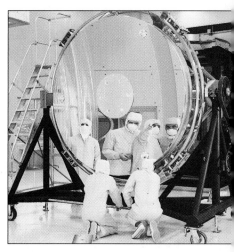

Figure 3-23 The 2.4-m (94-inch) mirror of the Hubble Space Telescope, after it was ground and polished and covered with a reflective coating. This single telescope has already made numerous discoveries but is raising as many questions as it answers. So we need still more ground-based telescopes as partners in the enterprise, in addition to more telescopes in space.

sign (Fig. 3-21) folds the light, so that the telescope is relatively short, making it easier to transport and set up. (www)

THE HUBBLE SPACE TELESCOPE

The first moderately large telescope to be launched above the Earth's atmosphere is the Hubble Space Telescope (HST or Hubble, for short) (Fig. 3-22), built by NASA with major contributions from the European Space Agency. The set of instruments on board is sensitive not only to visible light but also to ultraviolet and infrared radiation that don't pass through the Earth's atmosphere. (www)

When launched, the Hubble was supposed to provide images with about 10 times better resolution than ground-based images; now, over narrow fields of view, the latest advances in adaptive optics reduce its advantage in resolution to a factor of 3 or so. The Hubble saga is a dramatic one, since the main mirror (Fig. 3-23) turned out to be made with slightly the wrong shape. Apparently, an optical system used to test it was made slightly the wrong size, and it indicated that the mirror was in the right shape when it actually wasn't. The result is some amount of spherical aberration, which blurred the images and caused great disappointment when it was discovered soon after the April 1990 launch.

A mission launched in 1993 carried a replacement for the main camera and correcting mirrors for other instruments, and brought the telescope to full operation. Fortunately, the telescope was designed so that space-shuttle astronauts could visit it every few years to make repairs. A second generation of equipment, installed in 1997, includes detectors sensitive to the infrared. A mission in December 1999 replaced the gyroscopes that hold the telescope steady and made certain other repairs and improvements, such as installing better computers. A mission for late 2000 is to install an Advanced Camera for Surveys and a cooler (refrigerator) to make the infrared camera work again. Still another upgrade, to replace its main camera, is planned for about 2003.

Hubble's high resolution is able to concentrate the light of a star into an extremely small region of the sky—the star isn't blurred out. This, plus the very dark background sky at high altitude, allows us to see fainter objects than we could formerly (Fig. 3-24). The combination of resolution and sensitivity is

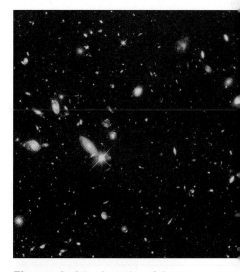

Figure 3-24 A section of the Hubble Deep Field, a tiny region of the sky in the constellation Ursa Major viewed for 10 days over and over with the Hubble Space Telescope. Many faint and distant galaxies are revealed. A Hubble Deep Field—South has since been observed.

leading to great advances toward solving several basic problems of astronomy. As we shall see later, we are able to pin down our whole notion of the size and age of the Universe much more accurately.

The Hubble Space Telescope should last until about 2010. A Next Generation Space Telescope is being planned, with an 8-m mirror, for launch in about 2008. (www) It will be optimized to work at infrared wavelengths, which will make it especially useful in studying the origins of planets and in looking for extremely distant objects in the Universe.

SOLAR TELESCOPES

Telescopes that work at night usually have to collect a lot of light. Solar telescopes work during the day, and often have far too much light to deal with. They have to get steady images of the Sun in spite of viewing through air turbulence caused by solar heating. So solar telescopes are usually designed differently from nighttime telescopes.

Solar observatories in space have taken special advantage of their position above the atmosphere to make x-ray and ultraviolet observations. The Japanese Yohkoh ("Sunbeam") spacecraft, launched in 1991, carries x-ray and ultraviolet telescopes for study of solar activity. The Solar and Heliospheric Observatory (SOHO), a joint European Space Agency/NASA mission, is located a million kilometers upward toward the Sun, for constant viewing of all parts of the Sun. The Transition Region and Coronal Explorer (TRACE) gives even higher resolution images of loops of gas at the edge of the Sun (Fig. 3–25). All these telescopes should give especially interesting data during the maximum of sunspot and other solar activity of 2000–2001.

OUTSIDE THE VISIBLE SPECTRUM

We can describe a light wave by its wavelength (Fig. 3–26). But a large range of wavelengths is possible, and visible light makes up only a small part of this broader spectrum (Fig. 3–27). **Gamma rays, x-rays,** and **ultraviolet** light have shorter wavelengths than visible light, and **infrared** and **radio waves** have longer wavelengths.

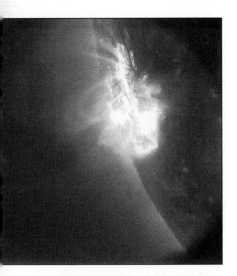

Figure 3–25 Loops of hot coronal gas from the Transition Region and Coronal Explorer (TRACE) spacecraft imaged in the ultraviolet.

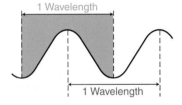

Figure 3–26 Waves of electric and magnetic fields travelling across space are called "radiation." The wavelength is the length over which a wave repeats.

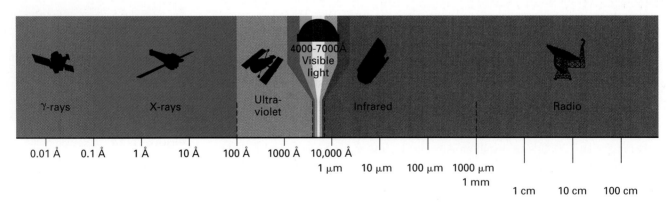

Figure 3–27 The spectrum. The silhouettes represent telescopes or spacecraft used or planned for observing that part of the spectrum: the Gamma Ray Observatory, the Chandra X-ray Observatory, the Hubble Space Telescope, the dome of a ground-based telescope, the Infrared Space Observatory, and a ground-based radio telescope.

Figure 3–28 An outside view of the nested Chandra X-ray Observatory mirrors.

Figure 3–29 One of the cylindrical mirrors from the Chandra X-ray Observatory. Its inside is polished to reflect x-rays by grazing incidence.

X-Ray and Gamma-Ray Telescopes

The shortest wavelengths would pass right through the glass or even the reflective coatings of ordinary telescopes, so special imaging devices have to be made to study them. And x-rays and gamma rays do not pass through the Earth's atmosphere, so they can be observed only from satellites in space. NASA's series of three High-Energy Astronomy Observatories (HEAOs) was tremendously successful in the late 1970s. One of them even made detailed x-ray observations of individual objects with resolution approaching that of ground-based telescopes working with ordinary light.

NASA's best x-ray telescope is called the Chandra X-ray Observatory, named after a scientist who made important studies of white dwarfs and black holes. It was launched in 1999 (see the Chapter Opener). It makes its high-resolution images with a set of nested mirrors made on cylinders (Figs. 3–28 and 3–29). Ordinary mirrors could not be used because x-rays would pass right through them. However, x-rays bounce off mirrors at low angles, just as stones can be skipped across a lake at low angles (Fig. 3–30). Chandra joins Hubble as one of NASA's "Great Observatories."

In the gamma-ray part of the spectrum, the Compton Gamma-Ray Observatory was launched in 1991, also as part of NASA's Great Observatories program. NASA planned to destroy it in June 2000, since the loss of one more of its gyros would make it more difficult to control where its debris would land on Earth.

⬤ TELESCOPES FOR SHORT WAVELENGTHS

Ultraviolet wavelengths are longer than x-rays but still shorter than visible light. All but the longest wavelength ultraviolet light does not pass through the Earth's atmosphere, so must be observed from space. For about two decades, the 20-cm telescope on the International Ultraviolet Explorer spacecraft sent back valuable ultraviolet observations. Overlapping it in time, the 2.4-m Hubble Space Telescope was launched, and has a much larger mirror and so is much more sensitive to ultraviolet radiation.

Several ultraviolet telescopes have been carried aloft for brief periods aboard space shuttles, and brought back to Earth at the end of the shuttle mission. At present, NASA's Far Ultraviolet Spectrographic Explorer (FUSE) is taking high-

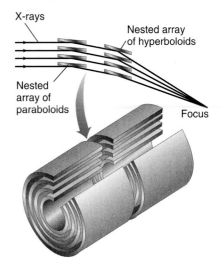

Figure 3–30 Four similar paraboloid/hyperboloid mirror arrangements, used at low angles to incoming x-rays and all sharing the same focus, are nested within each other to increase the area of telescope surface that intercepts x-rays in NASA's Chandra X-ray Observatory.

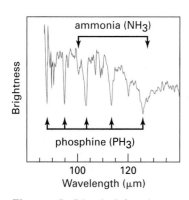

Figure 3-31 An Infrared Space Observatory spectrum of Saturn in a region of infrared radiation with wavelengths over 10 times longer than those of visible light.

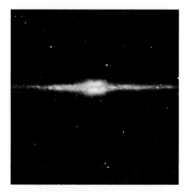

Figure 3-32 The sky, mapped by the Cosmic Background Explorer spacecraft in 1990, reveals mainly the Milky Way and thus the shape of our galaxy. The image here is false color, a translation into visible wavelengths of three infrared wavelengths that penetrate interstellar dust.

resolution spectra largely in order to study the origin of the elements in the Universe.

Infrared Telescopes

From high-altitude sites such as Mauna Kea, parts of the infrared can be observed from the Earth's surface. From high aircraft altitudes, even more can be observed, and NASA is refitting an airplane named SOFIA with a 2.5-m infrared telescope. Cool objects such as planets and dust around stars in formation emit most of their radiation in the infrared, so studies of planets and of how stars form have especially benefited from infrared observations.

An international observatory, the Infrared Astronomical Satellite (IRAS), mapped the sky in the 1980s, and then was followed by the European Infrared Space Observatory (ISO) in the mid-1990s. (www) Since the telescopes and detectors themselves, because of their warmth, emit enough infrared radiation to overwhelm the faint signals from space, the telescopes had to be cooled way below normal temperatures using liquid helium. These telescopes mapped the whole sky, and discovered a half dozen comets, hundreds of asteroids, hundreds of thousands of galaxies, and many other objects. ISO took the spectra of many of these objects (Fig. 3–31). During 1990–1994, the Cosmic Background Explorer (COBE) spacecraft mapped the sky in a variety of infrared and radio wavelengths in order to make cosmological studies. We shall discuss its cosmological discoveries later on. Since infrared penetrates the haze in space, its whole-sky view (Fig. 3–32) reveals our own Milky Way Galaxy.

A large, sensitive infrared mission, the Space Infrared Telescope Facility, is on NASA's plans as the infrared Great Observatory, with launch hoped for in 2002. (www)

Radio Telescopes

Since Karl Jansky's 1930s discovery (Fig. 3–33) that astronomical objects give off radio waves, radio astronomy has advanced greatly. Huge metal "dishes" are giant reflectors that concentrate radio waves onto antennae that enable us to detect faint signals from objects in outer space. A new radio telescope at the National Radio Astronomy's site at Green Bank, West Virginia, has a reflecting surface 100 m × 100 m in an unusual design (Fig. 3–34). (www) The world's largest telescope that can point anywhere in the sky, it replaces in that role a 100-m-diameter radio telescope near Bonn, Germany (Fig. 3–35).

Figure 3-33 *(A)* Karl Jansky *(inset)* and the full-scale model of the rotating antenna with which he discovered radio waves from space. *(B)* A sculpture honoring the original Jansky telescope, erected at the original site in Holmdel, New Jersey, in 1998.

A   *B*

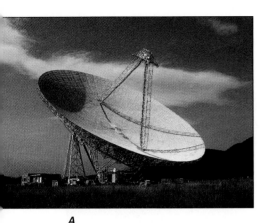

A   *B*

Figure 3–34   The 300-foot (91-m) radio telescope at Green Bank, West Virginia, *(A)* before and *(B)* after its dramatic collapse in 1988. *(C)* The new 100-m Green Bank Telescope, which went into use in 2000.

C

A still larger dish in Arecibo, Puerto Rico, is 1000 feet (330 m) across, but points only more or less overhead. Still, all the planets and many other interesting objects pass through its field of view. This telescope and one discussed below starred in the movie *Contact*.

Astronomers almost always convert the incoming radio signals to graphs or intensity values in computers and print them out, rather than converting the radio waves to sound with amplifiers and loudspeakers. If the signals are converted to sound, it is usually only so that the astronomers can monitor them to make sure no radio broadcasts are interfering with the celestial signals.

Radio telescopes were originally limited by their very poor resolution. The resolution of a telescope depends only on the telescope's diameter, but we have to measure the diameter relative to the wavelength of the radiation we are studying. So for a radio telescope studying waves 10 cm long, even a 100-m telescope is only 1000 wavelengths across. Even a 10-cm optical telescope studying ordinary light is 200,000 wavelengths across, so is effectively much larger and gives much finer images (Fig. 3–36). (See *Figure It Out: Angular Resolution of a Telescope*.)

Figure 3–35   The 100-m (330-ft) radio telescope near Bonn, Germany, the largest fully steerable radio telescope in the world.

Radio wave   **Light wave**

← 1¾ λ →   ← 8 λ →

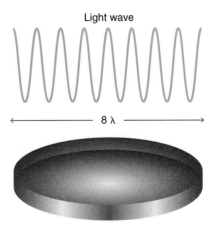

Radio telescope dish   Light telescope mirror

Figure 3–36   For familiarity's sake, we often refer to telescopes by their physical size, though this is not the most significant measure. It is more meaningful to measure the diameter of telescope mirrors in terms of the wavelength of the radiation that is being observed than it is to measure it in terms of units like centimeters that have no particularly relevant significance. The radio dish at left is only 1¾ wavelengths across, whereas the mirror at right is 8 wavelengths across, making it effectively much bigger. The wavelengths are greatly exaggerated in this diagram relative to the size of any actual reflectors. For the Bonn radio telescope used to observe 10-cm waves, 100-m diameter divided by 0.1 m per wave = 1000 wavelengths.

Figure 3–38   The Very Large Array (VLA) is laid out in a giant "Y" with a diameter of 27 km, and sprawls across a plain in New Mexico. It contains 27 telescopes (plus a spare). It is so large that even the weather is sometimes different from one side to another. Its imaging capability has been important for understanding radio sources.

Figure 3–37   The millimeter interferometer of the Berkeley-Illinois-Maryland Association (BIMA).

New technology has made it possible to observe radio waves well at relatively short wavelengths, those measured as a few millimeters. Molecules in space are especially well studied at these wavelengths (Fig. 3–37). The new Green Bank telescope will be useful for studying such molecules.

A breakthrough in providing higher resolution has been the development of arrays of radio telescopes that operate together and give the resolution of a single telescope spanning kilometers or even continents. The **Very Large Array (VLA)** is a set of 27 radio telescopes each 26 m in diameter, and was also seen in *Contact* (Fig. 3–38). All the telescopes operate together, and powerful computers analyze the joint output to make detailed pictures of objects in space. These telescopes are linked to allow the use of "interferometry" to mix the signals; analysis later on gives images of very high resolution. The VLA's telescopes are spread out over dozens of square kilometers on a plain in New Mexico.

To get even higher resolution, astronomers have built the **Very Long Baseline Array (VLBA)**, spanning the whole United States (Fig. 3–39). Its images are many times higher in resolution than even those of the VLA. Astronomers often use the technique of "very-long-baseline interferometry" to link telescopes at such distances, or even distances spanning continents, but the VLBA dedicates telescopes full-time to such high-resolution work.

Figure 3–39   *(A)* The telescopes in the VLBA—the Very-Long-Baseline Array. *(B)* The VLBA antenna on Kitt Peak.

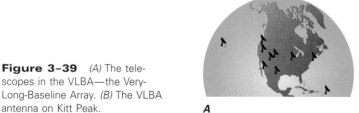

A

B

Figure It Out

Angular Resolution of a Telescope

Telescopes improve the clarity with which objects are seen: they have higher *angular resolution* (the ability to see fine detail) than the human eye.

Angular measure is important in astronomy. The full circle is divided into 360 degrees (360°). The Moon and the Sun each subtend (cover) about ½°. Each degree consists of 60 arc minutes (60′). Each minute of arc consists of 60 arc seconds (60″). A second of arc is very small—approximately the angle subtended by a dime viewed from a distance of 3.7 km. One can also use radians for angular measure. There are 2π radians in 360°, so $1″ = 1/206265$ radian.

If two point-like objects are closer together than 1–2 arc minutes, the unaided eye will perceive them as only one object because their individual "blur circles" merge together. With a telescope, the size of the individual blur circles decreases, and the objects become resolved (see the figure). The angular diameter of the blur circle is proportional to λ/D, where λ is the observation wavelength and D is the diameter of the lens or mirror. Hence, in principle, large telescopes are able to resolve finer details than small telescopes (at a given wavelength).

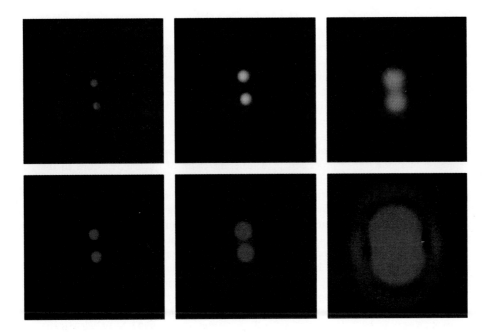

In addition to their primary use for gathering light from faint objects, telescopes are often used to increase the resolution, our ability to distinguish details in an image. We see a pair of point sources (sources so small or so far away that they appear as points), in the right-hand column, that are not quite resolved in *(top row)* violet light (about 4000 Å) and *(bottom row)* red light (about 6500 Å). Note that the rightmost violet image is substantially better resolved than the rightmost red image. In this series, the images were obtained with lenses (or mirrors) of different diameter; from right to left, the diameter is larger by a factor of 2 in successive images. In astronomy, as long as we aren't limited by shimmering (turbulence) in the Earth's atmosphere, a smaller image disk is produced by a larger telescope or at a shorter wavelength.

A rule of thumb found by the amateur astronomer Dawes in the 19th century is that the resolution (in arc seconds) equals $0.002\lambda/D$, if λ is given in Å and D is given in cm. This resolution is the angular separation at which a double star is detectable as having two separate components (if they have equal brightness).

As an example, the resolution of a telescope 1 meter (i.e., 100 cm) in diameter for green light of approximately 5000 Å is given by $0.002(5000/100) = 0.1$ arc second.

In practice, Earth's atmosphere blurs starlight: Layers of air with different densities move in a turbulent way relative to each other, and the rays of light bend in different directions. This is related to the twinkling of starlight. The angular resolution of telescopes larger than 20–30 cm is limited by the blurring effects of the atmosphere (typically 1 arc second, and rarely smaller than ⅓ arc second), not by the size of the mirror or lens. Thus, even bigger ground-based telescopes do not give clearer images, unless advanced techniques such as adaptive optics are used.

A Closer Look

A Night at Mauna Kea

An "observing run" with one of the world's largest telescopes is a highlight of the work of many astronomers. The construction of several of the world's largest telescopes at the top of Mauna Kea in Hawaii has made the mountain the site of one of the world's major observatories. Mauna Kea was chosen because of its clear and steady air. These conditions stem from the fact that the summit is so high—4200 m (13,800 ft) above sea level—and the mountain is surrounded by ocean. Though there are taller mountains in the world, none has such favorable conditions for astronomy.

Let us imagine you have applied for observing time and have been chosen. During the months before your observing run, you make lists of objects to observe, prepare charts of the objects' positions in the sky based on existing star maps and photographs, and plan the details of your observing procedure. The air is very thin at the telescope. You may not think as clearly or rapidly as you do at lower altitudes. Therefore, it is wise and necessary to plan each detail in advance.

The time for your observing run comes, and you fly off to the island of Hawaii, the largest of the islands in the state of Hawaii. You rent a car and drive up the mountain, as the scenery changes from tropical to relatively barren, crossing dark lava flows. First stop is the mid-level facility (Hale Pohaku) at an altitude of 2750 m (9000 ft). The Onizuka Center for International Astronomy at Hale Pohaku is named after Ellison Onizuka, the Hawaiian astronaut who died in the Challenger space-shuttle explosion in 1986.

The astronomers and technicians sleep and eat at Hale Pohaku. The top of Mauna Kea is too high for people to sleep and work comfortably for too long. The rules say that you must spend a full night acclimatizing to the high altitude before you can begin your own telescope run.

The next night is yours. During the afternoon, you test the instrument with which the data will be obtained and determine the exact configuration in which it is to be used. For example, if you will be obtaining optical spectra of objects, the spectrograph needs to be focused properly and you must choose specific wavelength ranges to observe. Often this can be done remotely from Hale Pohaku, using computers, but for some telescopes an afternoon trip to the summit might be necessary.

After dinner you return to the summit to start your observing. You dress warmly, since the nighttime temperature approaches freezing at this altitude, even in the summertime. Although most telescope domes are equipped with warm rooms in which astronomers sit when collecting data, occasionally you might step outside to absorb the magnifi-

cence of the dark sky—or monitor the motion of encroaching clouds.

During the night you are assisted by a Telescope Operator who knows the telescope and its systems very well, and who is responsible for the telescope, its operation, and its safety. The operator points the telescope to the object you have chosen to observe and makes sure the telescope is properly compensating for the Earth's rotation. A television camera mounted on the telescope shows objects within the telescope's field of view. You examine that region of the sky, carefully comparing it with your chart to identify the right object among many random, unwanted stars. That object is then centered at the proper position to take data.

Nearly everything else is computer-controlled. You measure the brightness of the object at the wavelength you are observing, or begin an exposure to get a spectrum, by starting a computer program on a console provided for the observer. The data are stored in digital form on computer disks and tapes, and you might even transfer them over the Internet to your university computer.

Suppose you are interested in what kinds of chemical elements are produced and ejected by supernovae (exploding stars). Spectra of a given type of supernova, but at various stages of its development, could reveal that certain heavy elements are closer to the center of the explosion, while others are farther out. But another type of supernova observed that night might show a different set of heavy elements, processed by a different chain of nuclear reactions in the stellar furnace before and during the explosion. You labor through the night, occasionally seeing an obviously exciting result, but more often just getting tantalizing hints that demand much more detailed and time-consuming analysis.

At some time during the night, you may go into the telescope dome to ponder the telescope at work. The telescope all but blocks your view of the sky. Hardly anyone ever actually looks through the telescope; indeed, there is usually no good way to do so. Still, you may sense a deep feeling for how the telescope is looking out into space.

You drive down the mountain to Hale Pohaku in the early morning sun. Over the next few days and nights, you repeat your new routine. When you leave Mauna Kea, it is a shock to leave the pristine air above the clouds. But you take home with you data about the objects you have observed. It may take you months or years to study the data fully that you gathered in a few brief days and nights at the top of the world. But you hope that your data will allow a more complete understanding of some aspect of the Universe. Sometimes major conceptual breakthroughs are made.

Concept Review

Refracting telescopes use lenses and **reflecting telescopes** use mirrors to collect light. **Spherical aberration** occurs when parallel light is not reflected or refracted to a good **focus,** the place where the image is formed; it is solved for on-axis light with a mirror in the shape of a **paraboloid.** Chromatic aberration, a property only of lenses, occurs when light of different colors does not reach the same focus. **Newtonian telescopes** reflect the light from the main mirror diagonally out to the side; **Cassegrain** telescopes use a secondary mirror to reflect the light from the main mirror through a hole in the primary. The larger the area of the telescope's main lens or mirror, the better its light-gathering power, the ability to detect faint objects. Also, the larger the diameter, the higher the theoretical **resolution,** the ability to distinguish detail. In practice, the shimmering of the Earth's atmosphere limits the resolution of large telescopes, but a new technique called **adaptive optics** allows astronomers to achieve very clear images over small fields of view.

Schmidt telescopes have a very wide field of view. The Sloan Digital Sky Survey is carrying out an electronic sky survey. The Hubble Space Telescope gives some observations of resolution about ten times better than most ground-based observations. Solar telescopes have technical differences from nighttime telescopes.

Telescopes to observe **x-rays,** gamma rays, and most **ultraviolet** light must be above the Earth's atmosphere. The Chandra X-ray Observatory is a Hubble-class instrument. Some parts of the **infrared** can be observed with telescopes on the ground, but telescopes in airplanes or in orbit are also necessary. Radio telescopes on the ground observe a wide range of wavelengths covered by **radio waves.** Arrays of radio telescopes, such as the **Very Large Array (VLA)** in New Mexico and the continent-spanning **Very Long Baseline Array (VLBA),** give high resolution through a process known as interferometry.

Questions

1. What are three discoveries that immediately followed the first use of the telescope for astronomy?
2. What advantage does a reflecting telescope have over a refracting telescope?
3. What limits a large ground-based telescope's ability to see detail?
4. List the important criteria in choosing a site for an optical observatory meant to study stars and galaxies.
5. Describe a method that allows us to make large optical telescopes more cheaply than simply scaling up designs of previous large telescopes.
6. Name some of the world's largest telescopes and their locations.
7. What are the similarities and differences between making radio observations and using a reflector for optical observations? Compare the path of the radiation, the detection of signals, and limiting factors.
8. Why is it sometimes better to use a small telescope in orbit around the Earth than it is to use a large telescope on a mountaintop?
9. Why is it better for some purposes to use a medium-size telescope on a mountain instead of a telescope in space?
10. Describe and compare the Hubble Space Telescope and the Chandra X-ray Observatory. Mention their uses, designs, and other relevant factors.
11. What are two reasons why the Hubble Space Telescope can observe fainter objects than we can now study from the ground?
†12. What is the light-gathering power of a telescope that is 3 meters in diameter, relative to a 1-m telescope? What is it relative to a dilated human eye (pupil 6 mm in diameter)?
†13. What mirror diameter gives 0.1 arc second resolution for infrared radiation of wavelength 2 micrometers?
†14. What mirror diameter gives 1 arc second resolution for radio radiation of wavelength 1 m? Compare this with the size of existing optical telescopes.

†This question requires a numerical solution.

People in Astronomy

JEFF HOFFMAN

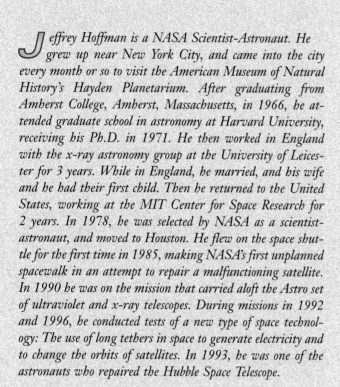

Jeffrey Hoffman is a NASA Scientist-Astronaut. He grew up near New York City, and came into the city every month or so to visit the American Museum of Natural History's Hayden Planetarium. After graduating from Amherst College, Amherst, Massachusetts, in 1966, he attended graduate school in astronomy at Harvard University, receiving his Ph.D. in 1971. He then worked in England with the x-ray astronomy group at the University of Leicester for 3 years. While in England, he married, and his wife and he had their first child. Then he returned to the United States, working at the MIT Center for Space Research for 2 years. In 1978, he was selected by NASA as a scientist-astronaut, and moved to Houston. He flew on the space shuttle for the first time in 1985, making NASA's first unplanned spacewalk in an attempt to repair a malfunctioning satellite. In 1990 he was on the mission that carried aloft the Astro set of ultraviolet and x-ray telescopes. During missions in 1992 and 1996, he conducted tests of a new type of space technology: The use of long tethers in space to generate electricity and to change the orbits of satellites. In 1993, he was one of the astronauts who repaired the Hubble Space Telescope.

How did you enjoy being in space?

Answering as an astronomer first, astronomers are used to working on mountaintops, and space is the ultimate mountaintop. Actually, for me, working on the Astro mission [a set of ultraviolet telescopes carried on a space shuttle] was unique, because my professional work had been x-ray astronomy using satellites, so I had never actually done anything with traditional telescopes. The first time I guided an actual optical-type telescope in my life was in space—but my three astronomer-astronaut colleagues let me do it anyway.

Since I had spent most of my professional career building x-ray telescopes to fly in rockets and satellites, it was gratifying to fly with some telescopes on board. Of course, the fun of being in space goes beyond what your actual mission is. No matter what you are doing up there, it is an incredible view, an incredible feeling. But working as an astronomer on one of my space flights gave me a lot of professional satisfaction. And of course

I was delighted to be able to make an important contribution to astronomy when I fixed the Hubble Space Telescope, which was undoubtedly the most significant of all my five space missions.

What did it feel like to repair the Hubble Space Telescope?

Fixing the Hubble Space Telescope was the most important and challenging task of my entire astronaut career. It is easy in retrospect to forget the incredible shock produced by the discovery of Hubble's initial optical flaw, but in many ways the whole future of NASA's human spaceflight program rested on our ability to show that astronauts working in space suits could fix the problem. Had we not succeeded, it is quite possible that Congress would not have given NASA the go-ahead to build the new International Space Station. We worked extremely hard training for the mission, spending over four hundred hours underwater and countless weeks in simulators. When we finally took off, I can honestly say that we had done everything we could think of to ensure the success of the mission. Of course, we knew that there were many unexpected things that could go wrong no matter how hard we trained. In fact, as things turned out, the most surprising thing about the mission was how few unpleasant surprises we had. The most critical problem was our initial inability to close one of the instrument compartment doors during our first spacewalk, but after analyzing the problem we figured out how to adapt a special tool to allow us to do the job. I felt pleased at having solved the problem, which shows the ability of human beings to adapt to unforeseen difficulties. At the end of the fifth and final spacewalk, we were elated at having been able to accomplish every one of the tasks that we had set out to do. Of course, we would not know for several weeks whether the new optics we installed had actually corrected Hubble's vision. It was New Year's Eve when I finally got the news from some astronomer friends working at the Space Telescope Science Institute that Hubble was finally working flawlessly. What a great way to celebrate the new year!

How did you get to be an astronaut?

How did I originally decide I wanted to be an astronaut? That's been going on for a long time, ever since I was a little kid. I first got interested in astronomy back in the Hayden Planetarium in New York City. But I got interested not only in astronomy but also in anything having to do with space, including rockets. Of course, there was no such thing as a real space program back then, and I wanted to be a scientist. The first astronauts were all jet pilots, and that didn't appeal to me, though I was excited by the rocket part of it. But when they announced that they needed scientists to be astronauts on the shuttle program, I always knew that that was something that I wanted to do, so I applied.

> ". . . my inspiration to become an astronomer and to become an astronaut spring from the same fascination as looking beyond where we are now, looking out from the Earth."

What do you do for NASA now?

I represent NASA in Europe. This is a bit like being an ambassador. Space flight is increasingly an international activity, and it is useful to have someone "on the scene" to work with our partner space agencies. My office is in the American Embassy in Paris, since the headquarters of the European Space Agency is located in Paris. My technical background and experience with international projects during my various space missions gave me the qualifications for this job. In addition, I speak several European languages, which helps a lot.

What would you most like to see NASA do next?

Exploration in all its aspects is NASA's primary mission. We need to develop space telescopes even more powerful than Hubble to continue our exploration of the astronomical universe. I think the search for extrasolar planetary systems and the search for life in the Universe is one of the most exciting scientific goals of the new millennium. Closer to home, we have a lot of exploring to do in our own Solar System. I am particularly excited about searching for signs of life on Mars and Europa. And of course I am interested in expanding human capability to travel and work in space. The International Space Station is in the next step in this development. Making all this happen requires more reliable and cheaper space transportation, which is also one of NASA's main goals.

What message do you have for students?

First of all, I like to try to spread an ecological message that we have to take care of the Earth, our planetary home in space. We get a lot of responses to pictures that we take of the Earth from space, particularly where we can show the environmental changes taking place on the planet. Kids really seem to respond to that. We get disturbing sequences of pictures taken over the last 15 years showing the deforestation of the Amazon, the encroaching desert in sub-Saharan Africa, and land erosion in Madagascar—one environmental disaster after the other, which you can see better from space than from anywhere else.

I also like to talk to young people about the fact that you can study physics and astronomy and apply it in numerous different ways other than just becoming a professional astronomer. For instance, I can show my younger son, who is thinking about what he wants to do after university, two examples of friends of mine from graduate school. One of them started in physics and moved to biology and one was in applied mathematics and also moved to biology, and both do a lot of work in environmental science. Both developed the skills of mathematical analysis and facility with computers, which we use all the time to model complex systems in astronomy. I often find that my training as a physicist allows me to cut to the heart of problems in a way that some people who were trained as engineers sometimes don't do.

The other thing that I often stress is the fact that my inspiration to become an astronomer and to become an astronaut spring from the same fascination at looking beyond where we are now, looking out from the Earth. I hope that we can keep the dream alive for the next generation so that they will be able to live out some of their dreams as well, whether they are studying through telescopes or travelling outside the Earth.

Observing the Stars and Planets: Clockwork of the Universe

ORIGINS *Throughout their existence, humans have used the apparent positions of celestial objects to define the day, month, year, and the seasons, as well as for navigation.*

The Sun, the Moon, and the stars rise every day in the eastern half of the sky and set in the western half. If you leave your camera on a tripod with the lens open for a few minutes or hours, you will photograph the star trails, the trails across the photograph left by the individual stars. In this chapter, we will discuss the phases of the Moon and planets, and how to find stars and planets in the sky. Stars twinkle; planets don't twinkle much (Fig. 4–1). We will also discuss the motions of the Sun, Moon, and planets as well as of the stars in the sky.

THE PHASES OF THE MOON AND PLANETS

From the simple observation that the apparent shapes of the Moon and planets change, we can draw conclusions that are important for our understanding of the mechanics of the Solar System. In this section, we shall see how the positions of the Sun, Earth, and other Solar-System objects determine the appearance of these objects.

The **phases** of moons or planets are the shapes of the sunlighted areas as seen from a given vantage point. The fact that the Moon goes through a set of such phases approximately once every month is perhaps the most familiar astronomical observation, aside from the day/night cycle and the fact that the stars come out at night (Fig. 4–2). In fact, the name "month" comes from the word "moon." The actual period of the phases, the interval between a particular phase of the Moon and its next repetition, is approximately 29½ Earth days (Fig. 4–3).

AIMS

To learn about the basic motions of celestial objects, and how they make us see and experience phases, rising and setting, seasons, time, and calendars.

•

To learn about how the sky looks in different seasons.

Figure 4–1 A camera was moved steadily from left to right *(bottom)*, with the bright star Sirius in view. The star trail shows twinkling. Also, the camera was moved from left to right *(top)* when pointed at Jupiter, which left a solid, non-twinkling trail. Notice the contrast between the nontwinkling planet trail and the twinkling star trail. The fact that the star trails break up into bits of different colors was caused by part of the twinkling effect of the Earth's atmosphere, given that Sirius was low in the sky when the picture was taken.

49

Figure 4-2 The phases of the Moon depend on the Moon's position in orbit around the Earth. Here we visualize the situation as if we were looking down from high above the Earth's orbit. Each of the Moon images shows how the Moon is actually lighted. Note that the Sun always lights half the Moon.

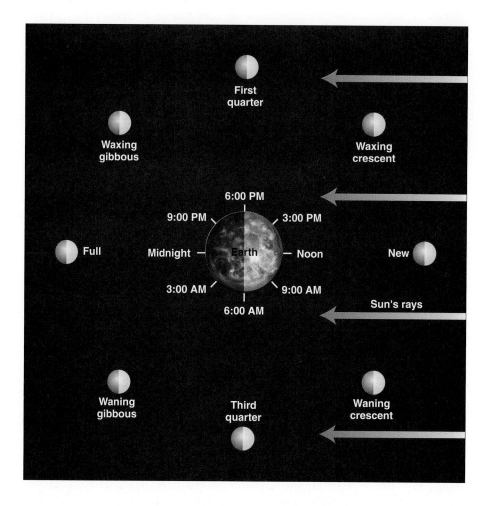

The explanation of the phases is quite simple: the Moon is a sphere that shines by reflecting sunlight, and at all times the side that faces the Sun is lighted and the side that faces away from the Sun is dark. The phase of the Moon that we see from the Earth, as the Moon revolves (orbits) around us, depends on the relative orientation of the three bodies: Sun, Moon, and Earth. The situation is simplified by the fact that the plane of the Moon's revolution around the Earth is nearly, although not quite, the plane of the Earth's revolution around the Sun.

When the Moon is almost exactly between the Earth and the Sun, the dark side of the Moon faces us. We call this a "new moon." A few days earlier or later we see a sliver of the lighted side of the Moon, and call this a "crescent." As the month progresses, the crescent gets bigger (a "waxing crescent"), and about seven days after a new moon, half the face of the Moon that is visible to us is lighted.

Figure 4-3 The phases of the Moon.

Waxing crescent

Waxing crescent

First quarter

Waxing gibbous

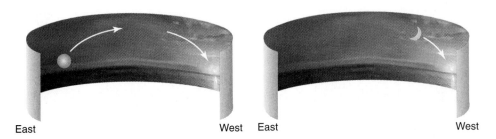

East   West   East   West

Figure 4–4 Because the phase of the Moon depends on its position in the sky with respect to the Sun, a full moon always rises at sunset. A crescent moon is either setting shortly after sunset, as shown here, or rising shortly before sunrise.

We are one quarter of the way through the cycle of phases, so we have the "first-quarter moon."

When over half the Moon's disk is visible, it is called a "gibbous moon." As the sunlighted portion visible to us grows, the gibbous moon is said to be "waxing." One week after the first-quarter moon, the Moon is on the opposite side of the Earth from the Sun, and the entire face visible to us is lighted. This is called a "full moon." Thereafter we have a "waning gibbous moon." One week after full moon, when again half the Moon's disk that we see appears lighted, we have a "third-quarter moon." This phase is followed by a "waning crescent moon," and finally by a "new moon" again. The cycle of phases then repeats.

Note that since the phase of the Moon is related to the position of the Moon with respect to the Sun, if you know the phase, you can tell approximately when the Moon will rise. ☼ For example, since the Moon is 180° across the sky from the Sun when it is full, a full moon is always rising just as the Sun sets (Fig. 4–4). Each day thereafter, the Moon rises an average of 50 minutes later. The third-quarter moon, then, rises near midnight, and is high in the sky at sunrise. The new moon rises with the Sun in the east at dawn, and sets with the Sun in the west at dusk. The first-quarter moon rises near noon and is high in the sky at sunset.

It is natural to ask why the Earth's shadow doesn't generally hide the Moon during full moon, and why the Moon doesn't often block the Sun during new moon. These phenomena are rare because the Moon's orbit is tilted by about 5° relative to the Earth–Sun plane, making it difficult for the Sun, Earth, and Moon to become exactly aligned. However, when they do reach the right configuration, eclipses occur, as discussed in Chapter 9.

The Moon is not the only object in the Solar System that goes through phases. Mercury and Venus both orbit inside the Earth's orbit, and so sometimes we see the side that faces away from the Sun and sometimes we see the side that faces toward the Sun. Thus at times Mercury and Venus are seen as crescents, though it takes a telescope to observe their shapes. Spacecraft to the outer planets have looked back and seen the Earth as a crescent and the other planets as crescents (Fig. 4–5) as well.

Figure 4–5 Uranus in its crescent phase photographed as the Voyager 2 spacecraft looked back on it.

Full moon

Waning gibbous

Third quarter

Waning crescent

A

Figure 4–6 *(A) But there is no East Star or West Star, since the Earth rotates on its axis. Only the positions of the north and south celestial poles are steady, and there is no South Star. (Cartoon by Charles Schulz. © 1970 United Feature Syndicate, Inc.) (B) Lucy is still making things up. You can never see the north and south celestial poles in the sky at the same time, since they are 180° from each other in the sky. Further, there is no South Star, that is, there is no obvious star near the south celestial pole. (Cartoon by Charles Schulz. © 1970 United Feature Syndicate, Inc.)*

TWINKLING

If you look up at night in a place far from city lights, you may see hundreds of stars with the naked eye. They will seem to change in brightness from moment to moment, that is, to **twinkle.** This twinkling comes from moving regions of air in the Earth's atmosphere. The air bends starlight, just as a glass lens bends light. As the air moves, the starlight is bent by different amounts and the strength of the radiation hitting your eye varies, making the point-like stars seem to twinkle.

Unlike stars, planets are close enough to us that they appear as tiny disks when viewed with telescopes, though we can't quite see these disks with the naked eye. As the air moves around, even though the planets' images move slightly, there are enough points on the image to make the average amount of light we receive keep relatively steady. (At any given time, some points are brighter than average and some are fainter.) So planets, on the whole, don't twinkle. As a rule, a bright object in the sky that isn't twinkling is a planet.

But when a planet is low enough on the horizon, as Venus often is when we see it, it too can twinkle because there is then so much turbulent air along the line of sight. Since the bending of light by air is different for different colors, we sometimes even see Venus or a bright star turning alternately reddish and greenish. On those occasions, professional astronomers sometimes get calls that UFOs have been sighted—especially when the crescent moon is nearby, drawing attention to the unusual configuration.

RISING AND SETTING STARS

Though stars seem to rise, move across the sky, and set each night, the Earth is actually turning on its axis and the stars are holding steady. If you extend the Earth's axis beyond the north pole and the south pole, these extensions point to the **celestial poles.** Stars appear to traverse circles or arcs around the celestial poles.

Since the orientation of the Earth's axis doesn't change (at least on timescales of years), the celestial poles don't appear to move in the course of the night. From our latitudes (the United States ranges from about 20° north latitude for Hawaii, to about 49° north latitude for the northern continental United States, to 65° for Alaska), we can see the north celestial pole but not the south celestial pole. A star named Polaris happens to be near the north celestial pole, only about 1° away, so we call Polaris the **pole star.** If you are navigating at sea or in a desert at night, you can always go due north by heading straight toward Polaris (Fig. 4–6).

Polaris is conveniently located at the end of the handle of the Little Dipper, and can easily be found by following the "Pointers" at the end of the bowl of the Big Dipper (Fig. 4–6*A*). Polaris isn't especially bright, but you can find it if city lights have not brightened the sky too much.

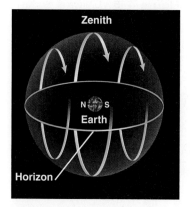

Figure 4–7 Viewed from the equator, the stars rise straight up, pass right across the sky, and set straight down.

B

The north celestial pole is the one fixed point in our sky, since it never moves. To understand the motion of the stars, let us first consider two simple cases. If we were at the Earth's equator, then the two celestial poles would be on the horizon (due north and due south), and stars would rise in the eastern half of the sky, go straight up and across the sky, and set in the western half (Fig. 4–7). Only a star that rose due east of us would pass directly overhead and set due west. If, on the other hand, we were at the Earth's north pole, then the north celestial pole would always be directly overhead. No stars would rise and set, but they would all move in circles around the sky, parallel to the horizon (Fig. 4–8).

We live in an intermediate case, where the stars rise at an angle (Fig. 4–9). Close to the celestial pole, we can see that the stars are really circling the pole (Fig. 4–10). Only the pole star itself remains relatively fixed in place, although it too traces out a small circle of radius about 1° around the north celestial pole.

As the Earth spins, it wobbles slightly, like a giant top, because of the gravitational pulls of the Sun and the Moon. As a result of this **precession,** the axis

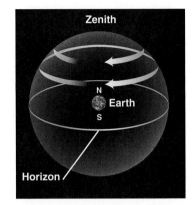

Figure 4–8 Viewed from the pole, the stars move around the sky in circles parallel to the horizon, never rising or setting.

Figure 4–9 Near the celestial equator, the star circles are so large that they appear almost straight. Here we are looking east past the twin Keck telescopes on Mauna Kea.

Figure 4–10 When we look toward the north celestial pole, stars appear to move in giant circles about the pole. Here, we are looking past one of the telescopes on Kitt Peak.

A Closer Look

Photographing the Stars

It is easy to photograph the stars if you have a dark sky far from city lights. You should use a camera that allows the option of taking a long exposure (normally the "bulb" setting); most "point-and-shoot" cameras won't work. Place your camera on a tripod, disable your flash unit, make the lens wide open (perhaps f/1.4 or f/2), and set the focus to infinity. Use a cable release to open the shutter for 8 minutes or more; don't take exposures of over 15 minutes without first testing shorter times to make sure that background skylight doesn't fog the film. You will have a picture of star trails. If the North Star is roughly centered in your field of view, you will see the circles that the stars take as a result of the Earth's spinning on its axis.

With the new fast color films, you can record the stars in a constellation with exposures of a few seconds. Use a 50-mm or 135-mm lens and try a series: 1, 2, 4, 8, and 16 seconds. The stars will not noticeably trail on the shorter exposures. The constellation Orion, visible during the winter, is a particularly interesting constellation to photograph, since your photograph will show the reddish Orion Nebula in addition to the stars. Your eyes are not sensitive to color when viewing such faint objects, but film will record them.

HINT: Take a picture of a normal scene at the beginning of the roll, and one at the end as well, so that the photofinisher will know where to cut apart the slides or how to make the prints. Be prepared to send back negatives for printing (or request that they be printed automatically), in spite of the photofinisher's note that they didn't come out. The photofinisher probably didn't notice the tiny specks the stars made.

---

actually traces out a large circle in the sky with a period of 26,000 years (Fig. 4–11). So Polaris is the pole star at the present time, and generally there isn't a prominent star near either celestial pole.

⬤ APPARENT MAGNITUDE

To describe the brightness of stars in the sky, astronomers—professionals and amateurs alike—use a scale that stems from the ancient Greeks. Over two millennia ago, Hipparchus described the typical brightest stars in the sky as "of the first magnitude," the next brightest as "of the second magnitude," and so on. The faintest stars were "of the sixth magnitude."

We still use a similar scale, the **magnitude scale,** though now it is on a mathematical basis. Each difference of 5 magnitudes is a factor of 100 times in brightness. A 1st-magnitude star is exactly 100 times brighter than a 6th-magnitude star—that is, we receive exactly 100 times more energy in the form of light. Sixth

Figure 4–11 The Earth's axis precesses with a period of 26,000 years. The two positions shown are separated by 13,000 years. As the Earth's pole precesses, the equator moves with it (since the Earth is a rigid body). The celestial equator and the ecliptic—the Sun's apparent path through the stars—will always maintain the 23½° angle between them, but the points of intersection, the equinoxes, will change. Thus, over the 26,000-year cycle, the vernal equinox will move through all signs of the zodiac. It is now in the constellation Pisces and approaching Aquarius (and thus the celebration in the musical *Hair* of the "Age of Aquarius").

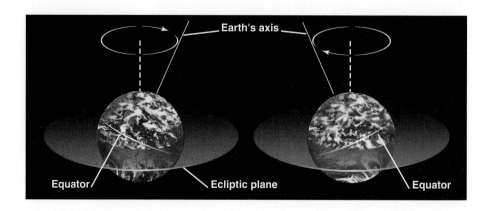

magnitude is still the faintest that we can see with the naked eye, though because of urban sprawl the dark skies necessary to see such faint stars are harder to find these days than they were long ago. Objects too faint to see with the naked eye have magnitudes greater than 6th. A few stars and planets are brighter than 1st magnitude, so the scale has also been extended in the opposite direction into negative numbers. For example, Sirius (the brightest star in the sky) has a magnitude of −1.4, and Venus can reach magnitude −4.4.

Since each difference of 5 magnitudes is a factor of 100 times, each one magnitude is a number that, when five of them are multiplied together, equals 100. No integer has this property. After all, $1 \times 1 \times 1 \times 1 \times 1 = 1$, $2 \times 2 \times 2 \times 2 \times 2 = 32$, which is less than 100, and $3 \times 3 \times 3 \times 3 \times 3 = 243$, which is greater than 100, so the number we want is somewhere between 2 and 3. A number with this property is simply called "the fifth root of 100." So, each difference of 1 magnitude is a factor of the fifth root of 100 (which is approximately equal to 2.512, or even just 2.5) in brightness. This choice is necessary to make a difference of 5 magnitudes ($1 + 1 + 1 + 1 + 1$, an additive process) equal to a factor of 100 ($2.5 \times 2.5 \times 2.5 \times 2.5 \times 2.5$, a multiplicative process).

The magnitude scale—**apparent magnitude,** since it is how bright the stars appear—is fixed by comparison with the historical scale (Fig. 4–12). If you read about a 13th-magnitude quasar, you should know that it is much too faint to see with the naked eye. If you read about a new telescope on the ground or in space observing a 29th-magnitude galaxy, you should know that the galaxy is one of the faintest objects we can currently study.

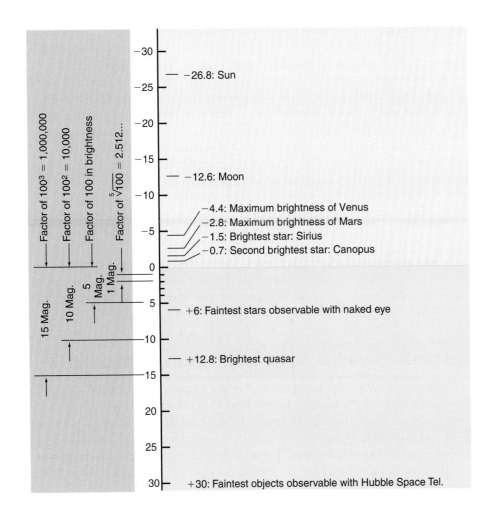

Figure 4–12 The apparent magnitude scale is shown on the vertical axis. At the left, sample intervals of 1, 5, 10, and 15 magnitudes are marked and translated into multiplicative factors of 2.512; 100; 10,000; and 1,000,000. For example, a difference of 1 magnitude is a factor of about 2.5 times in brightness; a difference of 5 magnitudes is a factor of 100 times in brightness.

Figure 4–13 The celestial equator is the projection of the Earth's equator onto the sky, and the ecliptic is the Sun's apparent path through the stars in the course of a year. The vernal equinox is one of the intersections of the ecliptic and the celestial equator, and is the zero-point of right ascension. From a given location at the latitude of the United States, the stars nearest the north celestial pole never set and the stars nearest the south celestial pole never rise above the horizon. Right ascension is measured along the celestial equator. Each hour of right ascension equals 15°. Declination is measured perpendicularly (−10°, −20°, etc.).

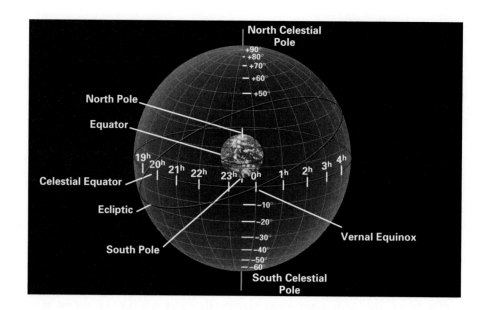

REDSHIFT
http://www.harcourtcollege.com/astro/cosmos/rsce

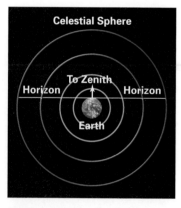

Figure 4–14 The zenith is the point over your head. The horizon marks as far as you can see. If the celestial sphere were nearby, it is clear that much less than half of it would be above the horizon. We can see progressively more of the celestial sphere at any given time if it is farther away, as shown. In fact, it is so far away that we see essentially half of it at a time.

CELESTIAL COORDINATES

Geographers divide the surface of the Earth into a grid, so that we can describe locations. The equator is the line half-way between the poles. Lines of constant *longitude* run from pole to pole, crossing the equator perpendicularly. Lines of constant *latitude* circle the Earth, parallel to the equator.

Astronomers have a similar coordinate system in the sky. They assume, for this purpose, that the celestial objects are on an imaginary sphere, the **celestial sphere,** that surrounds the Earth. Stars are so distant that we have no depth perception: they all seem to be glued to the enormously large sphere.

Imagine that we are at the center of the celestial sphere, looking out at the stars. ☼ The **zenith** is the point directly over our heads. The **celestial equator** circles the sky on the celestial sphere, half-way between the celestial poles. It lies right above the Earth's equator. Lines of constant **right ascension** run between the celestial poles, crossing the celestial equator perpendicularly. They are similar to terrestrial longitude. Lines of constant **declination** circle the celestial sphere, parallel to the celestial equator, similarly to the way that terrestrial latitude circles our globe (Fig. 4–13). The right ascension and declination of a star are essentially unchanging, just as each city on Earth has a fixed longitude and latitude. (Precession actually causes the celestial coordinates to change very slowly.)

From any point on Earth that has a clear horizon, one can see only half of the celestial sphere at any given time. This can be visualized by extending a plane that skims the surface so that it intersects the very distant celestial sphere (Fig. 4–14).

OBSERVING THE CONSTELLATIONS

When we look outward into space, we see stars that are at different distances. But though some stars are much farther away than others of the same brightness, our eyes do not show us these differences. People have long made up stories about groups of stars that appear in one part of the sky or another. The major star groups are called **constellations.** These constellations were given names, occasionally because they resembled something (for example, Scorpius, the Scorpion), but mostly to honor the object.

Figure It Out

Sidereal Time

Since a whole circle is divided into 360°, and the sky appears to turn completely once every 24 hours, the sky appears to turn 15 degrees per hour (15°/hr). Astronomers therefore divide right ascension (r.a.) into hours, minutes, and seconds (of time) instead of degrees. At the equator, one hour of r.a. = 15°. Dividing by 60, 1 minute of r.a. = 1/4° = 15 minutes of arc (since there are 60 minutes of arc in a degree). Dividing by 60 again, 1 second of r.a. = 1/4 minute of arc = 15 arc seconds (since there are 60 seconds of arc in a minute of arc).

We can keep "star time" by noticing the right ascension of a star that is passing due south of us; we say that such a star is "crossing the meridian," where the meridian is a line extending due north-south and passing through the zenith. This star time is called sidereal time; astronomers keep special clocks that run on sidereal time to show them which celestial objects are crossing the meridian, that is, going due south of us. A sidereal day (a day by the stars) is about 4 minutes shorter than a solar day (a day by the Sun), such as the one we usually keep track of on our watches.

The International Astronomical Union put the scheme of constellations on a definite system in 1930. The sky was officially divided into 88 constellations (see Appendix 9) with definite boundaries, and every star is now associated with one and only one constellation. But the constellations give only the directions to the stars, and not the stars' distances. Individual stars in a given constellation can have quite different distances from us (Fig. 4–15); the stars aren't physically associated with each other.

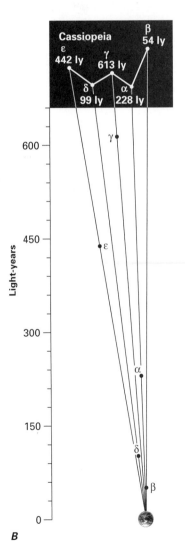

A

Figure 4–15 *(A)* The constellation Cassiopeia is easily found in the sky from its distinctive **W**. *(B)* The stars we see as a constellation are actually at different distances from us. In this case, we see the true relative distances of the stars in the **W** of Cassiopeia, as determined from the Hipparcos spacecraft. The stars' appearance projected on the sky is shown in the upper part.

B

starparties

USING THE SKY MAPS

Since the sidereal day is 4 minutes shorter than a solar day, the parts of the sky that are "up" after dark change slightly each day. A given star rises (and crosses the meridian) 4 minutes earlier each day. By the time a season has gone by, the sky has apparently slipped a quarter of the way around at sunset as the Earth has moved a quarter of the way around the Sun in its yearly orbit. Some constellations are lost in the afternoon and evening glare, while others have become visible just before dawn.

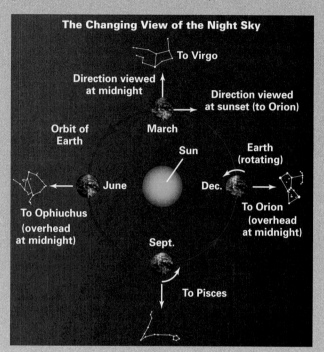

The Changing View of the Night Sky

To Virgo

Direction viewed
at midnight

Direction viewed
at sunset (to Orion)

Orbit of
Earth

March

Sun

Earth
(rotating)

June

Dec.

To Ophiuchus
(overhead
at midnight)

To Orion
(overhead
at midnight)

Sept.

To Pisces

In December of each year, the constellation Orion crosses the meridian at midnight. Three months later, in March, when the Earth has moved through one quarter of its orbit around the Sun, the constellation Virgo crosses the meridian at midnight, when Orion is setting. Orion crosses the meridian at sunset (that is, 6 hours earlier than in December—consistent with 4 minutes/day × 90 days = 360 minutes = 6 hours). Another three months later, in June, Orion crosses the meridian an additional 6 hours earlier—that is, at noon. Hence, it isn't then visible at night. Instead, the constellation Ophiuchus crosses the meridian at midnight.

Because of this seasonal difference, we have included at the end of this book four Sky Maps, one of which is best for the date and time at which you are observing. Suitable combinations of date and time are marked. Note also that if you make your observations later at night, it is equivalent to observing later in the year. Two hours later at night is the same as shifting over by one month.

Hold the map above your head while you are facing north or south, as marked on each map, and notice where your zenith is in the sky and on the map. The horizon for your latitude is also marked. Try to identify a pattern in the brightest stars that you can see. Finding the Big Dipper, and using it to locate the pole star, often helps you to orient yourself. Don't let any bright planets confuse your search for the bright stars—remember that planets usually appear to shine steadily instead of twinkling like stars.

Come back and look at the following sections when it is that time of year—even after you have finished with this course.

The Autumn Sky

As it grows dark on an autumn evening, you will see the Pointers in the Big Dipper point upward toward Polaris. Almost an equal distance on the other side of Polaris is a W-shaped constellation named Cassiopeia. Cassiopeia, in Greek mythology, was married to Cepheus, the king of Ethiopia (and the subject of the constellation that neighbors Cassiopeia to the west). Cassiopeia appears sitting on a chair (Fig. 4–15*A*).

Continuing across the sky away from the Pointers, we next come to the constellation Andromeda, who in Greek mythology was Cassiopeia's daughter. In Andromeda, you might see a faint, hazy patch of light; this is actually the center of the nearest large galaxy to our own, and is known as the Great Galaxy in Andromeda, or the Andromeda Galaxy. Though it is one of the nearest galaxies, about 2.4 million light-years away, it is much farther away than any of the individual stars that we see, since they are all in our own galaxy.

Southwest in the sky from Andromeda, but still high overhead, are four stars that appear to make a square known as the Great Square of Pegasus. One of its corners is actually in Andromeda.

If it is really dark out (which probably means that you are far from a city and also that the Moon is not full or almost full), you will see the Milky Way crossing the sky high overhead. It will appear as a hazy band across the sky, with ragged edges and dark patches and rifts in it. The Milky Way passes right through Cassiopeia.

Moving southeast from Cassiopeia, along the Milky Way, we come to the constellation Perseus; he was the Greek hero who slew the Medusa. (He flew off on Pegasus, the winged horse, who is conveniently nearby in the sky, and saw Andromeda, whom he saved.) On the edge of Perseus nearest to Cassiopeia, with a small telescope or binoculars we can see two hazy patches of light that are really clusters of hundreds of stars called "open clusters." This "double cluster in Perseus," also known as h and χ (chi) Persei, provides two of the star clusters (a type called "open clusters") that are easiest to see with small telescopes. In 1603, Johann Bayer assigned Greek letters to the brightest stars and lower-case Latin letters to less-bright stars. Although h and χ Persei are clusters, they were named using Bayer's system for labelling stars (Fig. 4–16). One cluster was confused with the star chi Persei—"chi of Perseus"—when the cluster was named, and the other was confused with the star h Persei.

Along the Milky Way in the other direction from Cassiopeia (whose W is relatively easy to find), we come to a cross of bright stars directly overhead. This "Northern Cross" is in the constellation Cygnus, the Swan (Fig. 4–17). In this direction, spacecraft detect x-rays of varying brightness, and we think a black hole is located there. Also in Cygnus is a particularly dark region of the Milky Way, called the northern Coalsack. Dust in space in that direction prevents us from seeing as many stars as we do in other directions in the Milky Way.

Slightly to the west is another bright star, Vega, in the constellation Lyra (the Lyre). And farther westward, we come to the constellation Hercules, named for the mythological Greek hero who performed twelve great labors, of which the most famous was bringing back the golden apples. In Hercules is an older, larger

Figure 4–16 Bayer, in 1603, used Greek letters to mark the brightest stars in constellations; he also used lower-case Latin letters. Here we see Cetus.

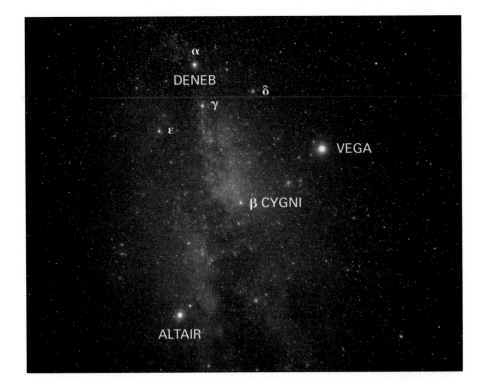

Figure 4–17 The Northern Cross, composed of the brightest stars in the constellation Cygnus, the Swan. Deneb (alpha Cygni), gamma Cygni, and beta Cygni make the long bar; epsilon, gamma, and delta Cygni make the crossbar. The bright star Vega, in Lyra, is nearby.

A

Figure 4-18 *(A)* The Pleiades, the Seven Sisters, in the constellation Taurus, the Bull. *(B)* The constellation Orion, marked by reddish Betelgeuse, bluish Rigel, and a belt of three stars in the middle. The Orion Nebula is the reddish object below the belt. It is M42 on the Winter Sky Map at the end of the book. Sirius, the brightest star in the sky, is to the lower left of Orion in this view.

type of star cluster called a "globular cluster." It is known as M13, the globular cluster in Hercules. It is visible as a fuzzy mothball in even small telescopes; larger telescopes have better resolution and so show the individual stars.

The Winter Sky

As the autumn proceeds and the winter approaches, the constellations we have discussed appear closer and closer to the western horizon for the same hour of the night. By early evening on January 1, Cygnus is setting in the western sky, while Cassiopeia and Perseus are overhead.

To the south of the Milky Way, near Perseus, we can now see a group of six stars close together in the sky (Fig. 4–18*A*). The grouping can catch your attention as you scan the sky. It is the Pleiades (pronounced "plee'a-deez"), traditionally the Seven Sisters of Greek mythology, the daughters of Atlas. (We can usually see six stars with the eye now; either one of the stars has faded over the millennia or it was never visible and the association with the Pleiades myth was loose.) These stars are another example of an open cluster. Binoculars or a small telescope will reveal dozens of stars there; a large telescope will ordinarily show too small a region of sky for you to see the Pleiades well at all. So a bigger telescope isn't always better.

Farther toward the east, rising earlier every evening, is the constellation Orion, the Hunter (Fig. 4–18*B*). Orion is perhaps the easiest constellation of all to pick out in the sky, for three bright stars close together in a line make up its belt. Orion is warding off Taurus, the Bull, whose head is marked by a large V of stars. A reddish star, Betelgeuse ("bee'tl-juice" would not be far wrong for pronunciation, though some say "beh'tl-jouz"), marks Orion's armpit, and symmetrically on the other side of his belt, the bright bluish star Rigel (rye'jel) marks his heel. Betelgeuse is an example of a red supergiant star; it is hundreds of millions of kilometers across, far bigger itself than the Earth's orbit around the Sun. Orion's sword extends down from his belt. A telescope, or a photograph, reveals a beautiful region known as the Great Nebula in Orion, or the Orion Nebula. Its shape can be seen in even a smallish telescope; however, only photographs clearly reveal its vivid colors—though whether it is reddish or greenish on an image depends on what film is used. It is a site where new stars are forming.

Rising after Orion is Sirius, the brightest star in the sky. Orion's belt points directly to it. Sirius appears blue-white, which indicates that it is very hot. Sirius is so much brighter than the other stars that it stands out to the naked eye. It is part of the constellation Canis Major, the Great Dog. (You can remember that it is near Orion by thinking of it as Orion's dog.)

Back toward the top of the sky, between the Pleiades and Orion's belt, is a group of stars that forms the V-shaped head of Taurus. This open cluster is known as the Hyades (hy'a-deez). The stars of the Hyades mark the bull's face, while the stars of the Pleiades ride on the bull's shoulder. In a Greek myth, Jupiter turned himself into a bull to carry Europa over the sea to what is now called Europe.

The Spring Sky

We can tell that spring is approaching when the Hyades and Orion get closer and closer to the western horizon each night, and finally are no longer visible shortly after sunset. Now the twins (Castor and Pollux), a pair of stars, are nicely placed for viewing in the western sky. Pollux is the twin closer to Procyon, while Castor is the twin closer to Capella; they are about equally bright. Castor and

Pollux were the twins in the Greek Pantheon of gods. The constellation is called Gemini, the twins.

On spring evenings, the Big Bear is overhead, and anything in the Big Dipper—which is part of the Big Bear—would spill out. Leo, the Lion, is just to the south of the zenith (follow the Pointers backward). Leo looks like a backward question mark, with the bright star Regulus at its base. Regulus marks the lion's heart. The rest of Leo, to the east of Regulus, is marked by a bright triangle of stars. Some people visualize a sickle-shaped head and a triangular tail.

If we follow the arc made by the stars in the handle of the Big Dipper, we come to a bright reddish star, Arcturus, another supergiant. It is in the kite-shaped constellation Boötes, the Herdsman.

Sirius sets right after sunset in the springtime; however, another bright star, Spica, is rising in the southeast in the constellation Virgo, the Virgin. It is farther along the arc of the Big Dipper through Arcturus. Vega, a star that is almost as bright, is rising in the northeast. And the constellation Hercules, with its notable globular cluster (visible in small telescopes), is rising in the east in the evening at this time of year.

The Summer Sky

Summer, of course, is a comfortable time to watch the stars because of the warm weather. Spica is over toward the southwest in the evening. A bright reddish star, Antares, is in the constellation Scorpius, the Scorpion, to the south. ("Antares" means "compared with Ares," another name for Mars, because Antares is also reddish.)

Hercules and Cygnus are high overhead, and the star Vega is prominent near the zenith. Cassiopeia is in the northeast. The center of our galaxy is in the dense part of the Milky Way that we see in the constellation Sagittarius, the Archer, in the south (Fig. 4–19).

Around August 12 every summer is a wonderful time to observe the sky, because that is when the Perseid meteor shower occurs. In a clear, dark sky, with not much moonlight, one bright meteor a minute may be visible at the peak of the shower. Just lie back and watch the sky in general—don't look in any specific direction. The rate of meteors tends to be substantially higher after midnight than before midnight. Although the Perseids is the most observed meteor shower, partly because it occurs at a time of warm weather in the northern part of the country, many other meteor showers occur during the year.

The summer is a good time of year for observing a variable star, Delta Cephei; it appears in the constellation Cepheus, which is midway between Cassiopeia and Cygnus. Delta Cephei varies in brightness with a 5.4-day period. As we will see in Chapters 11 and 16, studies of its variations have helped us figure out the distances to galaxies. This fact reminds us of the real importance of studying the sky—which is to learn what things are and how they work, and not just where they are. The study of the sky has led us to understand the Universe, and this is the major value and excitement of astronomy.

⬤ THE MOTIONS OF THE SUN, MOON, AND PLANETS

Though the stars appear to turn above the Earth at a steady rate, the Sun, the Moon, and the planets appear to slowly drift among the stars in the sky. The planets were long ago noticed to be "wanderers" among the stars, slightly deviating from the daily apparent motion of the entire sky overhead.

Figure 4–19 Halley's Comet (*left middle*, with the tail) near the Milky Way on March 13, 1986. It will be back in 2061.

The Seasons and the Path of the Sun

The path that the Sun follows through the stars in the sky is known as the **ecliptic.** We can't notice this path readily because the Sun is so bright that we don't see the stars when it is up, and because the Earth's rotation causes another more rapid daily motion (the rising and setting of the Sun), but the ecliptic is marked with a dotted line on the Sky Maps.

The Earth and the other planets revolve around the Sun in more or less a flat plane. So from Earth, the paths among the stars of the other planets are all close to the ecliptic. But the Earth's axis is not perpendicular to the ecliptic. It is, rather, tipped from perpendicular by $23\frac{1}{2}°$.

The ecliptic is therefore tipped with respect to the celestial equator. The two points of intersection are known as the **vernal equinox** and the **autumnal equinox,** respectively. The Sun is at those points at the beginning of our northern-hemisphere spring and autumn, respectively. On the equinoxes, the Sun's declination is zero. Three months after the vernal equinox, the Sun is on the part of the ecliptic that is farthest north of the celestial equator. The Sun's declination is then $+23\frac{1}{2}°$, and we say it is at the summer **solstice.** The summer is hot because the Sun is above our horizon for a longer time and because it reaches a higher angle above the horizon when it is at high declinations (Fig. 4–20). A consequence of the latter effect is that a given beam of light intercepts a smaller area than it does when the light strikes at a glancing angle, so the heating per unit area is greater.

The seasons (Fig. 4–21), thus, are caused by the variation in the declination of the Sun. ☼ This variation, in turn, is caused by the fact that the Earth's axis of spin is tipped by $23\frac{1}{2}°$. Many if not most people misunderstand the cause of the seasons. Note that the seasons are *not* caused by the changing distance between the Earth and the Sun. If they were, then seasons would not be opposite in the northern and southern hemispheres, and there would be no seasonal changes in the number of daytime hours. In fact, the Earth is closest to the Sun each year on or about January 4, which falls in the northern-hemisphere winter.

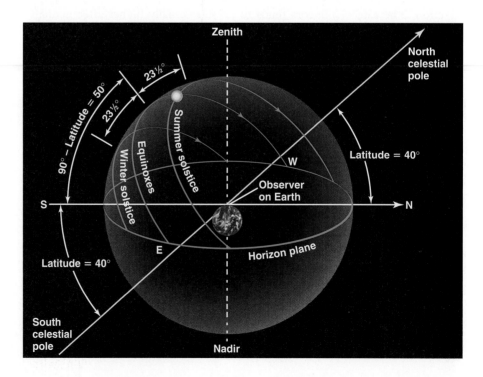

Figure 4–20 The blue arcs from the equator up and over the top of the sphere shown are the path of the Sun at different times of the year. Around the summer solstice, June 21, the Sun (shown in yellow) is at its highest declination, rises highest in the sky, stays up longer (because, as shown, more of its path is above the horizon), and rises and sets farthest to the north. The opposite is true near the winter solstice, December 21. The diagram is drawn for latitude 40°.

starparties

THE PATHS OF THE MOON AND PLANETS

The Moon goes through the full range of right ascension and its full range of declination each month as it circles the Earth. The motion of the planets is much more difficult to categorize. Mercury and Venus circle the Sun more quickly than does Earth, and have smaller orbits than the Earth. They wiggle in the sky around the Sun over a period of weeks or months, and are never seen in the middle of the night. They are the "evening stars" or the "morning stars," depending on which side of the Sun they are on. Mars, Jupiter, and Saturn (as well as the other planets too faint to see readily) circle the Sun more slowly than does Earth. So they drift across the sky with respect to the stars, and don't change their right ascension or declination rapidly.

The U.S. and U.K. governments jointly put out a volume of tables each year, *The Astronomical Almanac*. It is an "ephemeris," a list of changing things; the word comes from the same root as "ephemeral." The book includes tables of the positions of the Sun, Moon, and planets. One of us (J.M.P.) has written the *Field Guide to the Stars and Planets*, which includes graphs and less detailed tables.

The word "equinox" means "equal night," implying that in theory the length of day and night is equal on those two occasions each year. The equinoxes mark the dates at which the center of the Sun crosses the celestial equator. But the Sun is not just a theoretical point, which is what is used to calculate the equinoxes. Since the top of the Sun obviously rises before its middle, the daytime is actually a little longer than the nighttime on the day of the equinox. Also, bending of sunlight by the Earth's atmosphere allows us to see the Sun when it is really a little below our horizon, thereby lengthening the daytime. So the days of equal daytime and nighttime are displaced by a few days from the equinoxes.

Because of its apparent motion with respect to the stars, the Sun goes through the complete range of right ascension and between $+23\frac{1}{2}°$ and $-23\frac{1}{2}°$ in declination each year. As a result, its height above the horizon varies from day to day. If we were to take a photograph of the Sun at the same hour each day, over the year the Sun would sometimes be relatively low and sometimes relatively high.

If we were at or close to the north pole, we would be able to see the Sun whenever it was at a declination sufficiently above the celestial equator. This phenomenon is known as the **midnight sun** (Fig. 4–22). Indeed, from the north or south poles, the Sun is continuously visible for about six months of the year, followed by about six months of darkness.

Figure 4–21 A schematic view from space of the Sun and the Earth (not to scale). The seasons occur because the Earth's axis is tipped with respect to the plane of the orbit in which it revolves around the Sun. The dotted line is drawn perpendicularly to the plane of the Earth's orbit. When the northern hemisphere is tilted toward the Sun, it has its summer; at the same time, the southern hemisphere is having its winter. At both locations of the Earth shown, the Earth rotates through many 24-hour day-night cycles before its motion around the Sun moves it appreciably.

Summer in northern hemisphere
Winter in southern hemisphere

Winter in northern hemisphere
Summer in southern hemisphere

Figure 4-22 In this series taken in June from northern Norway, above the Arctic Circle, one photograph was taken each hour for an entire day. The Sun never set, a phenomenon known as the midnight sun. Since the site was not at the north pole, the Sun and stars move somewhat higher and lower in the sky in the course of a day.

● TIME AND THE INTERNATIONAL DATE LINE

Every city and town on Earth used to have its own time system, based on the Sun, until widespread railroad travel made this inconvenient. In 1884, an international conference agreed on a series of longitudinal time zones. Now all localities in the same zone have a standard time (Fig. 4–23). (www)

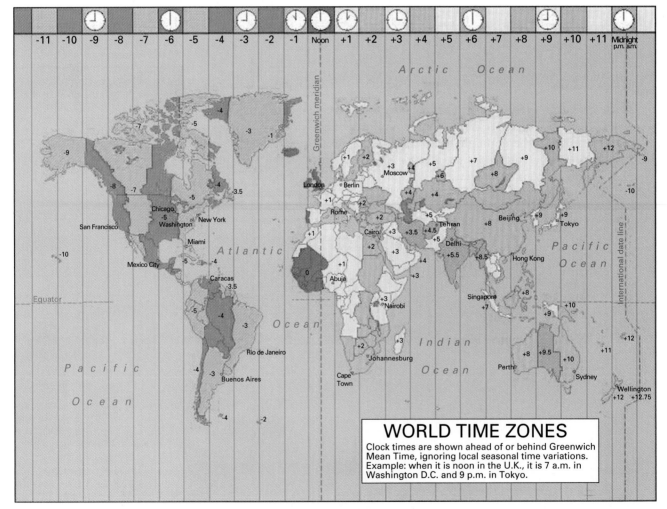

Figure 4-23 International time zones for standard time.

Since there are twenty-four hours in a day, the 360° of longitude around the Earth are divided into 24 standard time zones, each 15° wide. Each time zone is centered on a meridian of longitude exactly divisible by 15. Because the time is the same throughout each zone, the Sun is not directly overhead at noon at each point in a given time zone, but in principle is less than about a half-hour off. Standard time is based on a **mean solar day,** the average length of a solar day. As the Sun seems to move in the sky from east to west, the time in any one place gets later.

We can visualize noon, and each hour, moving around the world from east to west, minute by minute. We get a particular time back 24 hours later, but if the hours circled the world continuously the date would not be able to change at midnight. So we specify a north–south line and have the date change there. We call it the **international date line.** With this line present on the globe, as we go eastward the hours get later and we go into the next day. At some time in our eastward trip, we cross the international date line, and we go back one day. Thus we have only 24 hours on Earth at any one time.

A hundred years ago, England won for Greenwich, then the site of the Royal Observatory, the distinction of having the basic line of longitude, 0° (Fig. 4–24). Realizing that the international date line would disrupt the calendars of those who crossed it, that line was put as far away from the populated areas of Europe as possible—near or along the 180° longitude line. The international date line passes from north to south through the Pacific Ocean and actually bends to avoid cutting through continents or groups of islands, thus providing them with the same date as their nearest neighbor (Fig. 4–25).

In the summer, in order to make the daylight last into later hours, many countries have adopted "daylight-saving time." Clocks are set ahead 1 hour on a certain date in the spring. Thus if darkness falls at 7 p.m. Eastern Standard Time (E.S.T.), that time is called 8 p.m. Eastern Daylight Time (E.D.T.), and most people have an extra hour of daylight after work (but, naturally, one fewer hour of daylight in the morning). In most places, clocks are set back one hour in the fall, though some places have adopted daylight-saving time all year. The phrase to remember to help you set your clocks is "fall back, spring forward." Of course, daylight-saving time is just a bookkeeping change in how we name the hours, and doesn't result from any astronomical changes.

CALENDARS

The period of time that the Earth takes to revolve once around the Sun is called, of course, a year. This period is about 365¼ mean solar days.

Figure 4–24 The international zero circle of longitude at the Royal Observatory Greenwich, England, is marked by a telescope that looked, historically, up and down along the north-south line.

Figure 4-25 This top view of the Earth, looking down from above the north pole, shows how days are born and die at the international date line. When you cross the international date line, your calendar changes by one day. To be among the first locations to see the new millennium, countries have even changed their time zones, which shows how timekeeping can be arbitrary.

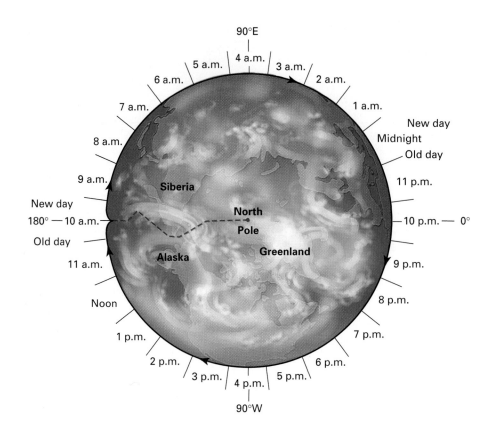

Roman calendars had, at different times, different numbers of days in a year, so the dates rapidly drifted out of synchronization with the seasons (which follow solar years). Julius Caesar decreed that 46 B.C. would be a 445-day year in order to catch up, and defined a calendar, the **Julian calendar,** that would be more accurate. This calendar had years that were normally 365 days in length, with an extra day inserted every fourth year in order to bring the average year to 365¼ days in length. The fourth years were, and are, called leap years.

The name of the fifth month, formerly Quintillis, was changed to honor Julius Caesar; in English we call it July. The year then began in March; the last four months of our year still bear names from this system of numbering. (For example, October was the eighth month, and "oct" is the Latin root for "eighth.") Augustus Caesar, who carried out subsequent calendar reforms, renamed August after himself. He also transferred a day from February in order to make August last as long as July.

The Julian calendar was much more accurate than its predecessors, but the actual **solar year** is a few minutes shorter than 365¼ days. By 1582, the calendar was about 10 days out of phase with the date at which Easter had occurred at the time of a religious council 1250 years earlier, and Pope Gregory XIII issued a proclamation to correct the situation: he simply dropped 10 days from 1582. Many citizens of that time objected to the supposed loss of the time from their lives and to the commercial complications. Does one pay a full month's rent for the month in which the days were omitted, for example? "Give us back our fortnight [14 days]," they cried.

In the **Gregorian calendar,** years that are evenly divisible by four are leap years, except that three out of every four century years—the ones not divisible evenly by 400—have only 365 days. Thus 1600 was a leap year; 1700, 1800, and 1900 were not; and 2000 was again a leap year. Although many countries immediately adopted the Gregorian calendar, Great Britain (and its American colonies) did not adopt it until 1752, when 11 days were skipped. As a result, we celebrate George Washington's birthday on February 22, even though he was born on

February 11 (Fig. 4–26). Actually, since the year had begun in March instead of January, the year 1752 was cut short. Washington was born on February 11, 1731, then often written February 11, 173 1/2, but we now refer to his date of birth as February 22, 1732. The Gregorian calendar is the one in current use. It will be over 3000 years before this calendar is as much as one day out of step.

When does the new millennium start? It is tempting to say that it was on "January 1, 2000." That marks the beginning of the thousand years whose dates start with a "2." But, if you are trying to count millennia, since there was no year zero, two thousand years after the beginning of the year one would be January 1, 2001. (Nobody called it the year one then, however; the dating came several hundred years later.) If the first century began in the year 1, then the 21st century should begin in the year 2001! Having big New Year's eve parties both years is okay with us.

Figure 4–26 In George Washington's family bible, his date of birth is given as "11th day of February 173 1/2." Some contemporaries would have said 1731; we now say 1732.

Concept Review

Planets and moons have detectable disks that show **phases,** depending on how much of the sunlighted side we see. Stars, whose images are almost always points, **twinkle** because we see them through the Earth's turbulent atmosphere. Stars appear to rise and set each day because the Earth turns. We see the stars as though they were on a giant **celestial sphere,** with **celestial poles** over the Earth's pole. Polaris, now the **pole star,** marks the north celestial pole, but **precession,** the changing tilt of the Earth's axis, will change that. The **celestial equator** is the line on the celestial sphere above the Earth's equator. Astronomers measure the brightness of stars with the **magnitude scale,** in which each factor of 100 in brightness corresponds to a difference of 5 magnitudes. A star's **apparent magnitude** is how bright it looks to us.

Astronomical coordinate systems include one based on the Earth, in which **right ascension** corresponds to terrestrial longitude and **declination** corresponds to terrestrial latitude. The meridian is the north–south line extending through the **zenith,** the point overhead. Sidereal time, time by the stars, differs from solar time. The sky is divided into 88 regions known as **constellations.** The planets "wander" into different constellations as their coordinates change. They travel close to the path of the Sun across the sky, the **ecliptic.** The ecliptic meets the celestial equator at the **vernal equinox** and at the **autumnal equinox.** The Sun's highest declination corresponds to the summer **solstice.** The seasons are caused by the tilt of the Earth's axis. When you are far enough north or south on the Earth so that the Sun never sets at the current season, you see the **midnight sun.**

Standard time is based on a **mean solar day.** Days change at the **international date line.** The **Julian calendar** included leap years, but the seasons drifted because the actual **solar year** is not exactly 365¼ days. We now use the **Gregorian calendar,** in which leap years are sometimes omitted to adjust for the fact that the year is not composed of exactly 365¼ days.

Questions

†1. On the picture opening the chapter, estimate the angle covered by the star trails and deduce how long the exposure lasted.

2. If you look toward the horizon, are the stars you see likely to be twinkling more or less than the stars overhead? Explain.

3. Is the planet Saturn, which is farther away and physically smaller than Jupiter, likely to twinkle more or less than Jupiter?

4. Explain how it is that some stars never rise in our sky, and others never set.

5. Using the Appendices and the Sky Maps, comment on whether Polaris is one of the twenty brightest stars in the sky.

†6. Compare a 6th-magnitude star and an 11th-magnitude star in brightness. Which is brighter, and by how many times?

†7. (a) Compare a 16th-magnitude star with an 11th-magnitude star in brightness, as in Question 6. (b) Now compare the 16th-magnitude with the 6th-magnitude star, specifying which is brighter and by how many times.

†8. Since the sky revolves once a day, how many degrees does it appear to revolve in 1 hour?

9. Use the suitable Sky Map to list the bright stars that are near the zenith during the summer.

10. Explain why Christmas comes in the summer in Australia.

11. Are the lengths of day and night exactly equal at the vernal equinox? Explain.

12. Does the Sun pass due south of you at noon on your watch each day? Explain why or why not.

13. Explain why for an observer at the north pole the Sun changes its altitude in the sky over the year, but the stars do not.

†14. When it is 5 p.m. on November 1 in New York City, what time of day and what date is it in Tokyo?

15. When it is 9 a.m. on February 21 in California, describe how to use Figure 4–23 to find the date and time in England. Carry out the calculation both by going eastward and by going westward, and describe why you get the same answer.

†16. List the leap years between 1800 and 2200. How many will there be?

---

†This question requires a numerical solution.

Gravity and Motion: The Early History of Modern Astronomy

5

ORIGINS *We explore the formation of the Solar System from a giant cloud of gas and dust, a process that appears to be common throughout the Universe.*

Ancient peoples knew five planets—Mercury, Venus, Mars, Jupiter, and Saturn. When we observe these planets in the sky, we join the people of long ago in noticing that the positions of the planets vary from night to night with respect to each other and with respect to the stars. In fact, our word "planet" comes from the Greek word for "wanderer." Of course, resulting from Earth's rotation about its axis, both stars and planets move together across our sky essentially once every 24 hours; by the "wandering" of the planets we mean that the planets appear to move at a very slightly different rate. Only over weeks or months do we notice them changing position with respect to the fixed stars.

In this chapter, we will not only discuss the motions of the planets but also learn how major figures in the history of astronomy explained these motions. The explanations have led to today's conceptions of the Universe and of our place in it.

AIMS

To summarize theories of the Universe developed two millennia ago, and to follow how modern astronomy progressed through the ideas of Copernicus, Kepler, Galileo, and Newton.

•

To discuss the formation of our Solar System.

SURVEY OF THE SOLAR SYSTEM

Our ideas of the Solar System took shape before even the telescope was invented, but nowadays we have visited almost all the Solar System with spacecraft (Fig. 5–1). Let us first give an overview of the Solar System before discussing the historical figures who started us on our current path. We will discuss the individual Solar-System objects further in later chapters.

Since 1957, when the first spacecraft were launched into orbit around the Earth, we have made giant strides. Satellites routinely study the Earth and even the Moon, mapping them in detail and watching how they change (Fig. 5–2). The series of crewed landings on the Moon during 1969–1972 revolutionized our understanding of that relatively nearby object (Fig. 5–3).

◀ The frontispiece of Galileo's *Dialogue on the Two Great World Systems,* published in 1632. According to the labels, Copernicus is to the right, with Aristotle and Ptolemy to the left; Copernicus was drawn with Galileo's face, however.

Figure 5–1 A composite of spacecraft images of the 8 innermost planets: Mercury, Venus, Earth, Mars, Jupiter, Saturn, Uranus, and Neptune.

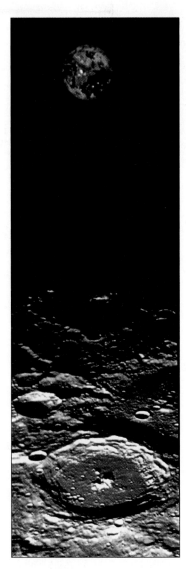

Figure 5-2 The Earth, with a lunar crater in the foreground, imaged by the Clementine spacecraft in orbit around the Moon. The angular distance between the images of the Earth and Moon was electronically reduced in this image.

Figure 5-3 Buzz Aldrin climbing down from the Lunar Module of Apollo 11 on July 20, 1969, in a photo taken during the Apollo 11 mission by Neil Armstrong. Armstrong and Aldrin were the first people on the Moon.

The closest planet to Earth is Venus, only 70 per cent our distance from the Sun. Venus is covered with clouds, but radar has mapped its surface (Fig. 5–4). Beyond Venus, Mercury orbits so close to the Sun that it has essentially no atmosphere left, given its small mass. No spacecraft has visited Mercury for decades.

Farther outward than the Earth's orbit is Mars, a fascinating planet now the subject of extensive study (Fig. 5–5). Its seasonal changes have long made it an object of interest, and the signs now photographed that water once flowed on its surface makes it all the more interesting, especially since flowing water is thought to be necessary for life to arise.

Figure 5-4 A global view of the surface of Venus as imaged with radar from NASA's Magellan spacecraft. The high volcano Maxwell appears near the top.

Figure 5-5 Mars, from the Hubble Space Telescope.

Lives in Science
GALILEO

Galileo.

Galileo was born in Pisa, in northern Italy, in 1564. In the 1590s, he became a professor at Padua, in the Venetian Republic. At some point in this decade, he adopted Copernicus's heliocentric theory. After hearing that a device to magnify far-off objects had been made, he figured out the principles himself and made his own. In late 1609 or early 1610, Galileo was among the first to turn a telescope toward the sky.

In 1613, Galileo published his acceptance of the Copernican theory. The book in which he did so was received and acknowledged with thanks by the Cardinal who, significantly, was Pope at the time of Galileo's trial. Just how Galileo got out of favor with the Church has been the subject of much study. His bitter quarrel with a Jesuit astronomer, Christopher Scheiner, over who deserved the priority for the discovery of sunspots did not put him in good stead with the Jesuits. Still, the disagreement on Copernicanism was more fundamental than personal, and the Church's evolving views against the Copernican system were undoubtedly more important than personal feelings. Also, the Church apparently objected to his belief that matter consists of atoms.

In 1615, Galileo was denounced to the Inquisition, and was warned against teaching the Copernican theory. There has been much discussion of how serious the warning was and how seriously Galileo took it.

Over the following years, Galileo was certainly not outspoken in his Copernican beliefs, though he also did not cease to hold them. In 1629, he completed his major manuscript, *Dialogue on the Two Great World Systems*, these systems being the Ptolemaic and Copernican. With some difficulty, clearance from the censors was obtained, and the book was published in 1632. It was written in contemporary Italian, instead of the traditional Latin, and could thus reach a wide audience. The book was condemned and Galileo was again called before the Inquisition.

Galileo's trial provided high drama, and has indeed been transformed into drama on the stage. He was convicted, and sentenced to house arrest. He lived nine more years, until 1642, his eyesight and his health failing. He devoted his time to the experimental study of the motions of falling bodies. Indeed, most of the principles in Newton's laws of motion are based on experimental facts determined by Galileo and documented in his important book on mechanics, written during this period. In 1992, Pope John Paul II acknowledged that Galileo had been correct to favor the Copernican theory, an implicit though not an explicit pardon.

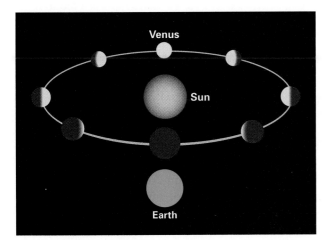

Figure 5–24 When Venus is on the far side of the Sun as seen from the Earth, it is more than half illuminated. We thus see it go through a whole cycle of phases, from thin crescent (or even "new") through full. Venus thus matches the predictions of the Copernican system.

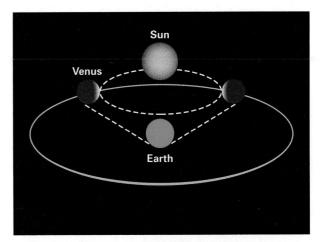

Figure 5–25 In the Ptolemaic theory, Venus and the Sun both orbit the Earth. However, because Venus is never seen far from the Sun in the sky, Venus's epicycle must be restricted to always fall on the line joining the center of the deferent and the Sun. In this diagram, Venus could never get farther from the Sun than the dotted lines. It would therefore always appear as a crescent.

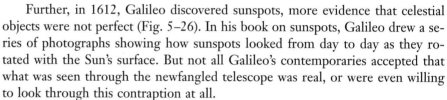

starparties

GALILEO'S OBSERVATIONS

With a small telescope, it is easy to recreate many of Galileo's observations. Try these with the telescope on your campus.

1. Observe Venus over a period of several months. This is most easily done when it is visible in the evening, but at some times of the year the morning is more suitable. You should be able to see it go through different phases, like the Moon.

2. Observe Jupiter over the course of several nights. You should be able to see its four bright moons, the "Galilean satellites." Their relative positions change daily. If you draw a picture each night for a few weeks, you'll even be able to determine their orbital periods and at least roughly verify that they obey Kepler's third law.

3. Observe the Moon. You should see craters and maria ("seas"), as did Galileo. The views will be best during the crescent, quarter, or somewhat gibbous phases, especially near the "terminator" (the border between the sunlit and dark parts of the Moon). During full moon, the features tend to appear washed out because of the lack of shadows.

4. Point the telescope to a bright region of the Milky Way and notice that many stars are visible. Use the eyepiece that gives the widest field of view for this observation. Do you think Galileo was surprised to find that the Milky Way consists of many faint stars?

5. A small telescope can be used to view sunspots, as did Galileo. But do *not* look directly at the Sun through the telescope. A better technique is to project the Sun's image onto a sheet of cardboard. You can look directly through the telescope *only if* you are certain that you have the right kind of filter at the top end of the telescope, and that it is securely fastened. Never use a filter in the eyepiece of the telescope, since it will get very hot and can break, leading to possible blindness.

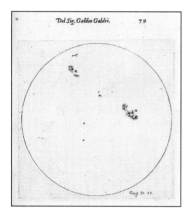

Figure 5-26 Sunspots observed by Galileo in 1612, from his book of the following year.

Further, in 1612, Galileo discovered sunspots, more evidence that celestial objects were not perfect (Fig. 5–26). In his book on sunspots, Galileo drew a series of photographs showing how sunspots looked from day to day as they rotated with the Sun's surface. But not all Galileo's contemporaries accepted that what was seen through the newfangled telescope was real, or were even willing to look through this contraption at all.

In Lives in Science, we discuss Galileo's controversy with the Roman Catholic Church, during which he was sentenced to house arrest (Fig. 5–27). The controversy echos down to today, and some of the truth about the real reasons behind the fight may still be hidden in the Vatican archives. But by now, almost four hundred years since Galileo made his discoveries and four hundred years since his near contemporary Giordano Bruno was burned at the stake at least in part because of his view that there were worlds beyond our Solar System, peace reigns between the Church and scientists, and the Vatican supports a modern observatory and respected contemporary astronomers.

◖ ISAAC NEWTON

Kepler discovered his laws of planetary orbits by trial and error, and he had no physical understanding of them. Only with the work of Isaac Newton 60 years later did we find out why the laws existed. Newton was born in England in 1642, the year of Galileo's death. He became the greatest scientist of his time and perhaps of all time. His work on optics, his invention of the reflecting telescope, and his discovery that visible light can be broken down into a spectrum would merit him a place in astronomy texts. But still more important was his work on motion and on gravity.

Newton set modern physics on its feet by deriving laws showing how objects move on Earth and in space, and by finding the law that describes the force of gravity. In order to work out the law of gravity, Newton had to invent calculus! Newton long withheld publishing his results, until Edmond Halley (whose name we associate with the famous comet) convinced him to publish his work. Newton's *Principia* (pronounced "prin-kip'ee-a"), short for the Latin form of *Mathematical Principles of Natural Philosophy*, appeared in 1687.

Figure 5-27 Galileo's house, in Florence, Italy.

Figure It Out

Newton's Version of Kepler's Third Law

Astronomers nowadays often make rough calculations to test whether physical processes under consideration could conceivably be valid. Astronomy has also had a long tradition of exceedingly accurate calculations. Pushing accuracy to one more decimal place sometimes leads to important results.

For example, Kepler's third law, in its original form—the period of a planet squared is proportional to its distance from the Sun cubed ($P^2 = \text{constant} \times R^3$)—holds to a reasonably high degree of accuracy and seemed completely accurate when Kepler did his work. But now we have more accurate observations, and they verify Newton's version.

Newton derived the formula for Kepler's third law, taking account of his own formula for the law of universal gravitation. He found that

$$P^2 = \frac{4\pi^2}{G(m_1 + m_2)} R^3,$$

where m_1 and m_2 are the masses of the two bodies. For planetary orbits, if m_1 is the mass of the Sun, then m_2 is the mass of the planet.

Newton's formulation of Kepler's third law shows that the constant of proportionality is determined by the mass of the Sun. Thus, we can use the formula to find out the Sun's mass. Newton's version of Kepler's third law, as applied to any planet's orbit around the Sun, also shows that the planet's mass contributes to the value of the proportionality constant, but its effect is very small because the Sun is so much more massive. (The effect can be detected even today only for the most massive planets—Jupiter and Saturn.) Newton therefore actually predicted that the constant in Kepler's third law depends slightly on the mass of the planet, being different for different planets. The observational confirmation of Newton's prediction was a great triumph.

Kepler's third law, and its subsequent generalization by Newton, applies not only to planets orbiting the Sun but also to any bodies orbiting other bodies under the control of gravity. Thus it also applies to satellites orbiting planets. We determine the mass of the Earth by studying the orbit of our Moon, and determine the mass of Jupiter by studying the orbits of its moons. Until recently, we were unable to reliably determine the mass of Pluto because we could not observe a moon in orbit around it. The discovery of a moon of Pluto in 1978 finally allowed us to determine Pluto's mass, and we found that our previously best estimates (based on Pluto's now-discredited gravitational effects on Uranus) were way off. The same formula can be applied to binary stars to find their masses.

The Principia contains Newton's three laws of motion. The first is that bodies in motion tend to remain in motion, in a straight line with constant speed, unless acted upon by a force. It is a law of inertia, which was really discovered by Galileo. Newton's second law relates a force with its effect on accelerating a mass. A larger force will make the same mass accelerate faster ($F = ma$, where m is the mass and a is the acceleration). Newton's third law is often stated "For every action, there is an equal and opposite reaction." The flying of jet planes is only one of the many processes explained by this law. *The Principia* also contains the law of gravity. One application of Newton's law of gravity is weighing the planets (see *Figure It Out: Newton's Version of Kepler's Third Law*).

One of the most popular tales of science is that an apple fell on Newton's head, leading to his discovery of the concept of gravity. Though no apple actually hit Newton's head, the story that Newton himself told, years later, is that he saw an apple fall and realized that just as the apple fell to Earth, the Moon is

Lives in Science
ISAAC NEWTON

Isaac Newton in 1702. He was knighted three years later. (Detail of the portrait by courtesy of the National Portrait Gallery, London.)

▮saac Newton was born in Woolsthorpe, Lincolnshire, England, on December 25, 1642. Even now, about three hundred sixty years later, scientists still refer regularly to "Newton's laws of motion" and to "Newton's law of gravitation."

Newton's most intellectually fertile years were those right after his graduation from college when he returned home to the country because fear of the plague shut down many cities. There, he developed his ideas about the nature of motion and about gravitation. In order to derive them mathematically, he invented calculus. Newton long withheld publishing his results, possibly out of shyness. Finally Edmond Halley—who became known by applying Newton's law of gravity to show that a series of reports of comets all really referred to this single comet that we now know as Halley's Comet—persuaded Newton to publish his results. A few years later, in 1687, the *Philosophiae Naturalis Principia Mathematica (Mathematical Principles of Natural Philosophy)*, known as *The Principia*, was published with Halley's practical and financial assistance. In it, Newton showed that the motions of the planets and comets could all be explained by the same law of gravitation that governed bodies on Earth. In fact, he derived Kepler's laws on theoretical grounds.

Newton was professor of mathematics at Cambridge University and later in life went into government service as Master of the Mint in London. He was knighted in 1705 (for his governmental rather than his scientific contributions) and lived until 1727. His tomb in Westminster Abbey bears the epitaph: "Mortals, congratulate yourselves that so great a man has lived for the honor of the human race."

falling to Earth, though its forward motion keeps it far from us. When he calculated that the acceleration of the Moon fit the same formula as the acceleration of the apple, he knew he had the right method. Newton was the first to realize that gravity was a force that acted in the same way throughout the Universe. Many pieces of the original apple tree have been enshrined (Fig. 5–28). There is no knowing which—if any—of the apple trees now growing in front of Newton's house, still standing at Woolsthorpe, were actually descended from the famous original (Fig. 5–29).

Figure 5–28 A piece of Newton's apple tree, now in the Royal Astronomical Society, London. The donor's father was present in about 1818 when some logs were sawed from this famous apple tree, which had blown over.

Figure 5–29 Newton's house at Woolsthorpe, Lincolnshire, England. It is now a museum.

Figure It Out

Orbital Speed of Planets

The more distant a planet is from the Sun, the slower is its orbital speed. We can derive this from Kepler's third law. Since most of the planets have nearly circular orbits, a reasonable approximation for our purposes here is to assume that the distance travelled by a planet in one full orbit is $2\pi R$, the circumference of a circle of radius R.

If the planet travels at a constant speed (which it would if the orbit were exactly circular), the distance travelled would equal its speed multiplied by time: $d = vt$. Applying this to one full orbit, we find that $2\pi R = vP$, where P is the planet's orbital period. Thus, $P = 2\pi R/v$, so

$$P \propto R/v, \text{ where the symbol "} \propto \text{" means "proportional to."}$$

But Kepler's law states that $P^2 \propto R^3$. When we substitute the expression $P \propto R/v$ for P into Kepler's law, we get $(R/v)^2 \propto R^3$. Rearranging gives $v^2 \propto 1/R$, and hence

$$v \propto (1/R)^{1/2}.$$

In words, the speed of a planet is inversely proportional to the square root of its orbital radius; distant planets move more slowly than those near the Sun. For example, a planet whose orbital radius is 9 times that of the Earth's has an orbital speed that is $(1/9)^{1/2} = 1/3$ as fast as the Earth's orbital speed, which itself is about 30 km/sec.

This "inverse-square-root" relationship is characteristic of systems in which a large central mass dominates over the masses of orbiting particles. Later, we will see that a different result is found for rotating galaxies.

Newton's most famous literary quotation is "If I have seen so far, it is because I have stood on the shoulders of giants." In fact, a whole book has been devoted to all the other usages of this phrase, which turns out to have been in wide use before Newton (and since). As of the 20th century, the pace of science has been so fast that it has even been remarked, "Nowadays we are privileged to sit beside the giants on whose shoulders we stand."

THE ROTATION AND REVOLUTION OF THE PLANETS

The scientists just discussed provided the basis of our current understanding of our Solar System, which in their time was essentially equivalent to the Universe. Studying the orbits of the planets shows that the Solar System has some basic properties. All the planets, for example, move in the same direction around the Sun and are more or less in the same plane. Considering these regularities gives important clues as to how the Solar System may have formed. As we study the other planetary systems now being discovered, we are eager to see how many things about our own Solar System are truly fundamental, and how many may be specific to our case.

As we have seen, the motion of a planet around the Sun in its orbit is its **revolution.** The spinning of a planet is its **rotation.** The Earth, for example, revolves around the Sun in one year and rotates on its axis in one day. (These motions define the terms "year" and "day.") According to Kepler's third law, planets far from the Sun have longer periods of revolution (orbital periods) than planets close to the Sun. The distant planets also move more slowly (see *Figure It Out: Orbital Speeds of Planets*).

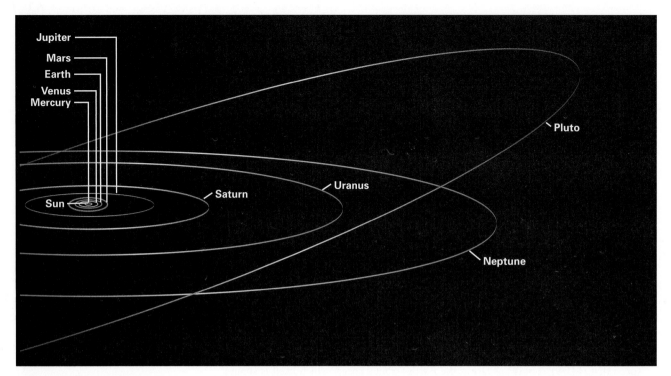

Figure 5–30 The orbits of the planets, with the exception of Pluto, have only small inclinations to the plane of the Earth's orbit. The orbits of all the planets other than Pluto are nearly circular, but look more extended and obviously elliptical here because of the perspective.

The orbits of the planets take up a disk rather than a full sphere (Fig. 5–30). The **inclination** of a planet's orbit is the angle that the plane of its orbit makes with the plane of the Earth's orbit.

The fact that the planets all orbit the Sun in roughly the same plane is one of the most important generalities about the Solar System. Spinning or rotating objects have a property called **angular momentum.** This angular momentum is larger if an object is more massive, if it is spinning faster, or if mass is farther from the axis of spin. One important property of angular momentum is that the total angular momentum of a system doesn't change (we say that it is "conserved"), unless the system interacts with something from outside that brings in or takes up angular momentum. Thus if one body were to spin more and more slowly, or were to begin spinning in the opposite direction, the angular momentum would have to be taken up by another body. In the next section, we will apply this idea to see how the planets undoubtedly all revolve around the Sun in the same direction because of how they were formed.

● THE FORMATION OF THE SOLAR SYSTEM

Many scientists studying the Earth and the planets are particularly interested in an ultimate question: How did the Earth and the rest of the Solar System form? We can accurately date the formation by studying the oldest objects we can find in the Solar System and allowing a little more time. For example, astronauts found rocks on the Moon older than 4.4 billion years. We think the Solar System formed about 4.6 billion years ago.

Our best current idea is that 4.6 billion years ago a huge cloud of gas and dust in space collapsed, pulled together by the force of gravity. What triggered the collapse isn't known.

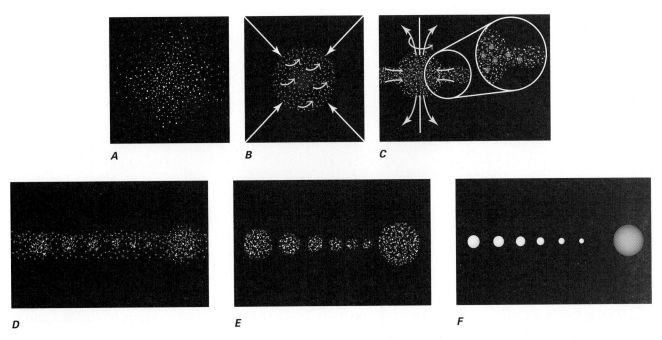

A   *B*   *C*

D   *E*   *F*

Figure 5–31 *(A)* The leading model for the formation of the Solar System. The protosolar nebula condenses because of the force of its own gravity *(B)*. Between states C and D, it contracts to form the protosun and a large number of small bodies called planetesimals. The planetesimals clumped together to form protoplanets *(D),* which, in turn, contracted to become planets *(E)* and *(F)*. Some of the planetesimals may have become moons or asteroids.

You have undoubtedly noticed that ice skaters spin faster when they pull their arms in. Similarly, the Solar System began to spin faster as it collapsed. (The original gas and dust may have had some spin, and this spin is magnified in speed by the collapse, because the total amount of angular momentum doesn't change.) Objects that are spinning around tend to fly off, and this force eventually became strong enough to counteract the effect of gravity pulling inward. Thus the Solar System stopped collapsing in one plane. Perpendicular to this plane, there was no spin to stop the collapse, so the solar nebula wound up as a disk.

In the disk of gas and dust, we think that the material began to clump (Fig. 5–31). Smaller clumps joined together to make larger ones, and eventually **planetesimals,** bodies a few hundred kilometers across, were formed. Gravity subsequently pulled many planetesimals together to make **protoplanets** (pre-planets) orbiting a **protosun** (pre-Sun). The protoplanets then contracted and cooled to make the planets we have today, and the protosun contracted to form the Sun (Fig. 5–32). Some of the planetesimals may still be orbiting the Sun; that is why we are so interested in studying small bodies of the Solar System like comets, meteorites, and asteroids. Most of the unused gas and dust, however, was blown away by a strong solar wind when the Sun was young and very active.

In one of the models for the formation of the outer planets, the solar nebula first collapsed into several large blobs. These blobs then became the outer planets. If this model is correct, the outer planets should all contain about the same relative amounts of the different elements as each other and as the space between the stars from which the gas came. In another model, a solid core condensed first for each of the outer planets. The gravity of this core then attracted the gas from around it. For this second model, the relative amounts of elements in the rocks and in the gases could be different from planet to planet. The Voyager spacecraft found that Jupiter, Saturn, Uranus, and Neptune have different

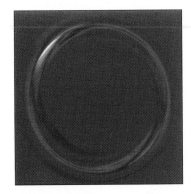

Figure 5–32 A false-color representation of temperature in the inner 10 A.U. of a model of the early solar nebula. A cross-section through the nebula along the rotation axis is shown. Red corresponds to the highest density, and white to the highest temperature.

Figure 5-33 The region around the southern star beta Pictoris, with the star itself blocked out in the telescope. This Hubble Space Telescope image with its Space Telescope Imaging Spectrograph clearly shows a side view of a disk of dust, which astronomers think is a planetary system in formation. The dips in the disk may indicate the positions of orbiting planets.

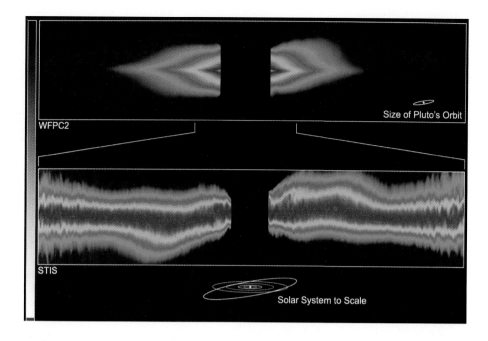

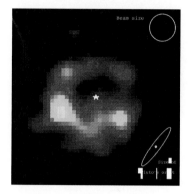

Figure 5-34 A dust ring around the star epsilon Eridani, imaged at 0.85 mm in the submillimeter part of the spectrum from a telescope in Hawaii. The peak of the dust ring is 60 A.U. from the star. The total amount of mass in it is about the same as that in comets orbiting in our Solar System. The ring may be analogous to the Kuiper belt (see Chapter 8) around our own Sun, with the inner region cleared by planetesimals.

relative amounts of some of the elements in their atmospheres. Also, the atmospheres of Jupiter and Saturn are very much more massive than the atmospheres of Uranus and Neptune. On the other hand, their cores all contain about the same amount of mass—10 to 20 times the mass of the Earth. Thus it seems that the outer planets assembled in several stages, with the cores forming first.

It is interesting that nothing about our current model of planetary-system formation implies that the Solar System is unique. (www) We are increasingly finding systems of planets around other stars. Disks have even been seen around many of the youngest nearby stars (Fig. 5-33 and Fig. 5-34), which may indicate that planetary systems are now forming there. We will discuss the newly discovered planets later on.

Concept Review

The Solar System has been largely explored with spacecraft over the last four decades. Long before that, and even before the telescope was invented, it was known that planets somehow undergo **retrograde motion** besides their generally **prograde motion.** Aristotle's cosmological **model,** a description of reality, had the Sun, Moon, and planets orbiting the Earth. Ptolemy enlarged on Aristotle's **geocentric** theory and explained retrograde motion by having the planets orbit the Earth on **epicycles,** small circles on their **deferent** that can be offset to be centered on the **equant.** Copernicus, in 1543, advanced a **heliocentric** theory that explained retrograde motion as a projection effect. Copernicus still used circular orbits and a few epicycles, and so had not completely broken with the past.

Tycho Brahe made unprecedentedly accurate observations of planetary positions, which Johannes Kepler interpreted. Kepler formulated three laws: (1) The orbits are ellipses with the Sun at one focus. Both **foci** and the **perihelion,** the closest point on the orbit to the Sun, lie on the **major axis.** (2) (**The law of equal areas.**) The line joining the Sun and a planet sweeps out equal areas in equal times. (3) The orbital **period**

squared is proportional to the **semimajor axis** cubed. The **minor axis** (including the **semiminor axis**) is perpendicular to the major axis; the difference between the lengths of the major and minor axes depends on the **eccentricity** of the ellipse, how far it is out of round. When measured in **Astronomical Units** and Earth **years,** Kepler's third law takes a simple form.

Galileo first systematically turned a telescope on the sky and discovered features on the Moon, moons of Jupiter, and a full set of phases of Venus. His discoveries strongly endorsed the Copernican heliocentric theory. Newton derived Kepler's laws mathematically, and also discovered the law of gravity and basic laws of motion.

A planet **revolves** around the Sun and **rotates** on its axis. Its orbit is generally **inclined** by only a few degrees to the plane of the Earth's orbit. Spinning or rotating objects have **angular momentum.** The conservation of angular momentum explains why planetary systems in formation contract into a disk. We think that small clumps of gas joined to make **planetesimals,** and planetesimals combined to make **protoplanets** orbiting the **protosun.**

Questions

1. Describe how you can tell in the sky whether a planet is in prograde or retrograde motion.
2. Describe the difference between the Ptolemaic and the Copernican systems in explaining retrograde motion.
3. Can planets interior to the Earth's orbit around the Sun undergo retrograde motion? Explain with a diagram.
4. If Copernicus's heliocentric model did not give more accurate predictions than Ptolemy's geocentric model, why do we now prefer Copernicus's model?
5. Draw the ellipses that represent Neptune's and Pluto's orbits. Show how one can cross the other, even though they are both ellipses with the Sun at one of their foci.
6. The Earth is closest to the Sun in January each year. Use Kepler's second law to describe the Earth's relative speeds in January and July.
†7. Pluto orbits the Sun in about 250 years. From Kepler's third law, calculate its semimajor axis. Show your work. Estimate the roots by hand rather than with a calculator.
†8. Mars's orbit has a semimajor axis of 1.5 A.U. From Kepler's third law, calculate Mars's period. Show your work. Estimate the roots by hand rather than with a calculator.
9. Explain in your own words why the observation of a full set of phases for Venus backs the Copernican system.
10. Do you expect Venus to have a larger angular size in its crescent phase or gibbous phase? Explain.
11. Discuss Galileo's main scientific contributions.
12. Discuss Newton's main scientific contributions.
†13. Explicitly show from Newton's version of Kepler's third law that Kepler's "constant" actually depends on the planet under consideration. (Substitute values for the mass of the Sun and of various planets into the constant.) Does it vary much for different planets?
†14. If a hypothetical planet is 4 times farther from the Sun than Earth is, what is its orbital *speed* (not period) relative to Earth's?
15. What is the difference between a protoplanet and a planet?
16. What role do planetesimals play in the origin of the planets?
17. Discuss the choice between models for the formation of the outer planets.

---

†This question requires a numerical solution.

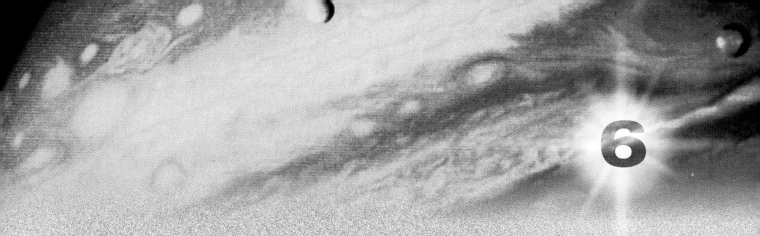

6

The Terrestrial Planets: Earth, Moon, and Their Relatives

ORIGINS *The small, rocky, terrestrial planets and the Moon show many similarities and differences among their interiors, surfaces, and atmospheres. By studying these objects, we can begin to understand the formation and evolution of the Earth, whose properties seem uniquely suitable for the development of complex life.*

AIMS

To describe the inner planets, four rocky worlds including the Earth.

Mercury, Venus, Earth, and Mars (Fig 6.1) share many similar features. Small compared with the huge planets beyond them, these inner planets also have rocky surfaces surrounded by relatively thin and transparent atmospheres, in contrast with the larger, gaseous/liquid planets. Together, we call these four the **terrestrial planets** (from the Latin "terra," meaning earth), which indicates their significance to us in our attempts to understand our own Earth. In this chapter, we discuss each of these rocky bodies, as well as their moons.

Venus and the Earth are often thought of as "sister planets," in that their sizes, masses, and densities are about the same (Fig. 6–2). But in many respects they are as different from each other as the wicked stepsisters were from Cinderella. The Earth is lush; it has oceans of water, an atmosphere containing oxygen, and life. On the other hand, Venus is a hot, foreboding planet with temperatures constantly over 750 K (900°F), a planet on which life seems unlikely to develop. Why is Venus like that? How did these harsh conditions come about? Can it happen to us here on Earth?

The one Solar-System body other than Earth that humans have visited is our Moon. It is so large relative to Earth that it joins us as a type of a "double-planet system." We will see how space exploration has revealed many of its secrets.

Mars is only 53 per cent the diameter of Earth and has 10 per cent of Earth's mass. Its atmosphere is much thinner than Earth's, too thin to breathe. But Mars has long been attractive as a site for exploration. We remain interested in Mars as a place where we may yet find signs of life or, indeed, where we might encourage life to grow. Mars's two tiny moons are but chunks of rock in orbit.

Mercury, the innermost planet, is more like our Moon than like our own Earth. Its atmosphere is negligible and its surface is seared by solar radiation.

Figure 6–1 Mars, viewed from one of the Viking spacecraft. The large volcanoes on its surface show clearly.

REDSHIFT
http://www.harcourtcollege.com/astro/cosmos/rsce

◀ A montage of space views of all the planets of the Solar System imaged from spacecraft: Mercury, Venus, Earth and Moon, Mars, Jupiter, Saturn, Uranus, and Neptune.

Figure 6-2 Earth and Venus at the same radar resolution.

Figure 6-3 The Moon viewed through green, blue, and ultraviolet filters from the Cassini spacecraft as it passed by our region of the Solar System in 1999 on its mission to Saturn. Mare Crisium is the dark circular region *(upper right).*

REDSHIFT
http://www.harcourtcollege.com/astro/cosmos/rsce

Figure 6-4 A composite of the Earth and Moon, viewed from the Galileo spacecraft as it passed by our region of the Solar System in 1992 on its mission to Jupiter. Structure on the Earth's surface, especially in South America and Central America, shows clearly.

OUR EARTH

On the first trip that astronauts ever took to the Moon, they looked back and saw for the first time the Earth floating in space. Nowadays we see that space view every day from weather satellites, so the view from the Saturn-bound Cassini spacecraft as it passed Earth in 1999 allowed us to test the instruments on known objects, the Earth and Moon (Fig. 6–3).

The realization that Earth is an oasis in space helped inspire our present concern for our environment. Until fairly recently, we studied the Earth only in geology courses and the other planets only in astronomy courses, but now the lines are very blurred. Not only have we learned more about the interior, surface, and atmosphere of the Earth but we have also seen the planets in enough detail to be able to make meaningful comparisons with Earth. The study of **comparative planetology** is helping us to understand weather, earthquakes, and other topics. This expanded knowledge will help us improve life on our own planet.

The Earth's Interior

The study of the Earth's interior and surface (Fig. 6–4) is called **geology.** Geologists study, among other things, how the Earth vibrates as a result of large shocks, such as earthquakes. Much of our knowledge of the structure of the Earth's interior comes from **seismology,** the study of these vibrations. The vibrations travel through different types of material at different speeds. From seismology and other studies, geologists have been able to develop a picture of the Earth's interior. The Earth's innermost region, the **core,** consists primarily of iron and nickel. The central part of the core may be solid, but the outer part is probably a very dense liquid. Outside the core is the **mantle,** and on top of the mantle is the thin outer layer called the **crust.** The upper mantle and crust are rigid and contain a lot of silicates, while the lower mantle is partially melted.

How did such a layered structure develop? The Earth, along with the Sun and the other planets, was probably formed from a cloud of gas and dust, and so was

A Closer Look

Comparative Data for the Terrestrial Planets and Their Moons

	Semimajor Axis	Orbital Period	Equatorial Radius (Earth = 100%)	Mass (Earth = 100%)
Planets				
Mercury	0.39 A.U.	88 days	38%	5.5%
Venus	0.72 A.U.	225 days	95%	82%
Earth	1 A.U.	365.25 days	100%	100%
Mars	1.52 A.U.	687 days	53%	11%
Earth's satellite				
The Moon	384,000 km	27⅓ days	27% (1738 km)	1.2%
Mars's satellites				
Phobos	9,378 km	7 h 39 m	$14 \times 11 \times 9 \ km^3$	1 millionth %
Deimos	23,459 km	1 d 6 h 18 m	$8 \times 6 \times 6 \ km^3$	0.3 millionth %

surrounded by a lot of debris in the form of dust and rocks. The young Earth was subject to constant bombardment from this debris. This bombardment heated the surface to the point where it began to melt, producing lava. Some of the original heat, though not enough to melt the Earth's interior, came from gravitational energy released as particles came together to form the Earth; such energy is released from gravity between objects when the objects move closer together and collide. The water at the base of a waterfall, for example, is slightly hotter than the water at the top; part of the falling water's energy of motion, gained by the pull of gravity, is converted to heat by the collision on the rocks at the waterfall's base.

The major source of the present-day energy in the interior is the natural radioactivity within the Earth. Certain forms of atoms are unstable—that is, they spontaneously change into more stable forms. In the process, they give off energetic particles that collide with the atoms in the rock and give some of their energy to these atoms. The rock heats up. The Earth's interior became so hot that the iron melted and sank to the center since it was denser, forming the core. Eventually other materials also melted. As the Earth cooled, various materials, because of their different densities and freezing points (the temperature at which they change from liquid to solid), solidified at different distances from the center. This process, called "differentiation," is responsible for the present layered structure of the Earth.

The rotation of the Earth's metallic core helps generate a magnetic field on Earth. The magnetic field has a north magnetic pole and a south magnetic pole that are not quite where the regular north and south geographic poles are—the Earth's magnetic north pole is near Hudson Bay, Canada.

Continental Drift

Some geologically active areas exist in which heat flows from beneath the surface at a rate much higher than average (Fig. 6–5). The outflowing **geothermal energy,** sometimes tapped as an energy source, signals what is below. The Earth's rigid outer layer is segmented into **plates,** each thousands of kilometers in extent but only about 50 km thick. Because of the internal heating, the top layers

Figure 6-5 A geyser and its surrounding steam, at Rotorua, New Zealand.

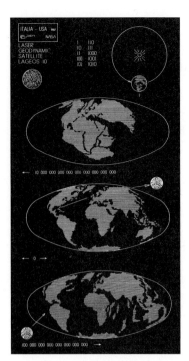

Figure 6–6 LAGEOS I and LAGEOS II (**La**ser **Geo**dynamic **S**atellites) bear plaques showing the continents in their positions 270 million years in the past *(top)*, at present *(middle)*, and 8 million years in the future *(bottom)*, according to our knowledge of continental drift. (The dates are given in the binary system.) Data on continental drift are gathered by reflecting laser beams from telescopes on Earth off the 426 retroreflectors that cover its surface. (Each retroreflector is a cube whose interior reflects incident light back in the direction from which it came.) Measuring the time until the beams return can be interpreted to show the accurate position of the ground station. We can now measure the position to an accuracy of about 1 cm.

float on an underlying hot layer (the mantle) where the rock is soft, though it is not hot enough to melt completely. The mantle beneath the rigid plates of the surface churns very slowly, thereby carrying the plates around. This theory, called **plate tectonics,** explains the observed **continental drift**—the drifting of the continents, over eons, from their original positions, at the rate of a few centimeters per year—about the speed your fingernails grow. ("Tectonics" comes from the Greek word meaning "to build.")

Although the notion of continental drift once seemed unreasonable, it is now generally accepted. The continents were once connected as two super-continents, which may have separated from a single super-continent called Pangaea ("all lands"). Over two hundred million years or so, the continents have moved apart as plates have separated. We can see from their shapes how they once fit together (Fig. 6–6). We even find similar fossils and rock types along two opposite coastlines, which were once adjacent but are now widely separated. More recently, studies have shown that Pangaea itself probably formed from the collision of previous generations of continents. The coming together and breaking apart of continents may have had many cycles in Earth's history.

In the future, we expect part of California to separate from the rest of the United States, Australia to be linked to Asia, and the Italian "boot" to disappear. The boundaries between the plates are geologically active areas (Fig. 6–7). (www) Therefore, these boundaries are traced out by the regions where earthquakes and most of the volcanoes occur. The boundaries where two plates are moving apart mark regions where molten material is being pushed up from the hotter interior to the surface, such as the mid-Atlantic ridge (Fig. 6–8). Molten material is being forced up through the center of the ridge and is being deposited as lava flows on either side, producing new seafloor. The motion of the plates relative to each other is also responsible for the formation of most great mountain ranges. When two plates come together, one may be forced under the other and the other raised. The great Himalayan chain, for example, was produced by the collision of India with the rest of Asia. The "ring of fire" volcanoes around the Pacific Ocean (including Mt. St. Helens in Washington) were formed when molten material made its way through gaps or weak points between plates.

Tides

It has long been accepted that **tides** are most directly associated with the Moon and to a lesser extent with the Sun. We know of their association with the Moon because the tides—like the Moon's passage across your meridian (the imaginary line in the sky passing from north to south through the point overhead)—occur

Figure 6–7 The San Andreas fault on Earth marks the boundary between the North American and Pacific plates. It is a strike-slip fault, in which two blocks of Earth move past each other. This view is paired with a Galileo-spacecraft image of a similar strike-slip fault on Jupiter's moon Europa, shown at the same scale and resolution.

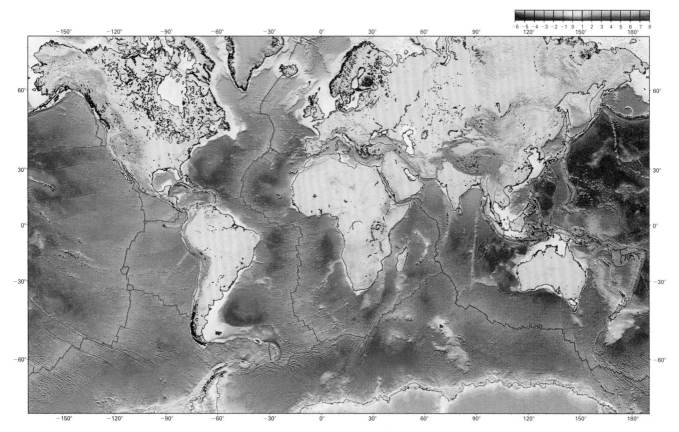

Figure 6-8 The Earth's ocean floor, mapped with microwaves bounced off the ocean's surface from spacecraft as well as with soundings from ships. Features on the seafloor with excess gravity attract the water enough to cause structure on the ocean's surface that can be mapped, though a 2000-meter-high seamount produces a bump on the surface only 2 meters high. High gravity is shown in yellow, orange, and red; low gravity is shown in blue and violet. The feature running from north to south in the middle of the Atlantic Ocean is the *mid-Atlantic ridge*. It marks the boundary between plates that are moving apart. Continents drift at about the rate that fingernails grow.

about an hour later each day. Tides result from the fact that the force of gravity exerted by the Moon (or any other body) gets weaker as you get farther away from it. Tides depend on the difference between the gravitational attraction of a massive body at different points on another body. ☼

To explain the tides in Earth's oceans, consider, for simplicity, that the Earth is completely covered with water. We might first say that the water closest to the Moon is attracted toward the Moon with the greatest force and so is the location of high tide as the Earth rotates. If this were the whole story, high tides would occur about once a day. However, two high tides occur daily, separated by roughly 12½ hours. To see why we get two high tides a day, consider three points, *A*, *B*, and *C*, where *B* represents the solid Earth (which moves all together as a single object and is marked by a point at its center), *A* is the ocean nearest the Moon, and *C* is the ocean farthest from the Moon (Fig. 6–9). Since the Moon's gravity weakens with distance, it is greater at point *A* than at *B*, and greater at *B* than at *C*. If the Earth and Moon were not in orbit about each other, all these points would fall toward the Moon, moving apart as they fell because of the difference in force. Thus the high tide on the side of the Earth that is near the Moon is a result of the water's being pulled away from the Earth. The high tide on the opposite side of the Earth results from the Earth's being pulled away from

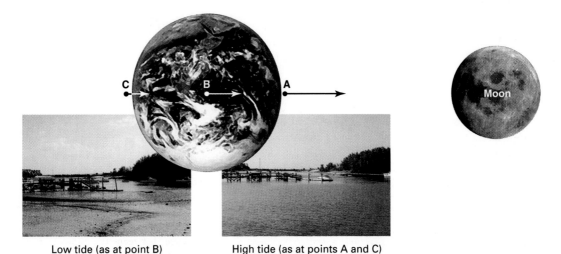

Low tide (as at point B)   High tide (as at points A and C)

Figure 6-9   *(top)* A schematic representation of the tidal effects caused by the Moon. The arrows represent the gravitational pull produced by the Moon (exaggerated in the drawing). The water at point A is pulled toward the Moon more than the Earth's center (deep behind point B) is; since the Earth is solid, the whole Earth moves with its center. (Tides in the solid Earth exist, but are much smaller than tides in the oceans.) Similarly, the solid Earth is pulled away from the water at point C. Water tends to collect at points A and C (high tide) and to be depleted in regions halfway between them on the Earth's surface, such as at point B (low tide). Note that although the tidal force results from gravity, it is the *difference* between the gravitational forces at two places and is therefore not the same as the "gravitational force." Tidal forces are important in many astrophysical situations, including accretion disks around neutron stars and black holes, and rings around planets. *(bottom)* The tidal range at Kennebunkport, Maine, where the shape of the ocean floor leads to an especially high tidal range.
Note that the Moon and the Earth are drawn to a different scale from their separation.

the water. In between the locations of the high tides the water has rushed elsewhere, so we have low tides. Since the Moon is moving in its orbit around the Earth, a point on the Earth's surface has to rotate longer than 24 hours to return to a spot nearest to the Moon. Thus a pair of tides repeats about every 25 hours, making 12½ hours between high tides.

The Sun's effect on the Earth's tides is only about half as much as the Moon's. Though the Sun exerts a greater gravitational force on the Earth than does the Moon, the Sun is so far away that its force does not change very much from one side of the Earth to the other. And it is only the difference in force from one place to another that counts for tides. Nonetheless, the Sun does matter. We tend to have very high and very low tides when the Sun, Earth, and Moon are aligned (as is the case near the time of full moon or of new moon), because their effects reinforce each other. Conversely, tides are less extreme when the Sun, Earth, and Moon form a right angle (as near the time of a first-quarter or third-quarter moon).

The Earth's Atmosphere

We name layers of our atmosphere (Fig. 6–10) according to the composition and the physical processes that determine its temperature. The Earth's weather is confined to the very thin **troposphere.** A major source of heat for the troposphere is infrared radiation emitted from the ground, so the temperature of the troposphere decreases as altitude increases. The rest of the Earth's atmosphere, as well as the Earth's surface, is heated mainly by solar energy from above.

A higher layer of the Earth's atmosphere is the **ionosphere,** where many of the atoms are ionized—that is, stripped of some of the electrons they normally contain—by the high temperature. The free electrons in the ionosphere reflect very-

A Closer Look

Density

The density of an object is its mass divided by its volume. The scale of mass in the metric system was set up so that 1 cubic centimeter of water would have a mass of 1 gram (symbol: g). Though the density of water varies slightly with the water's temperature and pressure, the density of water is always about 1 g/cm³. Density is always expressed in units of mass (grams, kilograms, and so on) per volume (cubic centimeters, cubic feet, and so on), where the slash means "per." In "customary units," the density of water is about 62 pounds/cubic foot (though, technically, pounds is a measure of weight—or gravitational force—rather than of mass). Always be careful to note if you are using metric or customary units. A spacecraft was lost just as it reached Mars in 1999 because one organization getting it there used metric units while another used customary units.

To measure the density of any amount of matter, astronomers must independently measure its mass and its size. They are often able to identify the materials present in an object on the basis of density. For example, since the density of iron is about 8 g/cm³, Saturn's density of only 0.7 g/cm³ shows that Saturn is not made largely of iron. On the other hand, the measurement that the Earth's density is about 5 g/cm³ shows that the Earth has a substantial content of dense elements, of which iron is a likely candidate.

Astronomical objects cover the extremes of density. In the space between the stars, the density is incredibly tiny fractions of a gram in each cubic centimeter (about 10^{-24} g/cm³), corresponding to just one hydrogen atom. And in neutron stars, the density exceeds a billion tons per cubic centimeter, comparable to that of an atomic nucleus.

---

long-wavelength radio signals. When the conditions are right, radio waves bounce off the ionosphere, which allows us to tune in distant radio stations. Observations from satellites have greatly enhanced our knowledge of Earth's atmosphere.

Scientists carry out calculations using the most powerful supercomputers to interpret the global data and to predict how the atmosphere will behave. The equations are essentially the same as those for the internal temperature and structure of stars, except that the sources of energy are different, with stars heated from below and the Earth's atmosphere mostly heated from above. Winds are caused partly by uneven heating of different regions. The rotation of the Earth

Densities

water	1	g/cm³
aluminum	3	g/cm³
typical rock	3	g/cm³
iron	8	g/cm³
lead	11	g/cm³

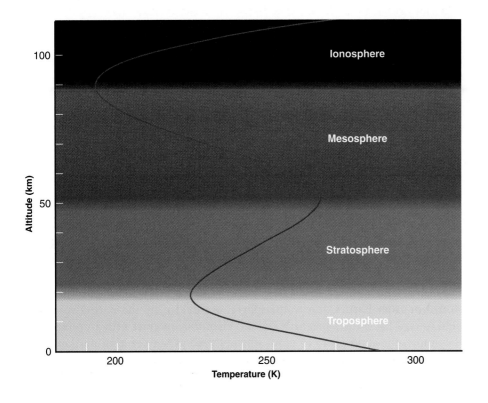

Figure 6-10 Temperature in the atmosphere as a function of altitude. In the troposphere, the energy source is the ground, so the temperature decreases as you go up in altitude. Higher temperatures in other layers result from ultraviolet radiation and x-rays from the Sun being absorbed.

Figure 6–11 The concentration of atmospheric carbon dioxide in parts per million (ppm) of dry air as it changed with time, observed at the Mauna Loa Observatory in Hawaii. In addition to the yearly cycle, there is an overall increase.

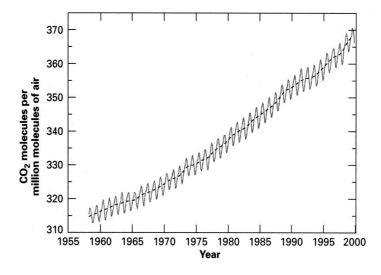

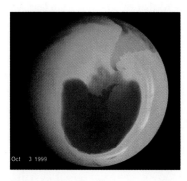

Figure 6–12 The amount of ozone (O_3) over Antarctica in the spring (September and early October for the Earth's southern hemisphere) is much lower than the amount in other seasons. Note the outline of the Antarctic continent in this display.

also has a very important effect in determining how the winds blow. Comparison of the circulation of winds on the Earth (which rotates in 1 Earth day), on slowly rotating Venus (which rotates in 243 Earth days), and on rapidly rotating Jupiter and Saturn (each of which rotates in about 10 Earth hours) helps us understand the weather on Earth. Our improved understanding allows forecasters of weather (day to day) and climate (long term) to be more accurate.

Comparison with the planet Venus led to our realization that the Earth's atmosphere traps radiation from the Sun, and that we are steadily increasing the amount of trapping. This "greenhouse effect" is caused largely by the carbon dioxide in Earth's atmosphere (Fig. 6–11), the amount of which is growing each year because of our use of fossil fuels. (See p. 113 for a comparison with Venus.)

Quite a separate problem is the discovery about the ozone in our upper atmosphere. This ozone (O_3) becomes thinner over Antarctica each Antarctic spring, a phenomenon known as the "ozone hole." The ozone hole apparently started forming only in the mid-1980s, but its maximum size has been larger almost every year since (Fig. 6–12). It is caused by the interaction in the cold upper atmosphere of sunlight with certain gases we give off in the lower atmosphere, such as chlorofluorocarbons used in air conditioners and refrigerators. International governmental meetings have arranged cutbacks in the use of these harmful gases. Some success has been achieved, but we have far to go to protect our future atmosphere.

The Van Allen Belts

> Every planet with a magnetic field — most notably Jupiter — has regions filled with charged particles. Earth's Van Allen belts were the first to be studied.

In 1958, the first American space satellite carried aloft, among other things, a device to search for particles carrying electric charge that might be orbiting the Earth. This device, under the direction of James A. Van Allen of the University of Iowa, detected a region filled with charged particles having high energies. Two such regions—the **Van Allen belts**—surround the Earth, like a small and a large doughnut, containing protons and electrons (Fig. 6–13). A more recently discovered innermost belt contains mainly ions of heavier elements from interstellar space. The particles in the Van Allen belts are trapped by the Earth's magnetic field. Charged particles preferentially move in the direction of magnetic-field lines, and not across the field lines. These particles, often from solar magnetic storms or from Earth's upper atmosphere, are guided by the Earth's magnetic field toward the Earth's magnetic poles. When they interact with air molecules, they cause our atmosphere to glow, which we see as the beautiful northern and southern lights—the **aurora borealis** and **aurora australis,** respectively (Fig. 6–14).

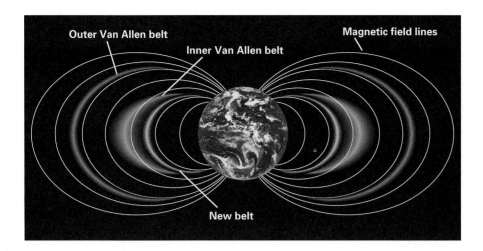

Figure 6–13 The outer Van Allen belt contains mainly electrons; the traditional "inner belt" contains mainly protons; and the newly discovered belt, which is within the other two belts, contains mainly ions of oxygen, nitrogen, and neon, for which the distribution of isotopes indicates that they are from interstellar space.

THE MOON

The Earth's nearest celestial neighbor—the Moon—is only 380,000 km (238,000 miles) away from us, on the average. At this distance, it appears sufficiently large and bright to dominate our nighttime sky. The Moon's stark beauty has captured our attention since the beginning of history. Now we can study the Moon not only as an individual object but also as an example of a small planet or a large planetary satellite, since spacecraft observations have told us that there may be little difference between small planets and large moons.

The Moon's Appearance

Even binoculars reveal that the Moon's surface is pockmarked with craters. Other areas, called **maria** (pronounced mar′ee-a; singular **mare,** pronounced mar′eyh), are relatively smooth. Indeed, the name comes from the Latin word for sea (Fig. 6–15). But there are no ships sailing on the lunar seas and no water in them; the Moon is a dry, airless, barren place.

The gravity at the Moon's surface is only one-sixth that of the Earth. You would weigh only 20 or 30 pounds there if you stepped on a scale! Gravity is so weak that any atmosphere and any water that may once have been present would long since have escaped into space.

Figure 6–14 The aurora with comet C/1996 B2 (Hyakutake).

Figure 6–15 A waning crescent moon showing dark lunar seas (maria) and lighter highlands.

The Moon rotates on its axis at the same rate that it revolves around the Earth, thereby always keeping the same face in our direction. The Earth's gravity has locked the Moon in this pattern, pulling on a bulge in the distribution of the lunar mass to prevent the Moon from rotating freely. As a result of this interlock, we always see essentially the same side of the Moon from our vantage point on Earth.

When the Moon is full, it is bright enough to cast shadows or even to read by. But a full moon is a bad time to try to observe lunar surface structure, for any shadows we see are short, and lunar features appear washed out. When the Moon is a crescent or even a quarter moon, however, the sunlit part of the Moon facing us is covered with long shadows. The lunar features then stand out in bold relief. Shadows are longest near the **terminator,** the line separating day from night. Note that nature photographers on Earth, concluding that views with shadows are more dramatic, generally take their best photos when the Sun is low.

The Moon revolves around the Earth every 27⅓ days with respect to the stars. But during that time, the Earth has moved part way around the Sun, so it takes a little more time for the Moon to complete a revolution with respect to the Sun (Fig. 6–16). The cycle of phases that we see from Earth repeats with this 29½-day period.

In some sense, before the period of exploration by the Apollo program, we knew more about bright stars than we did about the Moon. As a solid body, the Moon reflects the spectrum of sunlight rather than emitting its own spectrum, so we were hard pressed to determine even the composition or the physical properties of the Moon's surface (such as whether you would sink into it!).

The Lunar Surface

The kilometers of film exposed by the astronauts, the 382 kg of rock brought back to Earth, the lunar seismograph data recorded on tape, and other data have been studied by hundreds of scientists from all over the world. The data have led to new views of several basic questions, and have raised many new questions about the Moon and the Solar System.

The rocks that were encountered on the Moon are types that are familiar to terrestrial geologists (Fig. 6–17). Almost all the rocks are the kind that were formed by the cooling of lava, known as igneous rocks. Basalts are one example.

The Moon and the Earth seem to be similar chemically, though significant differences in overall composition do exist. Some elements that are rare on Earth—such as uranium and thorium—are found in greater abundances on the Moon. (Will we be mining on the Moon one day?) None of the lunar rocks contain any trace of water bound inside their minerals.

Meteoroids, interplanetary rocks that we will discuss later on, hit the Moon with such high speeds that huge amounts of energy are released at the impact. The effect is that of an explosion, as though TNT or an H-bomb had exploded. As a result of the Apollo missions, we know that almost all the craters on the Moon come from such impacts. One way of dating the surface of a moon or planet is to count the number of craters in a given area, a method that was used before Apollo. Surely those locations with the greatest number of craters must be the oldest. Relatively smooth areas—like maria—must have been covered over with volcanic material at some relatively recent time (which is still billions of years ago).

Obvious rays of lighter-colored matter splattered outward during the impacts that formed a few of the craters. Since these rays extend over other craters, the craters with rays must have formed more recently. The youngest rayed craters

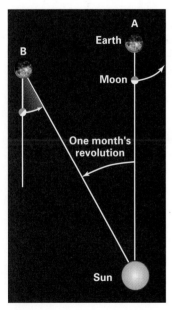

Figure 6–16 After the Moon has completed its 27⅓-day revolution around the Earth with respect to the stars, it has moved from A to B. But the Earth has moved one month's worth around the Sun. It takes about an extra two days for the Moon to complete its revolution with respect to the Sun.

A   **B**   **C**

Figure 6–17 Different types of lunar rocks brought back to Earth by the Apollo 15 astronauts. *(A)* A rock that is a compound of other types of rock. It weighs 1.5 kg. A 1-cm block appears for scale. *(B)* The dark part is dust that covered another rock to which this rock was attached. *(C)* A 4.5-kg compound rock.

may be very young indeed—perhaps only a few hundred million years. The rays darken with time, so rays that may have once existed near other craters are now indistinguishable from the rest of the surface.

Crater counts and the superposition of one crater on another give only relative ages. We found the absolute ages only when rocks were physically returned to Earth. Scientists worked out the dates by comparing the current ratio of radioactive forms of atoms to nonradioactive forms present in the rocks with the ratio that they would have had when they were formed. (Varieties of the chemical elements having different numbers of neutrons are known as isotopes, and radioactive isotopes are those that decay spontaneously; that is, they change into other isotopes even when left alone. Stable isotopes remain unchanged. For certain pairs of isotopes—one radioactive and one stable—we know the proportion of the two when the rock was formed. Since we know the rate at which the radioactive one is decaying, we can calculate how long it has been decaying from a measurement of what fraction is left.) The oldest rocks that were found on the Moon solidified 4.4 billion years ago. The youngest rocks found solidified 3.1 billion years ago.

The observations can be explained on the basis of the following general picture (Fig. 6–18): The Moon formed about 4.6 billion years ago. From the oldest rocks, we know that at least the top 100 km of the surface was molten about

> The superposition of one crater on another—when one crater lies wholly or partially over the other—indicates that the underlying crater formed first and is therefore older.

> This concept, known as "radioactive dating," is widely used, not only for dating rocks on the Moon, but also for dating rocks on Earth and for finding the ages of archaeological sites, of ancient pottery, and so on.

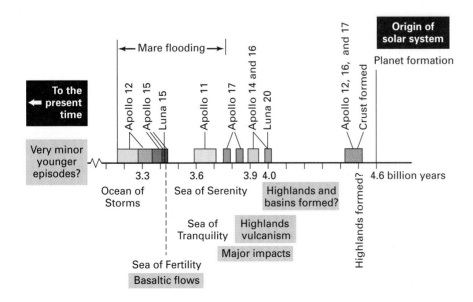

Figure 6–18 The chronology of the lunar surface. The ages of rocks found in 8 missions are shown. Note how many crewed American Apollo missions and uncrewed Soviet Luna missions it took to get a sampling of many different ages on the lunar surface.

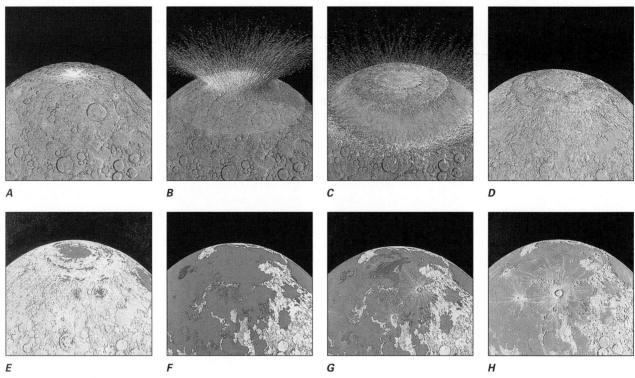

A   *B*   *C*   *D*

E   *F*   *G*   *H*

Figure 6–19 An artist's view of the formation of the Mare Imbrium region of the lunar surface. *(A)* An asteroid impact on the Moon, sometime between 3.85 and 4.0 billion years ago. *(B)* The shock of the asteroid impact began the Imbrium crater. *(C)* As the dust and heat subsided, the 1300-km Imbrium crater was left. *(D)* The molten rock (lava) flowed over outlying craters and cooled, leaving lunar mountains. *(E)* Lava welled up from inside the Moon 3.8 billion years ago. It filled the basin. *(F)* By 3.3 billion years ago, the lava flooding was nearly complete. *(G)* The final flow of thick lava came 2.5–3.0 billion years ago. *(H)* Subsequent cratering has left the Mare Imbrium of today's Moon.

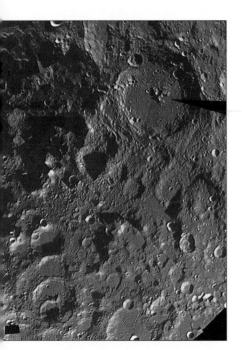

Figure 6–20 An oblique view of part of the lunar far side, from Clementine, shows how rough it looks.

200 million years later. Then the surface cooled. From 4.2 to 3.9 billion years ago, bombardment by interplanetary rocks caused most of the craters we see today. About 3.8 billion years ago, the interior of the Moon heated up sufficiently (from radioactive elements inside) that volcanism began. Lava flowed onto the lunar surface and filled the largest basins that resulted from the earlier bombardment, thus forming the maria (Fig. 6–19). By 3.1 billion years ago, the era of volcanism was over. The Moon has been geologically pretty quiet since then.

Up to this time, the Earth and the Moon shared similar histories. But active lunar history stopped about 3 billion years ago, while the Earth continued to be geologically active.

Almost all the rocks on the Earth are younger than 3 billion years of age; the oldest single rock ever discovered on Earth has an age of 4.5 billion years, but few such old rocks have been found. Erosion and the remolding of the continents as they move slowly over the Earth's surface have taken their toll. So we must look to extraterrestrial bodies—the Moon or meteorites—that have not suffered the effects of plate tectonics or erosion (which occurs in the presence of water or an atmosphere) to study the first billion years of the Solar System.

Not until the 1990s did spacecraft revisit the Moon. The Clementine spacecraft (named after the prospector's daughter in the old song, since the spacecraft was looking for minerals) took photographs and other measurements. Photographs of the far side of the Moon (Fig. 6–20) have shown us that the near and far hemispheres are quite different in overall appearance. The maria, which are so conspicuous on the near side, are almost absent from the far side, which is

A Closer Look

The First People on the Moon

On July 20, 1969, Neil Armstrong and Buzz Aldrin left Michael Collins orbiting the Moon and descended to the lunar surface. After Armstrong stepped out of the spacecraft first and took that first step on the Moon, he, the first man on the Moon, made the famous statement, "That's one small step for man, one giant leap for mankind."

At the time of the 30th anniversary of the Moon landing, looking back Aldrin said, "I was full of goose bumps when I stepped down onto the surface. I felt a bit disoriented because of the nearness of the horizon. On Earth when you look at the horizon it appears flat, but on the Moon, so much smaller and without hills, the horizon in all directions curved visibly down away from us."

"There was about every type of rock imaginable, all covered with a very fine powder. The rocks themselves actually had no color until you looked closely at the crystals. The thought briefly occurred to me that these rocks had been sitting there for hundreds of millions or billions of years, and that we were the first living beings to see them. But we were too busy to be philosophical for long or to study them closely, and we just grabbed what looked like an interesting assortment. I felt them crunch beneath my feet as we walked around." *(Griffith Observer interview, July 1999)*

REDSHIFT
http://www.harcourtcollege.com/astro/cosmos/rsce

cratered all over. We shall see in the next section that the difference probably results from the different thicknesses of the lunar crust on the sides of the Moon nearest Earth and farthest from Earth. The difference was first seen in the fuzzy photographs of the far side that were taken by the Soviet Lunik 3 Spacecraft in 1959.

In the 1990s, NASA's Lunar Prospector and Clementine spacecraft mapped the Moon (Figs. 6–21 and 6–22). Lunar Prospector confirmed indications from the Clementine spacecraft that there is likely to be water ice on the Moon, by detecting more neutrons coming from the Moon's polar regions than elsewhere. Clementine and Lunar Prospector scientists think that these neutrons are given off in interactions of particles coming from the Sun with hydrogen in water ice in craters near the poles, where they are shaded from the Sun's rays. But the detection is not of water directly, and Apollo 17 astronaut Harrison Schmitt, the only geologist ever to have walked on the Moon, told one of us (JMP) in 1999 that the neutrons may instead have come from the solar wind (Fig. 6–23). He

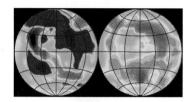

Figure 6–21 A Lunar Prospector map of the distribution of gravity on the Moon, showing hemispheres of the near side and the far side.

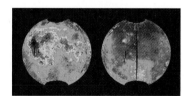

Figure 6–22 A Clementine map of the distribution of iron over the Moon.

Figure 6–23 Scientist–astronaut Harrison Schmitt, the only Ph.D. to have stood on the Moon, during the Apollo 17 mission. The giant rock indicates that the later Apollo missions went to less smooth sites than the earlier ones.

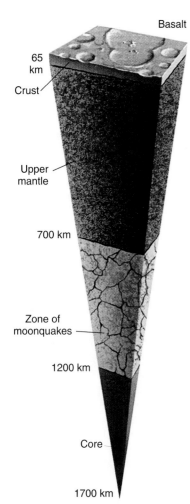

Basalt

65 km

Crust

Upper mantle

700 km

Zone of moonquakes

1200 km

Core

1700 km

Figure 6-24 The Moon's interior. The depth of solidified lava is greater under maria, which are largely on the side of the Moon nearest the Earth.

thinks the case for water on the Moon is not yet proved. He would love to find water there, because it would increase the chance that a crewed Moon base could be supplied on the Moon itself, much easier than bringing everything from Earth. In his estimation, it would take about 10 years to set up such a base, once basic funding is available on Earth. An attempt to test the idea of water in the crater was made by crashing Lunar Prospector into the most likely spot in 1999, but none of the spectrographs on Earth viewing the event detected any signal from the crash.

The Lunar Interior

Before the Moon landings, it was widely thought that the Moon was a simple body, with the same composition throughout. But we now know it to be differentiated (Fig. 6–24), like the planets. Most experts believe that the Moon's core is molten, but the evidence is not conclusive. The lunar crust is perhaps 65 km thick on the near side and twice as thick on the far side. This asymmetry may explain the different appearances of the sides, because lava would be less likely to flow through the far side's thicker crust.

The Apollo astronauts brought seismic equipment to the Moon (Fig. 6–25). One type of earthquake wave moves material to the side. Only if the Moon's core were solid would the material be moved back to where it started, continuing the wave. Since that type of wave doesn't come through the Moon, scientists deduce that the Moon's core is probably molten.

Tracking the orbits of the Apollo Command Modules and more recently the Clementine and Lunar Prospector spacecraft that orbited the Moon told us about the lunar interior. If the Moon were a perfect, uniform sphere, the spacecraft orbits would have been perfect ellipses. But they weren't. One of the major surprises of the lunar missions of the 1960s, refined more recently, was the discovery in this way of **mascons,** regions of **mas**s **con**centrations near and under most maria. The mascons may be lava that is denser than the surrounding matter, providing a stronger gravitational force on satellites passing overhead.

The Origin of the Moon

The leading models for the origin of the Moon that were considered at the time of the Apollo missions were

1. *Fission.* The Moon was separated from the material that formed the Earth; the Earth spun up and the Moon somehow spun off;
2. *Capture.* The Moon was formed far from the Earth in another part of the Solar System, and was later captured by the Earth's gravity; and
3. *Condensation.* The Moon was formed near to (and simultaneously with) the Earth in the Solar System.

But work over the last two decades has all but ruled out the first two of these and has made the third seem less likely. The model now strongly favored, especially because of computer simulations, is

4. *Ejection of a Gaseous Ring.* A planetesimal perhaps twice the size of Mars hit the proto-Earth, ejecting matter in gaseous (and perhaps some in liquid or solid) form (Fig. 6–26). Although some of the matter fell back to Earth, and part escaped entirely, a significant fraction started orbiting the Earth, probably in the same direction as the initial incoming planetesimal. The orbiting material eventually coalesced into the Moon.

Figure 6–25 Buzz Aldrin and the experiments he deployed on Apollo 11. In the foreground, we see the seismic experiment. The laser-ranging retroreflector is behind it.

Comparing the chemical composition of the lunar surface with the composition of the Earth's surface has been important in narrowing down the possibilities. The mean lunar density of 3.3 grams/cm^3 is close to the average density of the Earth's major upper region (the mantle), and the Moon seems especially deficient in iron. This fact favors the fission hypothesis; had the Moon condensed from the same material as Earth, as in the condensation scenario, it would contain much more iron. However, detailed examination of the lunar rocks and soils indicates that the abundances of elements on the Moon and Earth's mantle are sufficiently different from each other to indicate that the Moon did not form directly from the Earth. In the collision hypothesis, on the other hand, such collisions are expected.

Present calculations strongly suggest that the fission mechanism doesn't work. Also, the theory of plate tectonics now explains the formation of the Pacific Ocean basin. Before the ejection-of-a-ring theory was considered most probable, it seemed possible that the Pacific Ocean could be the scar left behind when the Moon was ripped from the Earth according to the fission hypothesis.

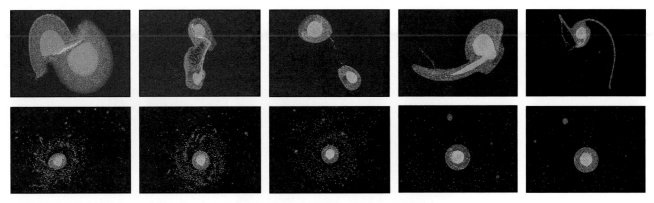

Figure 6–26 A computer simulation of a collision between the proto-Earth and an impactor. Each is composed of an iron core and a rock mantle. The internal energy in each increases with both temperature and pressure. Increasing internal energy is shown as dark red, light red, brown, and yellow for rock, and dark blue, light blue, dark green, and light green for iron. In the first images, the impactor hits the proto-Earth, separates, and then falls in again, pulling out a tail of matter. At the end of this sequence, which takes six days, several large clumps and many smaller clumps are left. This material then clumps together to form the Moon.

Figure 6-27 This rock is one of many meteorites found in the Earth's continent of Antarctica. Under a microscope, it seems like a sample from the lunar highlands and quite unlike any terrestrial rock. Eighteen meteorites are thought to have come from the Moon.

We are obtaining additional evidence about the capture model by studying the moons of Jupiter and Saturn. The outermost moon of Saturn, for example, is apparently a captured asteroid. However, the similarities in composition between the Moon and Earth's mantle argue against the capture hypothesis for the Earth–Moon system.

Thus as of now, ejection of a gaseous ring is the most accepted model.

Rocks from the Moon

A handful of meteorites found in Antarctica have been identified by their chemical composition as having come from the Moon (Fig. 6–27). They presumably were ejected from the Moon when craters formed. So we are still getting new moon rocks to study! A few other meteorites have even been found to come from Mars, as we shall discuss later on.

◖ MERCURY

Mercury is the innermost of our Sun's nine planets. Its average distance from the Sun is $4/10$ of the Earth's average distance, 0.4 A.U. Except for distant Pluto, its elliptical orbit around the Sun is the most elongated. Since we on the Earth are outside Mercury's orbit looking in at it, Mercury always appears close to the Sun in the sky (Fig. 6–28). At times Mercury rises just before sunrise, and at times it sets just after sunset, but it is never up when the sky is really dark. The Sun always rises or sets within an hour or so of Mercury's rising or setting. As a result, whenever Mercury is visible, its light has to pass obliquely through the Earth's atmosphere. This long path through turbulent air leads to blurred images. Thus astronomers have never had a really clear view of Mercury from the Earth, even with the largest telescopes. Even the best photographs taken from the Earth show Mercury as only a fuzzy ball with faint, indistinct markings. Most people have never seen it at all.

On rare occasions, Mercury goes into **transit** across the Sun; that is, we see it as a black dot crossing the Sun. One instance occurred in 1999; the next transits will be in 2003 and 2006.

The Rotation of Mercury

From studies of ground-based drawings and photographs, astronomers did as well as they could to describe Mercury's surface. A few features seemed to be barely distinguished, and the astronomers watched to see how long those features took

Figure 6-28 Since Mercury's orbit is inside that of the Earth *(left),* Mercury is never seen against a really dark sky. As a result, even Copernicus apparently never saw it. A view from the Earth appears at *right,* showing Mercury and Venus at their greatest respective distances from the Sun.

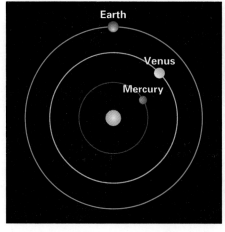

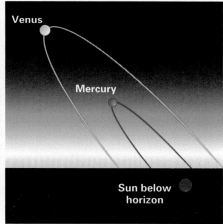

to rotate around the planet. From these observations they decided that Mercury rotates in the same amount of time that it takes to revolve around the Sun in its orbit, 88 Earth days. Thus they thought that one side always faces the Sun and the other side always faces away from the Sun. (Recall that one side of the Moon always faces the Earth, a similar phenomenon.) This discovery led to the fascinating conclusion that Mercury could be both the hottest planet and the coldest planet in the Solar System.

But when the first measurements were made of Mercury's radio radiation, the planet turned out to be giving off more energy than had been expected. This meant that it was hotter than expected. The dark side of Mercury was too hot for a surface that was always in the shade. (The light we see is merely sunlight reflected by Mercury's surface and doesn't tell us the surface's temperature. The radio waves are actually being emitted by the surface, as part of its "thermal" or nearly blackbody radiation.)

Later, we became able to transmit radar from Earth to Mercury. (Radar—radio detection and ranging—is when you send out radio waves so that they bounce off another object, allowing you to study their reflection.) Since one edge of the visible face of Mercury is rotating toward Earth, while the other edge is rotating away from Earth, the reflected radio waves were slightly smeared in wavelength according to the Doppler effect. This measurement allowed astronomers to determine Mercury's rotation speed, similarly to the way that radar is used by the police to tell if a car is breaking the speed limit. Knowing the rotation speed and Mercury's radius, we could determine the rotation period. The results were a surprise: it actually rotates every 59 days, not 88 days. Mercury's 59-day period of rotation with respect to the stars is exactly ⅔ of the 88-day orbital period, so the planet rotates three times for every two times it revolves around the Sun.

Mercury's rotation and revolution combine to give a value for the rotation of Mercury relative to the Sun (that is, a mercurian solar day) that is neither 59 nor 88 days long (Fig. 6–29). If we lived on Mercury we would measure each day (each day/night cycle) to be 176 Earth days long. We would alternately be fried for 88 Earth days and then frozen for 88 Earth days. Since each point on Mercury faces the Sun at some time, the heat doesn't build up forever at the place under the Sun, nor does the coldest point cool down as much as it would if it never received sunlight. The hottest temperature is about 700 K (800°F). The minimum temperature is about 100 K (−280°F).

No harm was done by the scientists' original misconception of Mercury's rotational period, but the story teaches all of us a lesson: we should not be too sure

Figure 6–29 Follow, from A to G, the arrow that starts facing to the right, toward the Sun in A. Mercury rotates once with respect to the stars in 59 days, when Mercury has moved only ⅔ of the way around the Sun (E). Note that after one full revolution of Mercury around the Sun (G), the arrow faces away from the Sun. It takes another full revolution, a second 88 days, for the arrow to again face the Sun. Thus Mercury's rotation period with respect to the Sun is twice 88, or 176, days.

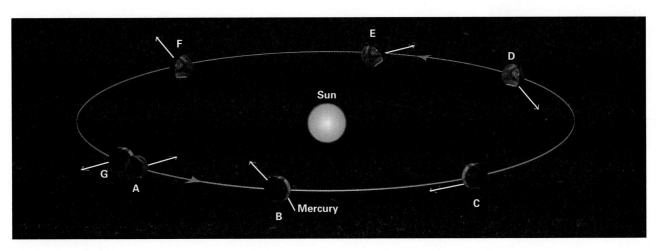

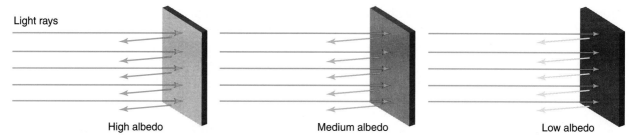

Figure 6-30   Albedo is the fraction of radiation reflected. A surface of low albedo looks dark.

of so-called "facts." Don't believe everything you read in this book, either. It should be fun for you to look back in twenty years and see how much of what we now think we know about astronomy actually turned out to be wrong.

Mercury Observed from the Earth

Even though the details of the surface of Mercury can't be seen very well from the Earth, other properties of the planet can be better studied. For example, we can measure Mercury's **albedo,** the fraction of the sunlight hitting Mercury that is reflected (Fig. 6–30). We can measure the albedo because we know how much sunlight hits Mercury (we know the brightness of the Sun and the distance of Mercury from the Sun). Then we can easily calculate at any given time how much light Mercury reflects, knowing how bright Mercury looks to us and its distance from the Earth. Comparing albedoes of materials on the Earth and on the Moon can teach us something of what the surface of Mercury is like.

Let us consider some examples of albedo. An ideal mirror reflects all the light that hits it; its albedo is thus 100 per cent. (The very best real mirrors have albedoes of as much as 96 per cent.) A black cloth reflects essentially none of the light that hits it; its albedo (in the visible part of the spectrum, anyway) is almost 0 per cent. Mercury's overall albedo is only about 10 per cent. Its surface, therefore, must be made of a dark (that is, poorly reflecting) material (though a few regions are very reflective, with albedoes up to 45 per cent). The albedoes of the Moon's maria are similarly low, 6 to 10 per cent. In fact, Mercury (or the Moon) appears bright to us only because it is contrasted against a relatively dark sky; if it were silhouetted against a bedsheet, it would look relatively dark.

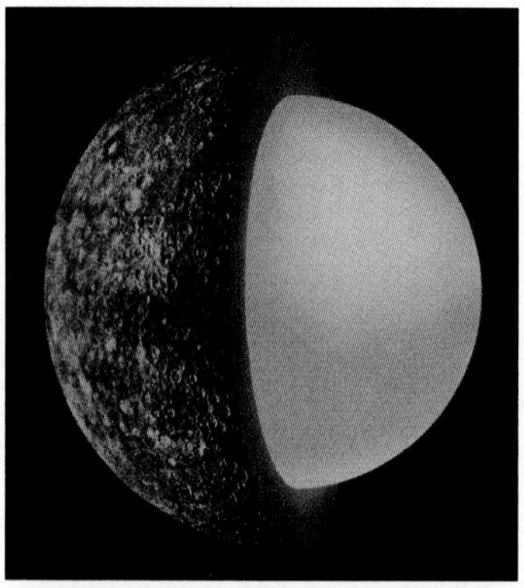

Mercury from Mariner 10

In 1974, we learned most of what we know about Mercury in a brief time. We flew right by. The tenth in the series of Mariner spacecraft launched by the United States went to Mercury. First it passed by Venus and had its orbit changed by Venus's gravity to direct it to Mercury. Tracking its orbit improved our measurements of the gravity of these planets and thus of their masses. Further, the 475-kg spacecraft had a variety of instruments on board.

The most striking overall impression is that Mercury is heavily cratered (Fig. 6–31). At first glance, it looks like the Moon! But there are several basic differences between the features on the surface of Mercury and those on the lunar surface. Mercury's craters seem flatter than those on the Moon, and have thinner rims (Fig. 6–32). Mercury's higher gravity at its surface may have caused the rims to slump more. Also, Mercury's surface may have been softer, more plastic-like, when most of the cratering occurred. The craters may have been eroded by any of a number of methods, such as the impacts of meteorites or microme-

Figure 6-31   A mosaic of photographs of Mercury from Mariner 10, showing its cratered surface. An artist has added a conception of Mercury's thick core *(yellow)* and thin crust/upper mantle *(red).*

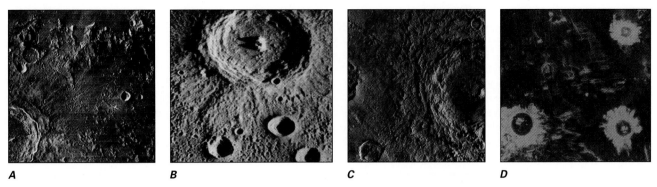

A   *B*   *C*   *D*

Figure 6–32   A comparison of *(A)* craters on Mercury, *(B)* lunar craters, *(C)* craters on Mars, and *(D)* craters on Venus. Though it is not noticeable on these images, material has been continuously thrown out less far on Mercury than on the Moon because of Mercury's higher gravity. The martian crater shows flow across its surface, possibly resulting from the melting of a permafrost layer under the surface. The photographs are dominated by Brahms on Mercury, 75 km in diameter; Copernicus on the Moon, 93 km in diameter; Cerulli on Mars, 115 km in diameter; and craters on Venus, 37 to 50 km in diameter.

teorites (large or small bits of interplanetary rock). Alternatively, erosion may have occurred during a much earlier period when Mercury may have had an atmosphere, undergone internal activity, or been flooded by lava.

Most of the craters seem to have been formed by impacts of meteorites. The Caloris Basin, in particular, is the site of a major impact. The secondary craters, caused by material ejected as primary craters were formed, are closer to the primaries than on the Moon, presumably because of Mercury's higher surface gravity. In many areas, the craters appear superimposed on relatively smooth plains. The plains are so extensive that they are probably volcanic. Their age is estimated to be 4.2 billion years, the oldest features on Mercury.

Smaller, brighter craters are sometimes, in turn, superimposed on the larger craters and thus must have been made afterward. Some craters have rays of higher albedo emanating from them (Fig. 6–33), just as some lunar craters do. The ray material represents relatively recent crater formation (that is, within the last hundred million years). The ray material must have been tossed out in the impact that formed the crater.

Lines of cliffs hundreds of miles long are visible on Mercury; on Mercury, as on Earth, such lines of cliffs are called **scarps.** The scarps are particularly apparent in the region of Mercury's south pole (Fig. 6–34). Unlike fault lines on the Earth, such as the San Andreas fault in California, on Mercury there are no

Figure 6–33   A field of rays on Mercury radiating from a crater off to the *top left.* The largest crater visible is 100 km in diameter.

Figure 6–34   A prominent scarp near Mercury's limb.

A Closer Look

Naming the Features of Mercury

Seeing features on the surface of Mercury led to a need for names. The scarps were named for historical ships of discovery and exploration, such as Endeavour (Captain Cook's ship), Santa Maria (Columbus's ship), and Victoria (the first ship to sail around the world, which it did from 1519 to 1522 under Magellan and his successors). Some plains were given the names of the gods equivalent to the Greek's Mercury, such as Tir (in ancient Persian), Odin (an ancient Norse god), and Suisei (Japanese). Craters were named for nonscientific authors, composers, and artists, in order to complement the lunar naming system, which honors scientists.

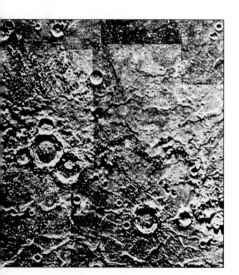

Figure 6–35 The fractured and ridged planes of Mercury's Caloris Basin.

signs of geologic tensions like rifts or fissures nearby. These scarps are global in scale, not just isolated. The scarps may actually be wrinkles in Mercury's crust. Part of Mercury's surface is very jumbled, probably from the energy released by an impact (Fig. 6–35).

Judging by the fact that Mercury's average density (its mass divided by its volume) is about the same as the Earth's, its core is probably iron and takes up perhaps 50 per cent of the volume, or 70 per cent of the mass, a much greater proportion than Earth's core. Perhaps it was once molten, and shrank by 1 or 2 km as it cooled. This shrinking would have caused the crust to buckle, creating the scarps in the quantity that we now observe.

Data from Mariner's infrared radiometer indicate that the surface of Mercury is covered with fine dust, as is the surface of the Moon, to a depth of at least several centimeters. Astronauts sent to Mercury, whenever they go, will leave footprints behind.

The biggest surprise of the mission was the detection of a magnetic field in space near Mercury. The field is weak, only about 1 per cent of the Earth's surface field. It had been thought that magnetic fields were generated by the rapid rotation of molten iron cores in planets, but Mercury is so small that its core should have quickly solidified. So the magnetic field is probably not now being generated. Perhaps the magnetic field has been frozen into Mercury since the time when its core was molten.

Mercury Research Rejuvenated

A dozen years after Mariner 10 sent back data about Mercury, an important new discovery about Mercury was made with a telescope on Earth. Mercury has an atmosphere! The atmosphere is very thin, but is still easily detectable in spectra.

Mercury's atmosphere contains more sodium than any other element—150,000 atoms per cubic centimeter compared with 4500 of helium and smaller amounts of oxygen, potassium, and hydrogen. At first, it appeared that the sodium was ejected into Mercury's atmosphere when particles from the Sun or from meteorites hit Mercury's surface. Newer evidence that the potassium and sodium are enhanced when the Caloris Basin is in view indicates, instead, that Mercury's atmosphere may have diffused up through Mercury's crust.

Mercury's surface features can be mapped from Earth with radar. The radar observations show altitudes and the roughness of the surface. The radar features show, in part, the half of Mercury not imaged by Mariner 10. It, too, is dominated by intercrater plains, though its overall appearance is different. The craters, with their floors flatter than the Moon's craters, show clearly on the radar maps. The scarps show clearly as well. Radio waves emitted from Mercury show conditions of Mercury's top layer (Fig. 6–36).

Mercury's History

Mercury may be the fragment of a giant early collision that nearly stripped it to its core. The stripping would account for the large proportion of iron in Mercury relative to Earth. Still early on, the core heated up and the planet's crust expanded. The expansion opened paths for molten rock to flow outward from the interior, producing the intercrater plains. As the core cooled, the crust contracted, and scraps resulted. Solar tides slowed Mercury's original rotation until its rotational period became ⅔ of the orbital period.

Back to Mercury, at Last

After decades of neglect, NASA hopes to send a spacecraft back to Mercury. The *ME*rcury *S*urface, *S*pace *EN*vironment, *GE*ochemistry and *R*anging (or MESSENGER) spacecraft is being planned for launch in 2004 and arrival at Mercury in 2008. MESSENGER would fly by Mercury twice in 2008 and then go into a year-long orbit around Mercury beginning in 2009.

⬤ VENUS

Venus orbits the Sun at a distance of 0.7 A.U. Although it comes closer to us than any other planet, we did not know much about it until recently because it is always shrouded in highly reflective clouds (Fig. 6–37).

Venus will be in transit (crossing the Sun's disk) in 2004 and 2012, for the first times since the pair of transits in 1874 and 1882. Previous transits of Venus were important for determining the scale of the Solar System, which we can now determine better by other means.

The Atmosphere of Venus

Studies from the Earth show that the clouds on Venus are primarily composed of droplets of sulfuric acid, H_2SO_4, with water droplets mixed in. Sulfuric acid may seem like a peculiar constituent of a cloud, but the Earth, too, has a significant layer of sulfuric acid droplets in its stratosphere, a higher layer of the atmosphere. However, the water in the lower layers of the Earth's atmosphere, circulating because of weather, washes the sulfur compounds out of these lower layers. Venus has sulfur compounds in the lower layers of its atmosphere in addition to those in its clouds.

Observations from Earth show a high concentration of carbon dioxide in the thick atmosphere of Venus. In fact, carbon dioxide makes up over 96 per cent of the mass of Venus's atmosphere (Fig. 6–38). The Earth's atmosphere, for comparison, is mainly nitrogen, with a fair amount of oxygen as well. Carbon dioxide makes up less than 1 per cent of the terrestrial atmosphere.

Because of the large amount of carbon dioxide in its atmosphere, which leads to the atmosphere being so massive, Venus's surface pressure is 90 times higher than the pressure of Earth's atmosphere. Carbon dioxide on Earth dissolved in seawater and eventually formed some types of terrestrial rocks, often with the help of life forms. (Limestone, for example, has formed from deceased marine life.) If this carbon dioxide were released from the Earth's rocks, along with other carbon dioxide

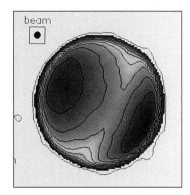

Figure 6–36 Mercury in radio waves, made with the Very Large Array. At the frequency of observation, the surface is translucent, allowing us to study the soil to a depth that depends on frequency. The maps show the temperature distribution. Red is hotter. This map, at a wavelength of 3.6 cm, shows the level 70 cm below the surface.

REDSHIFT
http://www.harcourtcollege.com/astro/cosmos/rsce

Figure 6–37 A crescent Venus, observed with a large telescope on Earth. In visible light images like this one, we see only a layer of clouds.

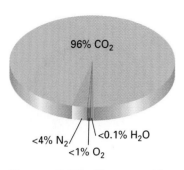

Figure 6-38 The composition of Venus's atmosphere.

Comparative planetology:
Atmospheres

Venus = 100 × Earth
Mars = Earth ÷ 100
Mercury—negligible

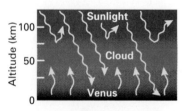

Figure 6-39 Though most sunlight is reflected from Venus's clouds, some sunlight penetrates them, so the surface is illuminated with radiation in the visible part of the spectrum. Venus's own radiation is mostly in the infrared. This infrared radiation is trapped, a phenomenon known as the greenhouse effect.

---

trapped in seawater, our atmosphere would become as dense and have as high a pressure as that of Venus. Venus, slightly closer to the Sun than Earth and thus hotter, had no oceans in which the carbon dioxide could dissolve to help take it up from the atmosphere. Thus the carbon dioxide remains in Venus's atmosphere.

The Rotation of Venus

In 1961, the radio waves used in radar astronomy penetrated Venus's clouds, allowing us to determine accurately how fast Venus rotates. Venus rotates in 243 days with respect to the stars in the direction opposite from the other planets. Venus revolves around the Sun in 225 Earth days. Venus's periods of rotation and revolution combine so that a solar day/night cycle on Venus corresponds to 117 Earth days; that is, the Sun returns to the same position in the sky every 117 days.

The notion that Venus rotates backward seems very strange to astronomers, since all the planets revolve around the Sun in the forward direction, and most of the other planets (except Uranus and Pluto, which are on their sides) and most satellites also rotate in that forward direction. Because the laws of physics do not allow the amount of spin to change on its own, and since the original material from which the planets coalesced was undoubtedly rotating, we expect all the planets to revolve and rotate in the same sense. Nobody knows definitely why Venus rotates "the wrong way." One possibility is that when Venus was forming, a large clump of material struck it at an angle that caused the merged resulting planet to rotate backward.

The slow rotation of Venus's solid surface contrasts with the rapid rotation of its clouds. The tops of the clouds rotate in the same sense as the surface rotates but about 60 times more rapidly, once every 4 days. Lower parts of the atmosphere, however, rotate very slowly.

The Temperature of Venus

We can determine the temperature of Venus's surface by studying its radio emission, since radio waves emitted by the surface penetrate the clouds. The surface is very hot, about 750 K (900°F), even on the night side. In addition to measuring the temperature on Venus, scientists theoretically calculate what the temperature would be if Venus's atmosphere were transparent to radiation of all wavelengths, both coming in and going out. This value—less than 375 K (215°F)—is much lower than the measured values. The high temperatures derived from radio measurements indicate that Venus traps much of the solar energy that hits it.

The process by which the surface is heated so much on Venus is similar to the process that is generally—though incorrectly—thought to occur in greenhouses here on Earth. The process is thus called the **greenhouse effect** (Fig. 6–39): The fraction of sunlight not reflected by the high clouds passes through the venusian atmosphere in the form of radiation in the visible part of the spectrum. The sunlight is absorbed by the surface of Venus, which heats up. At the temperatures that result, the thermal (nearly blackbody) radiation that the surface gives off is mostly in the infrared. But the carbon dioxide and other constituents of Venus's atmosphere are largely opaque to infrared radiation, so the energy is trapped. Thus Venus heats up far above the temperature it would reach if the atmosphere were transparent. The surface radiates more and more energy as the planet heats up. Finally, a balance is struck between the rate at which energy flows in from the Sun and the rate at which it trickles out (as infrared) through the atmosphere.

The situation is so extreme on Venus that we say a "runaway greenhouse effect" has taken place there. Understanding such processes involving the transfer of energy is but one of the practical results of the study of astronomy.

Greenhouses on Earth don't work quite this way. In actual greenhouses, closed glass on Earth prevents the mixing of air inside, heated by conduction from the warmed ground, with cooler outside air. The trapping of solar energy by the "greenhouse effect"—the inability of infrared radiation, once formed, to get out—is a less important process in an actual greenhouse or in a car when it is left in the sunlight.

The greenhouse effect can be beneficial; indeed, it keeps Earth's temperature at a comfortable level. (Earth would be quite chilly without it—about 33°C colder, on average.) Earth manages to achieve a comfortable greenhouse effect by recycling its greenhouse gases, primarily CO_2 and H_2O (water). As the oceans evaporate, water vapor builds up in the atmosphere, but then the rain comes down. The rain dissolves atmospheric CO_2, producing carbonic acid. This mild "acid rain" dissolves rocks and forms carbonates, and there are other forms of weathering as well. The carbonate solution eventually reaches the ocean. Various sea creatures then use the carbonates in their shells. When these organisms die, their carbonate shells form sediments on ocean bottoms. Because of plate tectonics, the sediments can later dive down under other plates, when they become part of the magma inside the Earth. The CO_2 is subsequently outgassed by volcanism, and the process repeats.

How did Venus reach its hellish state? It may have begun its existence with a pleasant climate, and possibly even with oceans. Perhaps rain and other processes were not quick enough in removing the atmospheric CO_2 outgassed by volcanoes. As CO_2 accumulated, the greenhouse effect increased, and temperatures rose; oceans evaporated faster, and a runaway greenhouse effect was in progress. The water vapor was gradually broken apart by ultraviolet radiation from the Sun, and the hydrogen (being so light) escaped from Venus. Eventually there were no oceans and no rain—but continued outgassing of CO_2 led to more greenhouse warming and extremely high temperatures.

It is also possible that Venus was essentially born in a hot state, or reached it very quickly. Without oceans, it is difficult for atmospheric CO_2 to become incorporated into rocks as it did on Earth. Extensive greenhouse heating is therefore maintained.

Studies of Venus are important for an overall understanding of Earth's climate and atmosphere. There is significant evidence for current global warming of Earth, although this is still controversial. Moreover, the atmospheric CO_2 concentration is increasing as a result of human activities. It is generally but not universally believed that the human-induced increase in CO_2 (from the burning of fossil fuels) is one of the main causes of global warming, if it is indeed occurring. However, it is unclear whether a "positive feedback" mechanism that leads to sustained global warming will necessarily occur; there are many variables, and their effects are not yet understood. For example, an increase in global temperatures leads to more evaporation of the oceans, but this can increase the cloud cover and hence Earth's reflectivity, thereby producing a decrease in the amount of visible sunlight reaching Earth's surface. This is a "negative feedback" mechanism that tends to stabilize temperatures. On the other hand, recent models suggest that the higher water-vapor content of the atmosphere actually increases greenhouse heating by an amount that more than compensates for the greater reflectivity of the Earth. Thus, the final outcome is far from clear; Earth's atmosphere is a very complicated system.

It is very unlikely that the Earth will experience a *runaway* greenhouse effect that makes the planet unfit for *all* forms of life. In the past, there have been times

Comparative planetology: Greenhouse effect

Earth: 33°C extra
Venus: 375°C extra
Mercury: negligible
Mars: negligible

Comparative planetology: Temperature at surface

Earth: about 0°C to 40°C (just right)
Venus: about 475°C (much too hot)
Mercury: about −175°C to 425°C (too cold to too hot)
Mars: about −125°C to 25°C (too cold to OK)

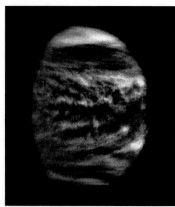

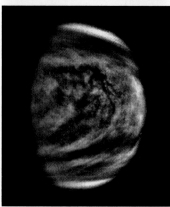

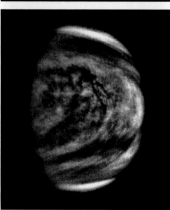

Figure 6–40 Changes in Venus's clouds over several hours, observed from the Galileo spacecraft.

when Earth was much warmer than now, yet a runaway did not occur. Moreover, life on Earth recovered from collisions with large asteroids and comets, some of which dumped enormous quantities of greenhouse gases into the atmosphere. On the other hand, even a rise of a few degrees in Earth's global average temperature over a time scale of a few decades, which is now generally predicted, would have terrible consequences for humans, including flooding of coastal cities and destruction of many agricultural areas. Also, since we cannot accurately predict what will happen under various circumstances, it is dangerous to act in ways that might disrupt the balance in our atmosphere and adversely affect life on Earth. Venus has shown us the importance of being very careful about how we treat the only home we have.

Venus's Atmosphere (and Lack of Magnetic Field) Observed from Space

Venus was an early target of both American and Soviet space missions. During the 1960s, American spacecraft flew by Venus, and Soviet spacecraft dropped through its atmosphere. In 1970, the Soviet Venera 7 spacecraft radioed 23 minutes of data back from the surface of Venus before it succumbed to the high temperature and pressure. Two years later, the lander from Venera 8 survived on the surface of Venus for 50 minutes. Both landers confirmed the Earth-based results of high temperatures, high pressures, and high carbon dioxide content.

Several United States spacecraft have observed Venus's clouds and followed the changes in them. The most recent view of changes in Venus's clouds came in 1990 from the Galileo spacecraft (Fig. 6–40). Structure in the clouds shows only when viewed in ultraviolet light. The clouds appear as long, delicate streaks, looking like terrestrial cirrus clouds.

Such studies of Venus have practical value. The principles that govern weather on Venus are similar to those that govern weather on Earth, though there are major specific differences such as the absence of oceans on Venus. The better we understand the interaction of solar heating, planetary rotation, and chemical composition in setting up an atmospheric circulation, the better we will understand our Earth's atmosphere. We then may be better able to predict the weather. The potential financial return from such knowledge is enormous: it would be many times the investment we have made in planetary exploration.

Not one of the spacecraft to Venus has detected a magnetic field. The absence of a magnetic field may indicate either that Venus does not have a liquid core or does not rotate fast enough.

Spacecraft have provided evidence that volcanoes may be active on Venus. The abundance of sulfur dioxide they found varied at different times by a factor of 10. The effect could come from eruptions at least ten times greater than that of even the largest recent terrestrial volcanic eruptions. Lightning that was detected is further evidence that those regions are sites of active volcanoes. On Earth, it is common for the dust and ash ejected by volcanoes to rub together, generating static electricity that produces lightning.

Probes that penetrated the atmosphere and went down to the surface found that high-speed winds at the upper levels are coupled to other high-speed winds at lower altitudes. The lowest part of Venus's atmosphere, however, is relatively stagnant. The probes detected three distinct layers of venusian clouds, separated from each other by regions of relatively low density.

One of the probes measured that only about 2 per cent of the sunlight reaching Venus filters down to the surface, making it like a dim terrestrial twilight. Thus most of the Sun's energy is absorbed in or reflected by the clouds, unlike the situation with Earth, and Venus's clouds are more important than Earth's for controlling weather. Most of the light that reaches the surface is orange, so photographs taken on the surface have an orange cast.

Venus's Surface

From Venus's size and from the fact that its mean density is similar to that of the Earth, we conclude that its interior is also probably similar to that of the Earth. This means that we expect to find volcanoes and mountains on Venus.

We can study the surface of Venus by using radar to penetrate Venus's clouds. Radars using huge Earth-based radio telescopes, such as the giant 1000-ft (305-m) dish at Arecibo, Puerto Rico, have mapped a small amount of Venus's surface with a resolution of up to about 20 km. (The **resolution** means the size of the finest details that could be detected, so we mean that only features larger than about 20 km can be detected.) Regions that reflect a large percentage of the radar beam back at Earth show up as bright, and other regions as dark.

NASA's Pioneer Venus and, in 1990 to 1993, Magellan were in orbit around Venus, allowing them to make observations over a lengthy time period. (www) They carried radars to study the topography of Venus's surface, mapping a much wider area than could be observed from the Earth.

From the radar maps (Fig. 6–41), we now know that 60 per cent of Venus's surface is covered by a rolling plain, flat to within plus or minus 1 km. Only about 16 per cent of Venus's surface lies below this plain, a much smaller fraction than the two thirds of the Earth covered by ocean floor.

Two large features, the size of small Earth continents, extend several km above the mean elevation. A northern "continent," Terra Ishtar, is about the size

Figure 6–41 The radar map of Venus's surface (omitting the polar regions) from the Magellan spacecraft. Red corresponds to high elevation and blue to low elevation in this false-color map.

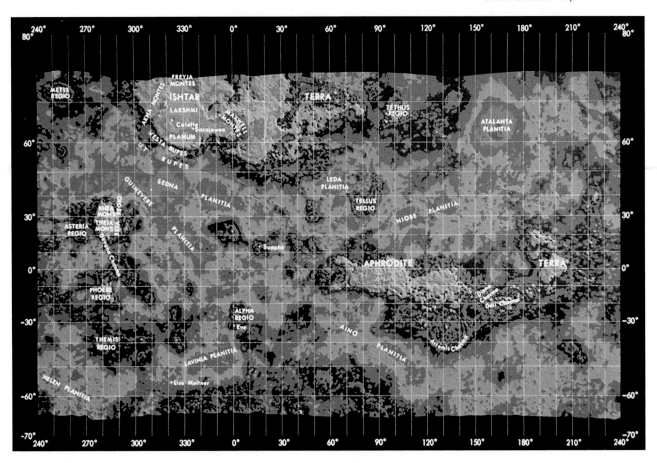

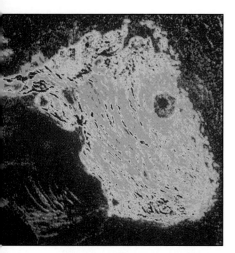

Figure 6–42 The Arecibo Observatory's high-resolution radar image of Maxwell Montes, with a resolution of about 2 km. Brighter areas show regions where more power bounces back to us from Venus, which usually corresponds to rougher terrain than darker areas.

of the continental United States. The giant chain of mountains on it known as Maxwell Montes (Fig. 6–42) is 11 km high, which is 2 km taller than Earth's Mt. Everest stands above terrestrial sea level. It had formerly been known only as a bright spot on Earth-based radar images. Ishtar's western part is a broad plateau, about as high as the highest plateau on Earth (the Tibetan plateau) but twice as large. A southern "continent," Aphrodite Terra, is about twice as large as the northern one and is much rougher.

From the radar maps, it appears that Venus, unlike Earth, is apparently made of only one continental plate. We observe nothing on Venus equivalent to Earth's mid-ocean ridges, at which new crust is carried upward. In particular, the high-resolution Magellan observations (Fig. 6–43) show no signs of crust spreading laterally on Venus. Venus may well have such a thick crust that any plate tectonics that existed in the distant past was choked off. Thus "venusquakes" are probably much less common than earthquakes.

Perhaps giant hot spots, like those that created the Hawaiian Islands on Earth, force mountains to form on Venus in addition to causing volcanic eruptions. The rising lava may also make the broad, circular domes that are seen. Venus got rid of its internal energy through widespread volcanism. Apparently, virtually the whole planet was resurfaced with lava about 500 million years ago.

In the 1970s and 1980s, a series of Soviet spacecraft landed on Venus and sent back photographs (Fig. 6–44). They also found that the soil resembles basalt in chemical composition and density, in common with the Earth, the Moon, and Mars. They measured temperatures of about 750 K and pressures over 90 times that of the Earth's atmosphere, confirming earlier, ground-based measurements.

Our recent space results, coupled with our ground-based knowledge, show us that Venus is even more different from the Earth than had previously been imagined. Among the differences are Venus's slow rotation, its one-plate surface, the absence of a satellite, the extreme weakness or absence of a magnetic field, the lack of water in its atmosphere, and its high surface temperature and pressure.

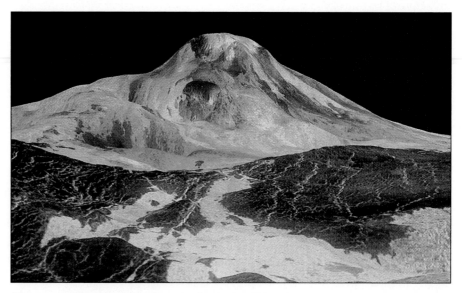

Figure 6–43 Maat Mons in a simulated perspective view based on radar data from Magellan and arbitrarily given the orange color that comes through Venus's clouds. Our viewpoint is 560 km north of Maat Mons at an elevation of 1.7 km. We see lava flows extending hundreds of kilometers across the foreground fractured plains. Maat Mons is 8 km high; its height here is exaggerated by a factor of about 20.

Figure 6-44 The view from a Soviet spacecraft on Venus's surface, showing a variety of sizes and textures of rocks. It survived on Venus for over 2 hours before it failed because of the high temperature and pressure. The camera first looked off to one side, and then scanned downward as though you were looking down toward your feet. Then it scanned up to the other side. As a result, the left horizon is visible as a slanted boundary at *upper left* and the right horizon is at *upper right*. The base of the spacecraft, Venera 13, and a lens cap are at *bottom center*.

MARS

Mars has long been the planet of greatest interest to scientists and nonscientists alike. Its unusual appearance as a reddish object in the night sky coupled with some past scientific studies have made Mars the prime object of speculation as to whether extraterrestrial life exists there.

In 1877, the Italian astronomer Giovanni Schiaparelli published the results of a long series of telescopic observations he had made of Mars. He reported that he had seen "canali" on the surface. When this Italian word for "channels" was improperly translated into "canals," which seemed to connote that they were dug by intelligent life, public interest in Mars increased. Percival Lowell, in particular, was fascinated by the prospect of an ancient civilization on Mars. In the late 19th century, he drew detailed maps of Mars that showed an elaborate system of canals, presumably built to bring water to arid regions.

Over the next decades, there were endless debates over just what had been seen. We now know that the channels or canals Schiaparelli and other observers reported are not present on Mars. They were an illusion. In fact, positions of the "canali" do not even always overlap the spots and markings that are actually on the martian surface (Fig. 6–45). But hope of finding life elsewhere in the Solar System springs eternal, and the latest studies have indicated the presence of considerable quantities of liquid water in Mars's past, a fact that leads many astronomers to hope that life could have formed during those periods.

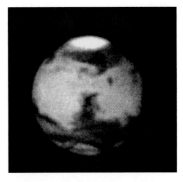

Figure 6–45 Mars, viewed from Earth with a CCD on a high-end amateur-grade telescope.

Figure 6–46 Mars's moons Deimos *(lower left)* and Phobos *(lower right)*, with the asteroid Gaspra *(top)* for comparison, shown to the same scale. Phobos is about 27 km across. Deimos measures about 15 km × 12 km × 11 km. Its surface is heavily cratered and thus presumably very old. The surface seems smoother than that of Phobos, though, because soil 10 m thick covers Deimos. The spectra of Phobos and Deimos are similar to those of one type of dark asteroid. The Phobos and Deimos images were from Viking; the Gaspra image was from Galileo.

Characteristics of Mars

Mars is a small planet, 6792 km across, which is only about half the diameter and one-eighth the volume of Earth or Venus, although somewhat larger than Mercury. It has two tiny (about 20 km in diameter), irregularly shaped moons named Phobos and Deimos (Fig. 6–46).

Mars's atmosphere is very thin, only 1 per cent of Earth's, but it might be sufficient for certain kinds of life. Unlike the orbits of Mercury or Venus, the orbit of Mars is outside the Earth's, so we can observe Mars in the late night sky.

Mars revolves around the Sun in 1.9 Earth years. The axis of its rotation is tipped at a 25° angle from perpendicular to the plane of its orbit, nearly the same as the Earth's 23½° tilt. Because the tilt of the axis causes the seasons, we know that Mars goes through its year with four seasons just as the Earth does.

We have watched the effects of the seasons on Mars over the last century. In the martian winter, in a given hemisphere, there is a polar cap. As the martian spring comes to the northern hemisphere, the north polar cap shrinks and material at more temperate zones darkens. The surface of Mars appears mainly

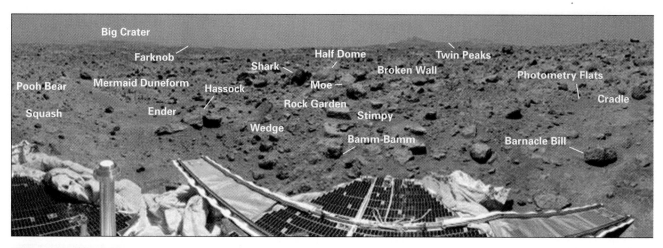

Figure 6–47 A Mars Pathfinder panorama.

reddish-orange when seen from Earth against dark space, probably due to the presence of iron oxide (rust), with darker gray areas that appear blue-green for physiological reasons, as the human eye and brain misjudge the color contrast. The color changes over time apparently result from dust that is either covering surface rock or is blown off by winds that can be hundreds of kilometers per hour. Sometimes, global dust storms occur, and then we can later watch the color change as the dust blows off the places where it first lands, exposing the dark areas underneath.

From Mars's mass and radius we can easily calculate that it has an average density about that of the Moon, substantially less than the density of Mercury, Venus, and Earth. The difference indicates that Mars's overall composition must be fundamentally different from that of these other planets. Mars probably has a smaller iron-rich core and a thicker crust than does Earth.

> Comparative planetology:
> Moons
>
> Mercury: none
> Venus: none
> Mars: 2, only 20 km diameter each
> Earth: 1, 3476 km diameter

Mars's Surface

Mars has been the target of several series of spacecraft, most notably several U.S. spacecraft, including Mariner 9 in 1971; a pair of Vikings that started orbiting in 1976 and which dropped landers onto the martian surface; Mars Pathfinder, which landed on July 4, 1997, and placed the Sojourner rover on Mars's surface (Fig. 6–47); and Mars Global Surveyor which is circling the planet. **www** Surface temperatures measured from the Viking landers ranged from a low of 150 K (−190°F) at the northern site of Lander 1 to over 300 K (80°F) at Lander 2. The temperature varied each day by 35 to 50°C (60 to 90°F).

The surface of Mars can be divided into four major types: volcanic regions, canyon areas, expanses of craters, and terraced areas near the poles. A chief surprise was the discovery of extensive areas of volcanism. The largest volcano—which corresponds in position to the surface marking long known as Nix Olympica, "the snow of Olympus"—is named Olympus Mons, "Mount Olympus." It is a huge volcano, 600 kilometers at its base and about 25 kilometers high (Fig. 6–48). It is crowned with a crater 65 km wide; Manhattan Island could be easily dropped inside. (The tallest volcano on Earth is Mauna Kea in the Hawaiian Islands, if we measure its height from its base deep below the ocean. Mauna Kea is slightly taller than Mount Everest, though still only 9 km high.) Perhaps the volcanic features on Mars can get so huge because continental drift is absent there, as on Venus. If molten rock flowing upward causes volcanoes to form, then on Mars the features just get bigger and bigger for hundreds of millions of years, since the volcanoes stay over the sources and do not drift away. (Each of the Hawaiian Islands was formed over a single "hot spot," but drifted away from it.)

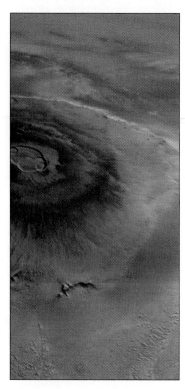

Figure 6–48 Olympus Mons, from Mars Global Surveyor.

Another surprise on Mars was the discovery of systems of canyons (Fig. 6–49). One tremendous canyon—about 5000 kilometers long—is as big as the continental United States and comparable in size to the Rift Valley in Africa, the longest geological fault on Earth.

Perhaps the most amazing discovery on Mars was the presence of sinuous channels. These are on a smaller scale than the "canali" that Schiaparelli and Lowell believed they had seen, and are entirely different phenomena. Some of the channels show the same characteristic features as streambeds on Earth. Even though liquid water cannot exist on the surface of Mars under today's conditions, it is difficult to think of ways to explain the channels satisfactorily other than to say that they were cut by running water in the past. Data from the Mars Pathfinder mission strengthen this conclusion: rocks are scattered over a large area in a manner reminiscent of flood plains. The high-resolution Mars Global Surveyor in 1998–1999 found additional signs, such as terracing, showing that water had flowed (Fig. 6–50).

This indication that water most likely flowed on Mars is particularly interesting because biologists feel that water is necessary for the formation and evo-

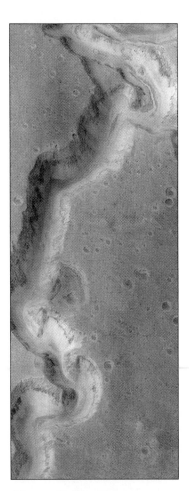

Figure 6–50 The past presence of water on Mars is revealed by features like this one, which appears as though it was cut by a flow of water for an extended period of time, similarly to the way that rivers on Earth are formed. We see a portion of the meandering canyons of a valley cutting through otherwise smooth, cratered plains. The image, showing a region 10 km by 28 km, is from the Mars Global Surveyor.

Figure 6–49 Mars, a mosaic of images from Viking. The valleys Valles Marineris appear horizontally across the center; together, they are as long as the United States is wide. At left we see three giant volcanoes.

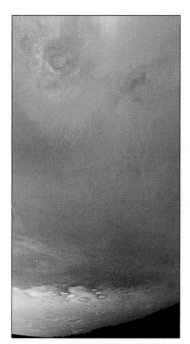

Figure 6–51 The martian south polar cap, observed from Mars Global Surveyor in 1999 just before the start of southern spring on Mars. We see the south polar cap as it retreats; wisps of a dust storm, caused by colder air blowing off the cap into warmer regions, appear just above it. The volcano Arsia Mons is at upper left.

lution of life as we know it. The presence of water on Mars, even in the past, may therefore indicate that life could have formed and may even have survived. Where has all the water gone? Most of the water is probably in a permafrost layer—permanently frozen subsoil—beneath middle latitudes and polar regions, although a substantial fraction may have escaped from Mars in gaseous form.

Some of the water is bound in the polar caps (Fig. 6–51). The large polar caps that extend to latitude 50° during the winter are carbon dioxide. But when a cap shrinks during its hemisphere's summer, a residual polar cap of water ice remains in the north, while the south has a residual polar cap of carbon dioxide ice, probably with water ice below.

Mars's Atmosphere

We have found that the martian atmosphere is composed of 95 per cent carbon dioxide with small amounts of carbon monoxide, oxygen, and water. The surface pressure is less than 1 per cent of that near Earth's surface. The current atmosphere is too thin to significantly affect the surface temperature, in contrast to the huge effects that the atmospheres of Venus and the Earth have on climate. But long ago, up to about a billion years after it formed, Mars had a thicker atmosphere and a hospitable climate. Perhaps partial loss of the atmosphere, as it escaped into space, led to colder overall temperatures and the freezing of CO_2 and water. There would consequently be less greenhouse heating and hence even colder temperatures. This mechanism may have led to the inverse of what happened on Venus.

Observations from the Viking orbiters showed Mars's atmosphere in some detail. The lengthy period of observation led to the discovery of weather patterns on Mars. The current spacecraft orbiting Mars, the Mars Global Surveyor, maps the weather there even better (Fig. 6–52). Additional information continues to come from the Hubble Space Telescope, which from time to time obtains a series of pictures of Mars (Fig. 6–53). With its rotation period similar to that of Earth, some features of Mars's weather are similar to our own. Studies of Mars's weather have already helped us better understand windstorms in Africa that affect weather as far away as North America.

Studies of the effect of martian dust storms on the planet led to the idea that the explosion of a nuclear bomb on Earth might lead to a "nuclear winter." Dust thrown into the air would shield the Earth's surface from sunlight for a lengthy period, with dire consequences for life on Earth. Improvement of computer models for the circulation of atmospheres is contributing to the investigations. The effect, at present, seems smaller than first feared, so it is perhaps better described as "nuclear fall," which would still be something to avoid. These models, and observations of Mars, also help us to understand the effects of smaller amounts of matter we are putting into the atmosphere from factories and fossil fuel power plants and by the burning of forests. Of course, the absence of oceans on Mars (and their associated effects) complicates direct comparisons with Earth.

The Search for Life

On July 20, 1976, exactly seven years after the first crewed landing on the Moon, Viking 1's lander descended safely onto a martian plain called Chryse. The views showed rocks of several kinds, covered with yellowish-brown material that is probably an iron oxide (rust) compound. (Many of the photographs released incorrectly show an orange-red color, but careful color balancing of the images revealed the true yellowish-brown.) Sand dunes were also visible. The sky on Mars also turns out to be yellowish-brown, almost pinkish (Fig. 6–54); the color is formed

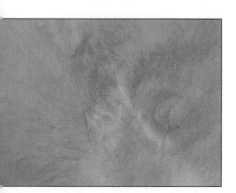

Figure 6–52 Mars, compiled from many photos taken with the Mars Global Surveyor. Atmospheric features show in addition to the background surface.

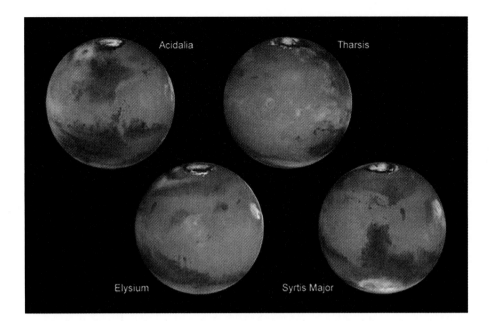

Figure 6-53 Mars from the Hubble Space Telescope's Wide Field and Planetary Camera 2 in 1999, during its closest approach to Earth in 8 years. Details as small as 20 km show. High cirrus clouds show especially on the limb. Though the larger bright and dark markings remain stable, some of the bright smaller markings seen from Viking 20 years earlier are now dark and vice versa as they are covered and uncovered by sand and dust.

as sunlight is scattered by dust suspended in the air as a result of one of Mars's frequent dust storms.

A series of experiments aboard the lander was designed to search for signs of life. A long arm was deployed and a shovel at its end dug up a bit of the martian surface. The soil was dumped into three experiments that searched for such signs of life as respiration and metabolism. The results were astonishing at first: The experiments sent back signals that seemed similar to those that would be caused on Earth by biological processes. But later results were less spectacular, and nonbiological chemical explanations seem more likely in all cases.

One important experiment gave much more negative results for the chance that there is life on Mars. It analyzed the soil and looked for traces of organic compounds. On Earth, many organic compounds left over from dead forms of life remain in the soil; living organisms themselves are only a tiny fraction of the organic material. Yet these experiments found no trace of organic material. Unfortunately, any ancient organic material on the surface would have been destroyed by solar ultraviolet radiation, so its absence isn't proof that Mars never had any life. Still, it is a strong argument against the presence of recent life on Mars. But even if the life signs detected by Viking come from chemical rather than biological processes, as seems likely, we have still learned of fascinating new chemistry going on.

In 1996, study of a martian meteorite found in the Earth's Antarctic revealed possible evidence for ancient primitive (microbial) organisms. The chunk of rock, estimated to be 4.5 billion years old, was blasted from the surface of Mars about 15 million years ago by a collision. There is little doubt that the rock is from Mars, based on analysis of gases trapped within it; also, calculations show the likelihood that rocks blasted off Mars can reach Earth. The meteorite landed in Antarctica about 13,000 years ago. It was the first meteorite found in 1984 in the Allan Hills, and so is called ALH 84001.

The meteorite contained carbonate globules (Fig. 6–55), which generally form in liquid water. Within them, certain types of organic compounds were found. Although these types of compounds can be formed in a number of ways other than by life, and are sometimes seen in normal meteorites, these particular ones were relatively unusual. A few other substances typical of life (such as

Figure 6-54 A Viking 2 view of Mars. The spacecraft landed on a rock and so was at an angle. The boom that supports Viking's weather station cuts through the center of the picture. ("Chance of precipitation," the local newscaster would say, "is 0 per cent.")

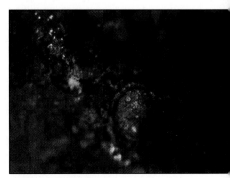

Figure 6-55 A microscopic view of a small (2.3 mm across) thin section from one of the Mars meteorites.

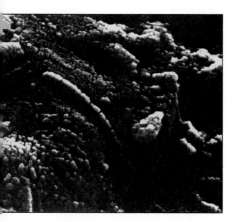

Figure 6-56 Structures in the "Mars meteorite," seen with a special kind of microscope, are thought by some scientists to be remnants of primitive life forms but by most other scientists to be natural formations, too small to be fossils of life forms.

magnetite, a mineral produced by some types of bacteria) were also seen. But photographs of tube-like structures resembling the tiniest Earth bacteria received the most publicity (Fig. 6–56).

Although none of these findings definitively imply the presence of life, taken together they are certainly intriguing (yet still not compelling). Some researchers have challenged the conclusions on a number of grounds. For example, they cite evidence for contamination of the meteorite by substances on Earth. The original research team has responded to these criticisms with reasonable counter-arguments, and stand by their original announcement. Still the results are highly controversial, and additional tests are necessary; most scientists are skeptical at this time. Stay tuned for further developments.

Regardless of whether the discovery of ancient microbial life is correct, it raises an interesting possibility: Earth life may have originated on Mars, and subsequently migrated to Earth in a meteorite. If so, we are the descendents of Martians!

Because of the tremendous interest in Mars and the possibility of finding out about the origin of life by studying it, space agencies in the United States and elsewhere are planning a whole set of spacecraft to Mars. But, the failure of two NASA spacecraft to orbit and land on Mars in 1999 has set back the American program severely. Eventually, martian rock and soil will be brought to Earth. Many people feel that only with this material to study using sophisticated Earth-based microscopes and other instruments can we really understand what Mars is like.

Some people are calling for a crewed mission to Mars. Note that though astronauts got to the Moon in only four days, Mars is so much farther away that a trip there would take a couple of years. Because of the cost and other problems, no such astronaut mission is likely until at least the year 2020.

Concept Review

The **terrestrial planets,** the four inner planets, are rocky worlds. Venus and the Earth have similar sizes, masses, and densities, but are otherwise very different. Mars is about half the diameter of the Earth and has an atmosphere $\frac{1}{100}$ the pressure of the Earth's. **Comparative planetology** has applications for understanding weather, earthquakes, and other topics of use to us on Earth.

Geology is the study of the Earth's interior and surface, and often uses **seismology,** the study of waves that pass through the Earth, to study the Earth's interior. The Earth has a layered structure, with a **core, mantle,** and **crust.** Radioactivity heats the interior, providing **geothermal energy.** Continental **plates** move around the surface, as explained by **plate tectonics,** the current theory that elaborates on the earlier notion of **continental drift. Tides** are a differential force caused by the fact that the near side of the Earth is closer to the Moon than the Earth's center, and so is subject to higher gravity, while the far side is farther than the center and is subject to lower gravity. The Earth's weather is confined in its atmosphere to the **troposphere.** In the **ionosphere,** atoms are ionized and the temperature is higher. The **Van Allen belts** of charged particles were a space-age discovery. Such particles lead to the **aurora borealis** and **aurora australis.**

Our own Moon is midway in size between Pluto and Mercury. Its volcanic surface shows smooth **maria** (singular: **mare**) and cratered highlands. Shadows reveal relief best along the **terminator,** the day–night line. Most of the craters are from meteoritic impacts. The oldest rocks are from 4.4 billion years ago; on the Earth, erosion and plate tectonics have erased most of the oldest rocks though a few equally old terrestrial rocks have been found. **Mascons,** discovered by the Apollo missions, are mass concentrations of high density under the Moon's surface. The current leading model for the Moon's formation is that a planetesimal perhaps twice the size of Mars hit the proto-Earth, ejecting matter into a ring that coalesced.

Mercury always appears close to the Sun in the sky; rarely it even goes into **transit** across the Sun's disk. Like the Moon, Mercury has a dark (low **albedo**) surface. Its rotation period, measured by radar, is linked to its orbital period. Mercury is heavily cratered, and shows **scarps** that resulted from a planet-wide shrinkage. It has a weak magnetic field, which is a surprise because we think Mercury's core has solidified. It is too close to the Sun to retain a thick atmosphere, but a very thin atmosphere does exist.

The highly reflective clouds that shroud Venus, giving it a high albedo, are mainly droplets of sulfuric acid, though the atmosphere consists primarily of carbon dioxide. Venus's atmospheric surface pressure is 90 times higher than Earth's. Venus's surface temperature is very hot, about 750 K, heated by the **greenhouse effect.** Earth escaped such a fate, but we should be careful about disrupting the balance in our atmosphere, lest conditions change too quickly. Radar penetrated Venus's clouds to show that Venus is rotating slowly backward and to map its surface. Higher **resolution** radar reveals surface features. Venus will be in transit across the Sun's disk in 2004 and 2012, for the first times in over 100 years.

Stories about life on Mars have long inspired study of this planet. We now know the seasonal surface changes to be the direct result of winds that arise as the sunlight hits the planet at different angles over a martian year. Mars's weather and dust storms help us understand our own weather. Mars's surface appears reddish because of rusty dust; careful calibration of photos taken on Mars shows that it is actually yellowish-brown. Mars boasts of giant volcanoes and a canyon longer than the width of the continental United States. There is strong evidence that Mars had liquid water flowing on its surface long ago. Studies of a martian meteorite provide suggestive, but not convincing, evidence for the presence of ancient, primitive life on Mars. A series of spacecraft to Mars should culminate in a sample return, in part to help settle this very controversial claim.

Questions

1. How did the layers of the Earth arise? Where did the energy that is flowing as heat come from?
2. What carries the continental plates around over the Earth's surface?
3. (a) Explain the origin of tides. (b) If the Moon were farther away from the Earth than it actually is, how would tides be affected?
4. Draw a diagram showing the positions of the Earth, Moon, and Sun at a time when there is the least difference between high and low tides.
†5. Calculate your weight if you were standing on the Moon.
6. Look at a globe and make a list or sketches of which pieces of the various continents probably lined up with each other before the continents drifted apart.
7. To what locations, relative to the Earth–Sun line, does the Earth's terminator correspond?
8. What does cratering tell you about the age of the surface of the Moon, compared to that of the Earth's surface?
9. Why are we more likely to learn about the early history of the Earth by studying the rocks from the Moon than those on the Earth?
10. Why may the near side and far side of the Moon look different?
11. Discuss one of the proposed theories to describe the origin of the Moon. List points both pro and con.
12. How can we get lunar material to study on Earth?
13. Assume that on a given day, Mercury sets after the Sun. Draw a diagram, or a few diagrams, to show that the height of Mercury above the horizon depends on the angle that the Sun's path in the sky makes with the horizon as the Sun sets. Discuss how this depends on the latitude or longitude of the observer.
14. If Mercury did always keep the same side toward the Sun, does that mean that the night side would always face the same stars? Draw a diagram to illustrate your answer.
†15. Explain why a day/night cycle on Mercury is 176 Earth day/night cycles long.
16. What did radar tell us about Mercury? How did it do so?
17. If ice has an albedo of 70–80%, and volcanic rocks typically have albedoes of 5–20%, what can you say about the surface of Mercury based on its measured albedo?

18. If you increased the albedo of Mercury, would its temperature increase or decrease? Explain.
19. How would you distinguish an old crater from a new one?
20. What evidence is there for erosion on Mercury? Does this mean there must have been water on the surface?
21. List three major findings of Mariner 10.
22. Make a table displaying the major similarities and differences between the Earth and Venus.
23. Why does Venus have more carbon dioxide in its atmosphere than does the Earth?
24. Why do we think that there have been significant external effects on the rotation of Venus?
25. Suppose a planet had an atmosphere that was opaque in the visible but transparent in the infrared. Describe how the effect of this type of atmosphere on the planet's temperature differs from the greenhouse effect.
26. Why do radar observations of Venus provide more data about the surface structure than a flyby with close-up optical cameras?
27. Why do we say that Venus is the Earth's "sister planet"?
28. Describe the most current radar observations of Venus.
29. What are some signs of volcanism on Venus?
30. Outline the features of Mars that made scientists think that it was a good place to search for life.
†31. Compare the tallest volcanoes on Earth and Mars relative to the diameters of the planets.
32. What evidence exists that there is, or has been, water on Mars?
33. Describe the composition of Mars's polar caps.
34. Consult an atlas and compare the sizes of the Grand Canyon in Arizona and the Rift Valley in Africa. How do they compare in size with the giant canyon on Mars?
35. Compare the temperature ranges on Venus, Earth, and Mars.
36. List the evidence from Viking for and against the existence of life on Mars.
37. Plan a set of experiments or observations that you, as a martian scientist, would have an unmanned spacecraft carry out on Earth to find out if life existed here. What data would your spacecraft radio back if it landed in a cornfield? In the Sahara? In the Antarctic? In Times Square?

---

†This question requires a numerical solution.

The Jovian Planets: Windswept Giants

ORIGINS *We find that the properties of giant planets differ from those of Earth because they formed much farther out from the center of the solar nebula. Discoveries of planetary systems around other stars are providing insights into the process by which planets form.*

Jupiter, Saturn, Uranus, and Neptune are **giant planets** also called the **jovian planets.** They are much bigger, more massive, and less dense than the inner planets. Their internal structure is entirely different from that of the four inner planets. In this chapter, we also discuss a set of moons of these giant planets, some of which range in diameter between ½ and ¼ the size of the Earth, as large as Mercury or Pluto. Close-up space observations have shown that these moons are themselves interesting objects for study.

Jupiter is the largest planet in our Solar System. Its moons make it a mini–solar-system of its own. Spacecraft that have gone to Jupiter and the other giant planets have revealed the nature not only of the planets themselves, but also of the moons.

Saturn has long been famous for its beautiful rings. We now know, however, that each of the other giant planets also has rings. When seen close-up, the astonishing detail in the rings is very beautiful.

Uranus and Neptune were known for a long time to us as mere points in the sky. Spacecraft views have transformed them into objects with more character.

The age of first exploration of the giant planets, with spacecraft that simply flew by the planets, is over. We now are in the stage of space missions to orbit the planets, with a Jupiter orbiter completing its mission and a Saturn orbiter en route. These missions study the planets, their rings, their moons, and their magnetic fields in a much more detailed manner.

The last few years have brought the exciting discovery of dozens of planets outside our Solar System, orbiting stars other than the Sun. All these planets are giant planets; most of the ones that have been discovered so far are several times as massive as Jupiter. After we discuss the giant planets in our own Solar System, we will summarize what we know of these other giant planets.

Crescent Saturn and its rings in close-up, as the Voyager I spacecraft looked back at them after its closest passage.

◀ The jovian planets and the Earth, to scale.

125

A Closer Look

Comparative Data for Some Major Worlds

	Semimajor Axis	Orbital Period	Equatorial Radius ÷ Earth's	Mass ÷ Earth's
Planets				
Jupiter	5.2 A.U.	12 years	11.2	318
Saturn	9.5 A.U.	29 years	9.4	95
Uranus	19.2 A.U.	84 years	4.0	15
Neptune	30.1 A.U.	164 years	3.9	17

JUPITER

Jupiter, the largest and most massive planet, dominates the Sun's planetary system. It alone contains two-thirds of the mass in the Solar System outside of the Sun, 318 times as much mass as the Earth (but only 0.001 times the Sun's mass). Jupiter has at least 16 moons of its own and so is a miniature planetary system (that is, several planetary objects orbiting a central object) in itself. It is often seen as a bright object in our night sky, and observations with even a small telescope reveal bands of clouds across its surface and show four of its moons, the Galilean satellites www (see *Starparties: Observing the Giant Planets*).

Jupiter is more than 11 times greater in diameter than the Earth. (See *Figure It Out: The Size of Jupiter*.) From its mass and volume, we calculate its density to be 1.3 g/cm^3, not much greater than the 1 g/cm^3 density of water. This low density tells us that any core of heavy elements (such as iron) makes up only a small fraction of Jupiter's mass. Jupiter, rather, is mainly composed of the light elements hydrogen and helium. Jupiter's chemical composition is closer to that of the Sun and stars than it is to that of the Earth (Fig. 7–1), so its origin can be traced directly back to the solar nebula with much less modification than the terrestrial planets underwent. Jupiter has no crust. At deeper and deeper levels, its gas just gets denser and denser, turning mushy and eventually liquefying about 20,000 km (15 per cent of the way) down. Jupiter's core, inaccessible to direct study, is calculated to be made of heavy elements and to be larger and perhaps 10 times more massive than Earth.

Jupiter's surface rotates in about 10 hours, though different latitudes rotate at slightly different speeds. Regions with different speeds correspond to different bands. The clouds are in constant turmoil; the shapes and distribution of bands can change within days.

The most prominent feature of Jupiter's surface is a large reddish oval known as the **Great Red Spot.** It is two to three times larger than the Earth. Other, smaller spots are also present.

Jupiter emits radio waves, which indicates that it has a strong magnetic field and strong "radiation belts" (actually, belts of magnetic fields filled with trapped energetic particles—large-scale versions of the Van Allen belts of Earth).

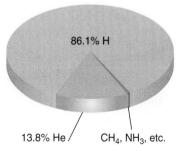

86.1% H

13.8% He / CH$_4$, NH$_3$, etc.

Figure 7-1 The composition of Jupiter.

A Closer Look

Comparative Data for Some Minor Worlds

	Semimajor Axis	Orbital Period	Equatorial Radius (Earth = 100%)	Mass (Earth = 100%)
Planets (for comparison)				
Mercury	0.39 A.U.	88 days	38%	5.5%
Pluto	39.5 A.U.	248 years	18%	0.2%
Earth's satellite (for comparison)				
The Moon	384,000 km	27 days	27%	1.2%

More detailed information appears in Appendices 3 and 4.

Spacecraft to Jupiter

Our understanding of Jupiter was revolutionized in the 1970s, when first Pioneer 10 and Pioneer 11 and then Voyager 1 and Voyager 2 flew by (Fig. 7–2). The Galileo spacecraft arrived at Jupiter in 1995, when it dropped a probe into Jupiter's atmosphere and went into orbit in the Jupiter system. Each spacecraft carried many types of instruments to measure properties of Jupiter, its satellites, and the space around them.

The Great Red Spot

The Great Red Spot is a gaseous island a few times larger across than the Earth (Fig. 7–3). It is the vortex of a violent, long-lasting storm, similar to large storms on Earth, and drifts about slowly with respect to the clouds as the planet rotates.

A

B

Figure 7–2 Jupiter from *(A)* Voyager 1, and *(B)* 4½ months later, from Voyager 2. Note how the white ovals have drifted around the Great Red Spot. Winds blow at different speeds at different latitudes, so clouds pass each other.

Figure It Out
The Size of Jupiter

It is easy and instructive to estimate the physical diameter of Jupiter from its angular size. Jupiter's disk appears to be about 50 arc seconds across when it is closest to the Earth. (You can actually measure this yourself, by looking through a telescope with an eyepiece whose field of view you have already calibrated—perhaps by looking at the Moon, which is 30 arc minutes in diameter.) Since Jupiter is roughly 5 A.U. from the Sun and the Earth is 1 A.U. from the Sun, Jupiter must be about 4 A.U. from Earth when at its closest.

Now we need to convert an angular diameter into a physical one, measured in kilometers. If the angular size (ϕ) of an object is measured in radians, and if the object appears small in the sky, then its physical size (*s*) is given by its distance (*d*) multiplied by its angular size. This relation, known as the *small-angle formula*, is expressed as $s = d\phi$; it is very useful in astronomy.

One radian is equal to 206,265 arc seconds, because in a full circle 2π radians correspond to 360°. Converting Jupiter's angular diameter of 50 arc seconds into radians, we have $\phi = 50/206,265 = 0.00024$ radians. Since 1 A.U. $= 1.5 \times 10^8$ km, the distance of Jupiter is 6.0×10^8 km. According to the small-angle formula, then, Jupiter's physical diameter must be $s = d\phi = (6.0 \times 10^8 \text{ km})(0.00024) = 1.4 \times 10^5$ km. This is about 11 times larger than Earth's diameter of 12,800 km.

From the sense of its rotation (counterclockwise rather than clockwise in its hemisphere), measured from time-lapse photographs, we can tell that it is a pressure high rather than a low. We also see how it interacts with surrounding clouds and smaller spots. The Great Red Spot has been visible for at least 150 years, and maybe even 300 years. Sometimes it is relatively prominent and colorful, and at other times the color may even disappear for a few years.

Why has the Great Red Spot lasted this long? Heat, energy flowing into the storm from below it, partly maintains its energy supply. The storm also contains more mass than hurricanes on Earth, which makes it more stable. Further, unlike Earth, Jupiter has no continents or other structure to break up the storm. Also, we do not know how much energy the Spot gains from the circulation of Jupiter's upper atmosphere and eddies (rotating regions) in it. Until we can sample lower levels of Jupiter's atmosphere, we will not be able to decide definitively. Studying the eddies in Jupiter's atmosphere helps us interpret features on Earth. For example, one theory of Jupiter's spots holds that they are similar to circulating rings that break off from the Gulf Stream in the Atlantic Ocean.

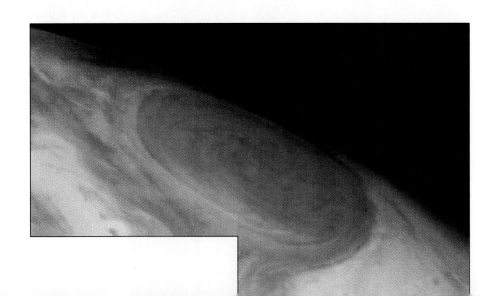

Figure 7-3 The Great Red Spot, in an image from the Galileo spacecraft. It rotates in the anticyclonic (counterclockwise) sense. The Great Red Spot is now about 26,000 km by 134,000 km and has been shrinking for 50 years. It is probably mostly ammonia gas and ice clouds, and goes 20 to 40 km deep.

Use a telescope, perhaps the one on your campus, to view the giant planets. Jupiter and Saturn are among the brightest objects in the sky, and are easy to locate, but Uranus and especially Neptune are more challenging. (To get current positions of planets, consult *Sky and Telescope*, *Astronomy*, or *Mercury* magazines, the *Field Guide to the Stars and Planets*, or RedShift.)

You should be able to see a few of the main bands of Jupiter, and maybe even the Great Red Spot, in addition to the bright Galilean satellites. Sometimes one or more of these moons is behind Jupiter or casts a shadow on its disk. Their relative positions change from night to night; the innermost moon, Io, takes only 1.8 days to complete an orbit.

Saturn's bands are less distinct that those of Jupiter, but the rings look magnificent. You may be able to see the dark gap in the rings known as Cassini's division, as well as the shadow of the rings on the disk of the planet. The rings appear edge-on to our line of sight twice during Saturn's 30-year orbital period, making them difficult to detect. Only one moon, Titan, is readily visible through a small telescope; it looks like a faint star.

Uranus and Neptune are too small and too far away to reveal detail, but you may be able to tell that they have resolved disks, unlike the much more distant, point-like stars. Try to decide whether Uranus and Neptune look greenish to your eyes.

http://www.harcourtcollege.com/astro/cosmos/rsce

Jupiter's Atmosphere

Heat emanating from Jupiter's interior churns the atmosphere. (In the Earth's atmosphere, on the other hand, most of the energy comes from the outside—from the Sun.) The bright bands on Jupiter are rising currents of gas driven by this process of convection (Fig. 7–4), described for the Sun in Chapter 9. The dark bands are falling gas. Their tops are somewhat lower (about 20 km) than the tops of the bright bands and so are about 10 K warmer.

Wind velocities show that each hemisphere of Jupiter has a half-dozen currents blowing eastward or westward. The Earth, in contrast, has only one westward current at low latitudes (the trade winds) and one eastward current at middle latitudes (the jet stream).

On December 7, 1995, the Galileo probe (Fig. 7–5A) transmitted data for 57 minutes as it fell through Jupiter's atmosphere. It gave us accurate measurements of Jupiter's composition; the heights of the cloud layers; and the variations of temperature, density, and pressure. It went through about 600 km of Jupiter's atmosphere, only about 1 per cent of Jupiter's radius. The probe found that Jupiter's winds were stronger than expected and increased with depth, which shows that the energy that drives them comes from below.

Extensive lightning storms, including giant-sized lightning strikes called "superbolts," were discovered from the Voyagers, as were giant aurorae (Fig. 7–5B). The Galileo spacecraft photographed giant thunderclouds on Jupiter, which indicates that some regions are relatively wet and others relatively dry. Probably, the probe found less water vapor than expected because it fell through a dry region.

Figure 7–4 Bright and dark bands on Jupiter, as seen from Voyager 1. Computer processing has unrolled the surface.

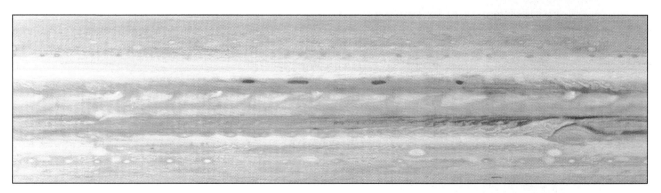

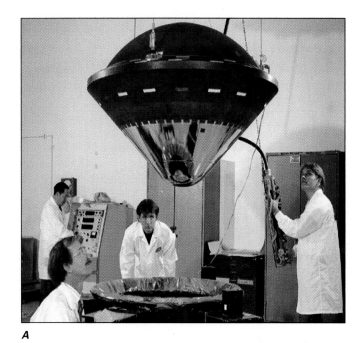

A

B

Figure 7-5 *(A)* The probe portion of Project Galileo. Six instruments inside the probe measured the composition of Jupiter's atmosphere, located clouds, studied lightning and radio emission, found where energy is absorbed, and measured precisely the ratio of the number of atoms of hydrogen and helium. *(B)* The northern and southern auroral ovals on Jupiter, imaged in the ultraviolet from the Hubble Space Telescope. They are about 500 km above the level in Jupiter's atmosphere where the pressure is the same as Earth's surface pressure.

Figure 7-6 The current model of Jupiter's interior.

Jupiter's Interior

Most of Jupiter's interior is in liquid form. Jupiter's central temperature may be between 13,000 and 35,000 K. The central pressure is 100 million times the pressure of the Earth's atmosphere measured at our sea level because of Jupiter's great mass pressing in. Because of this high pressure, Jupiter's interior is probably composed of ultracompressed hydrogen surrounding a rocky core consisting of perhaps 10 Earth masses of iron and silicates (Fig. 7–6). (The Earth's central pressure is 4 million times its atmosphere's pressure, and Earth's central temperatures are several thousand degrees.)

Jupiter radiates 1.6 times as much heat as it receives from the Sun. It must have an internal energy source—perhaps the energy remaining from its collapse from a primordial gas cloud 20 million km across. Jupiter is undoubtedly still contracting and this process also liberates energy. It lacks the mass necessary by a factor of about 75, however, to have heated up enough to become a star, generating energy by nuclear processes. It is therefore not "almost a star," contrary to some popular accounts.

Jupiter's Magnetic Field

The space missions showed that Jupiter's tremendous magnetic field is even more intense than many scientists had expected. At the height of Jupiter's clouds the magnetic field strength is 10 times that of the Earth, which itself has a strong field.

The inner field is shaped like a doughnut, containing several shells of charged particles, like giant versions of the Earth's Van Allen belts. The outer region of Jupiter's magnetic field interacts with the particles flowing outward from the Sun. When this solar wind is strong, Jupiter's outer magnetic field (shaped like a flattened pancake) is pushed in. When the high-energy particles interact with Jupiter's magnetic field, radio emission results.

Jupiter's Ring

Though Jupiter wasn't expected to have a ring, Voyager 1 was programmed to look for one just in case; Saturn's rings, of course, were well known, and Uranus's rings had been discovered only a few years earlier. The Voyager 1 photograph indeed showed a wispy ring of material around Jupiter at about 1.8 times Jupiter's radius, inside the orbit of its innermost moon. As a result, Voyager 2 was targeted to take a series of photographs of the ring. From the far side looking back, the ring appeared unexpectedly bright, probably because small particles in the ring scattered the light toward the spacecraft. Within the main ring, fainter material appears to extend down to Jupiter's cloud tops (Fig. 7–7). The ring particles may come from Jupiter's moon Io, or they may come from comet and meteor debris or from material knocked off the innermost moons by meteorites. Whatever their origin, the individual particles probably remain in the ring only temporarily.

Figure 7–7 Jupiter's ring (overexposed), with Jupiter's moon Europa behind it. The Galileo spacecraft was looking at light scattered forward through the ring, and at the side of Europa away from the Sun, illuminated by reflected light off Jupiter itself.

Jupiter's Galilean Satellites

Four of the innermost satellites were discovered by Galileo in 1610 when he first looked at Jupiter with his small telescope. These four moons (Io, Europa, Ganymede, and Callisto) are called the **Galilean satellites** (Fig. 7–8). One of these moons, Ganymede, 5276 km in diameter, is the largest satellite in the Solar System and is larger than the planet Mercury.

The Galilean satellites have played a very important role in the history of astronomy. The fact that these particular satellites were noticed to be going around another planet, like a solar system in miniature, supported Copernicus's Sun-centered model of our Solar System. Not everything revolved around the Earth! It was fitting to name the Galileo spacecraft after the discoverer of Jupiter's moons.

Jupiter also has another dozen satellites, some known from Earth and others discovered by the Voyagers. None of these other satellites is even 10 per cent the diameter of the smallest Galilean satellite.

Through first Voyager and then Galileo-spacecraft close-ups, the satellites of Jupiter have become known to us as worlds with personalities of their own. The four Galilean satellites, in particular, were formerly known only as dots of light. Not only the bigger satellites, which range between 0.9 and 1.5 times the size of our own Moon, but also the smaller ones that have been imaged in detail turn out to have interesting surfaces and histories.

Io, the innermost Galilean satellite, provided the biggest surprises. Scientists knew that Io gave off particles as it went around Jupiter, but Voyager 1 discovered that these particles resulted from active volcanoes on the satellite. Eight volcanoes were seen actually erupting, many more than erupt on the Earth at any one time. When Voyager 2 went by a few months later, most of the same volcanoes were still erupting. Though the Galileo spacecraft could not go close to Io for most of its mission, for fear of ruining itself because of Jupiter's strong radiation field in that region, it could get high-quality images of Io and its volcanoes (Fig. 7–9). Finally, it went within a few hundred kilometers of Io's surface, and found that 100 volcanoes were erupting simultaneously.

Io's surface (Fig. 7–10) has been transformed by the volcanoes, and is by far the youngest surface we have observed in our Solar System. Gravitational forces from the other Galilean satellites distort Io's orbit slightly, which changes the tidal force on it from Jupiter in a varying fashion. This changing tidal force flexes Io, creating heat from friction that heats the interior and leads to the volcanism. The surface of Io is covered with sulfur and sulfur compounds, including frozen

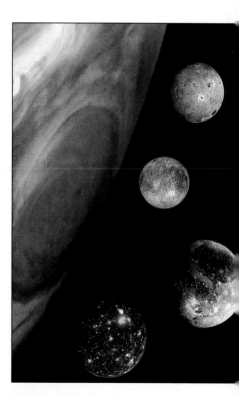

Figure 7–8 The four Galilean satellites of Jupiter, alongside Jupiter's Great Red Spot, imaged from Galileo.

A Closer Look

Jupiter and Its Satellites in Mythology

Jupiter was the supreme god in Roman mythology. He was also called Jove, so "jovian" refers to Jupiter — the jovian planets are those that are large like Jupiter, namely Jupiter, Saturn, Uranus, and Neptune.

All of Jupiter's moons except Amalthea are named in Greek mythology after lovers of Zeus, the Greek equivalent of Jupiter. Amalthea, a goat-nymph, was Zeus's nurse, and out of gratitude he made her into the constellation Capricorn, a goat. Zeus changed Io into a heifer to hide her from his jealous wife Hera; in honor of Io, Earth's crescent moon has horns, as do bulls.

Ganymede was a Trojan youth carried off by an eagle to be Jupiter's cup bearer (the constellation Aquarius). Callisto was changed into a bear as punishment for her affair with Zeus. She was then slain by mistake, and rescued by Zeus by being transformed into the Big Bear in the sky, the constellation Ursa Major. Jealous Hera persuaded the sea god to forbid Callisto to ever bathe in the sea, which is why Ursa Major never sinks below the horizon as seen from the latitude of Greece.

Europa was carried off to Crete on the back of Zeus, who took the form of a white bull. She became King Minos's mother. Pasiphae (also the name of one of Jupiter's moons) was the wife of Minos and the mother of the Minotaur.

sulfur dioxide, and the thin atmosphere is full of sulfur dioxide. It certainly wouldn't be a pleasant place to visit.

Io's surface, orange in color and covered with strange formations because of the sulfur, led Bradford Smith, the head of the Voyager imaging team, to remark that "It's better looking than a lot of pizzas I've seen."

Galileo imaging shows many mountains too tall to be supported by sulfur, so stronger types of rock must be involved with a crust at least 30 km thick above the molten regions. Also, Galileo's infrared observations show that some of the volcanoes are too hot to be sulfur volcanism. The surface changes substantially even as Galileo watches it.

Europa, Jupiter's Galilean satellite with the highest albedo, has a very smooth surface and is covered with narrow dark stripes. The lack of surface relief, mapped by the Galileo spacecraft to be no more than a couple of hundred meters high, suggests that the surface we see is ice. The markings may be fracture systems in

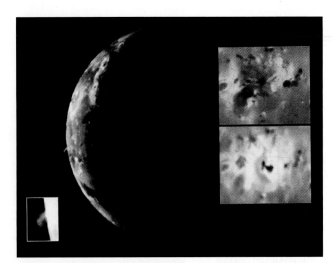

Figure 7-9 Io, from Galileo, showing a volcano erupting on its limb and changes in a volcano on its surface. Close fly-bys of Io in 1999 and 2000 have revealed over 100 erupting volcanoes. Many of the volcanic plumes active now are different from ones viewed by the Voyagers.

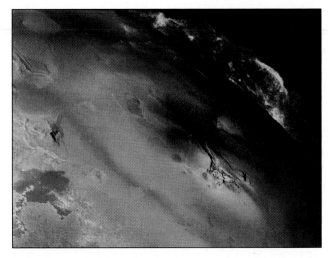

Figure 7-10 Io's surface has been transformed by the sulfur erupted from volcanoes like the one on the edge.

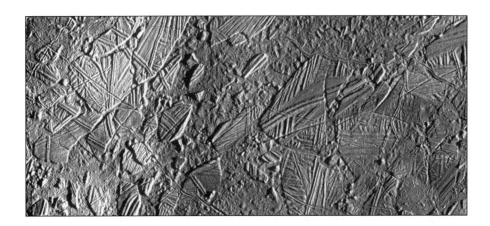

Figure 7-11 Close-ups of icy regions on Europa, almost certainly the crust of a global ocean.

the ice, like fractures in the large fields of sea ice near the Earth's north pole, as apparently verified in Galileo close-ups (Fig. 7–11). Some longer ridges can be traced far across Europa's surface. Few craters are visible, suggesting that the ice was soft enough below the crust to close in the craters. Either internal radioactivity or a gravitational tidal heating like that inside Io provides the heat to soften the ice. Because Europa possibly has a liquid-water ocean and extra heating, many scientists consider it a worthy location to check for signs of life. We can only hope that the ice crust, which may be about 10 km thick, is thin enough in some locations for us to be able to penetrate it to reach the ocean that may lie below— and see if life could have evolved there.

The largest satellite in the Solar System, Ganymede, shows many craters (Fig. 7–12) alongside weird grooved terrain (Fig. 7–13). Ganymede is larger than Mercury but less dense; it contains large amounts of water-ice surrounding a rocky core. But an icy surface is as hard as steel in the cold conditions that far from the Sun, so it retains the craters from perhaps 4 billion years ago. The grooved terrain is younger.

Ganymede shows many lateral displacements, where grooves have slid sideways, like those that occur in some places on Earth (for example, the San Andreas fault in California). It is the only place besides the Earth where such faults have been found. Thus further studies of Ganymede may help our understand-

Figure 7-12 Ganymede, Jupiter's largest satellite. Many impact craters, some with systems of bright rays, show. The large dark region has been named Galileo Regio. Low-albedo features on satellites are named for astronomers.

Figure 7-13 A computer-generated view, with exaggerated perspective, of ridges on Ganymede. Ganymede's icy surface has been fractured and broken into many parallel ridges and troughs. Such bright grooved terrain covers over half of Ganymede's surface.

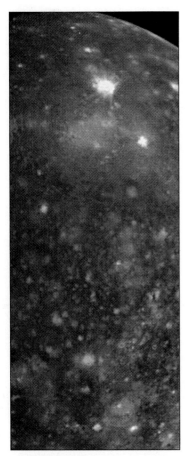

Figure 7–14 The side of Callisto that leads in its orbit around Jupiter, imaged from the Galileo spacecraft. The impact region Asgard *(center)* is surrounded by concentric rings up to 1700 km in diameter. The bright icy materials excavated by the younger craters contrast with the darker and redder coatings on older surfaces. Galileo images show, surprisingly, that Callisto has very few craters smaller than 100 m in diameter.

ing of terrestrial earthquakes. The Galileo spacecraft found a stronger magnetic field for Ganymede than expected, so perhaps Ganymede is more active inside than previously supposed.

Callisto, the outermost of Jupiter's Galilean satellites, has so many craters (Fig. 7–14) that its surface must also be the oldest. Callisto, like Europa and Ganymede, is covered with ice. A huge bull's-eye formation, Valhalla, contains about 10 concentric rings, no doubt resulting from a huge impact. Perhaps ripples spreading from the impact froze into the ice to make Valhalla. Callisto had been thought to be old and uninteresting, but observations from the Galileo spacecraft have revised the latter idea, by showing changes: There are fewer small craters than expected, so the small craters that must have once been there were probably covered by dust that meteorite impacts eroded from larger craters, or disintegrated by themselves through electrostatic changes. The spacecraft's measurements of Callisto's gravity from place to place show that its mass is concentrated more toward its center than had been thought. This concentration indicates that heavier materials inside have sunk, and perhaps even indicates that there is an ocean below Callisto's surface. Callisto's interactions with Jupiter's magnetic field have been interpreted to back up the idea of an ocean.

Studies of Jupiter's moons tell us about the formation of the Jupiter system, and help us better understand the early stages of the entire Solar System. We'll get some additional brief but close-up views of the Jupiter system when the Cassini spacecraft, en route to Saturn, passes through in late December 2000.

⬤ SATURN

Saturn, like Jupiter, Uranus, and Neptune, is a giant planet. Its diameter, without its rings, is 9 times greater than Earth's; its mass is 95 times greater. It is a truly beautiful object in a telescope of any size (Fig. 7–15). The glory of its system of rings makes it stand out even in small telescopes (see *Starparties: Observing the Giant Planets*).

The giant planets have low densities. Saturn's is only 0.7 g/cm³, 70 per cent the density of water (Fig. 7–16). The bulk of Saturn is hydrogen molecules and helium, reflecting Saturn's formation directly from the solar nebula. Saturn is thought to have a core of heavy elements, including rocky material, making up about the inner 20 per cent of its diameter.

Voyagers 1 and 2 flew by Saturn in 1980 and 1981, respectively. ⓦⓦⓦ A joint NASA/European Space Agency mission called Cassini is en route to Saturn, with arrival scheduled for 2004. ⓦⓦⓦ

Figure 7–15 A storm (bright equatorial region) that developed on Saturn was photographed with the Hubble Space Telescope.

Saturn's Rings

The rings extend far out in Saturn's equatorial plane, and are inclined to the planet's orbit. Over a 30-year period, we sometimes see them from above their northern side, sometimes from below their southern side, and at intermediate angles in between. When seen edge-on, they are almost invisible (Fig. 7–17).

The rings of Saturn were originally material that was torn apart by Saturn's gravity or material that failed to collect into a moon at the time when the planet and its moons were forming; the actual particles in the ring now are more newly arrived. Every massive object has a sphere, called its **Roche limit,** inside of which blobs of matter do not hold together by their mutual gravity. The forces that tend to tear the blobs apart from each other are tidal forces. They arise, like the Earth's tides, because some parts of an object are closer to the planet than others and are thus subject to higher gravity. The difference between the gravity force farther in and the gravity force farther out is the tidal force.

The radius of the Roche limit is usually 2½ to 3 times the radius of the larger body, closer to the latter for the relative densities of Saturn and its moons. The Sun also has a Roche limit, but all the planets lie outside it. The natural moons of the various planets lie outside the respective Roche limits. Saturn's rings lie inside Saturn's Roche limit, so it is not surprising that the material in the rings is spread out rather than collected into a single orbiting satellite. Artificial satellites that we send up to orbit the Earth are constructed of sufficiently rigid materials that they do not break up even though they are within the Earth's Roche limit; they are held together by forces much stronger than gravity.

Saturn has several concentric major rings visible from Earth. The brightest ring is separated from a fainter broad outer ring by an apparent gap called **Cassini's division.** Another ring is inside the brightest ring. We know that the rings are not solid objects, because their rotation speed decreases as their distances from the planet increase.

Radar waves bounced off the rings show that the particles in the rings are at least a few centimeters and possibly a meter across. Infrared studies show that at least their outsides are ice.

The images from the Voyagers (Fig. 7–18) revolutionized our view of Saturn, its rings, and its moons. Only from spacecraft can we see the rings from a vantage point different from the one we have on Earth. Backlighted views showed

REDSHIFT
http://www.harcourtcollege.com/astro/cosmos/rsce

Figure 7–16 Since Saturn's density is lower than that of water, it would float, like Ivory soap, if we could find a big enough bathtub. But it would leave a ring.

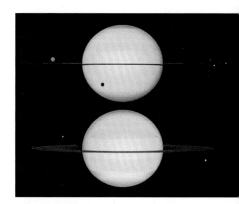

Figure 7–17 Saturn, from the Hubble Space Telescope, during the Earth's crossing of its ring plane.

Figure 7–18 Saturn and some of its moons, photographed from the Voyager 1 spacecraft. We see Enceladus *(foreground)*, Mimas *(in front of Saturn's disk)*, Dione *(upper left)*, Rhea *(top middle)*, Titan *(top right)*, Iapetus *(right center)*, and Tethys *(lower right)*.

Figure 7-19 Dark, radial spokes became visible in the rings as the Voyagers approached. They formed and dispersed within hours.

Figure 7-20 When sunlight scatters forward, the spokes appear bright. They are "radial" in that they point roughly from the center outward, along a radius.

that Cassini's division, visible as a dark (and thus apparently empty) band from Earth, appeared bright, so it must contain some particles.

The rings are thin, for when they pass in front of stars, the starlight easily shines through. Studies of the changes in the radio signals from the Voyagers when they went behind the rings showed that the rings are only about 20 m thick. Relative to the diameter of the rings, this is equivalent to a CD (compact disk) that is 30 km across though still its normal thickness.

The closer the spacecraft got to the rings, the more rings became apparent. Each of the known rings was actually divided into many thinner rings. The number of these rings (sometimes called "ringlets") is in the hundreds of thousands.

Everyone had expected that collisions between particles in Saturn's rings would make each ring perfectly uniform. But there was a big surprise. As Voyager 1 approached Saturn, we saw radial "spokes." The spokes look dark from the side illuminated by the Sun (Fig. 7–19), but look bright from behind (Fig. 7–20). This information showed that the particles in the spokes were very small, about 1 micron in diameter, since only small particles—like terrestrial dust in a sunbeam—reflect light in this way.

The outer major ring turns out to be kept in place by a tiny satellite orbiting just outside it. At least some of the rings are kept narrow by "shepherding" satellites that gravitationally affect the ring material, a concept that we can apply to rings of other planets.

A post-Voyager theory said that many of the narrowest gaps may be swept clean by a variety of small moons. These objects would be embedded in the rings in addition to the icy snowballs that make up most of the ring material. The tiny moon Pan has been observed clearing out the Encke Gap in just this way. Theorists suppose that smaller moons probably clear out several other gaps in Saturn's rings, but these bodies were not discovered by the Voyagers or found since.

Saturn's Atmosphere

Like Jupiter, Saturn rotates quickly on its axis. A complete period is only 10 hours, in spite of Saturn's diameter being over 9 times greater than Earth's. This makes Saturn look slightly "flattened"; it bulges out more at the equator than at its poles.

The structure in Saturn's clouds is of much lower contrast than that in Jupiter's clouds. It is not a surprise that the chemical reactions can be different; after all, Saturn is colder. Saturn has extremely high winds, up to 1800 km/hr, 4

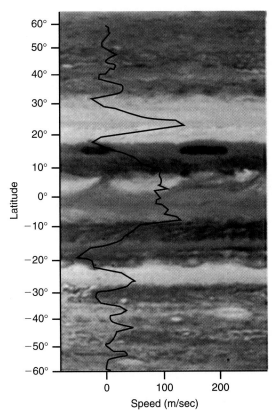

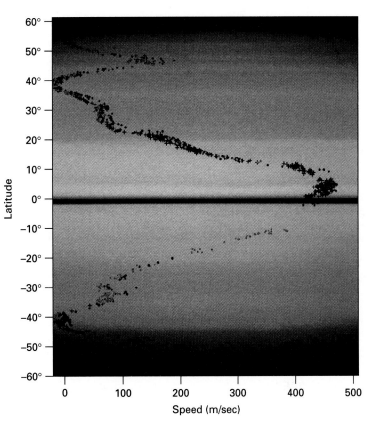

Figures 7–21 The winds of Jupiter, a graph of their speed on a background that shows Jupiter's surface and how it varies with latitude.

Figures 7–22 The winds of Saturn with Saturn's surface in the background. Compare Figure 7–21, at left. Jupiter's winds correspond to its bright and dark bands, but Saturn's do not.

times faster than the winds on Jupiter. On Saturn, the variations in wind speed do not seem to correlate with the positions of bright and dark bands, unlike the case with Jupiter (Figs. 7–21 and 7–22). Also as on Jupiter, but unlike the case for Earth, the winds seem to be driven by rotating eddies, which in turn get most of their energy from the planet's interior. Such differences provide a better understanding of storm systems in Earth's atmosphere.

Saturn's Interior and Magnetic Field

Saturn radiates about twice as much energy as it absorbs from the Sun, a greater factor than for Jupiter. One interpretation is that only ⅓ of Saturn's internal energy remains from its formation and from its continuing contraction under gravity. The rest would be generated by the gravitational energy released by helium sinking through the liquid hydrogen in Saturn's interior. The helium that sinks has condensed because Saturn, unlike Jupiter, is cold enough.

Saturn gives off radio signals, as does Jupiter, a pre-Voyager indication to earthbound astronomers that Saturn also has a magnetic field. The Voyagers found that the magnetic field at Saturn's equator is only ⅔ of the field present at the Earth's equator. Remember, though, that Saturn is much larger than the Earth and so its equator is much farther from its center. Saturn's magnetic field contains belts of charged particles (Van Allen belts), which are larger than Earth's but smaller than Jupiter's. (Saturn's surface magnetic field is 20 times weaker than Jupiter's.) These particles interact with the atmosphere near the poles and produce aurorae (Fig. 7–23).

Figure 7–23 Saturn's aurora, showing the planet's magnetic field. This ultraviolet image was taken with the Hubble Space Telescope.

A Closer Look

Saturn's Satellites in Mythology

Saturn's moons (Appendix 4) are named after the Titans. In Greek mythology, the Titans were the children and grandchildren of the early supreme god Uranus and of Gaea, the goddess of the Earth, who had been fertilized by a drop of Uranus's blood. The largest of Saturn's moons is named after Titan himself.

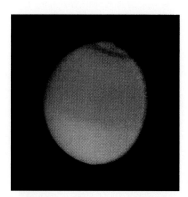

Figure 7–24 Titan was disappointingly featureless even to Voyager 1's cameras because of its thick, smoggy atmosphere. Its northern polar region was relatively dark. The southern hemisphere was lighter than the northern.

Saturn's Moon Titan

Saturn's largest moon, Titan, is larger than the planet Mercury. At 40 per cent the diameter of the Earth, Titan, however, is an intriguing body.

Titan has an atmosphere that was detected from Earth pre-Voyager. Studies of how the radio signals faded when Voyager 1 went behind Titan showed that Titan's atmosphere is denser than Earth's. The surface pressure on Titan is 1½ times that on Earth.

Titan's atmosphere is opaque, apparently because of the action of sunlight on chemicals in it, forming a sort of "smog" and giving it its reddish tint. Smog on Earth forms in a similar way. The Voyagers showed (Fig. 7–24) several layers of haze. They detected nitrogen, which makes up the bulk of Titan's atmosphere. Methane is a minor constituent, perhaps 1 per cent. A greenhouse effect is present, making some scientists wonder whether Titan's surface may have been warmed enough for life to have evolved there.

The temperature near the surface, deduced from measurements made with Voyager's infrared radiometer, is only about −180°C (93K), somewhat warmed by the greenhouse effect but still extremely cold. This temperature is near that of methane's "triple point," at which it can be in any of the three physical states—solid, liquid, or gas. So methane may play the role on Titan that water does on Earth. Parts of Titan may be covered with lakes or oceans of methane mixed with ethane, and other parts may be covered with methane ice or snow. Using filters in the near-infrared, first the Hubble Space Telescope and then ground-based telescopes have been able to penetrate Titan's haze to reveal some structure on its surface (Fig. 7–25), though no unambiguous lakes have been found.

Some of the organic molecules formed in Titan's atmosphere from lightning storms and other processes may rain down on its surface. Thus the surface, largely hidden from our view, may be covered with an organic crust about a kilometer thick, perhaps partly dissolved in liquid methane. These chemicals are similar to those from which we think life evolved on the primitive Earth. But it is probably too cold on Titan for life to begin.

Figure 7–25 (A) The Hubble Space Telescope has glimpsed Titan's surface through its atmospheric haze by observing at near-infrared wavelengths. The large, bright region—about the size of Australia—may be a continent, a huge cloud, or the mark of a meteor impact. (B) A Keck image made at a longer infrared wavelength using a special technique of combining many short exposures gives a view of Titan's surface with higher contrast and resolution twice as good. The infrared dark areas may be basins filled with liquid methane or ethane; the bright regions may be a mixture of water-ice and rock.

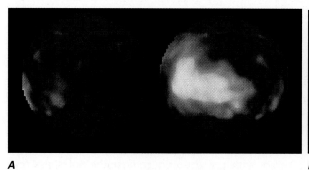

A

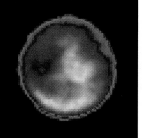

B

Titan is so intriguing, and potentially so important, that the lander of the Cassini mission is to plunge through its atmosphere. So while the lander of Jupiter's Galileo mission went into the planet itself, this lander, known as Huygens, will penetrate the clouds around Saturn's largest moon when it arrives in 2004. (The parts of the spacecraft are named after Giovanni Cassini, who discovered the main gap in Saturn's rings in 1675, and after Christian Huygens, who discovered Titan in 1655.)

⬤ URANUS

The two other giant planets beyond Saturn—Uranus (pronounced "U'ran us") and Neptune—are each about 4 times the diameter of and about 15 times more massive than the Earth. They reflect most of the sunlight that hits them, which indicates that they are covered with clouds. Like Jupiter and Saturn, Uranus and Neptune don't have solid surfaces. Their atmospheres are also mostly hydrogen and helium, but they have a higher proportion of heavier elements. Some of the hydrogen may be in a liquid mantle of water, methane, and ammonia. At the planets' centers, a rocky core contains mostly silicon and iron, probably surrounded by ices. From their average densities, we have deduced that the cores of Uranus and Neptune make up substantial parts of those planets, differing from the relatively more minor cores of Jupiter and Saturn.

Uranus was the first planet to be discovered that had not been known to the ancients. The English astronomer and musician William Herschel reported the discovery in 1781. Actually, Uranus had been plotted as a star on several sky maps during the hundred years prior to Herschel's discovery, but had not been singled out as other than an ordinary star.

Uranus revolves around the Sun in 84 years at an average distance of more than 19 A.U. Uranus appears so tiny that it is not much bigger than the resolution we are allowed by our atmosphere. Uranus is apparently surrounded by thick clouds of methane ice crystals (Fig. 7–26), with a clear atmosphere of molecular hydrogen above them. The trace of methane gas mixed in with the hydrogen makes Uranus look greenish.

Uranus is so far from the Sun that its outer layers are very cold. Studies of its infrared radiation give a temperature of −215°C (58 K). There is no evidence for an internal heat source, unlike the case for Jupiter, Saturn, and Neptune.

The other planets rotate such that their axes of rotation are very roughly parallel to their axes of revolution around the Sun. Uranus is different (as is Pluto), for its axis of rotation is roughly perpendicular to the other planetary axes, lying

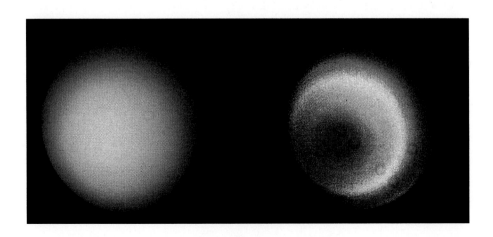

Figure 7-26 Uranus, as the Voyager 2 spacecraft approached. The picture on the left shows Uranus as the human eye would see it. On the right, false colors bring out slight contrast differences.

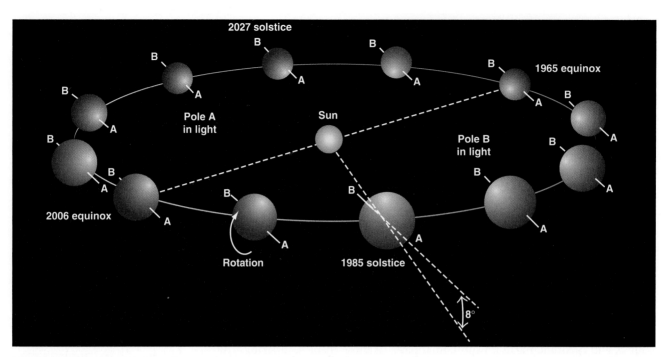

Figure 7-27 Uranus's axis of rotation lies roughly in the plane of its orbit. Notice how the planet's poles come with 8° of pointing toward the Sun, while ¼ of an orbit before or afterward, the Sun is almost over Uranus's equator.

Comparative planetology:

Jupiter: One major ring, discovered by Voyager.
Saturn: Glorious ring system with many rings and divisions.
Uranus: Nine narrow rings, discovered at a stellar occultation.
Neptune: Several rings discovered at a stellar occultation; one ring has several clumps.

only 8° from the plane of its orbit (Fig. 7–27). Sometimes one of Uranus's poles faces the Earth, 21 years later its equator crosses our field of view, and then another 21 years later the other pole faces the Earth. Polar regions remain alternately in sunlight and in darkness for decades. The strange seasonal effects that result on Uranus became obvious in 1999, when a series of Hubble Space Telescope views taken over the preceding years revealed the activity in the clouds of Uranus's once-every-84-year springtime (Fig. 7–28). When we understand just

Figure 7-28 Uranus and its rings, imaged in the near-infrared with the Hubble Space Telescope and displayed in exaggerated false color. Using a filter at the wavelength of the methane absorption from Uranus's disk diminishes the intensity of the disk. Since we see the rings by reflected sunlight, which does not have a diminished intensity at the methane wavelength, the rings appear relatively bright. We also see several of Uranus's moons.

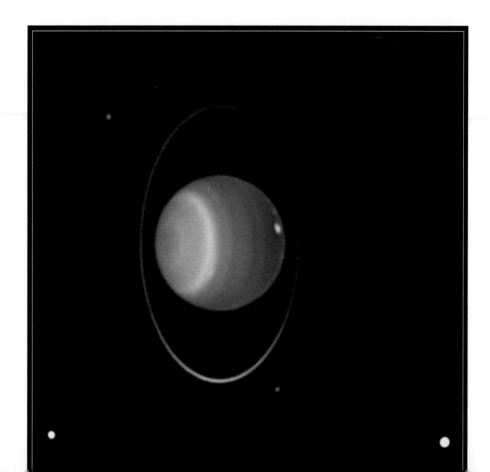

Uranus and Neptune in Mythology

In Greek mythology, Uranus was the personification of Heaven and ruler of the world, the son and husband of Gaea, the Earth. Neptune, in Roman mythology, was the god of the sea, and the planet Neptune's trident symbol reflects that origin.

how the seasonal changes in heating affect the clouds, we will be closer to understanding our own Earth's climate.

Voyager 2 reached Uranus in 1986. It revealed most of our current understanding of Uranus, its rings, and its moons. Its moon Miranda, for example, though relatively small, has a surface that was extremely varied and interesting (Fig. 7–29).

Uranus's Atmosphere

Even though Voyager 2 came very close to Uranus's surface, as close as 107,000 km (a quarter of the distance from the Earth to the Moon), it saw very little detail on it. Thus Uranus's surface is very bland. Apparently, chemical reactions are more limited than at Jupiter and Saturn because it is colder. Uranus's clouds form relatively deep in the atmosphere.

A dark polar cap was seen on Uranus, perhaps a result of a high-level photochemical haze added to the effect of sunlight scattered by hydrogen molecules and helium atoms. At lower levels, the abundance of methane gas (CH_4) increases. It is this gas that absorbs the orange and red wavelengths from the sunlight that hits Uranus. Thus most of the light that is reflected back at us is blue-green.

Two clouds were detected and monitored by Voyager 2, revealing the rotation period of their levels in Uranus's atmosphere. The larger cloud, at 35° latitude, rotated in 16.3 hours. The smaller, fainter cloud, at 27° latitude, rotated in 16.9 hours. Observations through color filters give evidence that these clouds are higher than their surroundings by 1.3 km and 2.3 km, respectively.

It was a surprise to find that both of Uranus's poles, even the one out of sunlight, are about the same temperature. The equator is nearly as warm. Comparing such a strange atmosphere with our own will help us understand Earth's weather and climate better.

Uranus's Rings

In 1977, astronomers on Earth watched as Uranus occulted (passed in front of) a faint star. Predictions showed that the occultation would be visible only from the Indian Ocean southwest of Australia. The scientists who went to study the occultation from an instrumented airplane turned on their equipment early, to be sure they caught the event. Surprisingly, about half an hour before the predicted time of occultation, they detected a few slight dips in the star's brightness (Fig. 7–30). They recorded similar dips, in the reverse order, about half an hour after the occultation. The dips indicated that Uranus is surrounded by several rings, some of which have since been photographed by Voyager 2 and with the Hubble Space Telescope. Each time a ring went between us and the distant star, the ring blocked some of the starlight, making a dip. Nine rings are now known. They are quite dark, reflecting only about 2 per cent of the sunlight that hits them.

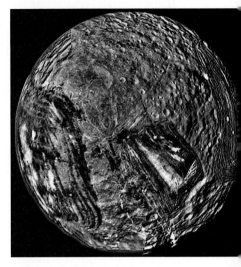

Figure 7–29 Uranus's moon Miranda, apparently broken up by impacts and solidified in new configurations in order to provide the jumbled surface that shows.

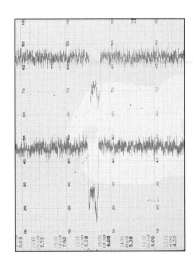

Figure 7–30 A graph of the intensity of starlight during an occultation of a star by one of Uranus's rings. Each color image, recorded through a different filter, is shown in a different color ink.

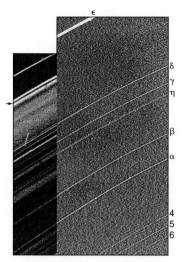

Figure 7–31 When the Voyager spacecraft viewed through the rings back toward the Sun, the backlighted view *(left)* revealed new dust lanes between the known rings, which also show in this view. The streaks are stars. The forward-lighted view *(right)* is provided for comparison. One of the new rings discovered from Voyager is marked with an arrow.

The rings have radii 1.7 to 2.1 times the radius of the planet. They are very narrow from side to side; some are only a few km wide. How can narrow rings exist, when we know that colliding particles tend to spread out? The discovery of Uranus's narrow rings led to the suggestion that a small unseen satellite (a "shepherd moon") in each ring keeps the particles together. As we saw, this model turned out to be applicable to at least some of the narrow ringlets of Saturn later discovered by the Voyagers.

Voyager provided detailed ring images (Fig. 7–31). Interpreting the small color differences is important for understanding the composition of the ring material. Quite significant was the single long-exposure, backlighted view taken by Voyager. Study of these data has shown that less of the dust in Uranus's rings is very small particles compared with the dust in the rings of Saturn and Jupiter.

The rings of Uranus are apparently younger than 100 million years of age, since the satellites that hold them in place are too small to hold them longer. Thus there must have been a more recent source of dust. That source may have been a small moon destroyed by a meteoroid or comet. The larger particles seen only when the rings were backlighted may have come from a different source. Perhaps they came from the surfaces of Uranus's current moons. We are now realizing that ring systems are younger and change more over time than had been thought.

Uranus's Interior and Magnetic Field

Voyager 2 detected Uranus's magnetic field. It is intrinsically about 50 times stronger than Earth's. Surprisingly, it is tipped 60° with respect to Uranus's axis of rotation. Even more surprising, it is centered on a point offset from Uranus's center by ⅓ its radius. Our own Earth's magnetic field is nothing like that! Since Uranus's field is so tilted, it winds up like a corkscrew as Uranus rotates. Uranus's magnetosphere contains belts of protons and electrons, similar to Earth's Van Allen belts.

Voyager also detected radio bursts from Uranus every 17.24 hours. These bursts apparently come from locations carried in Uranus's interior by the magnetic field as the planet rotates. Thus Uranus's interior rotates slightly more slowly than its atmosphere.

Memoranda.
1841, July 3. Formed a design, in the beginning of this week, of investigating, as soon as possible after taking my degree, the irregularities in the motion of Uranus, wh. are yet unaccounted for; in order to find whether they may be attributed to the action of an undiscovered planet beyond it; and if possible thence to determine the elements of its orbit, &c approximately, wh. wd. probably lead to its discovery.

Figure 7–32 From Adams's diary, kept while he was in college. "1841, July 3. Formed a design in the beginning of this week, of investigating, as soon as possible after taking my degree, the irregularities in the motion of Uranus which are yet unaccounted for."

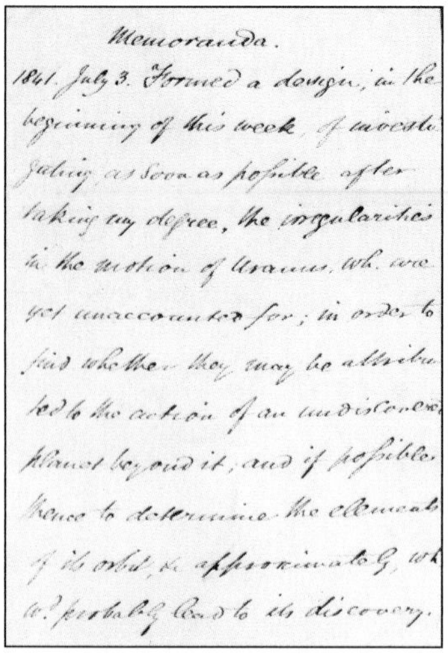

NEPTUNE

Neptune is even farther from the Sun than Uranus, 30 A.U. compared to about 19 A.U. Neptune takes 164 years to orbit the Sun. Its discovery was a triumph of the modern era of Newtonian astronomy. Mathematicians analyzed the amount that Uranus (then the outermost known planet) deviated from the orbit it would follow if gravity from only the Sun and the other known planets were acting on it. The small deviations could have been caused by gravitational interaction with another, as yet unknown, planet.

The first to work on the problem successfully was John C. Adams in England. In 1845, soon after he graduated from Cambridge University (Fig. 7–32), he predicted positions for the new planet. But neither of the two main astronomers in England made and analyzed observations to test this prediction quickly enough. A year later, the French astronomer Urbain Leverrier independently worked out the position of the undetected planet. The French astronomers didn't test his prediction right away either. Leverrier sent his predictions to an acquaintance at Berlin, where a star atlas had recently been completed. The Berlin observer, Johann Galle, discovered Neptune within hours by comparing the sky against the new atlas.

Neptune has not yet made a full orbit since it was located in 1846. But it now seems that Galileo inadvertently observed Neptune in 1613 and recorded its position, which more than doubles the period of time over which it has been studied. Galileo's observing records from January 1613 (when calculations indicate that Neptune had passed near Jupiter) show stars that were very close to Jupiter (Fig. 7–33), yet modern catalogues do not contain one of them. Galileo even once noted that one of the "stars" actually seemed to have moved from night to night, as a planet would. The object that Galileo saw was very close to but not quite exactly where our calculations of Neptune's orbit show that Neptune would have been at that time. Presumably, Galileo saw Neptune. We have used the positions he measured to improve our knowledge of Neptune's orbit.

Neptune's Atmosphere

Neptune, like Uranus, appears greenish in a telescope because of its atmospheric methane (Fig. 7–34). Some faint markings can be detected on Neptune even from Earth (Fig. 7–35). It was thus known before Voyager that Neptune's surface was more interesting than Uranus's. Still, given its position in the cold outer Solar System, nobody was prepared for the amount of activity that Voyager discovered.

As Voyager approached Neptune, active weather systems became apparent (Fig. 7–36). An Earth-sized region that was soon called the **Great Dark Spot** (Fig. 7–37) became apparent. Though colorless, the Great Dark Spot seemed analogous to Jupiter's Great Red Spot in several ways. For example, it was about the same size relative to its planet, and it was in the same general position in its

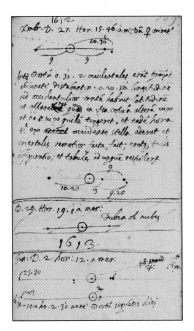

Figure 7–33 Galileo's notebook from late December 1612, showing a * marking an apparently fixed star to the side of Jupiter and its moons. The "star" was apparently Neptune. The episode shows the importance of keeping good lab notebooks, a lesson we should all take to heart.

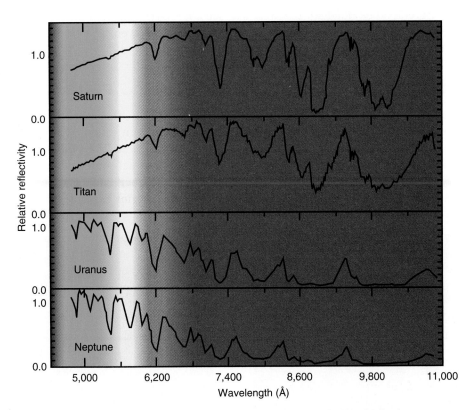

Figure 7–34 Each of the four graphs shows a spectrum divided by the Sun's spectrum. The spectra dip at wavelengths where there is a lot of absorption. The dark broad absorptions are from methane. Notice how strong the methane absorption is for Uranus and Neptune, and how it leaves mainly blue and green radiation for us to see.

Figure 7–35 Neptune, imaged in the near-infrared light (8900 Å) that is strongly absorbed by methane gas in Neptune's atmosphere, making most of the planet appear dark. The bright regions are clouds of methane ice crystals above the methane gas, reflecting sunlight. The northern hemisphere appeared brighter than the southern, a reversal of the earlier situation. Neptune's disk was 2.3 arc seconds across; the angular resolution observable was less than 1 arc second.

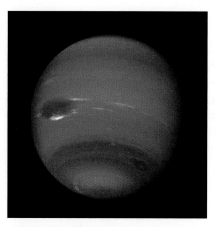

Figure 7–36 Neptune, from Voyager 2, with its Great Dark Spot, a giant storm in Neptune's atmosphere.

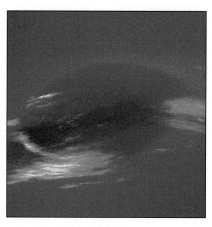

Figure 7–37 The Great Dark Spot, as seen by Voyager 2. It was about the size of the Earth, but has since disappeared.

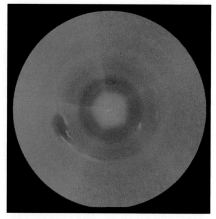

Figure 7–38 Cloud systems in Neptune's southern hemisphere. Neptune's south pole is at the center of this polar projection. The outer circle corresponds to 15° north latitude.

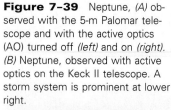

Comparative planetology:

Great Red Spot on Jupiter: lasts hundreds of years.
Great Dark Spot on Neptune: there when Voyager flew by and gone a few years later as shown by Hubble images.

Figure 7–39 Neptune, *(A)* observed with the 5-m Palomar telescope and with the active optics (AO) turned off *(left)* and on *(right)*. *(B)* Neptune, observed with active optics on the Keck II telescope. A storm system is prominent at lower right.

planet's southern hemisphere (Fig. 7–38). Putting together a series of observations into a movie, scientists discovered that it rotated counterclockwise, as does the Great Red Spot. Thus it was anticyclonic, which made it a high-pressure region. Clouds of ice crystals, similar to Earth's cirrus but made of methane, form at the edge of the Great Dark Spot as the high pressure forces methane-rich gas upward. But this Great Dark Spot had disappeared when the Hubble Space Telescope photographed Neptune a few years later, so it was much less long lived than Jupiter's Great Red Spot.

Several other cloud systems were also seen by Voyager 2. Clouds can also be viewed with the Hubble Space Telescope and can now even be followed with ground-based telescopes with active-optics systems (Fig. 7–39).

Neptune's Interior and Magnetic Field

Voyager 2 measured Neptune's average temperature: 59 K, that is, 59°C above absolute zero. This temperature, though low, is higher than would be expected on the basis of solar radiation alone. Neptune gives off about 2.7 times as much energy as it absorbs from the Sun. Thus there is an internal source of heating.

Why is the average density of Uranus and Neptune higher than that of Jupiter and Saturn? It may reflect slight differences in the origins of those planets.

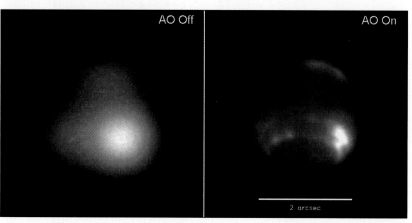

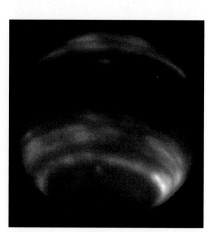

A

B

A Closer Look

Naming the Rings of Neptune

The rings of Neptune were named by a committee of the International Astronomical Union. The major rings are named after the two theoretical predictors and the actual observational discoverer of Neptune. The three subdivisions of the outer ring, at the suggestion of the French member of the committee, were named *Liberté*, *Egalité*, and *Fraternité* after the slogan of the French Revolution (1789). A fourth clump, named later, was called *Courage* (pronounced in the French way), in part because that is another worthwhile value that should be used to defend the other values. See Appendix 4 for details on the rings and clumps.

Their densities show that Uranus and Neptune have a higher percentage of heavy elements than Jupiter and Saturn. Perhaps the rocky cores they built up from the solar nebula were smaller, giving them less gravity and thus attracting less hydrogen and helium. Or, perhaps the positions of Uranus and Neptune farther out in the solar nebula put them in a region where either the solar nebula was less dense, or their slower orbital motion moved them through less gas.

Voyager detected radio bursts from Neptune every 16.11 hours. Thus Neptune's interior must rotate with this rate. On Neptune and Uranus, as on the Earth, equatorial winds blow more slowly than the interior rotates. By contrast, equatorial winds on Venus, Jupiter, Saturn, and the Sun blow more rapidly than the interior rotates. We now have quite a variety of planetary atmospheres to help us understand the basic causes of circulation.

Voyager discovered and measured Neptune's magnetic field. The field, as for Uranus, turned out to be both greatly tipped and offset from Neptune's center (Fig. 7–40). Thus the tentative explanation for Uranus that its field was tilted by a collision is not plausible; such a rare event would not be expected to happen twice. Astronomers favor the explanation that the magnetic field is formed in an electrically conducting shell outside the planets' cores. The fields of Earth and

Comparative planetology:

All the giant planets have atmospheres that are almost all hydrogen and helium. The small percentage of heavier elements is greater in Uranus and Neptune.

Figure 7–40 The magnetic fields for those planets with them. Only Mercury, whose magnetic field is only 1% that of Earth's, is not shown. The fields at the planets' equators are given in nanoteslas (nT). 10,000 nT = 1 gauss, the approximate strength of a toy magnet.

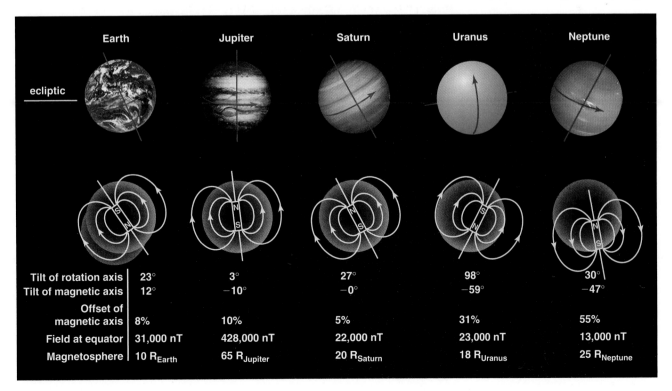

	Earth	Jupiter	Saturn	Uranus	Neptune
Tilt of rotation axis	23°	3°	27°	98°	30°
Tilt of magnetic axis	12°	−10°	−0°	−59°	−47°
Offset of magnetic axis	8%	10%	5%	31%	55%
Field at equator	31,000 nT	428,000 nT	22,000 nT	23,000 nT	13,000 nT
Magnetosphere	10 R_Earth	65 R_Jupiter	20 R_Saturn	18 R_Uranus	25 R_Neptune

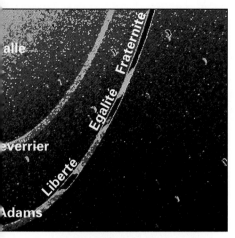

Figure 7–41 Three of Neptune's rings, showing the clumpy structure in one of them that led ground-based scientists to conclude from occultation data that Neptune had ring arcs. The Leverrier and Adams rings are 53,000 and 63,000 km, respectively, from Neptune's center. These photographs were taken as Voyager 2 left Neptune, 1.1 million km later, and so are backlighted. The main clumps are 6° to 8° long.

Jupiter, in contrast, are thought to be formed deep within the core. The tilted and offset magnetic fields are some of the biggest surprises found by Voyager 2.

Neptune's Rings

Before Voyager's arrival, astronomers wanted to know: Does Neptune have rings, like the other giant planets? There was no obvious reason why it shouldn't. Some observations from Earth detected dips at occultations of stars, similar to those produced by Uranus's rings, while others didn't. It was thought that perhaps Neptune had incomplete rings, "ring arcs."

As Voyager came close to Neptune, it radioed back images that showed conclusively that Neptune had narrow rings. Further, it showed the rings going all the way around Neptune. The material in the densest of Neptune's rings is very clumpy. The clumps had led to the incorrect idea of "ring arcs." The clumpy parts of the ring had blocked starlight, while the other parts and the other rings were too thin to do so (Figs. 7–41 and 7–42). The rings can now be studied from the Hubble Space Telescope.

The fact that Neptune's rings are so much brighter when seen backlighted tells us about the sizes of particles in them. The most detectable parts of the rings have at least a hundred times more dust-size grains than most of the rings of Uranus and Saturn. Since dust particles settle out of the rings, new sources must continually be active. Probably moonlets collide and are destroyed. Though much less dusty, Saturn's outer ring, a ring in one of Saturn's apparent ring gaps, and Uranus's rings are similarly narrow.

Neptune's Moon Triton

Neptune's largest moon, Triton, is a little larger than our Moon and has a retrograde orbit. It is named after a sea god who was a son of Poseidon. It is massive enough to have a melted interior. Its density is 2.07 grams/cm^3, so it is probably about 70 per cent rock and 30 per cent water-ice. It is denser than any jovian-planet satellite except Io and Europa.

Even before Voyager 2 visited, it was known that Triton has an atmosphere. Since Triton was Voyager 2's last objective among the planets and their moons, the spacecraft could be sent very close in. The scientists waited eagerly to see if Triton's atmosphere would be transparent enough to see the surface. It was. The atmosphere is mostly nitrogen gas, like Earth's.

Triton's surface is incredibly varied. Much of the region Voyager 2 imaged was near Triton's south polar cap (Fig. 7–43). The ice appeared slightly reddish. The color probably shows the presence of organic material formed by the action of solar ultraviolet light and particles from Neptune's magnetosphere hitting methane in Triton's atmosphere and surface. Nearer Triton's equator, nitrogen frost was seen.

Many craters and cliffs were seen. They could not survive if they were made of only methane-ice, so water-ice (which is stronger) must be the major component. Since Neptune's gravity captures many comets in that part of the Solar System, most of the craters are thought to result from collisions with comets.

Triton's surface showed about 50 dark streaks parallel to each other. They are apparently dark material vented from below. The material is spread out by winds. A leading model is that the Sun heats darkened methane ice on Triton's surface. This heating vaporizes the underlying nitrogen ice, which escapes through vents in the surface. A couple of these "ice volcanoes" were erupting when Voyager 2 went by. Since the streaks are on top of seasonal ice, they are all probably less than 100 years old.

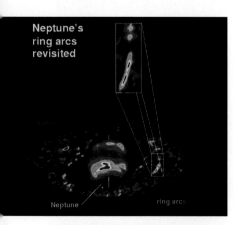

Figure 7–42 Neptune, its moon Galatea, and some of its ring arcs, imaged with the Hubble Space Telescope.

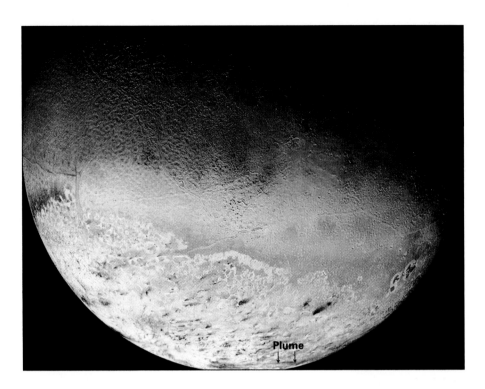

Plume

Figure 7–43 Triton's southern region, including its south polar ice cap. Most of the cap may be nitrogen ice. The pinkish color may come from the action of ultraviolet light and magnetospheric radiation upon methane in the atmosphere and surface. Most of the dark streaks represent material spread downwind from eruptions. One erupting plume is the long, thin, dark streak marked with arrows. Though the streaks are dark relative to other features on Triton, they are still ten times more reflective than the surface of our own Moon. The most detailed information on the photograph was taken in black and white through a clear filter; color information was added from lower-resolution images.

Much of Triton is so puckered that it is called the "cantaloupe terrain." It contains 30-km diameter depressions crisscrossed by ridges (Fig. 7–44).

Triton has obviously been very geologically active. Since it is in a retrograde (backward) orbit around Neptune, it was probably born elsewhere in the Solar System and later captured by Neptune. Tidal forces from Neptune would have kept Triton molten until its orbit became circular. While it was molten, the heavier rocky material would have settled to form a 2000-km-diameter core.

One of us (JMP) was fortunate to participate in a series of Earth-based expeditions to observe occultations of stars by Triton. Only rarely does Triton go in front of a star bright enough to study, but James Elliot, the discoverer of Uranus's rings, has organized a few such expeditions. In 1997, several collaborating expeditions of astronomers went to Australia and to Hawaii to try to get in Triton's shadow, as cast by the star. It turns out that on the second occasion he got the best observations with the Hubble Space Telescope, which happened to pass the right place at the right time. When Triton blocked out the star, which was about 6 times brighter, the brightness of the merged image dropped. From the way in which it dropped, he could analyze what Triton's atmosphere was like. The temperature turned out to be a few degrees warmer than it was when Voyager flew by, so our joint publication, in the journal *Nature*, was called "Global Warming on Triton."

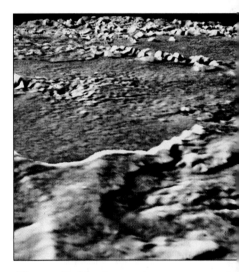

Figure 7–44 A computer-generated perspective view of one of Triton's caldera-like depressions.

🌑 EXTRA-SOLAR PLANETS

People have been looking for planets around other stars for decades. Many times in the last half century, astronomers have reported the discovery of a planet orbiting another star, but each of these reports has proven false. Finally, in the 1990s, the discovery of extra-solar planets (planets outside, "extra-," the Solar System) seems valid.

Planets around other stars would be so faint that they would be extremely difficult to see in the glare of the parent star with current technology. So the

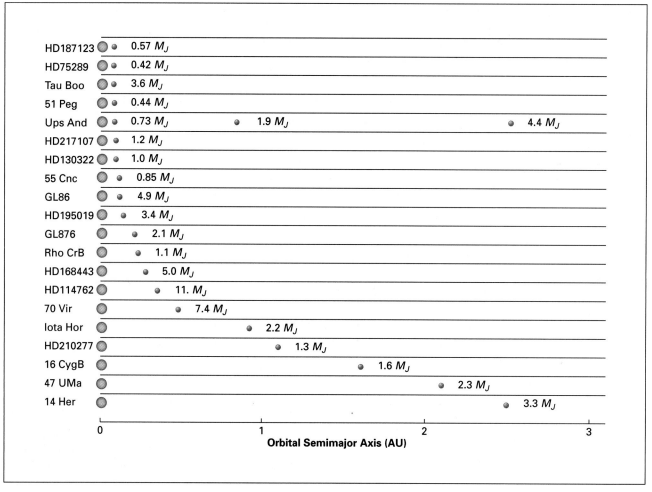

Figure 7–45 The characteristics of about 20 of the new planets discovered around nearby Sun-like stars.

search for planets has not concentrated on visible sightings of these planets. They have depended on watching for motions in the star that would have to be caused by something orbiting it.

The earlier reports, now rejected, were based on tracking the motion of the nearest stars across the sky with respect to other stars. If a star wobbled from side to side, it would reveal that a planet was wobbling invisibly the other way, so that the star/planet system was moving together in a straight line. (Technically, the "center of mass" of the system has to move in a straight line, unless its motion is distorted by some outside force.) Measurements have been made of star positions over the last hundred years or so, and a few of the nearby stars whose motions in the sky were followed seemed to show such wobbling. But the effects always turned out to be artifacts of the measuring process.

The first extra-solar planets were discovered around a pulsar, a weird kind of collapsed star (Chapter 13) that gives off extremely regular pulses of radio waves with a period that is a fraction of a second. The pulses came more frequently for a time and then less frequently in a regular pattern. So the planet around this pulsar must be first pulling the pulsar in our direction, making the pulses come more rapidly, and then pulling it in the opposite direction, making the pulses come more seldom. But this planet must have formed after the star collapsed; it could not have survived the explosion that preceded collapse. Thus it was not the kind of planet that is born at the same time as the star, like Earth. Even when the existence of a second planet orbiting that pulsar was established, the pulsar system seemed too unusual to think much about, except by specialists.

It didn't help that another pulsar planet, discovered slightly earlier, turned out to be a mistaken report.

In the 1990s, techniques were developed using the Doppler effect; in Chapter 2 we saw how the Doppler effect is a change in the wavelengths of light that has been emitted, caused by motion of the source or the receiver along the line of sight. The breakthrough came because new computer methods were found that measure the Doppler effect very precisely. On a computer, the star's spectrum can be simulated including changes in the wavelengths of light, just as happens in the Doppler effect. These simulated spectra can be compared to the observed spectrum of that star, until an excellent match is found. This method allows very small Doppler shifts to be detected, and the speeds of stars toward or away from us can be measured to a precision of 3 meters/sec, a fast walking speed.

The first surprising report came in 1995 from a Swiss astronomer and his student, Michel Mayor and Didier Queloz. They found a planet around a nearby star, 51 Pegasi. One strange thing about the planet is that it seemed to be a giant planet, at least half as massive as Jupiter, but with an orbit far inside what would be Mercury's orbit in our Solar System. The planet orbited 51 Peg in only 4.2 days, much faster than any planet in our Solar System.

Two American astronomers, Geoff Marcy and Paul Butler, then at San Francisco State University and the University of California at Berkeley, had been collecting similar data on dozens of other stars. But they had assumed, reasonably, that a planet like Jupiter around another star would take years to orbit, so they were collecting years of data while perfecting their analysis techniques to measure exceedingly small speeds. They hadn't run their spectra through the computer programs they were writing to measure the Doppler shifts. When they heard of the Mayor and Queloz results, they quickly examined their existing data and also observed 51 Peg. They soon verified the planet around 51 Peg and discovered planets around several other stars (Fig. 7-45). Almost all the approximately three dozen new planets, most of which were discovered by Marcy and Butler, turned out to be giant planets, either in extremely elliptical orbits or in circular orbits very close to the parent stars. **www** Most of these planets are orbiting stars within 50 light-years from us, not extremely far away but not the very closest few dozen stars. Stars this close are bright enough to carry out the extremely sensitive spectroscopic measurements.

One limitation of the method is that we generally don't know the angle of the plane in which the planets are orbiting the stars. The Doppler-shift method works only for the part of the star's velocity that is toward or away from us, and not for the part that is from side to side. So the planets we discover can be more massive than our measurements suggest; we are able to find only a minimum value to their masses. This problem left the nagging question whether the objects were really planets or whether they were merely low-mass companion stars or objects called "brown dwarfs," which didn't have quite enough mass to make it to star status. Nevertheless, most astronomers believed these objects are planets, because there is a large gap in mass between them and low-mass stars. Very few intermediate-mass objects have been found, yet they should have been easily detected if they existed. This gap suggested that the new objects are much more numerous than brown dwarfs.

The discovery, in 1999, of a system of three planets around the star Upsilon Andromedae clinched the case for planets (Fig. 7-46). It seems most unlikely that a system would have formed with four stars (or brown dwarfs) in it, while a system with one star and three planets is reasonable. One of the planets even has an orbit that would correspond to Venus's in our Solar System, not as elliptical as the orbits of the planets around other stars. This planet is even in a zone that

Comparative planetology:

Our Solar System: Massive planets in almost circular orbits beyond 5 A.U. and periods of many years.

Other Planetary Systems: Massive planets often in very elliptical orbita often much smaller than 1 A.U. and with periods of months.

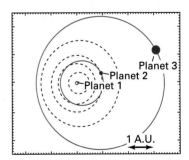

Figure 7–46 A diagram of the orbits of the 3 planets (*red dots*) detected around the star Upsilon Andromedae. The dashed circles show the orbits of Mercury, Venus, Earth, and Mars overplotted to give the scale of the orbits. Jupiter, at 5 A.U., would be outside the boundaries of this plot.

may be not too hot for life nor too cold for it. Though such a massive planet would be gaseous, and so not have a surface for life to live on, it could have a moon with a solid surface, just as Jupiter and the other giant planets in our Solar System have moons.

No telescope now in space has enough resolution to detect a planet orbiting another star. It will take an interferometry system, with two widely separated telescopes working together, to make such a detection. NASA's Space Interferometry Mission is on the drawing board, but it will be close to a decade before it is completed. Though we now detect only giant planets, we presume that small planets exist in those systems as well. It will take even longer to detect them; the current technology is inadequate. NASA's planned Terrestrial Planet Finder, with optical telescopes flying in space kilometers apart but used together, may be able to image small planets in the second decade of the century.

Other methods are also being pursued to find planets. For example, many stars are being monitored to see if their brightness changes slightly as a planet—perhaps terrestrial sized—passes in front of it. In late 1999, one large planet was indeed detected this way, although it was first discovered with the Doppler technique. Since the object's orbital plane is along our line of sight, its mass can be accurately determined, and this turns out to be 63 per cent of Jupiter's mass.

Also, stars are being monitored to see if their gravity focuses and brightens the light from other stars behind them, a process called "gravitational lensing" that we will discuss in Chapters 16 and 17. The hope is that not only a background star but then, slightly before or after the star passes, a planet will cause a brightness blip.

The discovery of several planetary systems instead of just our own will obviously change the models for how they are formed (Chapter 5). Many theorists are still sticking to their ideas of how giant planets form far away from their parent stars, just as Jupiter and the other giant planets are far out in our Solar System. Thus it follows that these planets must be jostled loose from their orbits and put in orbits that bring them closer to their parent stars. Perhaps their orbits shrink as the planets encounter debris in the dusty disk from which they formed. Similarly, the highly elliptical orbits probably didn't start out that way, but were produced by gravitational interactions that completely ejected some planets from the system. Whether these ideas survive, as new planetary systems are discovered, remains to be seen.

Concept Review

Jupiter (318 times more massive than Earth), Saturn (95 times more massive), Uranus (15 times more massive), and Neptune (17 times more massive) are **giant planets,** also known as **jovian planets.** They have extensive gaseous atmospheres, liquid interiors, and rocky cores. Each of them has many moons and at least one ring. All were visited by at least one Voyager spacecraft.

Jupiter's **Great Red Spot,** a giant circulating storm, is a few times larger than Earth. Jupiter's atmosphere shows bright and dark bands of rising and falling gas. Jupiter has an internal source of energy, perhaps the energy remaining from its collapse. Its magnetic field is very strong.

Jupiter's **Galilean satellites** are comparable in size to our Moon. They were studied in detail from the Voyagers and the Galileo spacecraft. Io has over 100 volcanoes, many of which are erupting at any given time. Europa is covered with smooth ice, perhaps over an ocean. Ganymede and Callisto are heavily cratered.

Saturn has a very low density, lower even than water. Saturn's rings were produced by tidal forces that prevented the formation of a moon, and are within Saturn's **Roche limit.** The rings are so thin that stars can be seen through them. **Cassini's division** is the major break between inner and outer rings. Saturn's atmosphere shows fewer features than Jupiter's, perhaps because it is colder. It has an internal source of heat, perhaps both energy from its collapse and energy from sinking helium. Saturn's moon Titan has a thick, smoggy atmosphere. The Cassini spacecraft, in 2004, will drop a probe down to its surface.

Uranus is covered with clouds. Its core makes up a substantial part of the planet, and there is no internal energy source. Uranus rotates on its side, so is heated by the Sun in a strange way over an 84-year period of revolution. Methane is a major constituent of Uranus's atmosphere and accounts for its greenish color. Uranus has several thin rings, discovered by an occultation of a star observed from Earth.

Neptune is also covered by clouds and has a substantial core. Neptune has an internal energy source. It, too, looks greenish because of its atmospheric methane. When Voyager 2 flew by in 1989, its surface showed a **Great Dark Spot,** a high-pressure region analogous to Jupiter's Great Red Spot. It disappeared between the Voyager visit and a later Hubble Space Telescope view. Neptune's magnetic field, like Uranus's, is greatly tipped and offset from the planet's center. These magnetic fields are thus probably formed in electrically conducting shells of gas. Neptune's rings are clumpy.

Neptune's moon Triton is also very large. Voyager 2 took high-resolution images that showed a varied surface with plumes from ice volcanoes. Triton has been very geologically active. Occultation studies show that it is gradually warming.

Planets around stars other than the Sun have been discovered since 1995 using the Doppler shift. All the ones discovered so far are giant planets, though terrestrial planets probably remain undiscovered in those planetary systems. The planets are often in elongated orbits, perhaps as a result of previous gravitational encounters with other planets. Some of the planets are in circular orbits very close to their parent stars, which may mean that they were formed elsewhere in the systems and later drifted inward. It remains to be seen whether our Solar System is typical of others.

Questions

1. Why does Jupiter appear brighter than Mars despite its greater distance from the Earth?
2. Even though Jupiter's atmosphere is very active, the Great Red Spot has persisted for a long time. How is this possible?
3. What advantages over the 5-m Palomar telescope on Earth did Voyagers 1 and 2 have for making images of the outer planets?
4. Other than photography, what are two types of observations made from the Voyagers?
5. How does the interior of Jupiter differ from the interior of the Earth?
†6. Given that Jupiter's diameter is 11.2 times that of Earth, about how many Earths could fit inside of Jupiter?
7. Why did we expect Io to show considerable volcanism?
8. Contrast the volcanoes of Io with those of Earth.
9. Compare the surfaces of Callisto, Io, and the Earth's Moon. Explain what this comparison tells us about the ages of features on their surfaces.
10. What are the similarities between Jupiter and Saturn?
11. What is the Roche limit and how does it apply to Saturn's rings?
12. How did we know, even before Voyager, that Saturn's rings are very thin.
13. What did the Voyagers reveal about Cassini's division?
14. How are narrow rings thought to be kept from spreading out?
15. What is strange about the direction of rotation of Uranus?
16. Which of the giant planets are known to have internal heat sources?
†17. What fraction of its orbit has Neptune traversed since it was discovered? Since it was first seen?
18. Compare Neptune's Great Dark Spot with Jupiter's Great Red Spot and with the size of the Earth.
19. Compare the rings of Jupiter, Saturn, Uranus, and Neptune.
20. Describe the major developments in our understanding of Jupiter, from ground-based observations to the Voyagers to Galileo.
21. Describe the major developments in our understanding of Saturn, from ground-based observations to the Voyagers, including what is expected from Cassini.
22. Describe the major developments in our understanding of Uranus, from ground-based observations to the Voyagers.
23. Describe the major developments in our understanding of Neptune, from ground-based observations to the Voyagers.
24. Compare Triton in size and surface with other major moons, like Ganymede and Titan.
25. Explain how planets have been discovered around other stars.
26. Why may it be that only giant planets and not terrestrial planets have been discovered so far around other stars?

†This question requires a numerical solution.

Pluto, Comets, and Space Debris

ORIGINS *We see how objects in the outer part of our Solar System, including comets, are left over from the origin of our Solar System and so can be studied to learn about conditions in our Solar System's early years. We find that asteroids, meteoroids, and comets may lead to our demise, just as they produced episodes of major extinction in the past when they collided with Earth.*

We have learned about our Solar System's giant planets, which range in size from about 4 to about 11 times the diameter of the Earth. We have seen that our Solar System has a set of terrestrial planets, which range in size from the Earth down to 40 per cent the diameter of the Earth. This size range includes four planets as well as seven planetary satellites. The remaining object that has long had the name "planet," Pluto, is only 20 per cent the diameter of Earth but is still over 2300 km across, so there is much room on it for interesting surface features. Recently, additional objects like it, but smaller, have been found in the outer reaches of the Solar System. We shall see how we determined Pluto's odd properties, and what the other, similar objects are.

Besides the planets and their moons, many other objects are in the family of the Sun. The most spectacular, as seen from Earth, are comets (Fig. 8–1). Bright comets have been noted throughout history, instilling great awe of the heavens. Comets have long been seen as omens. As Shakespeare wrote in *Julius Caesar,* "When beggars die, there are no comets seen; The heavens themselves blaze forth the death of princes."

Asteroids, which are minor planets, and chunks of rock known as meteoroids are other residents of our Solar System. We shall see how they and the comets are storehouses of information about the Solar System's origin.

Asteroids, meteoroids, and comets are suddenly in the news as astronomers are finding out that some come relatively close to the Earth. We are realizing more and more that collisions of these objects with the Earth can be devastating. Every few tens of millions of years, one large enough to do very serious damage should hit. Perhaps a comet or an asteroid caused the dinosaurs to become extinct some 65 million years ago. Should we be worrying about asteroid, meteoroid, or comet collisions? Should we be monitoring the sky around us better? Should we be planning ways of diverting an oncoming object if we were to find one?

AIMS

To learn about the space debris that makes up the outer part of our Solar System, including Kuiper belt objects (which may or may not include Pluto, historically called a planet), and comets.

•

To discuss some other minor constituents of our Solar System, such as asteroids and meteoroids.

Figure 8–1 Comet Hale-Bopp was so bright it was even visible over Central Park in the middle of New York City.

◀ Comet Hale-Bopp, very bright in the sky in 1997.

REDSHIFT
http://www.harcourtcollege.com/astro/cosmos/rsce

Figure 8-2 Small sections of the photographic plates on which Tombaugh discovered Pluto. On February 18, 1930, Tombaugh noticed that one dot among many had moved between January 23, 1930 *(left)*, and January 29, 1930 *(right).* He didn't have the arrows on the plates, of course!

⬤ PLUTO

Pluto, the outermost known planet, is a deviant. Its orbit is the most out of round (eccentric) and is inclined by the greatest angle with respect to the ecliptic plane, near which the other planets revolve.

Pluto's elliptical orbit is so eccentric that part lies inside the orbit of Neptune. Pluto was closest to the Sun in 1989 and moved farther away from the Sun than Neptune in 1999. So Pluto is still relatively near its closest approach to the Sun out of its 248-year period, and it appears about as bright as it ever does to viewers on Earth. It hasn't been as bright—about magnitude 13.5—for over 200 years. It is barely visible through a medium-sized telescope under dark-sky conditions.

The discovery of Pluto was the result of a long search for an additional planet that, together with Neptune, was believed to be slightly distorting the orbit of Uranus. ☼ Finally, in 1930, Clyde Tombaugh, hired at age 23 to search for a new planet because of his experience as an amateur astronomer, found the dot of light that is Pluto (Fig. 8–2). It took him a year of diligent study of the photographic plates he obtained at the Lowell Observatory in Arizona. From its slow motion with respect to the stars from night to night, he identified Pluto as a new planet.

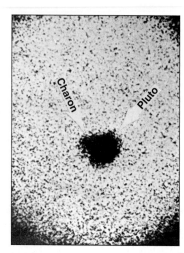

Figure 8-3 The image of Pluto and Charon on which Charon was discovered.

Pluto's Mass and Size

Even such basics as the mass and diameter of Pluto are very difficult to determine. It has been hard to deduce the mass of Pluto because to do so requires measuring Pluto's effect on Uranus, a far more massive body. (The orbit of Neptune is too poorly known to be of much use.) Moreover, Pluto has made less than one revolution around the Sun since its discovery, thus providing little of its path for detailed study. As recently as 1968, it was mistakenly concluded that Pluto had 91 per cent the mass of the Earth.

The situation changed drastically in 1978 with the surprise discovery (Fig. 8–3) that Pluto has a satellite. The moon was named Charon, after the boatman who rowed passengers across the River Styx to the realm of Pluto, god of the underworld in Greek mythology. (Its name is informally pronounced "Shar'on," similarly to the name of the discoverer's wife, Charlene, by astronomers working in the field.) The presence of a satellite allows us to deduce the mass of the planet by applying Newton's form of Kepler's third law (Chapter 5). Charon is 5 or 10 per cent of Pluto's mass, and Pluto is only $\frac{1}{500}$ the mass of the Earth, ten times less than had been suspected just before the discovery of Charon.

Pluto's rotation axis is nearly in the ecliptic, like that of Uranus. This is also the axis about which Charon orbits Pluto every 6.4 days. Consequently, there are two five-year intervals during Pluto's 248-year orbit when the two objects pass in front of (that is, occult) each other every 3.2 days, as seen from Earth. Such mutual occultations were the case during 1985 through 1990. When we measured their apparent brightness, we received light from both Pluto and Charon together. Their blocking each other led to dips in the total brightness we received. From the duration of fading, we deduced how large they are. Pluto is 2300 km in diameter, smaller than expected, and Charon is 1200 km in diameter. Charon is thus half the size of Pluto. Further, it is separated from Pluto by only about 8 Pluto diameters, compared with the 30 Earth diameters that separate the Earth and the Moon. So Pluto/Charon is almost a "double-planet" system.

The rate at which the light from Pluto/Charon faded also gave us information that revealed the reflectivities (albedoes) of their surfaces, since part of the surface of the farther object remained visible most of the time. The surfaces of both vary in brightness (Fig. 8–4). Pluto seems to have a dark band near its equator, some markings on that band, and bright polar caps.

In 1990, the Hubble Space Telescope took an image that showed Pluto and Charon as distinct objects for the first time (Fig. 8–5B), and they can now be viewed as separate by new telescopes on Mauna Kea (Fig. 8–5A). (www) The latest Hubble views show that Pluto has a dozen areas of bright and dark, the finest map ever made of Pluto (Fig. 8–6). But we don't know whether the bright areas are bright because they are high clouds near mountains or low haze and frost. We merely know that there are extreme contrasts on Pluto's surface.

If we were standing on Pluto, the Sun would appear over a thousand times fainter than it does to us on Earth. Consequently, Pluto is very cold; infrared measurements show that its temperature is less than 60 K. From Pluto, we would need a telescope to see the solar disk, which would be about the same size that Jupiter appears from Earth. NASA is considering a Pluto–Kuiper Express spacecraft, to give close-up images. (www) But the mission hasn't yet been approved, and the spacecraft wouldn't arrive at Pluto before 2010.

Pluto's Atmosphere

Pluto occulted—passed in front of and hid—a star on one night in 1988. Astronomers observed this occultation to learn about Pluto's atmosphere. If Pluto had no atmosphere, the starlight would wink out abruptly. Any atmosphere would make the starlight diminish more gradually. The observations showed that the

Figure 8–4 An image drawn to explain the observed variation of the total light as Pluto and Charon hide (occult) each other. The relative sizes, colors, and albedoes are approximately correct.

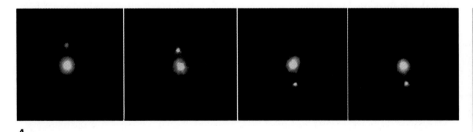

A

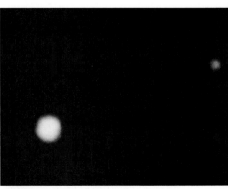

B

Figure 8–5 *(A)* This series of ground-based views of Pluto and Charon, showing them orbiting around each other, was taken in 1999 at the highest possible resolution at near-infrared wavelengths, with active optics (mirror-shape distortion in real time, to compensate for atmospheric turbulence) using the new Gemini North telescope on Mauna Kea. *(B)* Pluto and Charon, clearly distinguished with the Hubble Space Telescope.

Figure 8-6 Hubble Space Telescope views of Pluto, at the highest resolution ever obtained on this object. The Hubble image of Pluto is like standing in Los Angeles and looking at the spots on a soccer ball in San Francisco.

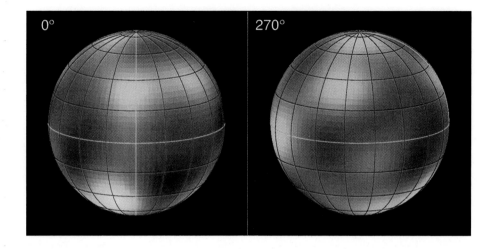

starlight diminished gradually and unevenly (Fig. 8–7). Thus Pluto's atmosphere has layers in it. Astronomers were also able to conclude that the bulk of Pluto's atmosphere is nitrogen. A trace of methane must also be present, since the methane ice on Pluto's surface, detected from its spectrum, must be evaporating. Still, Pluto's atmospheric pressure is very low, only $\frac{1}{100,000}$ of Earth's.

As Pluto goes farther from the Sun, as it is now doing, the atmosphere is predicted to freeze out and snow onto the surface. Though new calculations indicate that this might not be so, it is still possible that if we want to find out about the atmosphere, we had better get a spacecraft there within a decade or two, or we'll have to wait another 200 years for the atmosphere to form again.

What Is Pluto?

From Pluto's mass and radius, we calculate Pluto's density. It turns out to be about 2 g/cm^3, twice the density of water and less than half the density of Earth. Since ices have even lower densities than Pluto, Pluto must be made of a mixture of ices and rock. Its composition is more similar to that of the satellites of the giant planets, especially Neptune's large moon Triton, than to that of Earth or the other inner planets.

Ironically, now that we know Pluto's mass, we calculate that it is far too small to cause the deviations in Uranus's orbit that originally led to Pluto's discovery. The discrepancy probably wasn't real: The wrong mass had been assumed for Neptune when predicting the orbit of Uranus. The discovery of Pluto was purely the reward of hard work in conducting a thorough search in a zone of the sky near the ecliptic.

Pluto, with its moon and its atmosphere, has some similarities to the more familiar planets. Pluto remains strange in that it is so small next to the giants, and that its orbit is so eccentric and so highly inclined to the ecliptic. The new values of Pluto's mass and density revived the thinking that Pluto may be a for-

Figure 8-7 A series showing the merged images as Pluto occulted a star in 1988.

mer moon of one of the giant planets, probably Neptune, and escaped because of a gravitational encounter with another planet. But other lines of evidence, such as the existence of Charon close by, indicate that Pluto was never a moon.

Increasingly, Pluto is being identified with a newly discovered set of objects in the outer Solar System, which we will now study.

Kuiper-Belt Objects

Beyond the orbit of Neptune, a population of icy objects with diameters of a few tens or hundreds of kilometers is increasingly being found. The planetary astronomer Gerard Kuiper (pronounced koy′per) suggested a few decades ago that these objects would exist and should be the source of many of the comets that we see. As a result, these objects are now known as the **Kuiper-belt objects,** or **Trans-Neptunian Objects.**

The Kuiper belt is probably about 10 A.U. thick and extends from the orbit of Neptune about twice as far (Fig. 8–8). A few dozen objects have been found so far, but tens of thousands larger than 100 km across are thought to exist. The objects may be planetesimals left over from the formation of the Solar System. They are very dark, with albedos of only about 4 per cent. Pluto, by contrast, has an albedo of about 60 per cent. Still, Pluto could be the largest of the Kuiper-belt objects, so much larger than the others that it alone is covered with frost. Triton may have initially been a similar object, subsequently captured by Neptune.

David Jewitt of the University of Hawaii and Jane Luu of the University of Leiden have been the discoverers of most of the known Kuiper-belt objects. They found the first one in 1992 and continue to look for more.

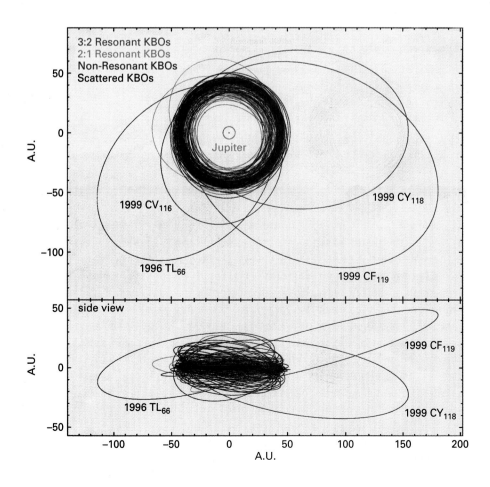

Figure 8-8 Orbits of Jupiter (*green*) and the Kuiper-belt objects.

A few objects may once have been Kuiper-belt objects but now come somewhat closer to the Sun, crossing the orbits of the outer planets. About 100 of these "centaur" objects a few hundred kilometers across may exist. Since they are larger and come closer to the Earth and Sun than most Kuiper-belt objects, we can study them better. On at least one, a coma (typical of comets, as we will soon see) was seen, so these centaurs are intermediate between comets and asteroids.

Is Pluto a Planet?

When Pluto was discovered, it was announced as the ninth planet. Its discovery resulted from a specific search for a ninth planet, though we have found that this did not validly come from the predictions. Since 1930, Pluto has had the status of planet. But now that we know that Pluto has only 1/500 the mass of Earth, some scientists think that Pluto should be demoted. Demoted to what? That is the question.

The discovery of a set of Kuiper-belt objects with orbits not very different from that of Pluto gives a possible home for categorizing Pluto. Pluto may be merely the largest and among the closest of the Kuiper-belt objects. But, given many decades of having nine planets, are we really prepared to go back to eight, especially since the case for demoting Pluto isn't too strong? If you remember the names of the planets as the initial letters of "My Very Educated Mother Just Sent Us Nine Pizzas," are you prepared to change to "My Very Educated Mother Just Sent Us Nothing"?

Word got around in 1999 that Pluto might get numbered in other systems of astronomical nomenclature. For example, as a Trans-Neptunian Object, it might be made TNO-1, or even, to single it out, TNO-0. Also, the asteroid numbers were at that time getting up close to 10,000, and Pluto could have been given that special, round number in the asteroid numbering scheme.

These leaks disturbed so many people that the official committee on nomenclature of the International Astronomical Union issued a statement in 1999 that Pluto would not be demoted. ⓦⓦⓦ A committee of the Division of Planetary Sciences of the American Astronomical Society also made a statement that Pluto would remain designated as a planet.

Given the history, the authors of this book are in favor of retaining Pluto's planethood. But if new searches turn up a couple of Kuiper-belt objects larger than Pluto, Pluto's status as a full-fledged planet will be harder to maintain.

◖◗ COMETS

Nearly every decade, a bright **comet** appears in our sky. From a small, bright area called the **head**, a **tail** may extend gracefully over one-sixth (30°) or more of the sky. The tail of a comet is always directed roughly away from the Sun, even when the comet is moving outward through the Solar System.

Although the tail may give an impression of motion because it extends out only to one side, the comet does not move noticeably with respect to the stars as we watch. With binoculars or a telescope, however, an observer can accurately note the position of the comet's head and after a few hours can detect that the comet is moving at a slightly different rate from the stars. Still, both comets and stars rise and set more or less together (Fig. 8–9). Within days, weeks, or (even less often) months, a bright comet will have become too faint to be seen with the naked eye, although it can often be followed for additional months with binoculars and then for additional months with telescopes.

Most comets are much fainter than the one we have just described. About a dozen new comets are discovered each year, and most become known only to as-

Figure 8–9 Comet Hyakutake as it moved with the stars; note the star trails.

tronomers. If you should ever discover a comet, and are among the first three people to report it to the International Astronomical Union Central Bureau for Astronomical Telegrams, at the Smithsonian Astrophysical Observatory in Cambridge, Massachusetts, it will be named after you. (www)

The Composition of Comets

At the center of a comet's head is its **nucleus,** which is composed of chunks of matter. The most widely accepted theory of the composition of comets, advanced in 1950 by Fred L. Whipple of the Harvard and Smithsonian Observatories, is that the nucleus is like a dirty snowball. It may be made of ices of such molecules as water (H_2O), carbon dioxide (CO_2), ammonia (NH_3), and methane (CH_4), with dust mixed in.

The nucleus itself is so small that we cannot observe it directly from Earth. Radar observations have verified in several cases that it is a few kilometers across. The rest of the head is the **coma** (pronounced coh′ma), which may grow to be as large as 100,000 km or so across (Fig. 8–10). The coma shines partly because its gas and dust are reflecting sunlight toward us and partly because gases liberated from the nucleus get enough energy from sunlight to radiate. The tail can extend 1 A.U. (150,000,000 km), so comets can be the largest objects in the Solar System. But the amount of matter in the tail is very small—the tail is a much better vacuum than we can make in laboratories on Earth.

Many comets actually have two tails (Fig. 8–11). The **dust tail** is caused by dust particles released from the ices of the nucleus when they are vaporized. The dust particles are left behind in the comet's orbit, blown slightly away from the Sun by the pressure of sunlight hitting the particles. As a result of the comet's orbital motion, the dust tail usually curves smoothly behind the comet. The **gas tail** is composed of gas blown out with high speed, more or less straight behind the comet by the "solar wind" of particles expanding outward from the Sun. As puffs of gas are blown out and as the solar wind varies, the gas tail takes on a structured appearance. Each puff of matter can be seen.

A comet—head and tail together—contains less than a billionth of the mass of the Earth. It has been said that comets are as close as something can come to being nothing.

Figure 8–10 The coma and inner tail of Halley's Comet in 1986.

Gas tail

Dust tail

Figure 8–11 Comet Hale-Bopp, technically C/1995 O1, was the brightest comet in decades at its peak in March and April 1997. Its 40-km-across nucleus gave off large quantities of gas and dust that made its long tails. The broad, yellowish, dust tail curved behind the comet's head. We see it by the sunlight that it reflects toward us. The much fainter, blue tail is made of ions (an ion is an atom that has lost an electron) given off by the comet's nucleus and is blown in the direction opposite to the Sun by the solar wind. We detect this "gas tail" by its own emission, though it was much less visible to the eye than it was to the camera. Comet Hale-Bopp was so bright that scientists discovered a third type of tail: a faint one from neutral sodium, not visible in this image.

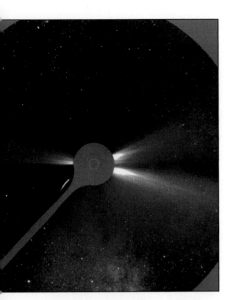

Figure 8–12 The outer solar corona extends outward near the solar equator in this view from the Solar and Heliospheric Observatory (SOHO) spacecraft. A comet plunging into the Sun is also seen. A disk that blocks the bright everyday Sun and the finger that holds it in place appear in silhouette; the size of the ordinary solar disk is drawn on. In the background, we see stars and the Milky Way.

The Origin and Evolution of Comets

It is now generally accepted that trillions of tail-less comets surround the Solar System in a sphere perhaps 50,000 A.U. (almost 1 light-year) in radius. This sphere, far outside Pluto's orbit, is the **Oort comet cloud** (named after the Dutch scientist Jan Oort). The total mass of matter in the cloud may be only 1 to 10 times the mass of the Earth. Occasionally one of these comets leaves the comet cloud, perhaps because the gravity of a nearby star has tugged it out of place. Generally, the comet gets directly ejected from the Solar System, but occasionally, the comet can approach the Sun. The comet's orbit may be altered, sometimes into an elliptical orbit, if it passes near a giant planet, most frequently Jupiter. Because the comet cloud is spherical, comets are not limited to the plane of the ecliptic, which explains why the longer-period comets come in randomly from all directions.

The short-period comets, those with periods of 200 years or less, are in orbits much more limited to the plane of the Solar System. They seem to come from what is now identified as the Kuiper belt, a flatter distribution of objects beyond the orbit of Neptune. Comets on highly eccentric orbits spend most of their time far away from the Sun, an excellent example of Kepler's second law (Chapter 5).

As a comet gets closer to the Sun than those distant regions, the solar radiation begins to vaporize the ice in the nucleus. The tail forms, and grows longer as more of the nucleus is vaporized. Even though the tail can be millions of kilometers long, it is still so tenuous that only 1/500 of the mass of the nucleus may be lost each time it visits the solar neighborhood. Thus a comet may last for many passages around the Sun before dwindling to nothing. But some comets may hit the Sun and be destroyed (Fig. 8–12).

We shall see in the following section that meteoroids can be left in the orbit of a disintegrated comet. Some of the asteroids, particularly those that cross the Earth's orbit, may be dead comet nuclei. In recent years, a handful of asteroids have shown comas or tails, making them comets; conversely, a few comets have died out and seem like asteroids. So we may have misidentified some of each in the past.

Because new comets come from the places in the Solar System that are farthest from the Sun and thus coldest, they probably contain matter that is unchanged since the formation of the Solar System. So the study of comets is important for understanding the birth of the Solar System. Moreover, some astronomers think that early in Earth's history, the oceans formed when an onslaught of water-bearing comets collided with Earth.

Halley's Comet

In 1705, the English astronomer Edmond Halley (Halley is pronounced to rhyme with "Sally," and not with "say'lee") (Fig. 8–13) applied a new method developed by his friend Isaac Newton to determine the orbits of comets from observations of their positions in the sky. He reported that the orbits of the bright comets that had appeared in 1531, 1604, and 1682 were about the same. Moreover, the intervals between appearances were approximately equal, so Halley suggested that we were observing a single comet orbiting the Sun. He predicted that it would again return in 1758. The reappearance of this bright comet on Christmas night of that year, 16 years after Halley's death, was the proof of Halley's hypothesis (and Newton's method). The comet has thereafter been known as Halley's Comet (Fig. 8–14). Since it was the first periodic comet known, it is officially called 1P, number 1 in the list of periodic (P) comets.

Figure 8–13 Edmond Halley.

Figure 8–14 Halley's Comet following its 1986 closest approach to the Sun. Both the dust tail and the gas tail show. The color image was composed from blue-, green-, and red-sensitive photographs taken at slightly different times, so the comet's motion makes each star appear three different colors.

It seems probable that the bright comets reported every 74 to 79 years since 240 B.C. were earlier appearances of Halley's Comet. The fact that it has been observed dozens of times endorses the calculations that show that less than 1 per cent of a cometary nucleus's mass is lost at each passage near the Sun.

Halley's Comet came especially close to the Earth during its 1910 return, and the Earth actually passed through its tail. Many people had been frightened that the tail would somehow damage the Earth or its atmosphere, but the tail had no noticeable effect. Even then, most scientists knew that the gas and dust in the tail were too tenuous to harm our environment.

The most recent close approach of Halley's Comet was in 1986. It was not as spectacular from the ground in 1986 as it was in 1910, for this time the Earth and comet were on opposite sides of the Sun when the comet was brightest. Since we knew long in advance that the comet would be available for viewing, special observations were planned for optical, infrared, and radio telescopes. For example, spectroscopy showed many previously undetected ions in the coma and tail.

When Halley's Comet passed through the plane of the Earth's orbit, it was met by an armada of spacecraft. The best was the European Space Agency's spacecraft Giotto (named after the 14th-century Italian artist who included Halley's Comet in a painting), which went right up close to Halley. Giotto's several instruments also studied Halley's gas, dust, and magnetic field from as close as 600 km from the nucleus. The most astounding observations were undoubtedly the photographs showing the nucleus itself (Fig. 8–15).

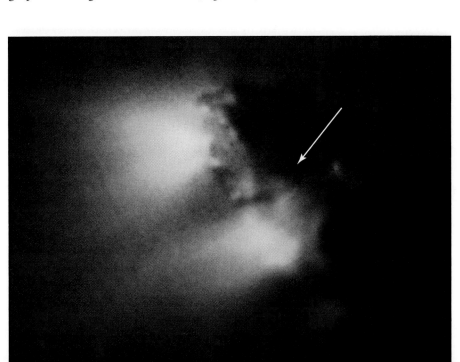

Figure 8–15 The nucleus of Halley's Comet from the Giotto spacecraft. The dark, potato-shaped nucleus is visible in silhouette. It is about 16 km by 8 km; the frame is 30 km across. Two bright jets of dust reflecting sunlight are visible to the left of Halley's nucleus. The small bright region in the center of the nucleus is probably a raised region of its surface that is struck by sunlight.

Figure 8-16 This potato looks remarkably like the nucleus of Halley's Comet, except is 100,000 times shorter.

The nucleus turns out to be potato-shaped (Fig. 8–16). It is about 16 km in its longest dimension, half the size of Manhattan Island. The "dirty snowball" theory of comets was confirmed in general, but the snowball is darker than expected. It is as black as velvet, with an albedo of only about 3 per cent. Further, the evaporating gas and dust is localized into jets that are stronger than expected. They come out of fissures in the dark crust. We now realize that comets may shut off not when they have given off all their material but when the fissures in their crusts close.

Giotto carried 10 instruments in addition to its camera. Among them were mass spectrometers to measure the types of particles present, detectors for dust, equipment to listen for radio signals that revealed the densities of gas and dust in the coma, detectors for ions, and a magnetometer to measure the magnetic field.

About 30 per cent of Halley's dust particles are made only of hydrogen, carbon, nitrogen, and oxygen (Fig. 8–17). This simple composition resembles that of the oldest type of meteorite. It thus indicates that these particles may be from the earliest years of the Solar System.

Many valuable observations were also obtained from the Earth. For example, radio telescopes were used to study molecules. Water vapor is the most prevalent gas, but carbon monoxide and carbon dioxide were also detected. The comet was bright enough that many telescopes obtained spectra (Fig. 8–18).

The next appearance of Halley's Comet, in 2061, again won't be spectacular. Not until the one after that, in 2134, will the comet show a long tail to Earthbound observers. Fortunately, though Halley's Comet is predictably interesting, a more spectacular comet appears every 10 years or so. When you read in the newspaper that a bright comet is here, don't wait to see it another time. Some bright comets are at their best for only a few days or a week. ☼

Comet Hale-Bopp

In 1995, Alan Hale and Thomas Bopp independently found a faint comet, which was soon discovered to be quite far out in the Solar System. Its orbit was to bring it into the inner Solar System, and it was already bright enough that it was likely

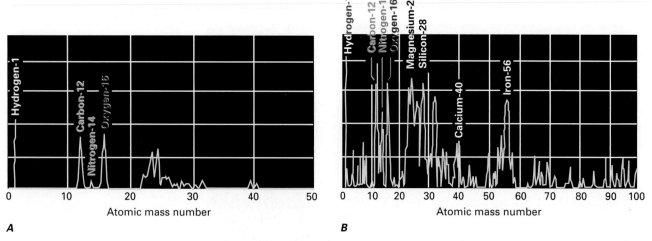

Figure 8-17 The mass spectrometer aboard the Vega spacecraft showed the number of nuclear particles (the atomic mass number) of each element in dust particles. *(A)* A CHON dust particle, composed only of carbon, hydrogen, oxygen, and nitrogen. *(B)* Other dust particles contain other elements as well.

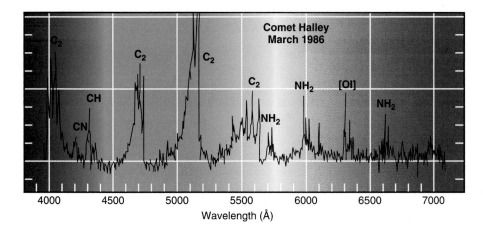

Figure 8–18 The visible spectrum of Halley's Comet on March 14, 1986, when it was at its brightest. We see several spectral lines and bands from molecules. Their relative strengths are approximately the same as they are in other comets, except for NH_2, which was slightly stronger in Halley.

to be spectacular when it came close to Earth in 1997. It lived up to its advance billing (Fig. 8–19).

Telescopes of all kinds were trained on Comet Hale-Bopp, and hundreds of millions of people were thrilled to step outside at night and see a comet just by looking up. Modern powerful radio telescopes were able to detect many kinds of molecules that had not before been recorded in a comet. The Hubble Space Telescope took high-resolution images of its head (Fig. 8–20).

Comet [Shoemaker-]Shoemaker-Levy 9

An earlier comet had also given thrills to people around the world. In 1993, Eugene Shoemaker, Carolyn Shoemaker, and David Levy had discovered their ninth comet in a search with a wide-field telescope at the Palomar Observatory. But the comet looked weird—it seemed squashed. Higher-resolution images taken with other telescopes, including the Hubble Space Telescope (Fig. 8–21), showed that the comet had broken into bits, forming a chain that resembled beads on a string. Even stranger, the comet was in orbit not around the Sun but around Jupiter, and would hit Jupiter a year later. Apparently, several decades earlier the comet was captured into a highly eccentric orbit around Jupiter, and in 1992, during its previous close approach, it was torn apart into more than 20 pieces by Jupiter's tidal forces. The authors of this book like to give both Shoemakers credit for the discovery, as in the heading, though the comet is generally called Shoemaker-Levy 9.

Figure 8–19 Comet Hale-Bopp (C/1995 O1) in the sky, showing both a whitish dust tail and a bluish gas tail.

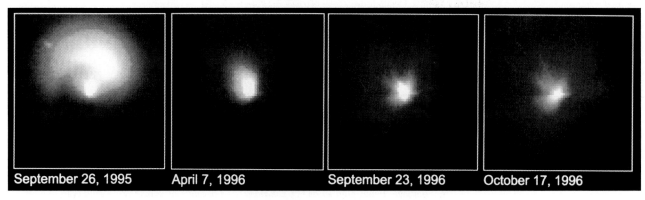

September 26, 1995 April 7, 1996 September 23, 1996 October 17, 1996

Figure 8–20 The head of Comet Hale-Bopp, imaged with the Hubble Space Telescope.

Figure 8-21 Comet Shoemaker-Levy 9 viewed from the Hubble Space Telescope. Over twenty fragments, each with its own tail, appear in this image taken six months before their collisions with Jupiter.

Telescopes all around the world and in space were trained on Jupiter when the first bit of comet hit. The site was slightly around the back side of Jupiter, but rotated to where we could see it from Earth after about 15 minutes. Even before then, scientists were enthralled by a plume rising above Jupiter's edge. When they could view Jupiter's surface, they saw a dark ring (Fig. 8–22). Infrared telescopes detected a tremendous amount of radiation from the heated gas. Over a period of almost a week, one bit of the comet after another hit Jupiter, leaving a series of Earth-sized rings and spots as Jupiter rotated. The dark spots could be seen for a few months even with small backyard telescopes.

The dark material showed us the hydrocarbon and other constituents of the comet. Spectra showed sulfur and other elements, presumably dredged up from lower levels of Jupiter's atmosphere than we normally see.

The biggest comet chunk released the equivalent of 6 million megatons of TNT—100,000 times more than the largest hydrogen bomb. Had any of the fragments hit Earth, they would have made a crater as large as Rhode Island, with dust thrown up to much greater distances. Had the entire comet (whose nucleus was 10 km across) hit Earth at one time, much of life could have been destroyed. So Comet Shoemaker-Levy 9 made us even more wary about what may be coming at us in space.

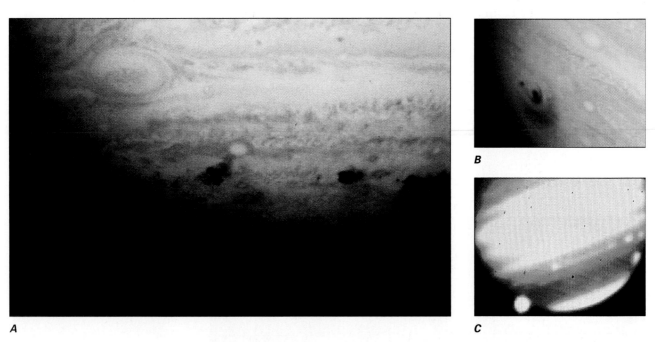

A   *C*

Figure 8-22 *(A)* Jupiter, with scars from the July 1994 crash of Comet Shoemaker-Levy 9. The image was taken with the Hubble Space Telescope. *(B)* The shape of the ring left on Jupiter's surface showed that the fragment of Comet Shoemaker-Levy 9 that caused it entered at an angle. The image was taken with the Hubble Space Telescope. *(C)* This infrared image shows bright plumes of hot gas from the collisions of comet fragments with Jupiter. This image was taken with the Keck Telescope.

Future Spacecraft to Comets

NASA's Stardust mission, launched in 1999, is travelling to Comet Wild-2. When it gets there in 2004, it will not only photograph the comet but also bring back to Earth some of its dust. It carries an extremely lightweight material called aerogel (Fig. 8–23), and will fly through the comet with the aerogel exposed so that the comet dust can stick in it. Stardust's orbit will bring it back near Earth in 2006, when it will parachute the aerogel down to the Utah desert. (WWW)

A more complicated spacecraft, Rosetta, is being prepared by the European Space Agency for launch in 2003. It is to reach a comet in 2011 and to orbit the Sun with it for a couple of years, also landing a probe on the comet's nucleus.

◖ METEOROIDS

There are many small chunks of matter orbiting in the Solar System, ranging up to tens of meters across. When these chunks are in space, they are called **meteoroids.** When one hits the Earth's atmosphere, friction heats it up—usually at a height of about 100 km—until all or most of it is vaporized. Such events result in streaks of light in the sky (Fig. 8–24), which we call **meteors** (popularly, and incorrectly, known as **shooting stars**). When a fragment of a meteoroid survives its passage through the Earth's atmosphere, the remnant that we find on Earth is called a **meteorite.**

Types and Sizes of Meteorites

Tiny meteorites less than a millimeter across, **micrometeorites,** are the major cause of erosion (what little there is) on the Moon. Micrometeorites also hit the Earth's upper atmosphere all the time, and remnants can be collected for analysis from balloons or airplanes or from deep-sea sediments. They are often sufficiently slowed down by Earth's atmosphere to avoid being vaporized before they reach the ground.

Space is full of meteoroids of all sizes, with the smallest being most abundant. Most of the small particles, less than 1 mm across, may come from comets.

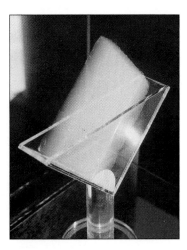

Figure 8-23 Aerogel, a frothy, extremely light substance. Even a large amount weighs very little, so a rocket can carry more of it into space than it could of a denser material.

Figure 8-24 Leonids (also see Figure 8–29). A time exposure of the 1998 Leonid meteor shower, showing fireballs that appeared over a 45-minute period. The Big Dipper is standing on its handle to the right, and the radiant in Leo is off the upper-right of the frame.

Figure 8–25 A 15 ton iron meteorite, the Willamette meteorite, at the Rose Center for Earth and Space, along with the Hayden Planetarium, at the American Museum of Natural History in New York City. Note how dense it has to be to have this high mass, given its size.

Figure 8–26 The largest mass of the Cape York meteorite, a 35-ton piece known as "Ahnighito" or "the Tent." This stony meteorite was discovered by Inuits in Greenland in the early 1800s and was brought to New York by William Peary and Matthew Henson in 1897. It is the largest on display in a museum (second in size only to a meteorite still in the ground in Namibia) and is in the American Museum of Natural History in New York City.

Figure 8–27 The Barringer "meteor crater" (actually a meteorite crater) in Arizona. It is 1.2 km in diameter. Dozens of other terrestrial craters are now known, many from aerial or space photographs. The largest may be a depression over 400 km across under the Antarctic ice pack, comparable with lunar craters. Another very large crater, in Hudson Bay, Canada, is filled with water. Most are either disguised in such ways or have eroded away.

The large particles, more than 1 cm across, may generally come from collisions of asteroids in the belt around the Sun in which most asteroids are found.

Most of the meteorites that are found (as opposed to most of those that exist) have a very high iron content (about 90 per cent); the rest is nickel. These iron meteorites are thus very dense—that is, they weigh quite a lot for their volume (Fig. 8–25).

Most meteorites that hit the Earth are stony in nature. Because they resemble ordinary rocks (Fig. 8–26) and disintegrate with weathering, they are not easily discovered unless their fall is observed. That difference explains why most meteorites discovered at random are made of iron. But when a fall is observed, most meteorites recovered are made of stone. Some meteorites are rich in carbon, and some of these even have complex molecules like amino acids. A large terrestrial crater that is obviously meteoritic in origin is the Barringer Meteor Crater in Arizona (Fig. 8–27). It resulted from what was perhaps the most recent large meteorite to hit the Earth, for it was formed only 50,000 years ago.

Every few years a meteorite is discovered on Earth immediately after its fall. The chance of a meteorite landing on someone's house is very small, but it has happened (Fig. 8–28)! Often the positions in the sky of extremely bright meteors are tracked in the hope of finding fresh meteorite falls. The newly discovered meteorites are rushed to laboratories in order to find out how long they have been in space by studying their radioactive elements. Many meteorites have recently been found in the Antarctic, where they have been well preserved as they accumulated over the years. A robot, successfully tested in Antarctica in 2000, has found some more. Some odd Antarctic meteorites are now known to have come from the Moon or even from Mars. Recall that in Chapter 6 we even discussed controversial evidence for ancient primitive life forms on Mars, found in one such meteorite. As of mid-2000, the conclusion hasn't been entirely ruled out, but few scientists accept it. As the late Carl Sagan said, "Extraordinary claims require extraordinary evidence," and the evidence about this meteorite is far from convincing.

A Closer Look

Meteor Showers

Shower	Date of Maximum	Approximate Duration	Limits	Number per Hour at Maximum	Parent Object
Quadrantids	Jan. 4	1 day	Jan. 1 – 6	100	—
Lyrids	Apr. 22	2 days	April 19 – 24	10	C/1861 G1 (Thatcher)
Eta Aquarids	May 5	3 days	May 1 – 8	20	1P/Halley
Delta Aquarids	July 27 – 28	7 days	July 15 – Aug. 15	30	—
Perseids	Aug. 12	5 days	July 25 – Aug. 18	70	107P/Swift-Tuttle
Orionids	Oct. 21	2 days	Oct. 16 – 26	30	1P/Halley
Taurids	Nov. 8	10 days	Oct. 20 – Nov. 30	10	2P/Encke
Leonids	Nov. 17	1 day	Nov. 15 – 19	10	55P/Tempel-Tuttle
Geminids	Dec. 14	3 days	Dec. 7 – 15	60	3200 Phaethon

Meteorites that have been examined were formed up to 4.6 billion years ago, the beginning of the Solar System. The relative abundances of the elements in meteorites thus tell us about the solar nebula from which the Solar System formed. In fact, up to the time of the Moon landings, meteorites were the only extraterrestrial material we could get our hands on. ☼

Meteor Showers

Meteors sometimes occur in **showers,** when meteors are seen at a rate far above average. The most widely observed—the Perseids—takes place each summer on about August 12 and the nights on either side of that date. The best winter show is the Geminids, which takes place around December 14. On any clear night a naked-eye observer with a dark sky may see a few **sporadic meteors** an hour— that is, meteors that are not part of a shower. (Just try going out to a field in the country and watching the sky for an hour.) During a shower, you may typically see one every few minutes. Meteor showers result from the Earth's passing

Figure 8-28 *(A)* A Peekskill, N.Y., woman heard a noise on October 9, 1992, and went outside to find extensive damage to her car. Under the car, she found *(B)* this just-fallen meteorite.

A

B

OBSERVING A METEOR SHOWER

Meteor showers can be fun to watch. You might see a "shooting star" nearly every minute or two. They are best viewed after midnight, when the Earth is rotated so that we are plowing into the stream of pebbles and ice chunks that produce the meteors. It is also very helpful if the Moon is down; a bright, moonlit sky tends to drown out the fainter meteors. A list of annual meteor showers is given in *A Closer Look–Meteor Showers*.

The Perseids on August 12 is the most commonly observed meteor shower. (For most others, bundle up against the cold.) Bring a reclining chair, blanket, or sleeping bag if possible, and find as dark a site as you can with a wide-open sky. Lie back, relax, and gaze up at the sky for at least half an hour. Any part of the sky is fine, but you will probably see the most meteors in the general direction of the "radiant," the constellation from which they seem to emerge (for example Perseus, in the case of the Perseids). Try to see how many meteors you can count. Note any especially bright ones. Sometimes they leave a vapor "trail" (or "train") for a few seconds. Occasionally, colors are seen in the streak of light.

Figure 8–29 A fireball–an extremely bright meteor–streaking away from the backward-question-mark shape of the constellation Leo, photographed during the 1999 Leonid meteor shower. Thousands of fainter meteors were seen in the peak hour.

through the orbits of defunct or disintegrating comets and hitting the meteoroids left behind. Though the Perseids and Geminids can be counted on each year, the Leonid meteor shower peaks every 33 years, when the Earth crosses the main clump of debris from Comet Tempel-Tuttle. ⓦⓦⓦ On November 17/18, 1998, one fireball (a meteor brighter than Venus) was visible each minute for a while (Fig. 8–29), and on November 17/18, 1999, about 3,000 meteors were seen in the peak hour by observers in Europe or the Middle East, though few fireballs appeared.

How visible meteors in a shower are depends in large part on how bright the Moon is. Meteors are best seen with the naked eye; using a telescope or binoculars merely restricts your field of view.

⬤ ASTEROIDS

The nine known planets were not the only bodies to result from the gas and dust cloud that collapsed to form the Solar System 4.6 billion years ago. Thousands of **minor planets,** called **asteroids,** also resulted. We detect them by their small motions in the sky relative to the stars (Fig. 8–30). Most of the asteroids have elliptical orbits between the orbits of Mars and Jupiter, in a zone called the **asteroid belt.** Most likely, Jupiter's gravitational tugs perturbed the orbits of asteroids, leading to collisions among them that were too violent to form a planet.

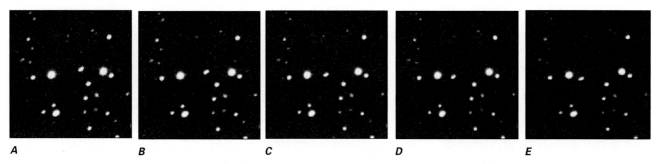

A B C D E

Figure 8–30 Asteroid 5100 Pasachoff moves significantly against the background of stars in this series of exposures spanning six minutes. The asteroid appears elongated only because of its motion during the individual exposures.

Asteroids are assigned a number in order of discovery and then a name: 1 Ceres, 16 Psyche, and 433 Eros, for example. Often the number is omitted when discussing well-known asteroids. Though the concept of the asteroid belt may seem to imply a lot of asteroids close together, asteroids rarely come within a million kilometers of each other. Occasionally collisions do occur, producing the small chips that make meteoroids. Only about 6 asteroids are larger than 300 km in diameter. Hundreds are over 100 km across (Fig. 8–31), roughly the size of some of the moons of the planets, but most are small, less than 10 km in diameter. Perhaps 100,000 asteroids could be detected with Earth-based telescopes; automated searches are now discovering asteroids at a prodigious rate. Yet all the asteroids together contain less mass than the Moon.

Spacecraft en route to Jupiter and beyond travelled through the asteroid belt for many months and showed that the amount of dust among the asteroids is not much greater than the amount of interplanetary dust in the vicinity of the Earth. So the asteroid belt will not be a hazard for space travel to the outer parts of the Solar System.

Asteroids are made of different materials from each other, and represent the chemical compositions of different regions of space. The asteroids at the inner edge of the asteroid belt are mostly stony in nature, while the ones at the outer edge are darker (because they contain more carbon). Most of the small asteroids that pass near the Earth belong to the stony group. Three of the largest asteroids belong to the high-carbon group. A third group is mostly composed of iron and nickel. The differences may be telling us about conditions in the early Solar System as it was forming and how the conditions varied with distance from the protosun.

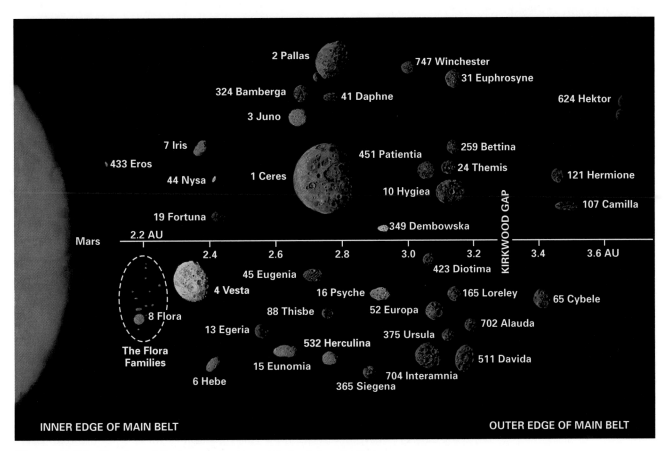

Figure 8–31   The sizes of the larger asteroids and their relative albedoes.

A Closer Look

The Extinction of the Dinosaurs

On Earth 65 million years ago, dinosaurs and about ⅔ of the other species died in a catastrophe known as the Cretaceous/Tertiary (K/T) extinction (see figure below). Evidence has recently been accumulating that this extinction was sudden and was caused by a Near-Earth Object or a comet hitting the Earth. Among the signs is the fact that the rare element iridium, which is more common in asteroids and comets, is widely distributed around the Earth in a thin layer laid down 65 million years ago, presumably after it was thrown up in the impact. The impact would have raised so much dust into the atmosphere that sunlight could have been shut out for months; plants and animals would not have been able to survive.

In 1991, a buried giant crater in Mexico was discovered that was probably the site of the impact (see figure below). It is one of the largest craters on Earth, about 200 kilometers across. Its age has been measured to be 65 million years — precisely coinciding with the extinction event. The impact of an asteroid or comet therefore seems compelling. Research is continuing in this exciting area.

The dinosaurs and many other species disappeared suddenly 65 million years ago, marking the end of the Cretaceous period, the third part of the Mesozoic era. In the movie *Jurassic Park* (the Jurassic—208 to 146 million years ago—was the second part of the Mesozoic era), dinosaurs were cloned from their DNA, something we can't do—yet.

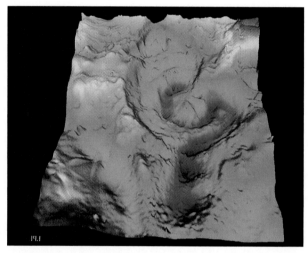

The impact structure known as Chicxulub, at the north end of the Yucatán. The blue areas are low-density rocks broken up by the impact. The green mound at the center is denser and probably represents a rebound at the point of impact. It is one of the largest craters known on Earth.

A Japanese space mission, MUSES-C, will be launched in 2002 and should return an asteroid sample to Earth in 2006. A NASA spacecraft, also to be launched in 2002, will do a Comet Nucleus Tour ("Contour") to three or more comets over a half-dozen years.

The path of the Galileo spacecraft to Jupiter sent it near the asteroid Gaspra in 1991 (Fig. 8–32). It detected a magnetic field from Gaspra, which means that the asteroid is probably made of metal and is magnetized. Galileo passed the asteroid Ida in 1993, and discovered that the asteroid has an even smaller satellite (Fig. 8–33), which was then named Dactyl. Other double asteroids have since been discovered, and astronomers newly recognize the frequency of such pairs. For example, ground-based astronomers found a 13-km satellite orbiting 200-km-diameter Eugenia every five days.

Near-Earth Objects

Some asteroids are far from the asteroid belt; their orbits approach or cross that of Earth. We have observed only a small fraction of these types of **Near-Earth Objects**, bodies that come within 1.3 A.U. of Earth.

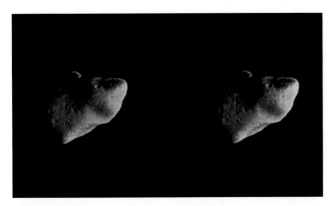

Figure 8-32 Gaspra, in this high-resolution image taken by the Galileo spacecraft, looks like a fragment of a larger body. The illuminated part seen here is 16 km by 12 km. Galileo discovered that Gaspra rotates every 7 hours.

We see a natural color image of Gaspra at *left*. At *right,* the color-enhanced image brings out variations in the composition of Gaspra's surface.

Figure 8-33 Ida and its moon (or satellite) Dactyl photographed from the Galileo spacecraft at a range of 10,500 km. Filters passing light at 4100 Å (violet), 7560 Å (near-infrared), and 9860 Å (infrared) are used, which enhances the color since the CCD is most sensitive in the infrared. Subtle color variations show differences in the physical state and composition of the soil. The bluish areas around some of the craters suggest a difference in the abundance or composition of iron-bearing minerals there. Dactyl differs from all areas of Ida in both infrared and violet filters.

The Near Earth Asteroid Rendezvous (NEAR Shoemaker, named in part after the planetary geologist Eugene Shoemaker) mission passed 253 Mathilde in 1997. NEAR Shoemaker went into orbit around 433 Eros on Valentine's Day, 2000 (Figs. 8–34 and 8–35). (Eros was the first near-Earth asteroid to be discovered.) Eros is 33 km by 13 km by 13 km. Craters, grooves, layers, house-sized boulders, and a 20-km-long surface ridge have been photographed. The existence of the craters and ridge, which indicates that Eros must be a solid body, disagrees with the previous suggestions of some scientists that all asteroids are mere rubble piles as Mathilde seems to be. Eros's density, 2.4 g/cm^3, is comparable to that of the Earth's crust, about the same as Ida's and twice Mathilde's. NEAR Shoemaker's infrared spectrometer is measuring how the minerals vary from place to place on Eros's surface.

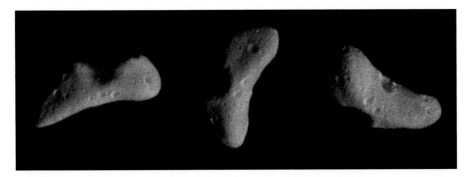

Figure 8-34 The rotation of asteroid 433 Eros, imaged from the Near Earth Asteroid Rendezvous mission (renamed NEAR Shoemaker), now in orbit around it. Eros's surface looks a uniform butterscotch color in visible light, but shows more variation in the infrared.

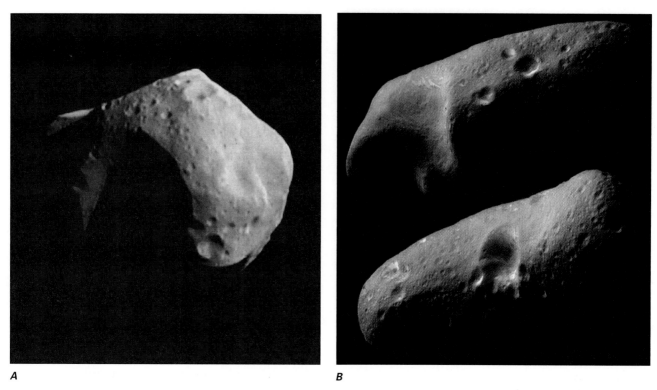

A   *B*

Figure 8-35 *(A)* Mathilde imaged from the Near Earth Asteroid Rendezvous mission (NEAR) in 1997. *(B)* Eros, from NEAR in orbit only 220 km above its surface. Surface detail down to 35 m is visible. Note how different its opposite hemispheres appear.

Figure 8-36 Antoine de Saint-Exupéry's Little Prince tending his asteroid.

Besides providing much detailed information, the close-up studies of these objects are allowing us to verify whether the lines of reasoning we use with ground-based asteroid observations give correct results.

Near-Earth asteroids may well be the source of most meteorites, which could be debris of collisions that occurred when these asteroids visit the asteroid belt. Eventually, most Earth-crossing asteroids will probably collide with the Earth. Measurements released in 2000 indicate that 1000 to 2000 of them are greater than 1 km in diameter, and none are known to be larger than 10 km across. Statistics show that there is a 1 per cent chance of a collision of this tremendous magnitude per millennium. This rate is pretty often on a cosmic scale. Such collisions would have drastic consequences for life on Earth. Even a collision with a small Near-Earth Object would do substantial damage (Fig. 8–36).

There are a hundred times more smaller objects, with a 1 per cent chance that an asteroid greater than 300 m in diameter would hit the Earth in the next century. Such a collision could kill hundreds of thousands or millions of people.

The question of how much we should worry about Near-Earth Objects hitting us is increasingly discussed, including a meeting sponsored by the United Nations. Even Hollywood movies have been devoted to the topic, though at present we can't send out astronauts to deflect or break up the objects the way the movies showed. Maybe we should be working on such capabilities. A first step should be to devote more resources to determining the depth of the problem, by setting up additional projects to find as many Near-Earth Objects as possible. Current plans are to map 90 per cent of them in the next couple of decades, and the pace of discovery is accelerating. A dedicated large telescope would speed up the search, but no funding is in view.

Concept Review

Pluto, which emerged in 1999 from inside Neptune's orbit, is so small and so far away that we know little about it. The discovery of its moon, Charon, allowed us to calculate that Pluto contains only $\frac{1}{500}$ the mass of the Earth. Mutual occultations of Pluto and Charon revealed sizes and surface structures of each. Pluto is hanging onto its designation as a planet, but is probably also the largest of the known **Kuiper-belt objects,** also known as **Trans-Neptunian Objects.**

In a **comet,** a long **tail** that points away from the Sun may extend from a bright **head.** The head consists of the **nucleus,** which is like a dirty snowball, and the gases of the **coma.** The **dust tail** contains dust particles released from the dirty ices of the nucleus. The **gas tail** is blown out behind the comet by the solar wind. Comets we see come from a huge spherical **Oort comet cloud** far beyond Pluto's orbit or from a flattened belt of comets (the Kuiper belt) just beyond Neptune's orbit. Spacecraft imaged the nucleus of Halley's Comet up close in 1986 and showed it to be an ellipsoidal snowball about 16 km long and 8 km across. Comet Hale-Bopp in 1997 was not only spectacular for the general public but also allowed many scientific studies. The impact of Comet Shoemaker-Levy 9 with Jupiter in 1994 produced temporary Earth-sized scars in Jupiter's atmosphere.

Meteoroids are chunks of rock in space. When one hits the Earth's atmosphere, we see the streak of light from its vaporization as a **meteor,** often called a **shooting star.** A fragment that survives and reaches Earth's surface is a **meteorite.** Most meteorites that are found are made of an iron-nickel alloy, but when a meteorite is seen to fall, stony meteorites are most often found. Bits of space dust are **micrometeorites.** The meteorites bring us primordial material to study and occasionally pieces of the Moon or Mars. Many meteors are seen in **showers,** which occur when the Earth crosses the path of a defunct or disintegrating comet. **Sporadic meteors** appear at the rate of a few each hour.

Asteroids are **minor planets.** Most asteroids are in the **asteroid belt** between the orbits of Mars and Jupiter. They range up to about 1000 km across, and were prevented from forming a planet by Jupiter's gravitational tugs. The Galileo and Near Earth Asteroid Rendezvous missions have studied asteroids up close, and NEAR is in orbit around the asteroid Eros. It is plausible that an impact of a large asteroid or comet threw so much dust into the Earth's atmosphere that it led to the extinction of many species, including dinosaurs, 65 million years ago. We worry about a **Near-Earth Object** or a comet hitting the Earth, which could lead to the demise of civilization, or of the more likely collision with the Earth of a smaller object, which could cause substantial damage, killing thousands or millions of people.

Questions

†1. What fraction of its orbit has Pluto traversed since it was discovered?

2. What evidence suggests that Pluto is not a "normal" planet?

3. Describe what mutual occultations of Pluto and Charon are and what they have told us.

4. Compare Pluto to the giant planets.

5. Compare Pluto to the terrestrial planets.

†6. If Pluto's surface is 60 K, use Wien's law (Chapter 2) to calculate the wavelength at which it gives off most of its thermal (black-body) radiation. In what part of the electromagnetic spectrum is this?

7. Which of Kepler's laws has enabled the discovery of Charon to help us find the mass of Pluto?

8. Describe the occultation observations of Pluto's atmosphere.

9. In what part of its orbit does a comet travel tail first?

10. What creates the tail of a comet? Is something burning?

11. Which part of a comet has the most mass?

12. From where are the comets that we see thought to come?

†13. How far is the Oort comet cloud from the Sun, relative to the distance from the Sun to Pluto?

14. Describe some of the Giotto spacecraft's discoveries about Halley's Comet.

†15. Use Kepler's third law (Chapter 5) and the known period of Halley's Comet (about 76 years) to deduce its semimajor axis. Which planet has a comparable semimajor axis?

16. Discuss the significance of the collision of Comet Shoemaker-Levy 9 with Jupiter.

17. What is the relation of meteoroids and asteroids?

18. What, physically, is a shooting star?

19. Why don't most meteoroids reach the Earth's surface?

20. Explain the process by which a meteor shower occurs.

21. Why might some meteor showers last only a day while others can last several weeks?

22. Why are meteorites important in our study of the Solar System?

23. Compare the sizes and surfaces of asteroids with the moons of the planets.

24. How might the occultation of a star by an asteroid tell us whether the asteroid has an atmosphere?

25. Why is the study of certain types of asteroids important for understanding our long-term survival on Earth?

26. What have the Galileo and NEAR Shoemaker spacecraft found out about asteroids?

---

†This question requires a numerical solution.

People in Astronomy

CAROLYN PORCO

*C*arolyn Porco is Associate Professor of Planetary Sciences at the Lunar and Planetary Laboratory of the University of Arizona. She is the leader of the imaging team for the Cassini mission to Saturn, due to arrive there in 2004.

Dr. Porco grew up in the Bronx, New York, and attended the State University of New York at Stony Brook. Wanting to specialize in planetary studies, she did her graduate work in the Division of Geological and Planetary Sciences at the California Institute of Technology in Pasadena. Armed with her Ph.D., she joined Professor Brad Smith, head of the Voyager imaging team, at the Lunar and Planetary Laboratory in Tucson, became a member of the Voyager imaging team, and headed the working group studying the rings during the Voyager Neptune encounter.

How is the Cassini spacecraft doing en route to Saturn?

On behalf of the Cassini Imaging Team, I am pleased to announce the successful imaging of asteroid 2685 Masursky with the Cassini cameras. It may be only a dot, but it's a very special dot to us!

The Masursky images represent the first time that Cassini has gathered information on a body not extensively studied from Earth. So far, the images reveal that the side of Masursky imaged by Cassini is roughly 15 to 20 kilometers (9 to 12 miles) across. This also was the first time that Cassini's automated object-targeting capabilities have been used and they functioned as expected.

Voyager was so wonderful; how can Cassini be better?

Voyager was indeed wonderful. It was the first spacecraft to visit the planets Uranus and Neptune, it did a much better job at Jupiter and Saturn than its predecessor Pioneer, and—it's been said so often it's become a cliche now—it rewrote the textbook on the outer Solar System. But you must keep in mind that the Voyager missions were flyby missions, so the spacecraft spent a very brief period of time in close proximity to the planets, the satellites, and the ring systems that they visited. And another thing to consider is when a spacecraft like Voyager flies by a spherical object such as a planet or satellite, it gets very close, but to only one hemisphere. So in some sense—particularly for the icy satellites of the giant planets—we have seen at close range only one-half of the outer Solar System. Taking all these factors into account, it leaves an awful lot left to do for a robotic spacecraft placed in orbit around a planet for a period of several years.

Cassini will encounter satellites of Saturn, some of them many times, at very close distances. It will be able to observe the rings over a reasonably large range of solar lighting conditions, and so we will be able to observe how the brightness of the rings changes with time as the Sun seen from Saturn changes in latitude. That will give us information about the way particles in Saturn's rings are distributed in the ring plane. Cassini will carry more capable instruments than did Voyager, so we will be able to get more information from these instruments than we could with Voyager's instruments. As one simple example, the camera system will now carry a charge-coupled device (a CCD), which is a detector that is far more sensitive to light than the vidicons that Voyager had in its camera system built during the mid-1970s.

So, quite simply, we will see more. We will be able to discover new satellites. We will be able to discover new rings. We should be able to see features in the atmospheres of Saturn and Titan that are at contrast levels far below what Voyager was able to detect, and I haven't even begun to mention what other instruments on the spacecraft will be able to do, because they too will have improved sensitivity over their Voyager counterparts. So I think Cassini will be just as exciting in an exploratory sense as was Voyager, and on top of that we will have the opportunity to answer all those questions left by the Voyager reconnaissance flybys.

How did you get into the business of space science in the first place?

I got into astronomy and planetary science in somewhat of a backward way. As a young teenager, I was interested in philosophy and Eastern religion and that got me thinking about the meaning of life and why we are here. By we, I mean why human beings are here, why I as an individual am here, and so on. From there, proceeding from internal to external, I began to question why the Universe was here, to ponder the whole scheme of things. Upon looking outward, I became very interested in astronomy and in the study of the Universe and the stars and the planets. By the time I decided to go to graduate school, my interest was focused on the exploration of planets—that was a big dream of mine. I knew that I wanted to be part of the American space program. So I decided to go to Caltech for graduate school because I knew it had an outstanding planetary science department, and also because it would give me the opportunity to get involved with the space missions going on at the nearby Jet Propulsion Laboratory.

> "Cassini will take images of Saturn and its rings and moons that will make Voyager images look second-rate in comparison. And, you know how beautiful and informative they were! In short, I'm absolutely thrilled to be leading the team responsible for our next imaging adventures at Saturn."

Does the fact that you study the Solar System mean that you studied more geology than physics?

No, I was a physics and astronomy double major at college. I chose the college I went to—the State University of New York at Stony Brook—because it had an astronomy department, but by the time I went through that program, I had decided that I wanted to study planets because I wanted to participate in the exploration of the Solar System. It was the intellectual and human adventure of space exploration that really grabbed me.

What followed graduate school?

Right out of graduate school, I went to work with Brad Smith at the University of Arizona. He was the leader of the Voyager imaging team at the time, so he was the one who added me to the team. I gained invaluable experience from being on the Voyager imaging team, and at Neptune was leader of the Rings Working Group of the imaging team. This experience gave me a taste of what it is like to lead a group of scientists through a planetary encounter. But I had to apply to be selected as imaging team leader for Cassini, just like everybody else.

Are you engaged in any other professional activities in addition to your on-going work on Cassini?

I am a professor here within the University of Arizona's Department of Planetary Sciences, so in this capacity I, of course, teach and advise graduate students. But I am also engaged in research: I'm still studying planetary rings (like those of Saturn and Neptune) using Voyager data as well as Earth-based telescopic observations of occultations of bright stars by these ring systems. There's a great deal yet to be learned about rings—like what causes all the finest scale structure, how quickly do these systems evolve, and how did they get there to begin with. It is questions like these, of course, which motivate us to return to Saturn and Cassini.

Can you tell us something of your life outside work?

I come from an Italian-American family; I have four brothers, my parents were immigrants. I like doing outdoorsy things—that's one thing I like about living in Arizona—the accessibility of the outdoors, though it gets too ungodly hot here in the summer. Of course, the sky here is oftentimes impeccable and it's nice to be able to go outdoors, look up at the stars, and "commune" with nature.

I like to ride bicycles, I like to play the guitar and sing. I love to sing.

As you look ahead to years of work for NASA, what do you think about the challenge that entails?

NASA is employing me to do a job I am ready and eager and able to carry out. The imaging experiment on any spacecraft mission is always the most visible, because it is the one experiment that the public can understand and relate to the best, and Cassini will take images of Saturn and its rings and moons that will make Voyager images look second rate in comparison. And you know how beautiful and informative they were! In short, I'm absolutely thrilled to be leading the team responsible for our next imaging adventures at Saturn. I feel enormously privileged.

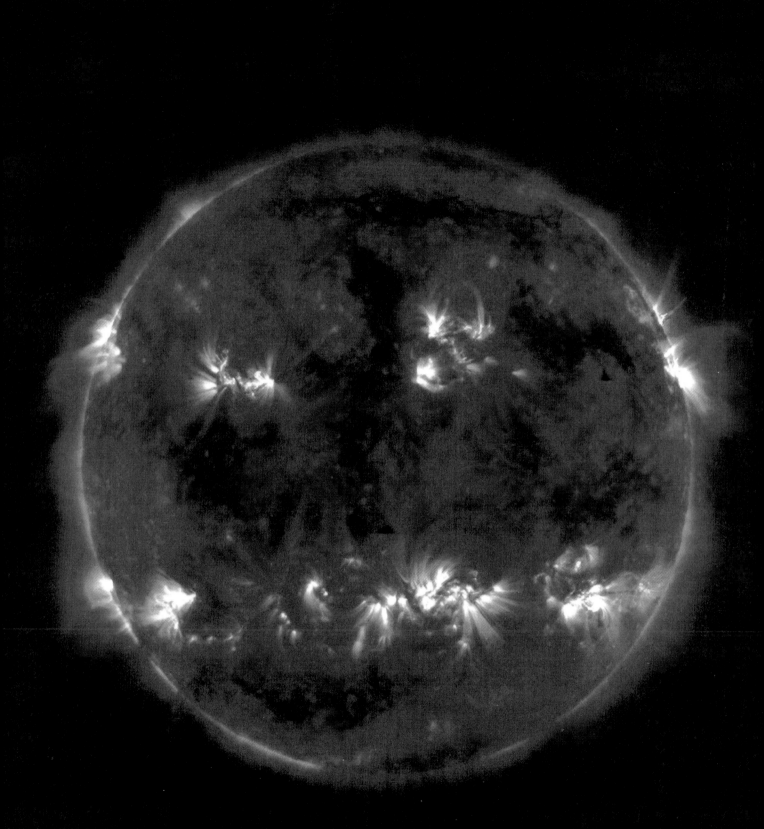

July 5,1998 171A

9

Our Star: The Sun

ORIGINS *The Sun, the source of light and heat on Earth, is critical to the existence of humans. Its properties are typical of stars.*

Not all stars are far away; one is close at hand. By studying the Sun, we not only learn about the properties of a particular star but also can study processes that undoubtedly take place in more distant stars as well. We will first discuss the **quiet Sun,** the solar phenomena that typically appear every day. Afterward, we will discuss the **active Sun,** solar phenomena that appear nonuniformly on the Sun and vary over time. We are now in a period of maximum solar activity, so it is particularly exciting to study the Sun. We study the Sun not only from telescopes permanently set up on Earth but also from temporary observation sites set up to observe total solar eclipses (Fig. 9–1).

⬤ BASIC STRUCTURE OF THE SUN

We think of the Sun as the bright ball of gas that appears to travel across our sky every day. We are seeing only one layer of the Sun, part of its atmosphere. The layer that we see is called the **photosphere** (Fig. 9–2), which simply means the sphere from which the light comes (from the Greek *photos*, "light"). The photosphere is about 110 Earth diameters across, so over a million Earths could fit inside the Sun!

As is typical of many stars, about 94 per cent of the atoms and nuclei in the outer parts are hydrogen, about 5.9 per cent are helium, and a mixture of all the other elements makes up the remaining one-tenth of one per cent (see *A Closer Look: The Most Common Elements in the Sun's Photosphere*). The overall composition of the Sun's interior is not very different.

The Sun is a typical star, in the sense that stars much hotter and much cooler, and stars intrinsically much brighter and much fainter, exist. Radiation from the photosphere peaks (is strongest) in the middle of the visible spectrum; after all, our eyes evolved over time to be sensitive to that region of the spectrum because the greatest amount of the solar radiation is emitted there. If we lived on a planet orbiting an object that emitted mostly x-rays, we, like Superman, might have x-ray vision. (Of course, unlike Superman, we would be only passively receiving x-rays, not sending them out.)

◄ The Sun in the ultraviolet, a mosaic of images from the Transition Region and Coronal Explorer (TRACE) spacecraft. The hottest gas, at millions of degrees, shows.

REDSHIFT
http://www.harcourtcollege.com/astro/cosmos/rsce

Figure 9–1 The 1999 total solar eclipse, observed from Romania. We see the solar corona in the sky, surrounding the dark disk of the Moon. Jupiter also shows.

A Closer Look

The Most Common Elements in the Sun's Photosphere

			Symbol	Atomic Number
for each				
1,000,000	atoms of	hydrogen, there are	H	1
98,000	atoms of	helium	He	2
850	atoms of	oxygen	O	8
400	atoms of	carbon	C	6
120	atoms of	neon	Ne	10
100	atoms of	nitrogen	N	7
47	atoms of	iron	Fe	26
38	atoms of	magnesium	Mg	12
35	atoms of	silicon	Si	14
16	atoms of	sulfur	S	16
4	atoms of	argon	Ar	18
3	atoms of	aluminum	Al	13
2	atoms of	calcium	Ca	20
2	atoms of	sodium	Na	11
2	atoms of	nickel	Ni	28

Note that the Sun isn't yellow, though it is often thought of as such. It only appears this color, or even orange or red, when it is close to the horizon: The blue and green light is selectively absorbed and scattered by Earth's atmosphere. (The scattered blue light produces the color of the daytime sky.) Also, on many photographs, the filter used to cut down the solar brightness to a safe level favors the yellow.

Beneath the photosphere is the solar **interior** (Fig. 9–3). All the solar energy is generated there at the solar **core,** which is about 10 per cent of the solar diameter at this stage of the Sun's life. The temperature there is about 15,000,000 K.

The photosphere is the lowest level of the **solar atmosphere.** Though the Sun is gaseous throughout, with no solid parts, we still use the term "atmosphere" for the upper part of the solar material because it is relatively transparent.

Just above the photosphere is a jagged, spiky layer about 10,000 km thick, only about 1.5 per cent of the solar radius. This layer glows colorfully pinkish when seen during an eclipse, when the photosphere is hidden, and is thus called the **chromosphere** (from the Greek *chromos,* "color"). Above the chromosphere, a ghostly white halo called the **corona** (from the Latin, "crown") extends tens of millions of kilometers into space (Fig. 9–4). The corona is continually expanding into interplanetary space and in this form is called the **solar wind.** The Earth is bathed in the solar wind.

The Photosphere

The Sun is a normal star with a surface temperature of about 5800 K, neither the hottest nor the coolest star. It is the only star close enough to allow us to study its surface in detail. We can resolve parts of the surface about 700 km across, about the distance from Boston to Washington, D.C.

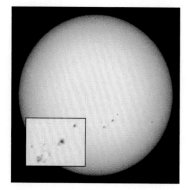

Figure 9–2 The solar photosphere in an image taken on the day of the 1999 total eclipse. The inset shows detail in the sunspots.

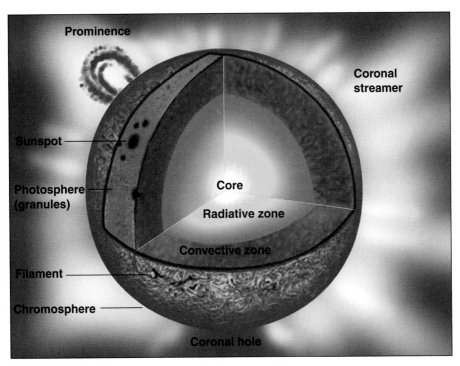

Figure 9-3 The parts of the solar atmosphere and interior. The solar surface is depicted as it appears through a hydrogen filter.

The interior has a core, where the energy is generated, surrounded by a zone in which radiation dominates and a zone in which convection (a boiling effect) dominates. On the surface, the photosphere, we find sunspots as well as loops of gas that are dark filaments when seen in projection against the disk and that are bright prominences when seen off the Sun's edge. The chromosphere, a thin layer above the photosphere, shows best in hydrogen-light and is composed of small spikes known as spicules. The white corona above it extends far into space. Its shape, including its huge streamers, is determined by the Sun's magnetic field.

Figure 9-4 The solar corona at the total solar eclipse of 1999, a composite image put together to show the shape of coronal streamers over a wide range of brightness.

When we study the solar surface in **white light**—all the visible radiation taken together—with typical good resolution (about 1 arc second), we see a salt-and-pepper texture called **granulation** (Fig. 9-5). The effect is similar to that seen in boiling liquids on Earth. But the granules are about the same size as the limit of our resolution, so are difficult to study. No spacecraft has yet exceeded the best resolution obtained from the ground.

On close examination, the photosphere as a whole oscillates—vibrating up and down slightly. The first period of vibration discovered was 5 minutes long, and for many years astronomers thought that 5 minutes was a basic duration for oscillation on the Sun. However, astronomers have since realized that the Sun simultaneously vibrates with many different periods, and that studying these periods tells us about the solar interior. The method works similarly to the way terrestrial geologists investigate the Earth's interior by measuring seismic waves on the Earth's surface; the studies on the Sun are thus called, by analogy, "solar seismology."

To find the longest periods, astronomers have observed the Sun from the Earth's south pole, where the Sun stays above the horizon for months on end. Now, even better, they use the Global Oscillation Network Group (GONG), a program centered at the National Solar Observatory that has erected a network of telescopes around the world to study solar oscillations. In addition, a NASA/European-Space-Agency spacecraft called SOHO (the Solar and Heliospheric Observatory) is stationed in space at a location from which it continu-

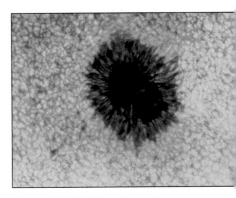

Figure 9-5 The solar surface, showing the salt-and-pepper granulation plus a sunspot.

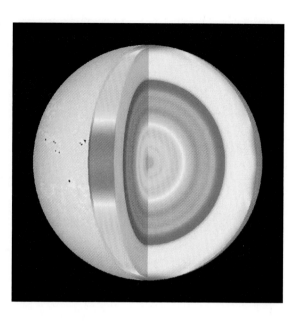

Figure 9-6 The rotation periods of different layers and regions of the Sun, based on helioseismology results. Red shows regions with the shortest rotation period (26 days), green corresponds to 28 days, and violet corresponds to 36 days. Thus the measurements show that the surface rotation periods, which vary from 26 days at the equator to 36 days at the poles, persist inward through the solar convection zone. Further toward the center, the data show that the Sun rotates as a solid body with a 27-day period.

Figure 9-7 Fraunhofer's original spectrum, shown on a German postage stamp. We see a continuous spectrum, from red on the left to violet on the right, crossed from top to bottom by the dark lines that Fraunhofer discovered, now known as "Fraunhofer lines" or "absorption lines."

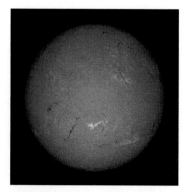

Figure 9-8 The Sun, photographed in hydrogen radiation, shows bright areas called plages and dark filaments around the many sunspot regions. We see an image from August 11, 1999, the day of a total eclipse.

ously views the Sun, and has also assembled long runs of observations lasting many months. Studies of solar vibrations thus far have told us about the temperature and density at various levels in the solar interior, and about how fast the interior rotates (Fig. 9–6). Since other stars must behave like the Sun, we are learning about the interiors of stars in general.

The spectrum of the solar photosphere, like that of almost all stars, is a continuous spectrum crossed by absorption lines (Fig. 9–7). Hundreds of thousands of these absorption lines, which are also called Fraunhofer lines, have been photographed and catalogued. They represent sets of spectral lines from most of the chemical elements. Iron has many lines in the spectrum. The hydrogen lines are strong but few in number. From the spectral lines, we can figure out the percentages of the elements.

The Chromosphere

When we look at the Sun through a filter that passes only the red light from hydrogen gas, the chromosphere is opaque. Thus through a hydrogen-light filter (Fig. 9–8), our view is of the chromosphere. The view has brighter and darker areas, with the brighter areas in the same regions as sunspots, but the chromosphere looks different from the photosphere below it.

An important theme valid throughout much of astronomy is that when you look at astronomical objects in detail, you see that they may vary in properties quite a lot from the average you see with the image blurred out. For example, under high resolution, we see that the chromosphere is not a spherical shell around the Sun but rather is composed of small "spicules." These jets of gas rise and fall, and have been compared in appearance to blades of grass or burning prairies. Spicules are more or less cylinders about 700 km across and 7000 km tall. They seem to have lifetimes of about 5 to 15 minutes, and there may be approximately half a million of them on the surface of the Sun at any given moment. They are about the same size as the granules, and are matter ejected into the chromosphere, presumably from the boiling effect that also makes the granules.

Chromospheric matter appears to be at a temperature of 7000 to 15,000 K, somewhat higher than the temperature of the photosphere. Ultraviolet spectra of stars recorded by spacecraft have shown unmistakable signs of chromospheres in Sun-like stars. Thus by studying the solar chromosphere we are also learning what the chromospheres of other stars are like.

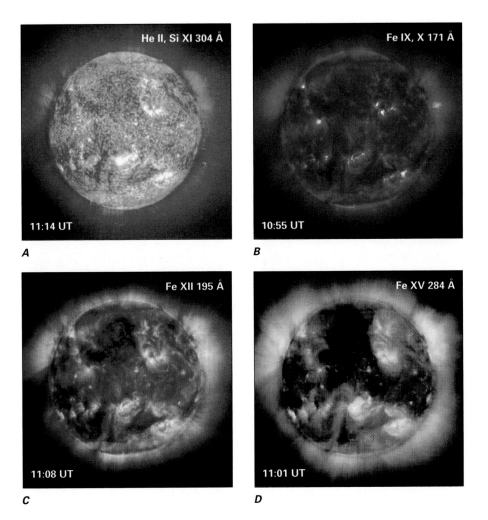

Figure 9–9 A set of images taken on August 11, 1999, the day of a total solar eclipse, through the four filters of the Extreme-ultraviolet Imaging Telescope (EIT) on the SOHO spacecraft. *(A)* Ionized helium gas at a temperature of about 60,000 K, with a small contribution from ionized silicon. *(B)* Ionized iron at a temperature of about 1,000,000 K. *(C)* Ionized iron at a temperature of about 1,500,000 K. *(D)* Ionized iron at a temperature of about 2,000,000 K to 2,500,000 K.

The Corona

During total solar eclipses, when first the photosphere and then the chromosphere are completely hidden from view, a faint white halo around the Sun becomes visible. This corona is the outermost part of the solar atmosphere, and technically extends throughout the Solar System. At the lowest levels, the corona's temperature is about 2,000,000 K. The Extreme-ultraviolet Imaging Telescope on the SOHO spacecraft images the corona every few minutes. By looking through filters that pass only light given off by gas at a very high temperature, the spacecraft can make images of the corona even in the center of the Sun's disk (Fig. 9–9). On the occasional dates of solar eclipses, these images can even be lined up with the eclipse images, to allow astronomers to trace the things seen during an eclipse back to their roots on the Sun's surface (Fig. 9–10).

Even though the temperature of the corona is so high, the actual amount of energy in the solar corona is not large. The temperature quoted is actually a measure of how fast individual particles (electrons, in particular) are moving. There aren't very many coronal particles, even though each particle has a high speed. The corona has less than one-billionth the density of the Earth's atmosphere, and would be considered a very good vacuum in a laboratory on Earth. For this reason, the corona serves as a unique and valuable celestial laboratory in which we may study gaseous "plasmas" in a near-vacuum. Plasmas are gases consisting of positively and negatively charged particles and can be shaped by magnetic fields. We are trying to learn how to use magnetic fields on Earth to control plasmas in order to provide energy through nuclear fusion.

Figure 9–10 The Sun at the time of the total solar eclipse of February 26, 1998. A composite image made from photographs of the solar corona made during the eclipse by one of us (J.M.P.) surrounds the solar disk. It shows the equatorial streamers and polar tufts in the corona over a wide range of intensity. Pasted over the dark lunar disk at the center of the image is a false-color image from the Extreme-ultraviolet Imaging Telescope on the Solar and Heliospheric Observatory in space. This image shows the coronal temperature based on observations of the corona in the ultraviolet taken at about the same time as the eclipse.

Figure 9–11 Loops of hot coronal gas from the TRACE spacecraft, imaged in the ultraviolet.

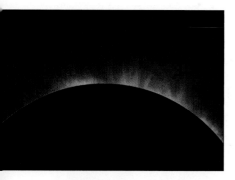

Figure 9–12 From a few mountain sites, the innermost corona can be photographed without need for an eclipse on many of the days of the year. The corona shows up best in its green emission lines from thirteen-times ionized iron, iron that has lost thirteen electrons.

Photographs of the corona show that it is very irregular in form. Beautiful long **streamers** extend away from the Sun in the equatorial regions. The shape of the corona varies continuously and is thus different at each successive eclipse. The structure of the corona is maintained by the magnetic field of the Sun. The TRACE (Transition Region and Coronal Explorer) spacecraft makes extremely high-resolution observations of the solar corona by observing in the ultraviolet. It shows clearly how the corona is made up of loops of gas, which are held in their shapes by the Sun's magnetic field (Fig. 9–11).

The corona is normally too faint to be seen except at an eclipse of the Sun because it is fainter than the everyday blue sky. But at certain locations on mountain peaks on the surface of the Earth, the sky is especially clear and dust-free, and the innermost part of the corona can be seen (Fig. 9–12) with special telescopes. The limited extent of the corona seen in this way shows how valuable the observations at eclipses and from space are.

Several crewed and uncrewed spacecraft have used devices that made a sort of artificial eclipse to photograph the corona hour by hour in visible light. One of the instruments aboard SOHO makes such observations very well at present. SOHO studies the corona to much greater distances from the solar surface than can be studied with coronagraphs on Earth. (SOHO's coronagraphs cannot see the innermost part, which we can still study best at eclipses.) Among the major conclusions of the research is that the corona is much more dynamic than we had thought. For example, many blobs of matter were seen to be ejected from the corona into interplanetary space, one per day or so. These "coronal mass ejections" sometimes even impact the Earth, causing surges in power lines and zapping—even occasionally destroying—satellites that bring you television or telephone calls. SOHO and other spacecraft far above the Earth in the direction of the Sun give us early warning when solar particles pass them.

The visible region of the coronal spectrum, when observed at eclipses, shows continuous radiation, absorption lines, and emission lines (Fig. 9–13). The emission lines do not correspond to any normal spectral lines known in laboratories on Earth or on other stars, and for many years their identification was one of the major problems in solar astronomy. The lines were even given the name of an element: coronium. (After all, the element helium was first discovered on the Sun.) In the late 1930s, it was discovered that the emission lines arose in atoms that had lost about a dozen electrons each. This was the major indication that the corona was very hot (and that coronium doesn't exist). The corona must be very hot indeed, millions of degrees, to have enough energy to strip that many electrons off atoms. This very hot corona also reflects the photospheric spectrum to us, but the Doppler shift broadens the absorption lines out so they are no longer visible and thus leaves only a continuum.

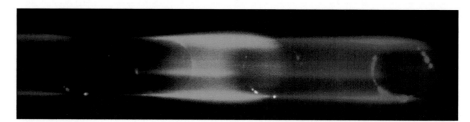

Figure 9–13 A spectrum of the prominences and corona at the 1999 total solar eclipse. The last traces of photospheric spectrum show as the band of color. The bright points are the emission lines from the chromosphere and prominences. The yellow helium D$_3$ line was the emission line from which helium was first identified over a hundred years ago. The element was named helium because at that time it was found only in the Sun (*helios* in Greek). Faintly visible in the green and in the red are complete circles that are emission lines in the spectrum of the corona.

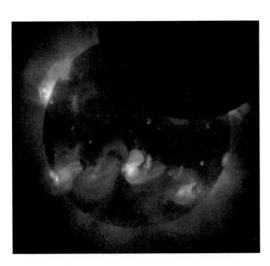

Figure 9–14 The corona in soft x-rays observed from the Yohkoh spacecraft. As the Sun was totally eclipsed on the ground on August 11, 1999, Yohkoh saw a partial eclipse.

While the emission lines tell us about the coronal gas, the visible absorption lines are mere reflections of the absorption lines in the spectrum of the Sun's photosphere. To provide these absorption lines, the photospheric spectrum is reflected toward us by dust in interplanetary space far closer to the Earth than to the Sun.

The gas in the corona is so hot that it emits mainly x-rays, photons of high energy. The photosphere, on the other hand, is too cool to emit x-rays. As a result, when photographs of the Sun are taken in the x-ray region of the spectrum (from satellites, since x-rays cannot pass through the Earth's atmosphere), they show the corona and its structure.

The x-ray image also reveals very dark areas at the Sun's north pole and extending downward across the center of the solar disk. These dark locations are **coronal holes,** regions of the corona that are particularly cool and quiet (Fig. 9–14). The density of gas in those areas is lower than the density in adjacent areas.

There is usually a coronal hole at one or both of the solar poles. Less often, we find additional coronal holes at lower solar latitudes. The regions of the coronal holes seem very different from other parts of the Sun. The solar wind flows to Earth mainly out of the coronal holes, so it is important to study the coronal holes to understand our environment in space.

The most detailed x-ray images support the more recent ultraviolet high-resolution images in showing that most, if not all, the radiation appears in the form of loops of gas joining separate points on the solar surface (Fig. 9–15). We must understand the physics of coronal loops in order to understand how the corona is heated. It is not sufficient to think in terms of a uniform corona, since the corona is obviously not uniform.

⬤ SUNSPOTS AND OTHER SOLAR ACTIVITY

Many solar phenomena vary with an 11-year cycle, which is called the **solar-activity cycle.** The most obvious are **sunspots** (Fig. 9–16), which appear relatively dark when seen in white light. Sunspots appear dark because they are giving off less radiation per unit area than the photosphere that surrounds them. Thus they are relatively cool (about 2000 K cooler than the photosphere), since cooler gas radiates less than hotter gas (recall Chapter 2). Actually, if we could somehow remove a sunspot from the solar surface and put it off in space, it would appear bright against the dark sky; a large one would give off as much light as the full moon seen from Earth. ☊

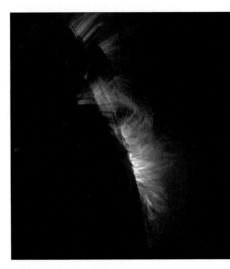

Figure 9–15 An ultraviolet image, taken from space on the day of the 1999 total solar eclipse, shows small coronal loops. The image was made with the TRACE spacecraft. The expedition of one of the authors (J.M.P.) observed this same group of loops from the ground during the eclipse, with lower spatial resolution but higher time resolution.

REDSHIFT
http://www.harcourtcollege.com/astro/cosmos/rsce

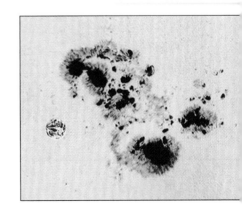

Figure 9–16 A sunspot, showing its dark umbra surrounded by its lighter penumbra. A photo of the Earth is superimposed to show its relative size.

A sunspot includes a very dark central region, called the **umbra** from the Latin for "shadow" (plural: **umbrae**). The umbra is surrounded by a **penumbra** (plural: **penumbrae**), which is not as dark.

To understand sunspots, we must understand magnetic fields. When iron filings are put near a simple bar magnet, the filings show a pattern (Fig. 9–17). The magnet is said to have a north pole and a south pole, and the magnetic field linking them is characterized by what we call **magnetic-field lines** (after all, the iron filings are spread out in what look like lines). The Earth (as well as some other planets) has a magnetic field that has many characteristics in common with that of a bar magnet. The structure seen in the solar corona, including the streamers, shows matter being held by the Sun's magnetic field.

The strength of the solar magnetic field is revealed in spectra. George Ellery Hale showed, in 1908, that the sunspots are regions of very high magnetic-field strength on the Sun, thousands of times more powerful than the Earth's magnetic field. Sunspots usually occur in pairs, and often these pairs are part of larger groups. In each pair, one sunspot will be typical of a north magnetic pole and the other will be typical of a south magnetic pole.

Magnetic fields are able to restrain charged matter—this is the property we are trying to exploit on Earth to contain superheated matter long enough to allow nuclear fusion to take place for energy production. The strongest magnetic fields in the Sun occur in sunspots. The magnetic fields in sunspots keep energy from being carried upward to the surface. As a result, sunspots are cool and dark.

Sunspots were discovered in 1610, independently by Galileo and by others. In about 1850, it was realized that the number of sunspots varies with an 11-year cycle (Fig. 9–18), the **sunspot cycle**. (www) Every 11-year cycle, the north magnetic pole and south magnetic pole on the Sun reverse; what had been a north magnetic pole is then a south magnetic pole and vice versa. So it is 22 years before the Sun returns to its original state, making the real period of the solar-activity cycle 22 years.

We are now going through the maximum of the sunspot cycle—the time when there is the greatest number of sunspots. After about 2001 or 2002, the number of sunspots should decrease.

Careful studies of the solar-activity cycle are now increasing our understanding of how the Sun affects the Earth. Although for many years scientists were skeptical of the idea that solar activity could have a direct effect on the Earth's weather, scientists currently seem to be accepting more and more the possibility of such a relationship.

Figure 9–17 Lines of force from a bar magnet are outlined by iron filings. One end of the magnet is called a north pole and the other is called a south pole. Similar poles ("like poles")—a pair of norths or a pair of souths—repel each other, and unlike (1 north and 1 south) attract each other. Lines of force go between opposite poles.

REDSHIFT
http://www.harcourtcollege.com/astro/cosmos/rsce

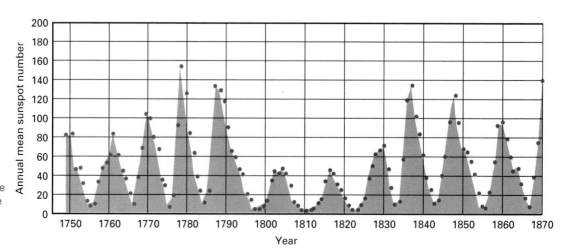

Figure 9–18 The 11-year sunspot cycle is but one manifestation of the solar-activity cycle.

An extreme test of the interaction may be provided by the interesting probability that there were essentially no sunspots on the Sun from 1645 to 1715! This period, called the **Maunder minimum,** was largely forgotten until recently. An important conclusion is that the solar-activity cycle may be much less regular than we had thought.

Much of the evidence for the Maunder minimum is indirect, and has been challenged, as has the specific link of the Maunder minimum with colder climate during that period. It would be reasonable for several mechanisms to affect the Earth's climate on this time scale, rather than only one.

Precise measurements made from spacecraft have shown that the total amount of energy flowing out of the Sun varies slightly, by up to 0.002 (0.2 per cent). On a short time scale, the dips in energy seem to correspond to the existence of large sunspots. Astronomers are now trying to figure out what happens to the blocked energy. ☀ On a longer time scale, the effect goes the other way. Spacecraft observations have shown that as sunspot minimum is reached, the Sun becomes overall slightly fainter. Fortunately, as the last maximum was approached, the Sun brightened back up, and so wasn't permanently fading away.

Flares

Violent activity sometimes occurs in the regions around sunspots. Tremendous eruptions called solar **flares** (Fig. 9–19) can eject particles and emit radiation from all parts of the spectrum into space. These solar storms begin over a few seconds and can last up to four hours. Temperatures in the flare can reach 5 million kelvins, even hotter than the quiet corona. Ultraviolet and x-rays that are given off reach Earth at the speed of light in 8 minutes and can disrupt radio communications, because they ionize the upper part of Earth's atmosphere. Flare particles that are ejected reach the Earth in a few hours or days and can cause the aurorae and even surges on power lines that lead to blackouts of electricity. Because of these solar-terrestrial relationships, high priority is placed on understanding solar activity and being able to predict it. The link between coronal mass ejections and solar flares is being debated; they often but not always occur together.

No specific theory for explaining the eruption of solar flares is generally accepted. But it is agreed that a tremendous amount of energy is stored in the solar magnetic fields in sunspot regions. Something unknown triggers the release of energy.

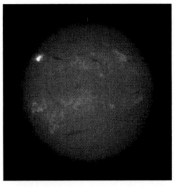

Figure 9–19 The brightening of regions of the Sun, seen in the hydrogen-alpha line, marking the position of a solar flare. To make a flare, magnetic field lines change the way their north and south magnetic poles are connected. This "reconnection" releases high-energy particles, which follow the magnetic field lines back down to lower levels of the solar atmosphere. We see these regions brightening drastically on this image of the February 5, 2000, solar flare.

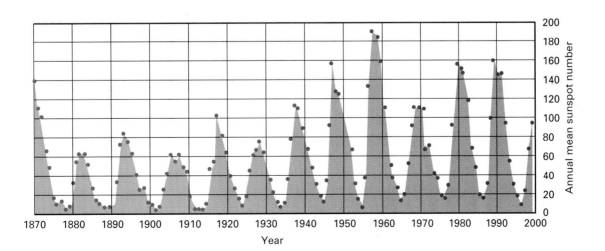

Figure 9-20 A solar prominence.

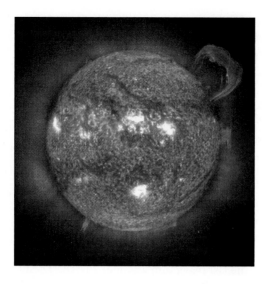

Filaments and Prominences

Studies of the solar atmosphere through filters that pass only hydrogen radiation also reveal other types of solar activity. Dark **filaments** are seen threading their way for up to 100,000 km across the Sun in the vicinity of sunspots. They mark the locations of zero magnetic field that separate regions of magnetic field pointing in opposite directions.

When filaments happen to be on the Sun's edge, they project into space, often in beautiful shapes; these are called **prominences** (Fig. 9–20). Prominences can be seen with the eye at solar eclipses and glow pinkish at that time because of their emission in hydrogen and a few other spectral lines (Fig. 9–21). They can be observed from the ground even without an eclipse, if a filter that passes only light emitted by hydrogen gas is used. Prominences appear to be composed of matter in a condition of temperature and density similar to matter in the quiet chromosphere, somewhat hotter than the photosphere. Sometimes prominences can hover above the Sun, supported by magnetic fields, for weeks or months. Other prominences change rapidly.

ECLIPSES

Because the Moon's orbit around the Earth and the Earth's orbit around the Sun are not precisely in the same plane (Fig. 9–22), the Moon usually passes slightly above or below the Earth's shadow at full moon, and the Earth usually passes slightly above or below the Moon's shadow at new moon. But up to seven times a year, full moons or new moons occur when the Moon is at the part of its orbit that crosses the Earth's orbital plane. At those times, we have a lunar eclipse or a solar eclipse (Fig. 9–23). Thus up to seven eclipses (mostly partial) can occur in a given year.

Many more people see a total lunar eclipse than a total solar eclipse when one occurs. At a total lunar eclipse, the Moon lies entirely in the Earth's full shadow and sunlight is entirely cut off from it (neglecting the atmospheric effects we will discuss below). So for anyone on the entire hemisphere of Earth for which the Moon is up, the eclipse is visible. In a total solar eclipse, on the other hand, the alignment of the Moon between the Sun and the Earth must be precise, and only those people in a narrow band on the surface of the Earth see the eclipse. Therefore, it is much rarer for a typical person on Earth to see a total solar eclipse—when the Moon covers the whole surface of the Sun—than a total lunar eclipse.

Figure 9-21 Prominences, especially at right and at lower left, as seen in a photograph of the Sun during the 1999 total eclipse. The chromosphere, as well as the last bit of the bright white diamond-ring effect, shows at upper left. The prominences and chromo-sphere appear pinkish because they radiate mainly the hydrogen-alpha emission line.

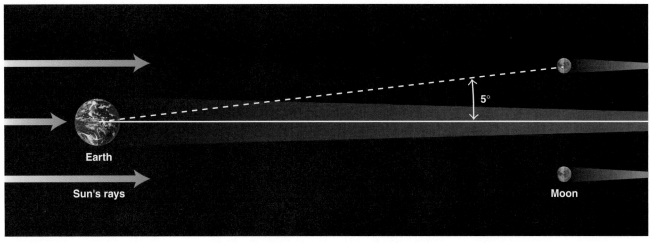

Figure 9–22 The plane of the Moon's orbit is tipped by 5° with respect to the plane of the Earth's orbit, so the Moon usually passes above or below the Earth's shadow. Therefore, we don't have lunar eclipses most months. (The diagram is not to scale.)

Lunar Eclipses

A total lunar eclipse is a much more leisurely event to watch than a total solar eclipse. The partial phase, when the Earth's shadow gradually covers the Moon, usually lasts over an hour, similar to the duration of the partial phase of a total solar eclipse. But then the total phase of a lunar eclipse, when the Moon is entirely within the Earth's shadow, can also last for over an hour, in dramatic contrast with the few minutes of a total solar eclipse. During the total lunar eclipse, the sunlight is not entirely shut off from the Moon: a small amount is refracted around the edge of the Earth by our atmosphere. Most of the blue light is scattered out during the sunlight's passage through our atmosphere; this explains how blue skies are made for the people part-way around the globe from places at which the Sun is high in the sky. The remaining light is reddish, as is the light from the Sun we see at sunset, and this is the light that falls on the Moon. Thus, the eclipsed Moon appears reddish.

The partial phases of a lunar eclipse will be visible from the United States on July 5, 2001. The total lunar eclipses of May 16, 2003, and of November 9, 2003, will be visible from the eastern United States. Try to see them! (www)

REDSHIFT
http://www.harcourtcollege.com/astro/cosmos/rsce

Solar Eclipses

The outer layers of the Sun, known as the corona, are fainter than the blue sky. To study them, we need a way to remove the blue sky while the Sun is up. A total solar eclipse does just that for us. Solar eclipses arise because of a happy circumstance: Though the Moon is 400 times smaller in diameter than the solar photosphere (the disk of the Sun we see everyday), it is also about 400 times

Figure 9–23 When the Moon is between the Earth and the Sun, we observe an eclipse of the Sun. When the Moon is on the far side of the Earth from the Sun, we see a lunar eclipse. The part of the Earth's shadow that is only partially shielded from the Sun's view is called the penumbra; the part of the Earth's shadow that is entirely shielded from the Sun is called the umbra. (The diagram is not to scale.)

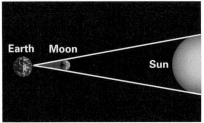

Solar eclipse

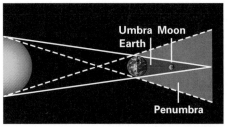

Lunar eclipse

Figure 9–24 An apple, the Empire State Building, the Moon, and the Sun are very different from each other in size, but here they cover the same angle because they are at different distances from us.

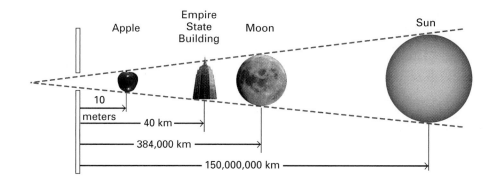

closer to the Earth. Because of this coincidence, the Sun and the Moon cover almost exactly the same angle in the sky—about ½° (Fig. 9–24).

Occasionally the Moon passes close enough to the Earth–Sun line that the Moon's shadow falls upon the surface of the Earth. The central part of the lunar shadow barely reaches the Earth's surface. This lunar shadow sweeps across the Earth's surface in a band up to 300 km wide. Only observers stationed within this narrow band can see the total eclipse.

From anywhere outside the band of totality, one sees only a partial eclipse. Sometimes the Moon, Sun, and Earth are not precisely aligned and the darkest part of the shadow—called the **umbra**—never hits the Earth. We are then in the intermediate part of the shadow, which is called the **penumbra.** Only a partial eclipse is visible on Earth under these circumstances. As long as the slightest bit of photosphere is visible, even as little as 1 per cent, one cannot see the important eclipse phenomena—the faint outer layers of the Sun—that are both beautiful and the subject of scientific study. Thus partial eclipses are of little value for most scientific purposes. After all, the photosphere is 1,000,000 times brighter than the outermost layer, the corona; if one per cent of the photosphere is showing, then we still have 10,000 times more light from the photosphere than from the corona, which is enough to ruin our opportunity to see the corona.

To see a partial eclipse or the partial phase of a total eclipse, you should not look at the Sun except through a special filter. (Alternatively, you can project the image of the Sun with a telescope or a pinhole camera onto a surface.) You need the filter to protect your eyes because the photosphere is visible throughout the partial phases before and after totality. Its direct image on your retina for an extended time could cause burning and blindness.

You still need the special filter to watch the final minute of the partial phases. As the partial phase ends, the bright light of the solar photosphere passing through valleys on the edge of the Moon glistens like a series of bright beads, which are called **Baily's beads.** At that time, the eclipse becomes safe to watch unfiltered, but only with the naked eye, not with binoculars or telescopes. The last bead seems so bright that it looks like the diamond on a ring—the **diamond-ring effect** (Fig. 9–25). ⓦⓦⓦ The phenomena of the darkening of the sky around you as totality approaches is dramatic. Watching on television loses most of the glory. For a few seconds, the chromosphere is visible as a pinkish band around the leading edge of the Moon. Then, as totality begins, the corona comes into view in all its glory (Fig. 9–26).

You see streamers of gas near the Sun's equator and finer plumes near its poles. The total phase may last a few seconds, or it may last as long as about 7 minutes. Spectrographs are operated; photographs are taken through special filters. Rockets may photograph the ultraviolet and x-ray spectrum from above the Earth's atmosphere, and tons of equipment are used to study the corona during this brief time of totality.

The next total solar eclipse will occur on June 21, 2001, and will be visible on land only from a path that crosses southern Africa. The total solar eclipse of December 4, 2002, will also be visible on land from Africa, though the end of the eclipse path reaches Australia. The next total eclipse visible from the United States isn't until 2017.

Should you go out of your way to observe a total solar eclipse? The entire event is indescribably beautiful and moving. Photographs and words simply do not convey the drama, beauty, and thrill of a total solar eclipse: It must be witnessed in person. In our opinion, everyone should see at least one!

Only during the total part of a total eclipse can one look at the Sun directly without special filters. (During the Baily's beads that mark the last two seconds or so before or after totality, it is also safe to observe the eclipse directly with the naked eye, though not with binoculars or a telescope.) During other phases and types of eclipses, you can observe the Sun safely only through special filters or by projecting the image so that you are not looking directly at the Sun.

Sometimes the Moon covers a slightly smaller angle in the sky than the Sun, because the Moon is on the part of its orbit that is relatively far from the Earth. When a well-aligned eclipse occurs in such a circumstance, the Moon doesn't quite cover the Sun. An annulus—a ring—of photospheric light remains visible, so we call this type an **annular eclipse** (see figure). An annular eclipse will cross Central America on December 14, 2001, and partial phases will be visible in most of the United States. People in the western United States can see partial phases of the annular eclipse of June 10, 2002.

During an annular eclipse, an annulus (ring) of photosphere remains visible. Here it is broken by the mountains of the Moon during the annular eclipse viewed from Australia in 1999.

Before then, maybe you can see the following eclipse: Sometimes, the central part of the Moon's shadow doesn't hit the Earth at all, and we on Earth never see more than a partial eclipse on that date. Such a partial eclipse will be visible from the United States on December 25, 2000.

At the end of the eclipse, the diamond ring appears on the other side of the Sun, followed by Baily's beads and then the more mundane partial phases. All too soon, the eclipse is over.

Figure 9–25 The diamond-ring effect marks the beginning and the end of the total phase of a solar eclipse.

Figure 9–26 The corona during a total solar eclipse.

Figure 9-27 Just as light is bent by space, a putted golf ball follows the warp of the green. Here Tiger Woods celebrates his successful accounting for the curvature of space.

THE SUN AND THE THEORY OF RELATIVITY

The intuitive notion we have of gravity corresponds to the theory of gravity advanced by Isaac Newton in 1687. We now know, however, that Newton's theory and our intuitive ideas are not sufficient to explain the Universe in detail. Theories advanced by Albert Einstein in the first decades of the 20th century now provide us with a more accurate understanding.

Einstein was a young clerk in the patent office in Switzerland when, in 1905, he published three scientific papers that revolutionized physics. One of the papers dealt with the effect of light on metals, another dealt with the irregular motions of small particles in liquids, and the third dealt with motion.

Einstein's work on motion became known as his "special theory of relativity." It is "special" in the sense that it does not include the effect of gravity or acceleration and is therefore not "general." "Special," thus, means "limited" in this case. Einstein's special theory of relativity assumes that the speed of light—300,000 km/sec—is an important constant that cannot be exceeded by real objects. This theory must be used to explain the motion of objects moving very fast, and has been thoroughly tested and verified.

In 1916, Einstein came up with his **"general theory of relativity,"** which explains gravity and also deals with accelerations. In this theory, we think objects have gravity but in truth they are simply moving freely in space that is curved. In fact, Einstein showed that time is a dimension, almost but not quite like the three spatial dimensions. Thus Einstein worked in four-dimensional space-time.

Picture a ball rolling on a golf green. If the green is warped, the ball will seem to curve (Fig. 9–27). If the surface could be flattened out, though, or if we could view it from a perspective in which the surface were flat, we would see that the ball is rolling in a straight line (the shortest path between two points). This analogy shows the effect of a two-dimensional space (a surface) curved into an extra dimension. Einstein's general theory of relativity treats mathematically what happens as a consequence of the curvature of space-time.

In Einstein's mathematical theory, the presence of a mass curves the space, just as your bed's surface ceases to be flat when you put your weight on it. Light travelling on Einstein's curved space tends to fall into the dents, just as a ball rolling on your bed would fall into its dents. Einstein, in particular, predicted that light travelling near the Sun would be slightly bent because the Sun's gravity must warp space (Fig. 9–28).

Figure 9-28 According to Einstein's general theory of relativity, the presence of a massive body warps the space nearby. The solid line shows the actual path of light. We on Earth, though, trace back the light in the direction from which it came. At any given instant we have no direct knowledge that the light's path had curved, so assume that the light has always been travelling straight.

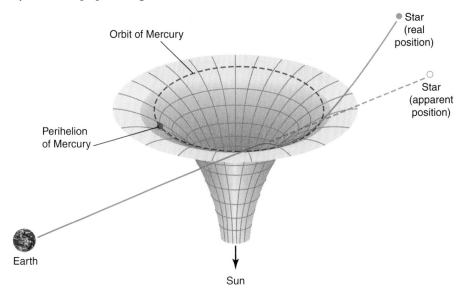

Figure 9-29 The prediction in Einstein's own handwriting of the deflection of starlight by the Sun. In this early version of his theory, Einstein predicted half the value he later calculated. The number he gives is slightly further wrong, because of an arithmetic error. We also see his signature with the name of his university.

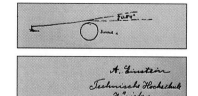

The Sun, as the nearest star to the Earth, has been very important for testing some of the predictions of Einstein's general theory of relativity. The theory could be checked by three observational tests that depended on the presence of a large mass like the Sun for experimental verification.

First, Einstein's theory showed that the closest point to the Sun of Mercury's elliptical orbit would move slightly around the sky over centuries. Such a movement had already been noted. Calculations with Einstein's theory accounted precisely for the amount of the movement. It was a plus for Einstein to have explained it, but the true test of a scientific theory is really whether it predicts new things rather than simply explaining old ones.

The second test arose from a major new prediction of Einstein's theory: Light from a star would act as though it were bent toward the Sun by a very small amount (Fig. 9–29). We on Earth, looking back past the Sun, would see the star from which the light was emitted as though it were shifted slightly away from the Sun. Only a star whose radiation grazed the edge ("limb") of the Sun would seem to undergo the full deflection; the effect diminishes as one considers stars farther away from the solar limb. To see the effect, one has to look near the Sun at a time when the stars are visible, and this could be done only at a total solar eclipse. The effect was verified at the total solar eclipse of 1919. Scientists hailed this confirmation of Einstein's theory, and from the moment of its official announcement, Einstein was recognized by scientists and the general public alike as the world's greatest scientist (Fig. 9–30).

Similar observations have been made at subsequent eclipses, though they are very difficult to make. Fortunately, the effect is constant through the spectrum and the test can now be performed more accurately by observing how the Sun bends radiation from radio sources, especially quasars (which are discussed in Chapter 17). The results agree with Einstein's theory to within 1 per cent, enough to make the competing theories very unlikely. As we shall see later on, the effect is now very well tested farther out in the Universe, and many studies are being made today taking advantage of the "gravitational lensing" effect.

The third traditional test of general relativity was to verify the prediction of Einstein's theory that gravity would cause the spectrum to be redshifted. This effect is very slight for the Sun, but has been detected. It has best been verified for extremely dense stars known as white dwarfs (see Chapter 13), in which mass is very tightly packed together.

As a general rule, scientists try to find theories that not only explain the data that are at hand, but also make predictions that can be tested. This is an important part of the scientific method. Because the bending of radiation was a prediction of the general theory of relativity that had not been anticipated, its verification provided more convincing evidence of the theory's validity than the theory's ability to explain the known shift in Mercury's orbit. Now, decades later, the general theory of relativity is a standard part of the arsenal of tools of a theoretical astronomer. We will meet it in several places later in this book.

ECLIPSE SHOWED GRAVITY VARIATION

Diversion of Light Rays Accepted as Affecting Newton's Principles.

HAILED AS EPOCHMAKING

British Scientist Calls the Discovery One of the Greatest of Human Achievements.

Copyright, 1919, by The New York Times Company.

Special Cable to THE NEW YORK TIMES.

LONDON, Nov. 8.—What Sir Joseph Thomson, President of the Royal Society, declared was " one of the greatest —perhaps the greatest—of achievements in the history of human thought " was discussed at a joint meeting of the Royal Society and the Royal Astronomical Society in London yesterday, when the results of the British observations of the total solar eclipse of May 29 were made known.

There was a large attendance of astronomers and physicists, and it was generally accepted that the observations were decisive in verifying the prediction of Dr. Einstein, Professor of Physics in the University of Prague, that rays of light from stars, passing close to the sun on their way to the earth, would suffer twice the deflection for which the principles enunciated by Sir Isaac Newton accounted. But there was a difference of opinion as to whether science had to face merely a new and unexplained fact or to reckon with a theory that would completely revolutionize the accepted fundamentals of physics.

The discussion was opened by the Astronomer Royal, Sir Frank Dyson, who described the work of the expeditions sent respectively to Sobral, in Northern Brazil, and the Island of Principe, off the west coast of Africa. At each of these places, if the weather were propitious on the day of the eclipse, it would be possible to take during totality a set of photographs of the obscured sun and a number of bright stars which happened to be in its immediate vicinity.

The desired object was to ascertain whether the light from these stars as it passed by the sun came as directly toward the earth as if the sun were not there, or if there was a deflection due to its presence. And if the deflection did occur the stars would appear on the

Figure 9-30 The first report of the eclipse results, showing how the eclipse captured the public's attention.

Lives in Science
ALBERT EINSTEIN

Albert Einstein.

Albert Einstein was born in Ulm, Germany, in 1879. He graduated from a university in Switzerland. Unable to get an academic position, he went to work in the Patent Office as an examiner. Out of this lemon, he made lemonade: he later stated that this job with its set working hours left him all the rest of his time to do physics unconstrained.

In 1905, Einstein published papers on three major ideas of physics. He explained the jiggling motion of tiny particles that a careful observer can see moving around inside fluids. The particles are hit, he said, time and again and driven from place to place. Next, he explained the mechanism by which light falling on a metal surface causes electrons to be given off. This work turned out to be basic to our current ideas of atoms and how they interact with light. (He later won the Nobel Prize in Physics for it.) It showed that particles of light exist, the particles we now call photons. And finally, he advanced his special theory of relativity, which described motion and gave a central role to the speed of light. He showed that time, space, and motion are all relative to the observer; absolute measurements of these quantities cannot be made.

Einstein's work made him well known in scientific circles, and he was offered university professorships. Over the next decade, he worked incessantly on incorporating gravity into his theories, and by 1914 had a prediction of the angle by which light passing near the Sun would be bent. He was then a professor in Berlin. German scientists went to Russia to try to observe this effect during a total solar eclipse, but were interned during the war and could not make their observations. The fact that they did not observe the eclipse in 1914 turned out to be a blessing, for Einstein had not yet found an adequate form of the theory, and his prediction was too low by a factor of 2. By 1916, Einstein had revised his theory, and his prediction was the value that we now know to be correct. When it was verified at the 1919 eclipse, Einstein triumphed.

The coming to power of the Nazis in the 1930s forced Einstein to renounce his German citizenship. He was persecuted as a Jew and his work was attacked as "Jewish physics," as though one's religion had some bearing on scientific truth. He accepted an offer to come to America to be the first professor at the Institute for Advanced Study, which was being set up in Princeton, New Jersey. During his years there, Einstein continued to work on scientific problems. However, his ideas were far from the mainstream and, partly as a result, his work on unifying the fundamental forces of nature did not succeed. Of course, Einstein had long worked far from the mainstream, but this time he was avoiding the basic physics of atoms—quantum mechanics—which he did not accept.

Einstein was a celebrity in spite of himself; he was so much photographed that he once gave his profession as "artist's model." He devoted himself to pacifist and Zionist causes and was very influential.

Einstein was a strong backer of the State of Israel and was even once asked to be its President. His understanding of the Nazi peril forced him to put aside his pacifism in that context. On one important occasion, scientists wishing to warn President Roosevelt that atomic energy could lead to a Nazi bomb enlisted Einstein to help them reach the president's ear.

Einstein died in 1955. A project is now well under way to publish all his papers in a uniform set of books. The volume containing his childhood documents has given insights into the formation of his thought. For his achievements and their influence on modern science and technology, Einstein was named "Person of the Century" in the December 31, 1999, issue of TIME magazine.

THE SCIENTIFIC VALUE OF ECLIPSES

In these days of orbiting satellites, why is it worth making an expedition to observe a total solar eclipse for scientific purposes? There is much to be said for the benefits of eclipse observing. Eclipse observations are a relatively inexpensive way, compared to space research, of observing the outer layers of the Sun. Artificial eclipses made by spacecraft hide not only the photosphere but also the inner corona. And for some kinds of observations, space techniques have not yet matched ground-based eclipse capabilities.

Concept Review

The everyday layer of the **quiet Sun** we see is the **photosphere;** the sunspots in it are signs of the **active Sun.** Beneath it is the solar **interior,** with the energy generated at the solar **core,** whose temperature is 15 million kelvins. The **solar atmosphere** above the photosphere contains the **chromosphere** and **corona.** The corona expands into space as the **solar wind,** which bathes the Earth.

The Sun's surface, observed in all the visible light together (which is known as **white light**) is covered with tiny **granulation.** Oscillations of the surface reveal the solar interior. The spectrum of the photosphere shows millions of Fraunhofer (absorption) lines.

The corona, best seen at total solar eclipses, contains gas at 2 million kelvins. At eclipses and from space we see how the magnetic field holds the coronal gas into **streamers.** Regions where the corona is less dense and cooler than average are **coronal holes.**

The **solar-activity cycle,** including the **sunspot cycle,** lasts about 11 years (or about 22 years, if magnetic polarity is in-cluded). **Sunspots** are regions of strong magnetic field and are cooler than the surrounding photosphere. Each sunspot has a dark **umbra** surrounded by a lighter **penumbra.** The **Maunder minimum** was a 17th and 18th century period when sunspots were essentially absent. Solar **flares** are eruptions of tremendous amounts of energy, and occur in sunspot regions. **Prominences** are **filaments** seen in silhouette off the edge of the Sun, following **magnetic-field lines.**

At a total solar eclipse, the **umbra** of the Moon's shadow hits the Earth. (People in the **penumbra** see only a partial eclipse.) Just before an eclipse is total, the last bits of photosphere appear as **Baily's beads.** The last bead makes the **diamond-ring effect.** When the Moon's disk appears too small to cover the Sun, we have an **annular eclipse.**

Some of the basic tests of Einstein's **general theory of relativity** involved the large mass of the Sun. The Sun was seen to bend starlight passing near it, and it was verified that the Sun's mass affects the orbit of Mercury.

Questions

1. Sketch the Sun, labelling the interior, the photosphere, the chromosphere, the corona, sunspots, and prominences.
2. Draw a graph showing the Sun's approximate temperatures, starting with the core and going upward through the corona.
3. Define and contrast a prominence and a filament.
4. Why are we on Earth particularly interested in coronal holes?
5. List three phenomena that vary with the solar-activity cycle.
6. Why can't we observe the corona every day from the Earth's surface?
7. How do we know that the corona is hot?
8. Describe relative advantages of ground-based eclipse studies and of satellite studies of the corona.
9. Describe the sunspot cycle.
†10. If the Moon were placed at twice its current distance from the Earth, how large would it need to be so that total solar eclipses could still occur?
†11. From the table in *A Closer Look: The Most Common Elements in the Sun's Photosphere,* calculate the percentage of helium atoms in the Sun and the percentage of iron atoms.
12. What are two ways that the Sun's corona can be studied from the Solar and Heliospheric Observatory (SOHO)?
13. What is the difference between the special theory of relativity and the general theory of relativity?
14. Why was the Sun useful for checking the general theory of relativity?
15. Explain in your own words how Einstein's general theory of relativity explains the Sun's gravity.

---

†This question requires a numerical solution.

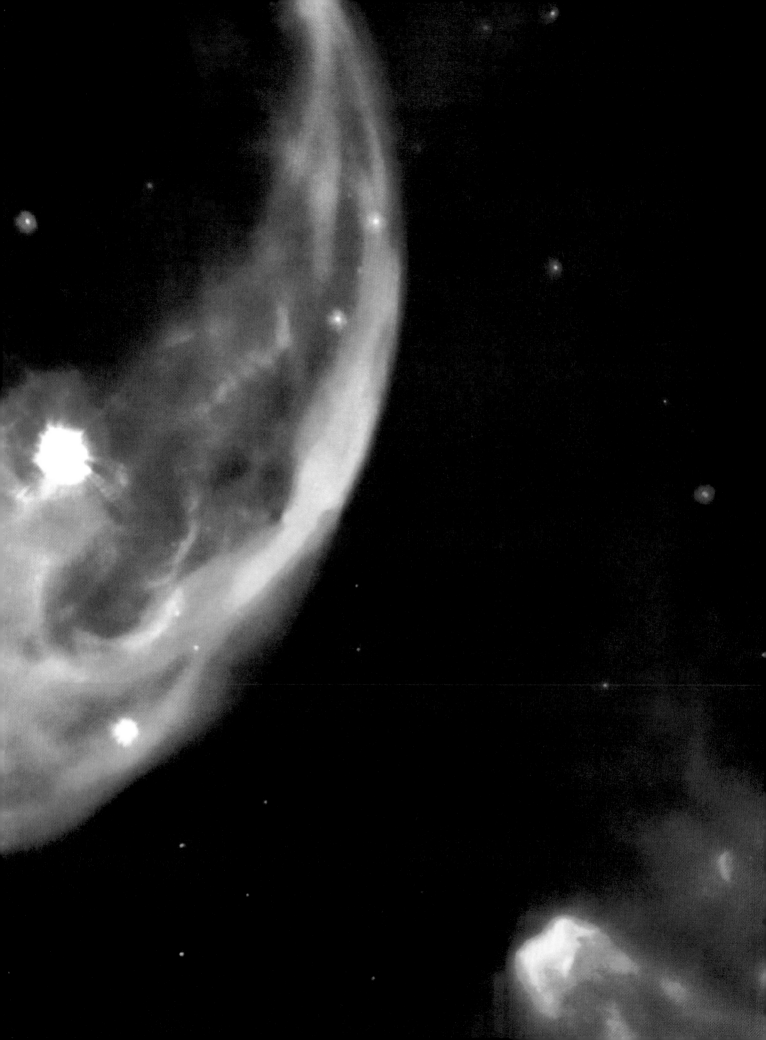

10

Stars: Distant Suns

ORIGINS *The stars are simply distant versions of our Sun. By studying their properties, such as surface temperature and intrinsic brightness, we can achieve a deeper understanding of the Sun and how its light and other energy originate.*

The thousands of stars in the sky that we see with our eyes, and the millions more that telescopes reveal, are glowing balls of gas. Their bright surfaces send us the light that we see. Though we learn a lot about a star from studying its surface, we can never see through to a star's interior where the important action goes on. In this chapter, we will discuss the surfaces of stars and what they tell us. First we discuss how we tell the temperatures of stars and what we observe to study them. We also discuss how stars move, and how we determine their distances. In the next chapter we will learn about stars that come with friends: other stars or groups of stars. Only when we finish these useful studies will we go on to discuss the stellar interiors.

AIMS

To learn how the color and spectral type of stars tell us their surface temperatures.

•

To explore the most fundamental way of measuring how far away stars are, and to see how their intrinsic brightnesses are determined.

•

To learn more about how the Doppler effect tells us a star's motion toward or away from us.

⬤ COLORS AND TEMPERATURES

When you heat an iron poker in a fire, it begins to glow and then becomes red hot. If we could make it hotter still, it would become white hot, and eventually bluish-white. To understand the temperature of the poker or of a hot gas, we measure its spectrum (Fig. 10–1). A dense, opaque gas or a solid gives off a continuous spectrum, that is, light changing smoothly in intensity from one color to the next.

Recall from Chapter 2 that a "black body" is a perfect emitter: its spectrum depends only on its temperature, not on chemical composition or other factors. We can approximate the spectrum of the visible radiation from the outer layer of a star as a **black-body curve.** The black-body curve is also called the Planck curve, in honor of the physicist Max Planck. Its derivation about 100 years ago was a triumph in the early development of quantum physics.

A different black-body curve corresponds to each temperature (Fig. 10–2). Note that as the temperature increases, the gas gives off more energy at every wavelength. Indeed, per unit of surface area, a hot black body emits much more energy per second than a cold one. Note also that the wavelength at which most energy is given off is farther and farther toward the blue as the temperature in-

◀ An expanding shell of glowing gas, the Bubble Nebula, around a hot, massive star. Winds from the star, the bright white object at left, are sculpting gas and dust into the bubble, which is 10 light-years across. The gas at lower right is set to glowing by radiation from the star. The image is from the Hubble Space Telescope.

Figure 10–1 When a narrow beam of light is dispersed (spread out) by a prism, we see a spectrum. Incoming light is usually passed through an open slit to provide the narrow beam.

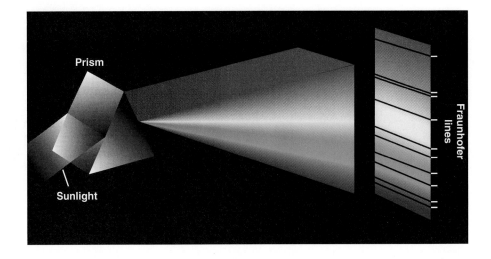

creases. The wavelength of this peak energy is shown with a dotted line. At temperatures of 4000 K, 5000 K, 6000 K, and 7000 K, the peak of the black-body curve is at wavelengths of 7200 Å, 5800 Å, 4800 Å, and 4100 Å, respectively. Thus, the hottest stars look blue and the coolest ones are red. Those of intermediate temperature (like the Sun) appear white, despite having spectra that peak around yellow or green wavelengths, because of the physiological response of our eyes. ☼

Astronomers rely heavily on the quantitative expression of the two radiation laws noted in the preceding paragraph to determine various aspects of stars (see Chapter 2, *Figure It Out: Black-Body Radiation*). For example, by simply measuring the wavelength of the peak brightness of a star's spectrum, astronomers can take the star's temperature.

The actual amount of radiation a star gives off depends both on how hot its surface is and on how large its surface is. (Again, see Chapter 2, *Figure It Out: Black-Body Radiation*.) But the shape of the black-body curve doesn't change with a star's size.

Figure 10–2 The brightness (intensity) of radiation at different wavelengths for different temperatures. Black bodies—ideally radiating matter—give off radiation that follows these curves. Stars follow these curves fairly closely.

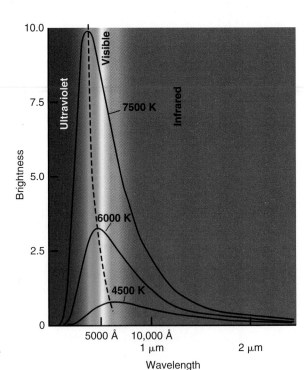

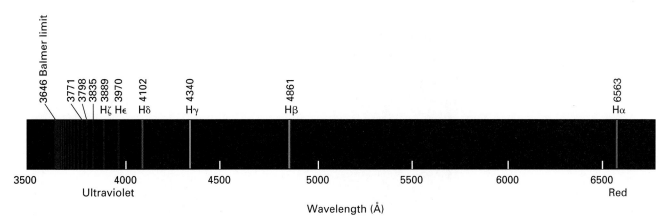

Figure 10–3 The series of spectral lines of hydrogen that appears in the visible part of the spectrum. The strongest line in this series, Hα (H-alpha), is in the red. The series is labelled with the first letters of the Greek alphabet.

THE SPECTRAL TYPES OF STARS

All the spectral lines of normal stars are absorption lines, also called dark lines. The absorption lines cause a star's spectrum to deviate from that of a black body, but not by much. Studying the absorption lines has been especially fruitful in understanding the stars and their composition.

We can duplicate on Earth many of the contributors to the spectra of the stars. Since hydrogen has only one electron, hydrogen's spectrum (Chapter 2) is particularly simple (Fig. 10–3). Atoms with more electrons have more possible energy levels, resulting in more choices for jumps between energy levels. Consequently, elements other than hydrogen have more complicated sets of spectral lines.

When we look at a variety of stars, we see many different sets of spectral lines, usually from many elements mixed together in the star's outer layers. Hydrogen is usually prominent, though often iron, magnesium, sodium, or calcium lines are also present.

Early in the 20th century, Annie Jump Cannon at the Harvard Observatory classified hundreds of thousands of stars by their visible-light spectra (Fig. 10–4). She first classified them by the strength of their hydrogen lines, defining stars with the strongest lines as "spectral type A," stars of slightly weaker hydrogen lines as "spectral type B," and so on. (Many of the letters wound up not being used.) It was soon realized that hydrogen lines were strongest at some particular temperature, and were weaker at hotter temperatures (because the electrons escaped completely from the atom) or at cooler temperatures (because the electrons were only on the lowest possible energy levels, which do not allow the visible-light hydrogen lines to form). Rearranging the spectral types in order of temperature—from hottest to coolest—gave O B A F G K M. Recently, even

Figure 10–4 Part of the "computing staff" of the Harvard College Observatory in 1917, when the word "computer" meant a person and not a machine. Annie Jump Cannon, 5th from the right, classified over 500,000 spectra in the decades following 1896. We discuss the work of Henrietta Leavitt, 5th from the left, when we discuss variable stars in Chapter 11.

Henrietta Leavitt   *Annie Cannon*

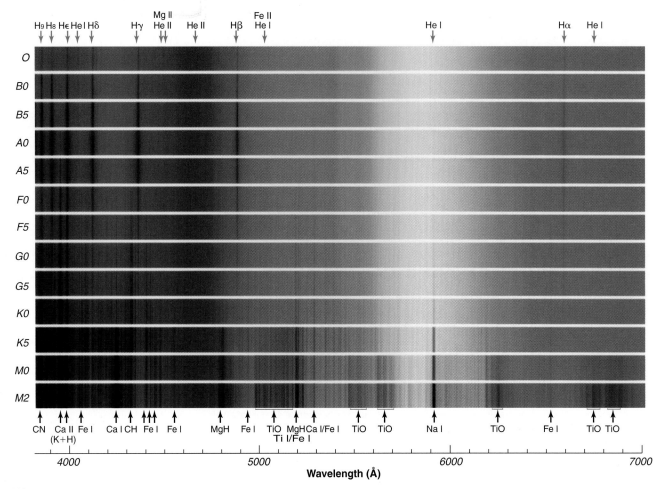

Figure 10–5 Computer simulation of the spectra of stars from a wide range of spectral types. The hottest stars are at the top and the coolest at the bottom. Notice how the hydrogen lines decline in darkness for types either above or below A stars. The original lettering was in order of the strength of the hydrogen lines. Only subsequently was the temperature ordering discovered.

cooler stars have been found, and are designated L-stars. Generations of astronomy students have memorized the order using the mnemonic "Oh, be a fine girl [or guy], kiss me." We invite you to think of your own mnemonic for O B A F G K M, and we will be glad to receive your entries at this book's Web site.

Looking at a set of stellar spectra in order (Fig. 10–5) shows how the hydrogen lines are strongest at spectral type A, which corresponds roughly to 10,000 K. A pair of spectral lines from calcium becomes strong in spectral type G (a type that includes the Sun), at about 6000 K (Fig. 10–6). In very cool stars, those of spectral type M, the temperatures are so low (only 3000 K) that molecules can survive, and we see complicated sets of spectral lines from them. At the other extreme, the hottest stars, of spectral type O, can reach 50,000 K. Some stars of type O have shells of hot gas around them and give off emission lines. They are known as Wolf-Rayet stars (Fig. 10–7).

Astronomers subdivide each spectral type into ten subtypes ranging from hottest (0) to coolest (9); thus, for example, we have A0 through A9, and then F0 through F9. The first thing an astronomer does when studying a star is to determine its spectral type and thus its surface temperature. The Sun is a type G2 star, corresponding to a temperature of 5800 K.

We have known that stars are mainly made of hydrogen and helium only since the 1920s. When Cecilia Payne (later Cecilia Payne-Gaposchkin) at Harvard first suggested the idea that stars were mainly made of hydrogen, based on her analyses of stellar spectra, it seemed impossible. But she was right. It was years before other astronomers accepted it.

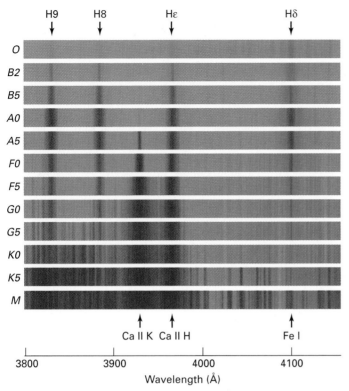

Figure 10-6 The violet and blue part of the previous image, displayed to show more clearly the change in visibility of the H and K lines of Ionized calcium (Ca II) versus the H-epsilon line of hydrogen as spectral types go from cooler to hotter.

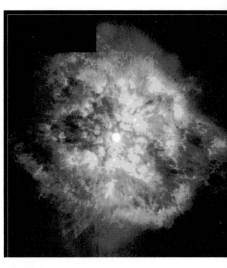

Figure 10-7 A Wolf-Rayet star *(center)* surrounded by hot clumps of gas that are being ejected into space at tremendously high speeds. This Hubble Space Telescope image also shows chaotic structures, extending 100 billion km into space around this hot star.

⬤ THE DISTANCES OF STARS

When you spot your friends on campus, you might use your binocular vision to judge how far away they are. Your brain interprets the slightly different images from your two eyes to give such nearby objects some three-dimensionality and to assess their distance. Also, we often unconsciously judge how far away an object is by assessing its apparent size compared with the sizes of other objects.

But the stars are so far away that they are all points, so we cannot judge their size (Fig. 10–8). And our eyes are much too close together to give us binocular vision at that great distance. Only for the nearest stars can we reliably measure their distance fairly directly even with our best methods, as we shall see next.

⬤ TRIANGULATING TO THE NEARBY STARS

The distances of nearby stars can be determined by triangulation—a sort of binocular vision obtained by taking advantage of our location on a moving platform (the Earth). The basic idea is that the position of a nearby object shifts relative to distant objects when viewed from different lines of sight.

For example, if you put a finger in front of your eyes, and close one eye, the finger will appear to be projected against a particular background object. If you close this eye and open the other one, the position of the finger will shift to a different background object. The amount of shift is smaller if the finger is farther away.

Figure 10-8 Stars near the center of our galaxy, in exaggerated color. They are so far away that they appear as merely points. Their colors reveal their temperatures, and these stars show a wide variety.

Figure It Out

Stellar Triangulation

The technique of measuring the distance of a nearby star by triangulation is also known as "trigonometric parallax." We photograph the star twice, about 6 months apart. We then measure the angular shift of the star relative to galaxies or very distant stars, whose positions are stationary or nearly so.

The "parallax" (*p*) of a star is defined to be *half* of the angular shift produced over a 6-month baseline (2 A.U., the diameter of Earth's orbit—see Figure 10–9). The distance of a star whose parallax is 1″ (1 second of arc) is called 1 parsec, abbreviated 1 pc. (This comes from a *par*allax of one *sec*ond of arc. See Chapter 4, *Figure It Out: Angular Resolution*, for a review of angular measure.) In the more familiar units of light-years, 1 pc is 3.26 ly.

It turns out that with this definition of a parsec, the relationship between distance and parallax is very simple. A star's distance *d* (in parsecs) is simply the inverse (reciprocal) of its parallax *p* (in seconds of arc): $d = 1/p$. For example, a star whose measured parallax is 0.5″ has a distance of $1/0.5 = 2$ pc, or about 6.5 ly. The nearest star, Proxima Centauri, has a parallax of 0.77″ (its full angular shift over a 6-month period is twice this value, 1.54″), and hence a distance of 1.3 pc (or 4.2 ly).

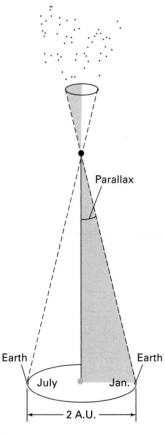

Figure 10–9 The nearer stars seem to be slightly displaced with respect to the farther stars when viewed from different locations in the Earth's orbit.

Here's how we apply this method to nearby stars (see *Figure It Out: Stellar Triangulation* for details). At six-month intervals, the Earth carries us entirely across its orbit, half-way around. Since the average radius of the Earth's orbit is 1 astronomical unit (A.U.), we move by 2 A.U. This distance is enough to give us a slightly different perspective on the nearest stars. These stars appear to shift very slightly against the background of more distant stars (Fig. 10–9). The nearest star—known as Proxima Centauri—appears to shift by only the diameter of a dime at a distance of 2.4 km! It turns out to be about 4.2 light-years away. (It is in the southern constellation of Centaurus, and is not visible from most of the United States.) By calculating the length of the long side of a giant triangle—by "triangulating"—we can measure distances in this way out to a few hundred light-years. But there are only a few thousand stars that are so close to us, and at the farthest distances the results are not very accurate.

The European Space Agency lofted a satellite, Hipparcos, in 1989 to measure the positions of stars. Based on its data, in 1997 the scientists involved released the Hipparcos catalogue, containing distances for 118,000 stars, relatively accurate to about 300 light-years away. A secondary list, the Tycho-2 catalogue, contains less-accurate distances for a million stars but provides accurate proper motions (motion across the sky, *see* p. 205) for a total of two and a half million stars. These catalogues greatly improve our fundamental knowledge of the distances of nearby stars. Moreover, our understanding of distant stars and even galaxies is often based on the distances to these closer stars, so the improvement is important for much of astronomy. NASA has selected the U.S. Naval Observatory's Full-sky Astrometric Explorer (FAME), to be launched in 2004, to send back accurate positions of 40 million stars out to about 6000 light-years away, about a quarter of the way from Earth to the center of our galaxy. (www)

But our galaxy is perhaps 100,000 light-years across, so even the 600 light-years (diameter) spanned by the Hipparcos catalogue takes up less than 1 per cent of the diameter of the galaxy. And for stars near the limit of the Tycho catalogue, our distances are still very uncertain.

◖◗ ABSOLUTE MAGNITUDES

Automobile headlights appear faint when they are far away but can almost blind us when they are up close. Similarly, stars that are inherently the same brightness appear to be of different brightnesses to us, depending on their distances. We have previously described (in Chapter 4) how we give their brightnesses in "apparent magnitude."

To tell how inherently bright stars are, astronomers have set a specific distance at which to compare stars. The distance, which is a round number in the special units astronomers use for triangulation (parsecs; see *Figure It Out: Stellar Triangulation*), comes out to be about 33 light-years. We calculate how bright a star would appear on the magnitude scale if it were 33 light-years away from us. We call this value its **absolute magnitude.** For a star that is actually at the standard distance, its absolute magnitude and apparent magnitude are the same. For a star that is farther from us than that standard distance, we would have to move it closer to us to get it to that standard distance. This would make its magnitude at the standard distance brighter than the magnitude we saw for it before it moved. Thus its absolute magnitude (the magnitude when it is at the standard distance) is brighter than its apparent magnitude. (A brighter star has a lower numerical value, 4th magnitude instead of 6th magnitude, for example.) The method links absolute magnitude, apparent magnitude, and distance, and so is particularly valuable, since astronomers have several ways of finding a star's absolute magnitude. Then by comparing its absolute magnitude and its apparent magnitude (which can easily be measured at a telescope), they calculate how far away the star must be.

The method works because we understand how a star's energy spreads out with distance (Fig. 10–10). The energy from a star changes with the square of the distance; the energy decreases as the distance increases. Since one value goes up as the other goes down, it is an inverse relationship. We call it the **inverse-square law.** Using this law, there is a simple way to express the luminosity (power, or intrinsic brightness) of a star in terms of its apparent brightness and distance (see *Figure It Out: The Inverse-Square Law*).

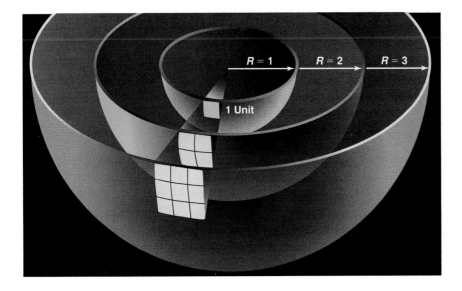

Figure 10–10 The inverse-square law of brightness. Radiation passing through a sphere twice as far away as another sphere from a central point has spread out so that it covers $2^2 = 4$ times the area. The central point thus appears 1/4 as bright to an observer located on the second sphere. If it were n times farther away, it would cover $1/n^2$ times the area, and appear $1/n^2$ as bright.

Figure It Out

The Inverse-Square Law

The magnitude scale is intuitive but from another era; still, astronomers often use it, and it appears in amateur astronomy magazines. It has the advantage of using small numbers that are relatively easy to remember. A different approach is to use the apparent brightness itself, b, instead of apparent magnitude. Its units are energy received per unit area, per unit time. For example, b can be measured in ergs/cm²/sec, where an "erg" is a small unit of energy (formally, one g-cm²/sec²). Similarly, instead of absolute magnitude, we can use luminosity L to denote a star's intrinsic brightness or power (the amount of energy it emits per unit time, ergs/sec).

A star's apparent brightness depends on its distance, because the light spreads out uniformly in all directions over a sphere whose radius is the distance. The larger the distance, the greater the surface area of the sphere, and so the smaller the amount of light passing through any small section of the sphere (see Figure 10–10). Since the surface area of a sphere is just 4π times the square of the radius, and in this case the radius is the star's distance d, the surface area is $4\pi d^2$. So, we find that the star's apparent brightness b is related to its luminosity L and distance d as follows: $b = L/(4\pi d^2)$. (Note that the units work out correctly: b is in ergs/cm²/sec, L is in ergs/sec, and d is in cm.) In other words, each unit area of area of the sphere (for example, 1 cm²) intercepts only a fraction $1/(4\pi d^2)$ of the total L.

This equation is just the well-known inverse-square law of light. For example, if a star's distance is doubled, it appears four times fainter, because a given amount of light has spread over a surface area four times larger (Figure 10–10). As an analogy, recall that if you are spraying water out of a hose by putting your thumb over part of the opening, you can either get many people slightly wet (by being relatively far from them) or one person very wet (by being close).

We now have the tools needed to determine the luminosity of a nearby star. First, measure its apparent brightness, b (for example, with a CCD); this procedure is called "photometry." Also measure its parallax, p (units: seconds of arc), and determine its distance in parsecs from the equation $d = 1/p$ (recall *Figure It Out: Stellar Triangulation*). Then convert parsecs to cm (the conversion factor is given in Appendix 2). Finally, use the inverse-square law to compute $L = 4\pi d^2 b$.

It is often convenient to express the luminosity in terms of the Sun's luminosity, $L_{Sun} = 3.83 \times 10^{33}$ ergs/sec. For example, a star whose luminosity is 7.66×10^{33} ergs/sec is said to have $L = 2L_\odot$. (In astronomy, the symbol $\odot$ with a dot in the middle of it denotes the Sun. Thus, the Sun's luminosity is often written as an L with this symbol as a subscript, $L_\odot$.) The Sun's luminosity is equivalent to an absolute magnitude of 4.8—that is, the Sun would just barely be visible to the naked eye if it were at a distance of about 33 light-years (10 pc).

TEMPERATURE-LUMINOSITY DIAGRAMS

If you plot a graph that has the surface temperature of stars on the horizontal axis (x-axis) and the luminosity (intrinsic brightness) of the stars on the vertical axis (y-axis), the result is called a **temperature-luminosity diagram.** The luminosity (vertical axis) can also be expressed in magnitudes, so the term temperature-magnitude diagram is synonymous. Sometimes the horizontal axis is labeled with spectral type, since temperature is often measured from spectral type or from a star's color.

If we plot such a diagram for the nearest stars (Fig. 10–11), we find that most of them fall on a narrow band that extends downward (fainter) from the Sun. If we plot such a diagram for the brightest stars we see in the sky, we find that the stars are more scattered, but that all are intrinsically brighter than the Sun.

The idea of such plots came to two astronomers in the early years of the 20th century. Henry Norris Russell (Fig. 10–12), at Princeton in the United States,

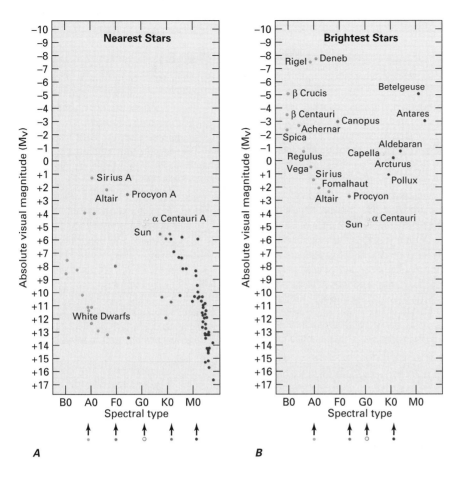

Figure 10–11 Temperature-luminosity (or temperature-magnitude) diagrams *(A)* for the nearest stars in the sky and *(B)* for the brightest stars in the sky. The brightness scale is given in absolute magnitude—the apparent magnitude a star would have at a standard distance of about 33 light-years. The upward arrow indicates increasing luminosity (intrinsic brightness). Because the effect of distance has been removed, the intrinsic properties of the stars can be compared directly on such a diagram.

Note that there are no stars near the top of the diagram to the left. Thus none of the nearest stars is intrinsically very luminous. Note further that there are no stars near the bottom of the diagram to the right. Thus none of the stars that appear brightest to us is intrinsically very faint.

Figure 10–12 Henry Norris Russell and his family, circa 1917.

plotted the absolute magnitudes (equivalent to luminosities) as his measure of brightness. Ejnar Hertzsprung (Fig. 10–13), in Denmark, plotted apparent magnitudes but did them for a group of stars that were all at the same distance. He could do this by considering a cluster of stars, since all the stars in the cluster are essentially the same distance away from us. The two methods came to the same thing, in that the magnitudes found for different stars could be directly compared. Temperature-luminosity (temperature-magnitude) diagrams are often known as Hertzsprung–Russell diagrams or as H–R diagrams.

When we graph quite a lot of stars, or put together both nearby and farther stars (Fig. 10–14), we can see clearly that most stars fall in a narrow band across the temperature-luminosity diagram. This band is called the **main sequence.** Normal stars in the longest-lasting phase of their lifetimes are on the main sequence. The position of a star on the main sequence turns out to be determined by its mass: Massive stars are hotter and more powerful than low-mass stars. The position of a star does not change much while it is on the main sequence: The Sun stays at more or less the same position on the main sequence for about 10 billion years. Stars on the main sequence are called **dwarfs,** so the Sun is a dwarf. ☼

A few stars are more luminous (intrinsically brighter) than main-sequence stars of the same surface temperature. Since the same surface area of gas at the same temperature gives off the same amount of energy per second, these stars must be bigger than the main-sequence stars (see *Figure It Out: A Star's Luminosity*). They are thus called **giants** or even **supergiants.** The reddish star Betelgeuse in the shoulder of the constellation Orion is a supergiant.

A few stars are intrinsically fainter than dwarfs (main-sequence stars) of the same color. These less luminous stars are called **white dwarfs.** Do not confuse

Figure 10–13 Ejnar Hertzsprung in the 1930s.

Figure It Out

A Star's Luminosity

We can easily relate the luminosity (intrinsic brightness, or power), radius, and surface temperature of a star. Recall (Chapter 2, *Figure It Out: Black-Body Radiation*) the Stefan-Boltzmann law, $E = \sigma T^4$, where E is the energy emitted per unit area (for example, cm^2) per second, T is the temperature, and σ (Greek lower-case "sigma") is a constant. To get the luminosity (ergs/sec) of a star, we must multiply E by the surface area, $4\pi R^2$ for a sphere of radius R, or $L = 4\pi R^2 \sigma T^4$.

Thus, if we know the luminosity and surface temperature of a star, we can derive its radius R. (The luminosity may have been derived with the inverse-square law by using the apparent brightness and distance, as in *Figure It Out: The Inverse-Square Law*. The surface temperature can be found from Wien's law, Chapter 2, *Figure It Out: Black-Body Radiation*.)

Suppose instead we have two stars, labelled by subscripts 1 and 2, with $L_1 = 4\pi R_1^2 \sigma T_1^4$ and $L_2 = 4\pi R_2^2 \sigma T_2^4$. If we divide one equation by the other, constants such as 4π and σ cancel, and we get $(L_1/L_2) = (R_1/R_2)^2 (T_1/T_2)^4$. So, for example, if Star 1 has twice the radius of Star 2, but only half the surface temperature of Star 2, then the luminosity of Star 1 will be 1/4 that of Star 2: $(L_1/L_2) = (2)^2 (1/2)^4 = (4)(1/16) = 1/4$. This example illustrates the advantage of using *ratios* when solving certain kinds of problems: Nowhere do we have to actually use the values of 4π and σ.

white dwarfs with normal dwarfs; white dwarfs are smaller than normal dwarfs. The Sun is 1.4 million kilometers across, while the white dwarf Sirius B (a companion of the bright star Sirius) is only about 10 thousand kilometers across, roughly the size of the Earth. We shall see later on that giants, supergiants, and white dwarfs are later stages of life for stars.

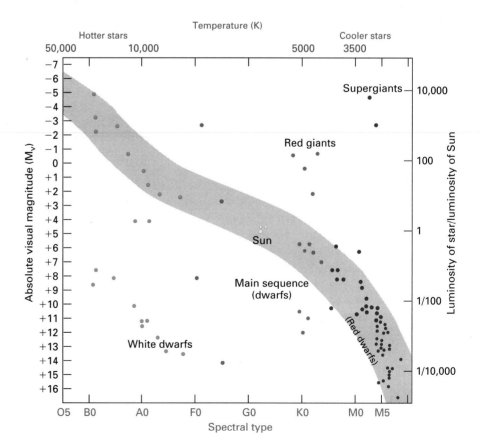

Figure 10–14 The temperature-luminosity (temperature-magnitude) diagram with both nearest and brightest stars included. The spectral-type axis *(bottom)* is equivalent to the surface temperature axis *(top)*. The absolute-magnitude axis *(left)* is equivalent to the luminosity axis *(right)*.

THE MOTIONS OF STARS

The stars are so far away that they hardly move across the sky relative to each other. (Don't confuse this relative motion of stars with the nightly motion of stars across the sky, which is merely a reflection of Earth's rotation about its axis, or with the seasonal changes of constellations, which is a reflection of Earth's orbit around the Sun.) Only for the nearest stars can we detect any such relative motion among the stars, which is called **proper motion.** Only for the most precise work do astronomers have to take the effect of the proper motions over decades into account. The Hipparcos satellite has improved our measurements of many stars' proper motions.

Astronomers can actually measure motions toward and away from us much better than they can measure motions from side to side. Recall (Chapter 2) that a motion toward or away from us on a line joining us and a star (a radius) is called a radial velocity. A radial velocity shows up as a **Doppler shift,** a change of the spectrum so that anything that originally appeared at a given wavelength now appears at another wavelength. We briefly discussed the Doppler shift in Chapter 2, but here we will give a more thorough explanation.

Doppler shifts in sound are more familiar to us than Doppler shifts in light. You can easily hear the pitch of a car's engine drop as the car passes us. We are hearing the wavelength of its sound waves increase as the object passes us and begins to recede. ☼ A similar effect takes place with light when we observe light that was emitted by an object that is moving toward or away from us. The effect, though, is not obvious to the eye; sensitive devices are necessary to detect the Doppler shift in light.

To explain the Doppler shift in light or sound, consider an object that emits waves of radiation (Fig. 10–15). The waves can be represented by spheres that show the peaks of the wave. Each sphere is centered on the object and is expanding. Consider an emitter moving in the direction shown by the arrow. In part A, we see that the peak emitted when the emitter was at point 1 becomes a sphere (labelled S_1) around point 1, though the emitter has since moved toward the left. In part B, some time later, sphere S_1 has continued to grow around point 1 (even though the emitter is no longer there). We also see sphere S_2, which shows the position of the peaks emitted when the emitter had moved to point 2. Sphere S_2 is thus centered on point 2, even though the emitter has continued to move on. In C, still later, yet a third peak of the wave has been emitted, sphere S_3, while spheres S_1 and S_2 have continued to expand.

For the case of the moving emitter, observers who are being approached by the emitting source (observers on the left side of C) see the three peaks coming past them bunched together. They pass at shorter intervals of time, as though the wavelengths were shorter. This situation corresponds to a color for each wavelength farther to the blue than it started out, so is called a **blueshift.** Observers from whom the emitter is receding (observers on the right side of C) see the three peaks at increased intervals of time, as though the wavelengths were longer. This situation corresponds to a color for each wavelength farther to the red than it started out, a **redshift.**

By contrast, for the stationary emitter, all the peaks are centered on the same point. No redshifts or blueshifts arise.

So whenever an object is moving away from us, its spectrum is shifted slightly to longer wavelengths. We say that the object's light is redshifted. When an object is moving toward us, its spectrum is blueshifted, that is, shifted toward shorter wavelengths. Even when an object's radiation is already beyond the red, we still say it is redshifted whenever the object is receding.

The fraction of its wavelength by which light is redshifted or blueshifted is the same as the fraction of the speed of light that the object is moving. (That is,

REDSHIFT
http://www.harcourtcollege.com/astro/cosmos/rsce

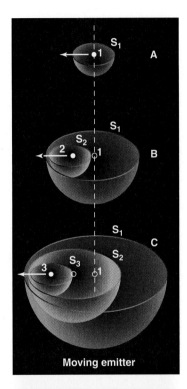

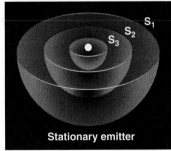

Figure 10–15 How Doppler shifts arise. When the emitter is moving toward us, the wave peaks arrive more frequently. We say that the radiation is blueshifted. When the emitter is moving away from us, the wave peaks arrive less frequently. We say that the radiation is redshifted. Radiation from a stationary emitter is not shifted in wavelength.

Figure It Out
Doppler Shifts

The wavelength emitted by a source of light that is not moving relative to the observer (it is "at rest") is known as its "rest wavelength" or "laboratory wavelength," λ_0 (pronounced "lambda nought"). If the source is moving relative to the observer, the observed wavelength λ will differ from this. For the Doppler shift of a moving source, we can write

$$\frac{\text{Change in wavelength}}{\text{Rest wavelength}} = \frac{\text{Speed of emitter}}{\text{Speed of light}},$$

or $\Delta\lambda/\lambda_0 = v/c,$

where v is the speed of the source toward or away from us, c is the speed of light (3×10^5 km/sec), and $\Delta\lambda = \lambda - \lambda_0$ is the change in wavelength (the upper-case Greek letter delta, Δ, usually stands for "the change in" or "the difference in"). This equation is valid only if v is much less than c; an expression from the special theory of relativity must be used at high speeds (say, $v > 0.2c$).

Note that when λ is larger than λ_0 (that is, $\lambda > \lambda_0$), the light is redshifted, so the source and observer are receding away from each other. Conversely, when $\lambda < \lambda_0$, the light is blueshifted, so the source and observer are approaching each other.

The procedure to use with a star, for example, is to obtain its spectrum, recognize a familiar pattern of lines (such as the hydrogen Balmer series), measure the observed wavelength of an absorption or emission line, and compare it with the known wavelength at rest to get $\Delta\lambda$.

For example, if the Hα absorption line is observed at $\lambda = 6565$ Å, and its rest wavelength is known to be $\lambda_0 = 6563$ Å, we find that $\Delta\lambda = \lambda - \lambda_0 = 2$ Å, so $\Delta\lambda/\lambda_0 = 2$ Å$/6563$ Å $= 3 \times 10^{-4}$. But we know that this result must be equal to v/c, so $v = (3 \times 10^{-4})c = (3 \times 10^{-4})(3 \times 10^5$ km/sec$) = 90$ km/sec. Thus, we and the star are moving away from each other at about 90 km/sec.

Figure 10–16 The Doppler effect in stellar spectra. In each pair of spectra, the position of a spectral line in the laboratory on Earth is shown at top, and the position observed in the spectrum of a star is shown below it. Lines from approaching stars appear blueshifted and lines from receding stars appear redshifted. Lines from stars that are not moving toward or away from us, even if they are moving transversely, are not shifted. A short, dashed, vertical line marks the unshifted position on the spectrum showing shifts.

the wavelength shift is proportional to the speed of the object; *see Figure It Out: Doppler Shifts*.) So astronomers can measure an object's radial velocity by measuring the wavelength of a spectral line and comparing the wavelength to a similar spectral line measured from a stationary source on Earth (Fig. 10–16).

Stars in our galaxy have only small Doppler shifts, which shows that they are travelling less than 1 per cent of the speed of light. Though they may be redshifted or blueshifted, the shifts are not enough to change the color that we see

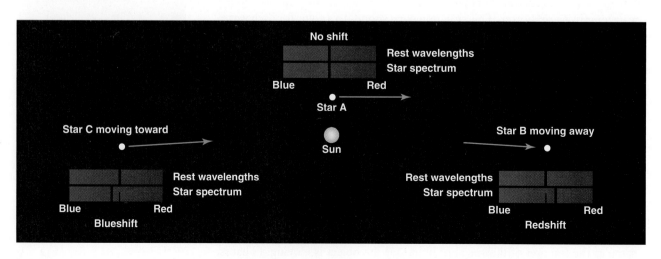

when we look at these stars. We shall find, when we discuss galaxies, that the situation is very different for the Universe as a whole. The observed redshifts of galaxies turn out to be the key to our understanding of the past and future of our Universe.

Concept Review

Heating dense, opaque matter causes it to grow much brighter and to have the maximum in its continuous spectrum shift to shorter wavelengths. The continuous radiation approximately follows a **black-body curve.** Classifying stars by their spectra led to O B A F G K M, with spectral-type A stars having the strongest hydrogen lines.

We find the distance to the nearest stars by triangulation. The Hipparcos spacecraft has improved the accuracy for a million stars, but only within about 1 per cent the diameter of our galaxy. **Absolute magnitudes** are the magnitudes that stars would appear to have if they were at a standard distance equivalent to 33 light-years from us. The energy from a star decreases with distance from it following the **inverse-square law.**

Plotting temperature (often as spectral type) on the horizontal axis and luminosity (intrinsic brightness) on the vertical axis gives a **temperature-luminosity diagram,** also known as a temperature-magnitude diagram. Most stars appear in a band known as the **main sequence,** which goes from hot, luminous stars to cool, dim stars. Stars on the main sequence are known as **dwarfs.** Some brighter stars exist and are **giants** or even **supergiants.** Less luminous stars than the main sequence for a given temperature are **white dwarfs.**

Stars have small motions across the sky known as **proper motions.** Studies of their **Doppler shifts** show their radial velocities as **blueshifts** or **redshifts.**

Questions

1. The Sun's spectrum reaches its maximum brightness at a wavelength of 5000 Å. Would the spectrum of a star whose surface temperature is higher than that of the Sun peak at a longer or a shorter wavelength?

†2. The Sun's surface temperature is about 5800 K and its spectrum peaks at 5000 Å. An O star's temperature may be 40,000 K. According to Wien's law, at what wavelength does its spectrum peak? In what part of the spectrum might that peak be? Can the peak be observed with the Keck telescopes? Explain.

3. One black body peaks at 2000 Å. Another, of the same surface area, peaks at 10,000 Å. Which gives out more radiation at 2000 Å? Which gives out more radiation at 10,000 Å?

†4. If a star has a surface temperature four times higher than that of the Sun, but the same surface area, by what factor is the star more luminous than the Sun?

5. Star A appears to have the same brightness through a red and blue filter. Star B appears brighter in the red than in the blue. Star C appears brighter in the blue than in the red. Rank these stars in order of increasing surface temperature.

6. What is the difference between continuous radiation and an absorption line? Continuous radiation and an emission line? Graph a spectrum that shows both continuous radiation and absorption lines. Can you draw absorption lines without continuous radiation? Can you draw emission lines without continuous radiation? Explain.

7. Make up your own mnemonic for the spectral types.

†8. What is the distance of a star whose parallax is 0.2″? What is the parallax of a star whose distance is 100 pc?

†9. If a star that is 100 light-years from us appears to be 10th magnitude, would its absolute magnitude be a larger or a smaller number?

†10. If the Sun were 10 times farther from us than it is, how many times less light would we get from it?

11. Sketch a temperature-luminosity diagram, and distinguish among giants, dwarfs, and white dwarfs. Be sure to label the axes.

12. Compare the surface temperatures of white dwarfs with those of dwarfs, and explain their relative brightness.

†13. If a red giant has half the Sun's surface temperature but 100 times its radius, what is the giant's luminosity relative to that of the Sun?

14. Two stars the same distance from us are the same temperature, but one is a giant while the other is a dwarf. Which appears brighter?

15. Does measuring Doppler shifts depend on how far away an object is? Explain.

16. Consider two stars of the same spectral type. Describe whether this information is sufficient to tell you how the stars differ in (a) surface temperature, (b) size, and (c) distance.

17. Suppose you find an object that has a Doppler shift corresponding to 20% the speed of light. Is the object likely to be a star in our galaxy? Why or why not?

18. Use the explanation of Doppler shifts to show what happens to the sound from a fire truck as the truck passes you.

19. Sketch a star's spectrum that contains two spectral lines. Then sketch the spectrum of the same star if the star is moving toward us. Finally, sketch the spectrum if the star is moving away from us.

†20. Suppose a star is moving toward you with a speed of 200 km/sec. Will its spectrum appear blueshifted or redshifted? At what wavelength will you observe the Hα line (rest wavelength 6563 Å) in the star's spectrum?

---

†This question requires a numerical solution.

People in Astronomy

GIBOR BASRI

Gibor Basri is Professor of Astronomy at the University of California at Berkeley. He grew up in Colorado and went to Stanford University for his B.Sc. Then he returned to Colorado for graduate school in the department then called Astrogeophysics, where he wrote his thesis about the chromospheres of supergiant stars. More recently, he was one of those astronomers who finally proved that brown dwarfs really exist. Professor Basri is a high-resolution spectroscopist and also does stellar model atmospheres on the computer. He works in the optical and ultraviolet (and soon also in the infrared). He recently was awarded a Miller Research Professorship to carry out a year's work on his current interest in brown dwarfs. He has recently summarized the latest in that exciting field in both Annual Review of Astronomy and Astrophysics *and in* Scientific American.

What are you most interested in now?

I am interested in the question of the formation of sub-stellar objects, namely giant planets and brown dwarfs. More generally, this involves the formation of stars. It's all part and parcel of the same process.

Were you very surprised that all of a sudden we have dozens of new planets outside our Solar System?

Mostly. I kind of expected it, but I didn't expect them to have the orbits they do. And the discovery of the brown dwarfs and planets at the same time was a very exciting moment. This is one of those convergences in science where people are looking and looking and being frustrated, when all of a sudden the dam breaks and it all falls out. It was very exciting to be part of that process.

Why do you think it took so long to find brown dwarfs?

It was partly a matter of technology development. They are very faint and people weren't looking quite hard enough, basically. And it was also a matter of develop-

ing the right tests to be sure that we have brown dwarfs. When Gliese 229B was found, it was a no-brainer since it was so cool that it couldn't possibly be a star. Just prior to that there were some brown dwarfs that we identified because they had lithium in them, which was more subtle but also convincing. Once people realized that you could find them, everybody went after them, and they just started dropping out of the sky, so to speak. Also, it happens that the all-sky infrared surveys started up around then, and that has been a major source of new brown dwarfs, too.

What excites you most about the new planets: astronomical reasons or the idea that they are a possible location for life?

In the end, it is the question of life that is exciting, but I am also excited about the fact that we now have techniques that allow us to attack this question empirically. I'm particularly excited about the new planet seen in transit in front of its star, since that is a more direct observation of a planet. I happen to believe that we will discover Earth-like planets first by transits, so this discovery opens people's eyes to the possibility. I am hopeful that this method will be exploited more aggressively now.

What is your personal method for finding brown dwarfs?

I got into this game mostly because of the advent of the Keck telescopes. I was privileged to be a Keck user and wanted to find something exciting to do with the new world's largest telescopes. The idea of the new lithium test for brown dwarfs had already been suggested by the Canary Islands group, but they were finding that their telescope was not quite up to the task. So I got involved in that and we were lucky enough to make the first discoveries and confirm that way that brown dwarfs exist.

The basic idea is that stars will destroy lithium when they start their hydrogen fusion. Most brown dwarfs will never get hot enough to destroy lithium, so they will retain it. So you can do a simple spectroscopic test to see if lithium is still present or not. If a star is a very faint red object and it shows lithium, then it is probably a brown dwarf.

Along the way, in applying that test carefully, we discovered that the age scale for young open clusters was off. It turns out they are all about 50% older than we thought. This is because the normal way for finding the ages of those clusters involves high-mass stars turning off the main sequence, and those stars have convective nuclear burning cores. Convective overshoot is a poorly understood process, but the cores of those stars can basically grab extra hydrogen and live longer. The stellar evolution people were aware of this potential problem but they didn't have a way of calibrating how much hydrogen would be grabbed. We know how long it takes stars to destroy lithium. You look down the pre-main sequence until you see that lithium is not yet destroyed; you see how bright those objects are and that gives you the age (since they get fainter with time in a known way).

When did you get interested in astronomy?

I think I was about 8 years old. I came to it through science fiction. When I started reading science fiction, I thought it was really cool stuff and I started learning about space. My father was a physicist and he encouraged me, and I never lost my interest after that. However, I did do a career report in the 8th grade and decided that astronomy was too small and esoteric a field to be a realistic career. Later, I was majoring in physics at Stanford, and realized that I only wanted to do it if I could do it as astrophysics. So I just thought I'd go to grad school, and thought I'd see how long my astronomical career lasted.

Can you tell me about some outside interests?

I'm an expert skier because I grew up in Colorado. We go up to the Lake Tahoe area. I don't go up as much as I'd like to, but I go up 6 or 7 times a year. I also like hiking a lot, and tennis. My wife is a psychologist and we have a 9-year-old son. He's a little interested in skiing but not so much in astronomy. He's more of a Drama person at the moment.

You teach both elementary and advanced classes. How do you contrast them?

I have everything from basic astronomy students to grad students. I find that almost everybody has some interest in astronomy. Of course, the grad students want to do it as a career so they have a deeper interest, but I never have trouble conversing about astronomy with a student at any level.

How exciting has astronomy been to you since you started your career?

It turned out to be a particularly good time to get into astronomy, just as I became a grad student. We had just opened up space astronomy. I was able, my last student year, to work with the IUE [the International Ultraviolet Explorer] to do hands-on real-time observations in space. Also, computers have gotten better and better and better. And detectors have become far more sensitive and versatile. It has been a particularly exciting time for the last twenty years or so, and I think it will go on for the next twenty years.

What do you think are some of the major discoveries that will be made in the next 20 years?

As I said, I think that we will discover Earth-like planets. We will also make the first measurements of atmospheres on other planets. I think we will make major progress toward the question of life on Mars and Europa (and perhaps other Solar System locations like Titan). Cosmology will resolve the question of the Hubble constant, the cosmological constant, and the fate of the Universe. We will understand the formation of stars and planets in great detail, and the formation of galaxies in much better detail than we have now. Ground-based telescopes will remove atmospheric blurring well, and from space we will measure parallaxes and motions of stars across our galaxy, giving us a precise distance scale and a real understanding of dynamics in our galaxy and beyond. Vast amounts of data at all wavelengths will be available over the Web for many to analyze in new and different ways. And computer models of very complicated systems with very great resolution and physical detail will provide a new means of observing the cosmos.

Social Stars: Binaries and Clusters

11

ORIGINS *The properties of all stars, including the Sun, depend mostly on their mass, which in special cases can be determined from studies of binary stars. Cepheid variable stars give celestial distances, which we will later see are important for studying the age and origin of our Universe. Star clusters can be used to determine the ages of stars, and to study the evolution of typical stars like our Sun.*

Though the Sun is an isolated star in space, most stars are part of pairs or groups. If our planet were part of such a system, we might see two or more Suns rising. And if our Sun were part of a cluster of stars, the nighttime sky would be ablaze with very bright points of light.

In this chapter, we will discuss stars that are different from our Sun in that either they have stellar companions or they vary considerably in brightness over time. Sometimes the apparent variation is caused by a companion, and sometimes it is something intrinsic to the star itself. At the end of the chapter, we will see how studying clusters of stars enables us to tell their ages.

AIMS

To learn about binary stars and star clusters, and to see how they are important for measuring stellar masses and ages.
•
To learn about variable stars and to see how some types can be used to measure distances.

BINARY STARS

Sometimes, a star that looks double appears so merely because two unconnected stars happen to appear in essentially the same direction. They are thus examples of **optical doubles,** chance apparent associations. For example, if you look up at the Big Dipper, you might be able to see, even with your naked eye, that the middle star in the handle (Mizar) has a fainter companion (Alcor). Native Americans called these stars a horse and rider.

However, over 50 per cent of the apparently single "stars" you see in the sky are actually multiple systems bound by gravity and orbiting each other. (More precisely, the stars orbit their mutual center of mass; see *Figure It Out: Binary Stars.*) Astronomers take advantage of such **binary stars** to find out how much mass stars contain.

When you look at Mizar through a telescope, you can see that it is split in two. In fact, Mizar was the first double star to have been discovered telescopically (in 1650). These two stars are revolving around each other, and are thus known as a **visual binary.** We see visual binaries best when the stars are relatively far away from each other (Fig. 11–1).

◄ The globular star cluster M80, imaged with the Hubble Space Telescope. It contains hundreds of thousands of stars, held together by their mutual gravity. The cluster is 28,000 light-years from us.

REDSHIFT
http://www.harcourtcollege.com/astro/cosmos/rsce

Figure 11–1 The double-star Albireo (also known as β [beta] Cygni) contains a B star and a K star. They make a particularly beautiful pair because of their contrasting colors. Albireo is high overhead on summer evenings.

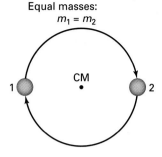

(CM = Center of Mass)
Equal masses:
$m_1 = m_2$

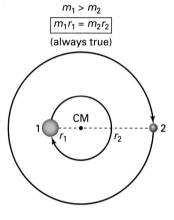

$m_1 > m_2$
$\boxed{m_1 r_1 = m_2 r_2}$
(always true)

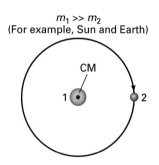

$m_1 \gg m_2$
(For example, Sun and Earth)

Figure It Out

Binary Stars

The two stars in a binary system generally have elliptical orbits. For simplicity, however, consider the orbits to be circular (see the figure). One can define the *center of mass* of the system with the equation $m_1 r_1 = m_2 r_2$, where r_1 and r_2 are the distances of Stars 1 and 2 (respectively) from the center of mass. Note that the center of mass must be along a line joining the two stars. This is analogous to two weights on opposite ends of a long plank that balances on a point, like a child's see-saw. The larger weight must be closer to the balance point than the smaller weight.

If the two stars have equal mass, they are the same distance from the center of mass. Their orbits have equal size, the stars move with equal speed, and their orbital periods are equal.

If $m_1 > m_2$, then Star 1 is closer to the center of mass than Star 2 is. The orbit of Star 1 is smaller than that of Star 2, and Star 1 moves around the center of mass with a lower speed than Star 2. Their orbital periods remain equal.

If m_1 is much, much larger than m_2 (that is, $m_1 \gg m_2$), then Star 1 is very close to the center of mass. The orbit of Star 1 is tiny, and Star 1 moves very slowly around the center of mass. Their orbital periods remain equal. A good example is the Sun-Earth system, ignoring the other planets. The Sun is nearly (but not quite) stationary. (This slight movement of a star makes it possible to detect planets indirectly, as discussed in Chapters 2 and 7.) We often say that the "Earth orbits the Sun," but a formally more correct statement is that the Sun and Earth each orbit the system's center of mass.

Most of these basic principles remain valid even when we consider elliptical orbits. The main difference is that the orbital speed of each star varies with position in the orbit; recall Kepler's second law (equal areas in equal times; Chapter 5).

The two components of Mizar are known as Mizar A and Mizar B. If we look at the spectrum of either one (Fig. 11–2), we can see that over a period of days, the spectral lines seem to split and come together again. The spectrum shows that two stars are actually present in Mizar A, and another two are present in Mizar B. They are **spectroscopic binaries,** since they are told apart by their spectrum. Even when only the spectrum of one star is detectable (Fig. 11–3), the other star being too faint, we can still tell that it is part of a spectroscopic bi-

Figure 11–2 Two spectra of this star taken 2 days apart show that it is a spectroscopic binary. The lines of both stars are superimposed in the upper stellar absorption spectrum *(red arrow)*. They are separated in the lower spectrum *(blue arrow)* by an amount that shows that the stars are then moving 140 km/sec with respect to each other.

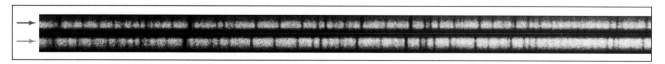

Figure 11–3 Two spectra of Castor B, taken at different times. The star's spectrum we see shows a change in the amount of Doppler shift caused by its changing speed toward us. Thus the star must be moving around another star. We conclude that it is a spectroscopic binary, even though only one of the components can be seen.

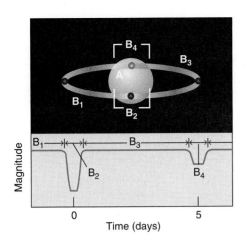 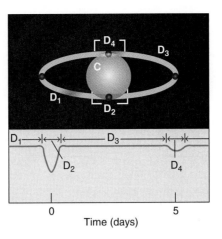

Figure 11–4 The shape of the light curve of an eclipsing binary depends on the sizes of the components and on the angle from which we view them. At *lower left,* we see the light curve that would result for star B orbiting star A, as pictured above it. When star B is in the positions shown at *top* with subscripts, the regions of the light curve marked with the same subscripts result. The eclipse at B_4 is total. At *right,* we see the appearance of the orbit and the light curve for star D orbiting star C, with the orbit inclined at a greater angle than at left. The eclipse at D_4 is partial.

nary if the spectral lines shift back and forth in wavelength over time. These shifts arise from the stars' motions. The radial velocity (that is, the speed along our line of sight) of each star at any given time can be determined from the Doppler formula (see Chapter 10, *Figure It Out: Doppler Shifts*).

Yet another type of binary star is detectable when the light from a star periodically dims because one of the components passes in front of (eclipses) the other (Fig. 11–4). From Earth, we graph the "light curves" of these **eclipsing binaries** by plotting their brightnesses over time. Eclipsing binaries are detected by us only because they happen to be aligned so that one star passes between the Earth and the other star. Thus all binaries would be eclipsing if we could see them at the proper angle. (At least one planet around a distant star has been detected in a similar fashion, by a dimming of the total light when it passes in front of its star.)

The same binary system might be seen as different types depending on the angle at which we happen to view the system (Fig. 11–5).

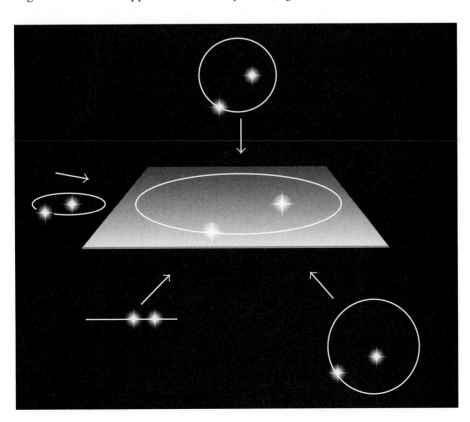

Figure 11–5 The appearance of the spectrum of a binary star and whether the lines shift in wavelength over time depend on the angle from which we view it. From far above or below the plane of the orbit, we might see a visual binary, as at *top* and at *lower right.* From close to the plane of the orbit, even if the stars are so close to each other in angle in the sky (closer than shown here *at left*), so that they can't be seen as separate stars, we might see only a spectroscopic binary, as at *left.* Similarly, for stars exactly in the plane of the orbit but too close in angle to see as separate objects, we could still see the system as an eclipsing binary, as at *lower left.*

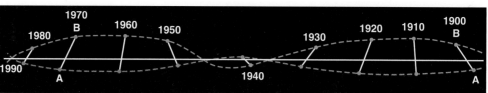

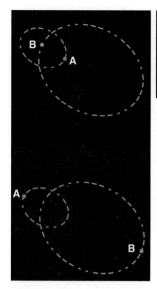

Figure 11–6 Sirius A and B, an astrometric binary. From studying the motion of Sirius A (usually called, simply, Sirius), astronomers deduced the presence of Sirius B before it was seen directly.

One more type of double star can be detected when we observe a wobble in the path of a star as its proper motion, its motion with respect to other stars in our galaxy, takes it slowly across the sky (Fig. 11–6). The laws of physics hold that a moving mass must move in a straight line unless affected by a force. The wobble off a straight line tells us that the visible star must be orbiting an invisible object, pulling the visible star from side to side. Measurement of the positions and motions of stars is known as **astrometry,** so these stars are called **astrometric binaries.** A similar method was used to try to detect planets around distant stars. The Hipparcos spacecraft, for example, was designed to measure such small motions of stars, and an even more precise telescope, the Full-sky Astrometric Explorer (FAME), is planned for launch in 2004. However, the new planets that have been located around other stars (Chapters 2 and 7) were found by using the Doppler effect rather than by astrometry.

A primary use of binary stars is the measurement of stellar masses. We can sometimes determine enough about the stars' orbits around each other that we can calculate how much mass the stars must have to stay in those orbits. The most accurate masses are derived from spectroscopic binaries in which the absorption lines of both stars are visible, and which are also eclipsing binaries. For these cases, we can also describe the radii of the two stars reliably. Such binary systems are relatively rare, so we know accurately only about one or two hundred stellar masses.

It turns out that the mass of a star is its single most important characteristic: Essentially everything else about its properties and evolution depends on it. In *Figure It Out: The Mass-Luminosity Relation* you can learn how the mass is connected to how intrinsically brightly the star shines. In particular, massive stars are far more luminous and have much shorter lives than low-mass stars. Though massive stars start with more fuel, they consume it at a huge rate.

When we discuss the evolution of stars, we will see that matter often flows from one member of a binary system to another. This interchange of matter can drastically alter a star's evolution. And the flowing matter can heat up by friction

Figure It Out

The Mass-Luminosity Relation

For main-sequence stars, we find that the luminosity L (intrinsic brightness) is roughly proportional to the fourth power of mass M, or $L \propto M^4$, the "mass-luminosity relation." This proportionality is only approximate; the exponent of M actually varies to some extent along the main sequence. However, it is sufficient for our purposes. Thus, for example, a star with twice the Sun's mass is a factor of $2^4 = 16$ times more luminous (powerful) than the Sun.

The mass-luminosity relation for main-sequence stars implies that massive stars have much shorter lives than low-mass stars. Massive stars have more fuel, but they consume it at a disproportionately rapid rate. For example, a star with twice the Sun's mass has twice as much fuel, so if all else were equal it would live twice as long. But it actually uses its fuel 16 times more rapidly! So, it can sustain itself for only $2/16$ (which is equal to $1/8$) as long as the Sun.

A Closer Look

A Sense of Mass: Weighing Stars

The amount of matter in a body is its *mass*. The standard kilogram resides at the International Bureau of Weights and Measures near Paris and is the ultimate standard of mass. To determine the mass of an object on Earth, we compare it, in some ultimate sense, with the standard kilogram or, more commonly, with secondary standard kilograms held by the various countries.

Kepler's third law, as modified by Newton (Chapter 5), gives us a way of determining the mass of an astronomical body from studying an object orbiting it. But this determination requires knowledge of the universal constant of gravitation, usually called *G*, that appears in Newton's law of gravity. This constant is measured by laboratory experiments, which limits the resulting accuracy. The first determination of the constant nearly two hundred years ago appeared in a paper whose title was "On Weighing the Earth." Note that weight is actually different from mass. An object's weight on Earth is the gravitational force of the Earth's mass on the object's mass. The object would have a different weight on another planet but the same mass.

The Earth's mass seems so large to us that we think we can jump up and down without moving the Earth. Actually when we jump up, the Earth moves away from us, according to Newton's third law of motion ("For every action, there is an equal and opposite reaction"). But the Earth's mass is 10^{23} times the mass of a person, so it doesn't move much!

The Earth and the rest of the Solar System may be important to us, but they are only minor companions to the Sun. In "Captain Stormfield's Visit to Heaven," by Mark Twain, the Captain races with a comet and gets off course.

He comes into heaven by a wrong gate, and finds that nobody there has heard of "the world." ("The world! H'm! there's billions of them!" says a gatekeeper.) Finally, someone goes up in a balloon to try to detect "the world" on a huge map. The balloonist rises into clouds and after a day or two of searching comes back to report that he has found it: an unimportant planet named "the Wart."

Indeed, the Earth is inconsequential in terms of mass next to Jupiter, which is 318 times more massive. And the Sun, an ordinary star, is 300,000 times more massive than the Earth. Some stars are up to 100 times more massive than the Sun, but others can be as little as $\frac{1}{12}$ as massive. We shall see that a star's mass is the key factor that determines how the star will evolve and how it will end its life.

Hundreds of billions of stars bound together by gravity make up a galaxy, the fundamental unit of the Universe as a whole. Our own Milky Way Galaxy contains about a trillion (a million million) times as much mass as the Sun.

Since there are billions of galaxies in the Universe, if not an infinite number, the total mass itself in the Universe is not a meaningful number. More often, we discuss the density of a region of the Universe—the region's mass divided by its volume. The Universe is so spread out, with so much almost empty space between the galaxies, that the Universe's average density is very low. In Chapter 18 we will see that determining that average density is important for knowing the future of the Universe, whether it will eventually collapse or expand forever. This choice, in turn, dictates whether time will ever end. Thus, mass is among the most important and intriguing quantities to know.

---

to the very high temperatures at which x-rays are emitted. In recent years, our satellite observatories above the Earth's atmosphere have detected many such sources of celestial x-rays, and the Chandra X-ray Observatory and the XMM spacecraft, both launched in 1999, are now the most powerful and sensitive x-ray telescopes ever available.

◖ VARIABLE STARS

Some stars vary in brightness over hours, days, or months. Eclipsing binaries are one example of such **variable stars,** but other types of variable stars are actually individually changing in brightness. Many professional and amateur astronomers follow the light curves of such stars, the graphs of how their brightness changes over time.

A very common type of variable star changes slowly in brightness with a period of months or up to a couple of years. The first of these **long-period variables** to be discovered was the star Mira in the constellation Cetus, the Whale (Fig. 11–7). At its maximum brightness, it is 3rd magnitude, quite noticeable to

Figure 11–7 The constellation Cetus, the Whale (or Sea Monster). It is the home of the long-period variable star Mira, also known as *o* (omicron) Ceti.

Figure 11–8 The light curve for Mira, the first known example of a long-period variable. The horizontal axis shows time and the vertical axis shows brightness. The Sun may become a Mira star in about 5 billion years.

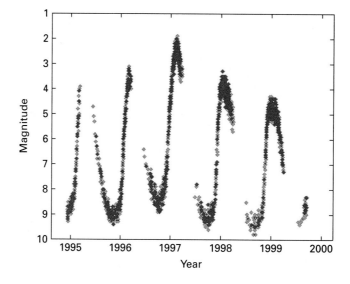

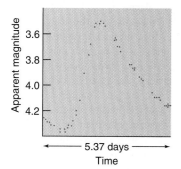

Figure 11–9 The light curve for δ (delta) Cephei, the first example of a Cepheid variable.

the naked eye. At its minimum brightness, it is far below naked-eye visibility (Fig. 11–8). Such stars are giants whose outer layers actually change in size and temperature.

A particularly important type of variable star for astronomers is the **Cepheid variable,** of which the star δ (delta) Cephei is a prime example (Fig. 11–9). A strict relation exists linking the period over which the star varies with the star's average luminosity (intrinsic brightness). This "period-luminosity relation" (Fig. 11–10) is the key to why Cepheid-variable stars are so phenomenally important, possibly the most important type of star known from the point of view of cosmology. After all, we can observe a star's light curve easily, with just a telescope, so we know how long its period is. From the period of a newly observed Cepheid, we know how bright it actually is (its luminosity). By comparing its average luminosity with how bright it appears on average, we can tell how far away the star is. We can do so because a given star would appear fainter if it were farther away, according to the inverse-square law of light (see Chapter 10, *Figure It Out: The Inverse-Square Law*).

Cepheids are supergiant stars, powerful enough that we can detect them in some of the nearby galaxies. Finding the distance to these Cepheids gives us the distances to the galaxies they are in. This chain of reasoning, linking the period of variation of Cepheid variables with their average luminosities, is perhaps the

Figure 11–10 The period-luminosity relation for Cepheid variables. The stars' average intrinsic brightnesses (luminosities) are shown on the vertical axis. The upward arrow shows increasing brightness.

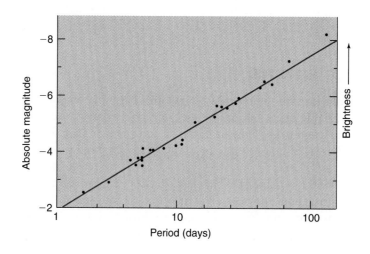

most accurate method we have of finding the distances to galaxies, and so is at the basis of most of our measurements of the size of the Universe.

The period-luminosity relation was discovered in the early years of the 20th century by Henrietta Leavitt in the course of her study of stars in the Magellanic Clouds (Fig. 11–11). She knew only that all the stars in each of the Magellanic Clouds were about the same distance from us, so she could compare their apparent brightnesses without worrying about distance effects. The Magellanic Clouds turn out to be the nearest galaxies to us.

The Hubble Space Telescope has been used to detect Cepheids in galaxies farther away than those in which Cepheids could be detected from ground-based telescopes (Fig. 11–12). The high resolution of Hubble has enabled Cepheids to be picked out and to have their brightnesses followed over time. The results, as we shall see in the cosmic distance scale discussion (Chapter 18), are at the basis of our knowledge of the size and age of the Universe. But we need a good way, now lacking, to measure the distances to a few Cepheids reliably in a manner independent of the period-luminosity relation. The FAME spacecraft, in the years following its 2004 launch, should be able to measure accurate, independent distances for 200 Cepheids.

Cepheid variables are rather massive stars, more massive than the Sun, at an advanced and unstable stage of their lives. They are giant stars, not on the main sequence. Cepheid variables are changing in size, which leads to their variations in brightness. They contract a bit, which makes them hotter, and because of a peculiarity in the structure of the star less energy escapes. Thus the pressure rises, and they expand. They overshoot, expanding too far. The expansion cools them. The energy easily escapes, the pressure decreases, and they respond by contracting again. A cycle is thus set up.

Stars of a related type, **RR Lyrae variables,** have shorter periods than regular Cepheids. RR Lyrae variables are mostly found in clusters of stars, group-

Figure 11–11 From the southern hemisphere, the Magellanic Clouds are high in the sky. They are not quite this obvious to the naked eye. Since all stars in a given Magellanic Cloud are essentially the same distance from us, their intrinsic brightnesses are related to each other as are their relative apparent brightnesses when we observe them from Earth.

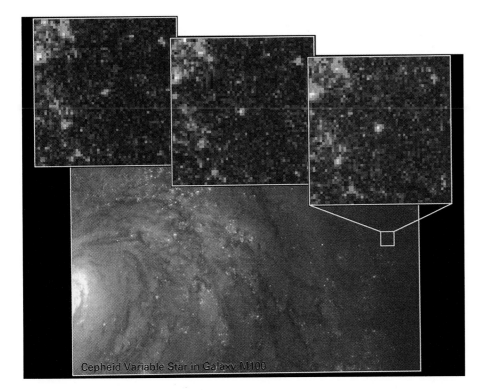

Cepheid Variable Star in Galaxy M100

Figure 11–12 Brightness variations over time (three epochs shown) in a Cepheid variable in a distant galaxy, imaged with the Hubble Space Telescope.

Figure 11-13 An open cluster of stars in the constellation Carina.

Figure 11-14 An open cluster, surrounded by the Rosette Nebula.

Figure 11-15 The globular cluster M15 in the constellation Pegasus.

ings of stars we will discuss next. Since all RR Lyrae stars have the same average luminosity (intrinsic brightness), observing an RR Lyrae star and comparing its average apparent brightness with its average luminosity enables us to tell how far away the cluster is.

STAR CLUSTERS

The face of Taurus, the Bull, is outlined by a "V" of stars, which are close together out in space. On Taurus's back rides another group of stars, the Pleiades (pronounced Plee'a-deez). Both are examples of **open clusters**, groupings of dozens or a few hundred stars. We can often see 6 of the Pleiades stars with the naked eye, but binoculars reveal dozens more and telescopes show the rest. Open clusters (Figs. 11–13 and 11–14) are irregular in shape. About 1000 open clusters are known in our galaxy. They are found near the Milky Way, which means that they are in the plane of our galaxy. Specifically, they are generally in or near our galaxy's spiral arms.

In some places in the sky, we see clusters of stars with spherical symmetry. Such a **globular cluster** looks like a faint, fuzzy ball when seen through a small telescope. Larger telescopes reveal many individual stars. Globular clusters (Fig. 11–15) contain tens of thousands or hundreds of thousands of stars, many more than typical open clusters. Globular clusters usually appear far above and below the Milky Way. About 180 of them are known, and they form a spherical **halo** around the center of our galaxy. We find their distances from Earth by measuring the apparent brightness of RR Lyrae stars in them.

When we study the spectra of stars in clusters to find out what elements are in them, we find out that all the stars are about 90 per cent hydrogen with almost all the rest helium (that is, 90 per cent of the number of atoms are hydrogen). Generally, under 1 per cent of the number of atoms are made of elements heavier than helium, even in the open clusters. But the abundances of these "heavy elements" in the stars in globular clusters are ten times lower. We now think that the globular clusters formed about 11 to 14 billion years ago, early in the life of the Galaxy, from the original gas in the galaxy. This gas had a very low abundance of elements heavier than hydrogen and helium. As time passed, stars in our galaxy "cooked" lighter elements into heavy elements through fusion in their interiors, died, and spewed off these heavy elements into space (see Chapters 12 and 13). The stars in open clusters formed later on from chemically enriched gas. Thus the open clusters have higher abundances of these heavier elements.

X-ray observations have detected bursts of x-rays coming from a few regions of sky, including a half-dozen globular clusters, and the Chandra X-ray Observatory and XMM are studying more. We think that among the many stars in those clusters, some have lived their lives, have died, and are attracting material from nearby companions to flow over to them. We are seeing x-rays from this material as it "drips" onto the dead star's surface and heats up. So globular clusters still have action going on inside them, in spite of their great age.

HOW OLD ARE STAR CLUSTERS?

Large star clusters contain many stars with a wide range of masses. We can tell the ages of such star clusters by looking at their temperature-luminosity (temperature-magnitude) diagrams. For the youngest open clusters, almost all the stars are on the main sequence. But the stars at the upper-left part of the main sequence—which are relatively hot, bright, and massive—live on the main sequence for much shorter times than the cooler stars, which are fainter and less massive.

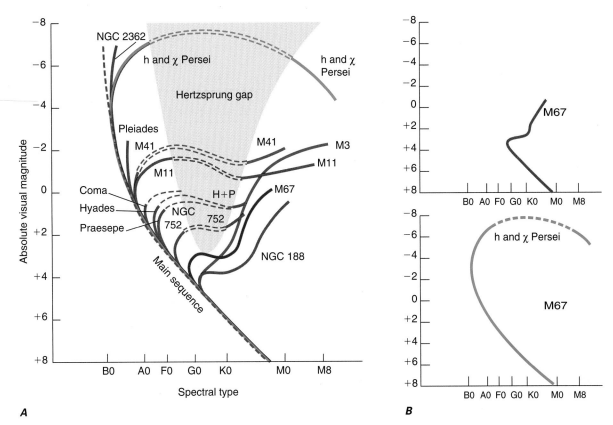

A

B

The points representing their temperatures and magnitudes begin to lie off the main sequence (Fig. 11–16).

Thus the length of the main sequence on the temperature-luminosity diagram of a cluster gets progressively shorter: it is like the wick of a candle burning down. Since we can calculate the main-sequence lifetimes of stars of different masses, we can tell the age of an open cluster by which stars do not appear on the main sequence. This calculation is similar to estimating how long a candle has been burning if the candle's original length and burn rate are known.

For example, suppose a cluster does not have any O and B main-sequence stars, but does have A, F, G, K, and M-type main-sequence stars. We deduce that the cluster is about 100 million years old, because that's how long it takes B-type stars to leave the main sequence. The cluster cannot be a billion years old, because A-type stars would have left the main sequence by that stage. As time passes, the end of the main sequence will move to F stars, then G, and so on—just as the wick of a candle gradually burns down.

The ages of open clusters range from "only" a few million years up to billions of years. The pair of clusters known as h and χ (the Greek letter "chi") Persei is among the youngest known (Fig. 11–17).

Figure 11–16 Temperature-luminosity (temperature-magnitude) diagrams of several open clusters. The positions of all the stars in one cluster make a curve, and several curves are put together here. Two of the individual curves are shown at *right*. Few stars appear in the region marked as the "Hertzsprung gap," because stars with these combinations of luminosity and temperature are apparently unstable, so do not stay in this region of the diagram for long.

If all stars were recently born, they would all lie on the leftmost curve. The older the stars are, the more likely they are to have evolved to the right. Thus the clusters with more stars matching the original main sequence, like h and χ (the Greek letter "chi") Persei, are younger.

The curve for M3, an old globular cluster, is shown for comparison.

Figure 11–17 The double cluster in Perseus, h and χ Persei, a pair of open clusters that is readily visible in a small telescope. Perseus is a northern constellation that is most prominent in the winter sky. In Greek mythology, Perseus slew the Gorgon Medusa and saved Andromeda from a sea monster.

A Closer Look

Star Clusters

Open Clusters	Globular Clusters
No regular shape	Shaped like a ball, stars more closely packed toward center
Many young stars	All old stars
Temperature-luminosity diagrams have long main sequences	Temperature-luminosity diagrams have short main sequences
Where stars leave the main sequence tells the cluster's age	All clusters have nearly the same temperature-luminosity diagram and thus the same age
Stars have similar composition to the Sun	Stars have much lower abundances of heavy elements than the Sun
Hundreds of stars per cluster	10,000–1,000,000 stars per cluster
Found in galactic plane	Found in galactic halo

Though different open clusters have different-looking temperature-luminosity diagrams because of their range of ages, all globular clusters have very similar temperature-luminosity diagrams (Fig. 11–18). Thus all the globular clusters in our galaxy are of roughly the same age, about 13 billion years based on the ab-

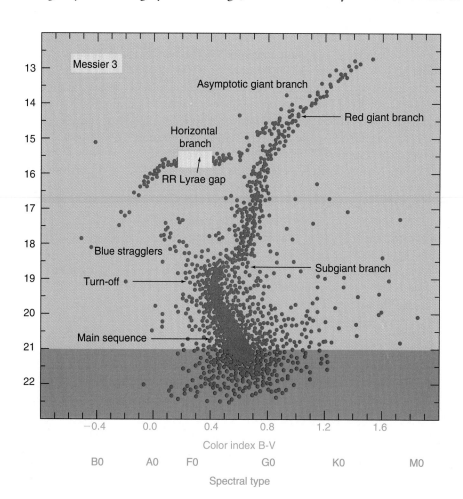

Figure 11–18 The temperature-luminosity diagram for the globular cluster M3. The main sequence is very short and stubby, so most of the stars have evolved off it. Thus this and other globular clusters are very old.

sence of main-sequence stars more massive than the Sun. Recent research, however, is finding a variation in age by two billion years or so, a greater range than had been thought.

The ages of globular clusters set minimum value for the age of the Universe: they must be younger than the Universe itself! The distances set from measurements made with the Hipparcos spacecraft have led to a revision in the calculated ages of some of the globular clusters, and the resulting calculation of their ages is a controversial part of the study of the age of the Universe.

Concept Review

Optical doubles appear close together by chance. Most stars are in multiple-star systems, **binary stars,** bound by gravity. **Visual binaries** are revolving around each other and can be detected as double from Earth. **Spectroscopic binaries** show their double status from spectra. **Eclipsing binaries** pass in front of each other, as shown in their light curves. **Astrometry** is the study of the positions and motions of stars. **Astrometric binaries** can be detected by their wobbling from side to side in their paths across the sky.

The primary use of binary stars is the determination of stellar masses. In the case of main-sequence stars, it is found that massive stars are much more luminous (intrinsically bright) than low-mass stars. Thus, massive stars use up their fuel faster, and have shorter lives, than low-mass stars.

Among **variable stars,** stars that change considerably in brightness over time, **long-period variables,** like Mira, are common. **Cepheid variables** are rare but valuable; the period of a Cepheid variable's variation is linked to its average intrinsic brightness (luminosity), so studying its variation shows its intrinsic brightness. Comparing intrinsic with apparent brightness gives stars' distances, and they are valuable tools for determining distances to other galaxies. **RR Lyrae variables** also can be used to find distances, particularly to globular clusters in our galaxy.

Open clusters contain up to a few thousand stars irregularly spread in a small region of sky. They are in the plane of our Galaxy. **Globular clusters,** consisting of 10,000 to a million stars, have spherical shapes and form a **halo** around the center of our Galaxy. They contain low abundances of heavy elements, so are very old.

The temperature-luminosity diagrams for open clusters show the ages of the clusters by how many stars have peeled off the main sequence. (The massive stars leave the main sequence faster than the low-mass stars.) The temperature-luminosity diagrams for globular clusters are all very similar, with short main sequences, indicating that the globular clusters are all about the same age—very old, providing a limit for the age of our Universe.

Questions

1. Sketch the orbit of a double star that is simultaneously a visual, an eclipsing, and a spectroscopic binary.
2. If the light in a binary star system is greatly dominated by one of the stars (the other star is very faint), how can you tell from its spectrum that the star is a binary?
3. **(a)** Assume that an eclipsing binary contains two identical stars. Sketch the intensity of light received as a function of time. **(b)** Sketch to the same scale another light curve to show the result if both stars were much larger while the orbit stayed the same.
†4. Suppose you find a binary system in which one star is 6 times as massive as the other star. Which one is closer to the center of mass, and by what factor?
5. Sketch the path in the sky of the visible (with a solid curve) and invisible (with a dotted curve) components of an astrometric binary.
6. Explain how astrometric methods were used to search for planets around other stars. How were such planets finally found?
7. Explain why massive main-sequence stars have shorter lives than low-mass main-sequence stars.
†8. **(a)** If one main-sequence star is 3 times as massive as another one, what is the ratio of their lumosities? **(b)** What is the ratio of their main-sequence lifetimes? Be sure to state which one is most luminous and which one lives longest.

9. Explain briefly how observations of a Cepheid variable in a distant galaxy can be used to find the distance to the galaxy. Why can't we use triangulation instead?
†10. If you find a Cepheid variable star in a star cluster, and the Cepheid appears 100 times fainter than another Cepheid with the same period but whose distance is known to be 400 light-years, what is the distance of the star cluster?
11. What ability of the Hubble Space Telescope is enabling astronomers to study Cepheid variables farther out in space than ever before?
12. What are the Magellanic Clouds? How did we find out the distances to them?
13. What are two distinctions between open and globular clusters?
14. Why is it more useful to study the temperature-luminosity diagram of a cluster of stars instead of one for stars in a field chosen at random?
15. Suppose you find two clusters, one whose main sequence doesn't have O, B, and A stars, and the other whose main sequence doesn't have any O, B, A, and F stars. Which is older?
16. Which are generally older: open clusters or globular clusters? How do we know?

†This question requires a numerical solution.

How Stars Shine: Cosmic Furnaces

12

ORIGINS *The formation of stars like the Sun is critical to our existence. With its large supply of fuel, the Sun exists for a long time as it produces energy through nuclear fusion, making possible the development of life on Earth.*

Even though individual stars shine for a relatively long time, they are not eternal. Stars are born out of the gas and dust that exist within a galaxy; they then begin to shine brightly on their own. Eventually, they die. Though we can directly observe only the outer layers of stars, we can deduce that the temperatures at their centers must be millions of kelvins. We can even figure out what it is deep down inside that makes the stars shine.

To determine the probable life history of a typical star, we observe stars having many different ages and assume that they evolve in a similar manner. However, we must take into account the different masses of stars; some aspects of their evolution depend critically on mass.

We start this chapter by discussing the birth of stars. We then consider the processes that go on inside a star during its life on the main sequence. Finally, we begin the story of the evolution of stars when they finish this stage of their lives. Chapters 13 and 14 will continue the story of what is called stellar evolution, all the way to the deaths of stars. ◯

Near the end of this chapter we will see that the most important experiment to test whether we understand how stars shine is the search for particles, called neutrinos, from the Sun. Over the past decades a search for them has been made, but only about a third to half of those expected have been found. Recent experiments have provided better ways of detecting neutrinos than we previously had. The results indicate that we did not understand neutrinos as well as we had thought. These astronomical results therefore have great potential for adding important knowledge about fundamental physics in addition to our understanding of the stars.

AIMS

To understand how stars form out of gigantic clouds of gas and dust.

•

To study the way in which stars generate their energy.

•

To learn how neutrino detection experiments test our most fundamental understanding of how stars shine.

REDSHIFT
http://www.harcourtcollege.com/astro/cosmos/rsce

A Hubble image of gas and dust around a young star.

◀ The Super-Kamiokande neutrino detector in Japan. We see the empty stainless steel vessel, a cube 40 m across, and some of the 13,000 photomultiplier tubes, each 50 cm across, surrounding it. Neutrinos interacting within molecules of the 50,000 tons of highly purified water (note the boat, photographed as the detector was partly filled) lead to the emission of blue flashes of light that are being detected.

Figure 12-1 The clouds of gas and dust around ρ (rho) Ophiuchi (in the blue reflection nebula). The bright star Antares is surrounded by yellow and red nebulosity. The complex is 500 light-years away. M4, the nearest globular cluster to us, is also seen.

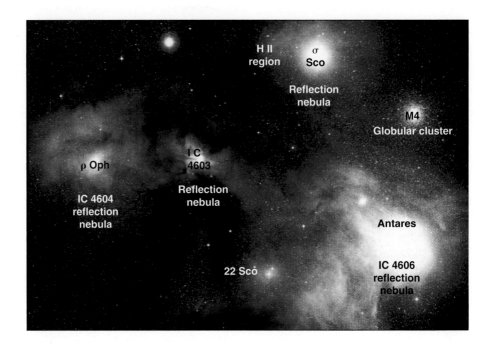

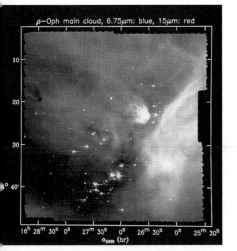

Figure 12-2 An infrared view of the region directly around ρ Ophiuchi, covering only about 10% of the field of view of the optical image. In this view made with the Infrared Space Observatory, infrared at 7 μm is reproduced as blue and infrared at 15 μm is reproduced as red. The brightest regions are dense clouds glowing in the infrared because their dust has absorbed the ultraviolet light from newly born, massive stars.

STARS IN FORMATION

The birth of a star begins with a nebula—a region of gas and dust (Fig. 12–1). The dust (tiny solid particles) may have escaped from the outer atmosphere of giant stars. The gas and dust from which stars are forming are best observed in the infrared and radio regions of the spectrum (Fig. 12–2).

Consider a region that reaches a higher density than its surroundings, perhaps from a random fluctuation in density or—in a leading theory of why galaxies have spiral arms—because a wave of compression passes by. Still another possibility is that a nearby star explodes (a "supernova"), sending out a shock wave that compresses the gas and dust. In any case, once the region (often called a cloud, or, specifically, a "giant molecular cloud") gains a higher-than-average density, the gas and dust continues to collapse due to gravity. Energy is released, and the material accelerates in free-fall.

Eventually dense cores, each with a mass comparable to that of a star, form and grow like tiny seeds within the vast cloud. These **protostars** (from the prefix of Greek origin meaning "primitive") continue to collapse, almost unopposed by internal pressure. But when a protostar becomes sufficiently dense, frequent collisions occur among its particles; hence, part of the gravitational energy released during subsequent collapse goes into heating the gas, increasing its internal pressure. (In general, compression heats a gas; for example, a bicycle tire feels warm after air is vigorously pumped into it.) The rising internal pressure, which is highest in the object's center and decreases outward, slows down the collapse until it becomes very gradual and more accurately described as contraction. The object is now called a **pre-main-sequence star.**

During the contraction phase, a disk tends to form because the original nebula was rotating slightly. We discussed this process when considering the "nebular hypothesis" for the formation of our Solar System (Chapter 5). Dusty disks have been found around young stars and pre-main-sequence stars in nebulae such as the Orion Nebula (Fig. 12–3). These are sometimes called protoplanetary disks ("proplyds"), and they support the theoretical expectation that planetary systems are common.

As energy is radiated from the surface of the pre-main-sequence star, its internal pressure decreases, and it gradually contracts. This release of gravitational energy heats the interior, thereby increasing the internal temperature and pressure. It is also the source of the radiated energy. Gravitational energy was released in this way in the early Solar System. As the temperature in the interior rises, the outward force resulting from the outwardly decreasing pressure increases, and eventually it balances the inward force of gravity. As we shall discuss later, this balance is the key to understanding stable stars.

Theoretical analysis shows that the dust surrounding the stellar embryo we call a protostar should absorb much of the radiation that the protostar emits. The radiation from the protostars should heat the dust to temperatures that cause it to radiate primarily in the infrared. Infrared astronomers have found many objects that are especially bright in the infrared but that have no known optical counterparts. These objects seem to be located in regions where the presence of a lot of dust, gas, and young stars indicates that star formation might still be going on. Imaging in the visible with the Hubble Space Telescope and in the infrared with the Infrared Space Observatory has shown how young stars are born inside giant pillars of gas and dust inside certain nebulae (Fig. 12–4). ☼

Astronomers have long studied some types of stars that they think are still forming or newly formed. For example, observations of many protostars show that they send matter out in oppositely directed beams. This "bipolar ejection" (Fig. 12–5) may imply that a disk of matter orbits such protostars, blocking an outward flow of gas in the equatorial direction. Thus the flow of gas is channelled toward the poles. Sometimes clumps of gas appear, though only recently have they been identified with ejections from stars in the process of collapsing.

Several classes of stars that vary erratically in brightness have been found. One of these classes, called T Tauri, contains pre-main-sequence stars as massive as or less massive than the Sun. Presumably, these stars are so young that they have not quite settled down to a steady and reliable existence on the main sequence.

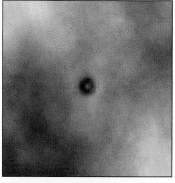
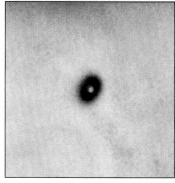

Figure 12–3 The Hubble Space Telescope has revealed protoplanetary disks around several stars.

REDSHIFT
http://www.harcourtcollege.com/astro/cosmos/rsce

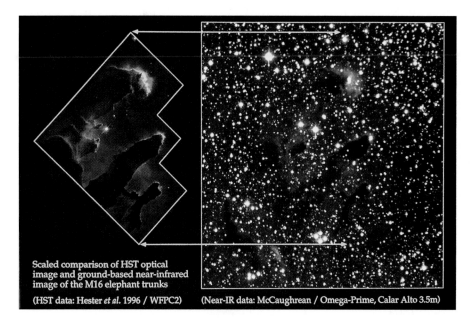

Scaled comparison of HST optical image and ground-based near-infrared image of the M16 elephant trunks

(HST data: Hester *et al.* 1996 / WFPC2) (Near-IR data: McCaughrean / Omega-Prime, Calar Alto 3.5m)

Figure 12–4 The orientation of the Hubble Space Telescope visible-light image of the Eagle Nebula *(opposite)* on a near-infrared image taken with a ground-based telescope.

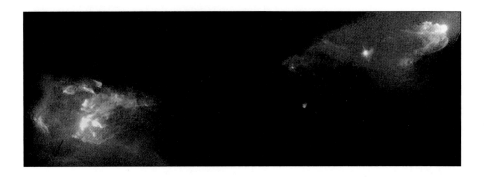

Figure 12-5 Hubble Space Telescope observations of Herbig-Haro objects #1 and #2 (HH-1 and HH-2), clouds of interstellar gas heated by shock waves from jets of high-speed gas. The jets are being ejected from a protostar, a star in the process of being born. Since the protostar is hidden by a dusty cocoon of gas, only the Herbig-Haro object reveals what is going on.

The jets of gas were formed as the protostar collapsed under the force of its own gravity. Because a thick disk of cool gas and dust sur-rounds the protostar, the gas squirts outward along the protostar's axis of rotation at speeds of perhaps 1 million km/hr. HH-1 and HH-2 are more irregular in shape than many other Herbig-Haro objects, perhaps because the bow shock wave we are seeing (a shock wave like those formed by the bow of a boat plow-ing through the water) has broken up. These objects are about 1500 light-years from us, in a star-forming region of the constellation Orion. The smallest features resolved are about the size of our Solar System, and the whole image is only about 1 light-year across. Herbig-Haro objects are named after George Herbig, formerly of the Lick Obser-vatory and now of the University of Hawaii, and Guillermo Haro of the Mexican National Observatory.

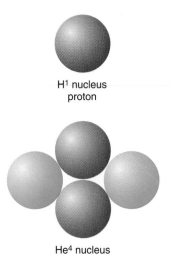

H¹ nucleus
proton

He⁴ nucleus
alpha particle

Figure 12-6 The nucleus of hydrogen's most common form is a single proton, while the nucleus of helium's most common form con-sists of two protons and two neu-trons. Protons and neutrons are made of smaller particles called quarks (Chapter 19).

ENERGY SOURCES IN STARS

If the Sun got all of its energy from contraction, it could have shined for only about 30 million years, not very long on an astronomical time scale. Yet we know that rocks 4 billion years old have been found on Earth, so the Sun has been around at least that long. Some other source of energy must hold the Sun and other stars up against their own gravitational pull.

Actually, a pre-main-sequence star will heat up until its central portions be-come hot enough for **nuclear fusion** to take place, at which time it reaches the main sequence of the temperature-luminosity (temperature-magnitude) diagram. Using this process, which we will soon discuss in detail, the star can generate enough energy to support it during its entire lifetime on the main sequence. The energy makes the particles in the star move around rapidly. Such rapid motions in a gas are the definition of high temperature. The **thermal pressure,** the force from these moving particles pushing on each area of gas, is also high. The vary-ing pressure, which decreases outward from the center, produces a force that pushes outward on any given pocket of gas. This outward force balances grav-ity's inward pull on the pocket, a condition known as "hydrostatic equilibrium."

The basic fusion process in main-sequence stars fuses four hydrogen nuclei into one helium nucleus, just as hydrogen nuclei are combined into helium in a hydrogen bomb here on Earth. In the process, tremendous amounts of energy are released.

A hydrogen nucleus is but a single proton. A helium nucleus is more com-plex; it consists of two protons and two neutrons (Fig. 12–6). The mass of the helium nucleus that is the final product of the fusion process is slightly less than the sum of the masses of the four hydrogen nuclei (protons) that went into it. A small amount of the mass, m, "disappears" in the process: 0.007 (0.7 per cent) of the mass of the four protons.

The mass difference does not really disappear, but rather is converted into energy, E, according to Albert Einstein's famous formula $E = mc^2$ (where c is the speed of light). Even though m is only a small fraction of the original mass, the amount of energy released is prodigious; in the formula, c is a very large num-ber. The loss of only 0.007 of the central part of the Sun, for example, is enough to allow the Sun to radiate as much as it does at its present rate for a period of about ten billion (10^{10}) years. This fact, not realized until 1920 and worked out in more detail in the 1930s, solved the longstanding problem of where the Sun and the other stars get their energy.

All the main-sequence stars are approximately 90 per cent hydrogen (that is, 90 per cent of the atoms are hydrogen), so there is a lot of raw material to fuel the nuclear "fires." We speak colloquially of "nuclear burning," although, of course, the processes are quite different from the chemical processes that are in-

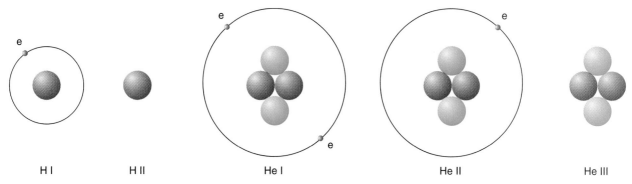

H I   H II   He I   He II   He III

Figure 12–7 Hydrogen and helium ions. The sizes of the nuclei are greatly exaggerated with respect to the sizes of the "orbits" of the electrons.

volved in the "burning" of logs or of autumn leaves. In order to be able to discuss these processes, we must first review the general structure of nuclei and atoms.

ATOMS AND NUCLEI

As we mentioned in Chapter 2, an atom consists of a small nucleus surrounded by electrons. Most of the mass of the atom is in the nucleus, which takes up a very small volume in the center of the atom. The effective size of the atom, the chemical interactions of atoms to form molecules, and the nature of spectra are all determined by the electrons.

The nuclear particles with which we need be most familiar are the **proton** and **neutron.** Both these particles have nearly the same mass, 1836 times greater than the mass of an electron, though still tiny (Appendix 2). The neutron has no electric charge and the proton has one unit of positive electric charge. The electrons, which surround the nucleus, have one unit each of negative electric charge. When an atom loses an electron, it has a net positive charge of 1 unit for each electron lost. The atom is now a form of **ion** (Fig. 12–7).

Since the number of protons in the nucleus determines the charge of the nucleus, it also determines the quota of electrons that the neutral state of the atom must have. To be neutral, after all, there must be equal numbers of positive and negative charges. Each **element** (sometimes called "chemical element") is defined by the specific number of protons in its nucleus. The element with one proton is hydrogen, that with two protons is helium, that with three protons is lithium, and so on.

Though a given element always has the same number of protons in a nucleus, it can have several different numbers of neutrons. (The number of neutrons is usually somewhere between 1 and 2 times the number of protons. The most common form of hydrogen, just a single proton, is the main exception to this rule.) The possible forms of the same element having different numbers of neutrons are called **isotopes.**

For example, the nucleus of ordinary hydrogen contains one proton and no neutrons. An isotope of hydrogen (Fig. 12–8) called deuterium (and sometimes "heavy hydrogen") has one proton and one neutron. Another isotope of hydrogen called tritium has one proton and two neutrons.

Most isotopes do not have specific names, and we keep track of the numbers of protons and neutrons with a system of superscripts and subscripts. The subscript before the symbol denoting the element is the number of protons (called the atomic number), and a superscript after the symbol is the total number of protons and neutrons together (called the mass number, or atomic mass). For example, $_1H^2$ is deuterium, since deuterium has one proton, which gives the sub-

Figure 12-8 Isotopes of hydrogen and helium. $_1H^2$ (deuterium) and $_1H^3$ (tritium) are much rarer than the normal isotope $_1H^1$. Similarly, $_2He^3$ is much rarer than $_2He^4$.

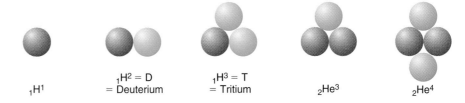

$_1H^1$ $_1H^2 = D$ = Deuterium $_1H^3 = T$ = Tritium $_2He^3$ $_2He^4$

script, and an atomic mass of 2, which gives the superscript. (Note that $_1^2H$ is also correct notation.) Deuterium has atomic number equal to 1 and mass number equal to 2. Similarly, $_{92}U^{238}$ is an isotope of uranium with 92 protons (atomic number = 92) and mass number of 238, which is divided into 92 protons and $238 - 92 = 146$ neutrons.

Each element has only certain isotopes. For example, most of the naturally occurring helium is in the form $_2He^4$, with a lesser amount as $_2He^3$. Sometimes an isotope is not stable, in that after a time it will spontaneously change into another isotope or element; we say that such an isotope is **radioactive.** ☼ The most massive elements, those past uranium, are all radioactive, and have average lifetimes that are very short. It has been theoretically predicted that around element 114, elements should begin being somewhat more stable again. The handful of atoms of element 114, 116, and 118 discovered in 1998 and 1999 are more stable than those of slightly lower mass numbers—lasting even about 5 seconds instead of a small fraction of a second. The known elements are listed in Appendix 11.

During certain types of radioactive decay, as well as when a free proton and electron combine to form a neutron, a particle called a **neutrino** is given off. A neutrino is a neutral particle (its name comes from the Italian for "little neutral one"). Neutrinos have a very useful property for the purpose of astronomy: they do not interact very much with matter. Thus when a neutrino is formed deep inside a star, it can usually escape to the outside without interacting with any of the matter in the star. A photon of electromagnetic radiation, on the other hand, can travel only about 1 cm in a stellar interior before it is absorbed, and it is over a hundred thousand years before a photon zigs and zags its way to the surface.

The elusiveness of the neutrino not only makes it a valuable messenger—indeed, the only possible direct messenger—carrying news of the conditions inside the Sun at the present time, but also makes it very difficult for us to detect on Earth. A careful experiment carried out over many years has found only about ⅓ the expected number of neutrinos, as we shall soon see.

◖◗ THE STELLAR PRIME OF LIFE

Let us now use our knowledge of atomic nuclei to explain how stars shine. For a pre-main-sequence star, the energy from the gravitational contraction goes into giving the individual particles greater speeds; that is, the temperature rises. When atoms collide at high temperature, electrons get knocked away from their nuclei, and the atoms become fully ionized. The electrons and nuclei can move freely and separately. For nuclear fusion to begin, atomic nuclei must get close enough to each other so that the force that holds nuclei together, the strong nuclear force, can play its part. But all nuclei have positive charges, because they are composed of protons (which bear positive charges) and neutrons (which are neutral). The positive charges on any two nuclei cause an electrical repulsion between them, which tends to prevent fusion from taking place.

However, at the high temperatures typical of a stellar interior, some nuclei have enough energy to overcome this electrical repulsion. They come sufficiently close to each other that they essentially collide, and the strong nuclear force (to be discussed in Chapter 19) takes over. Fusion on the main sequence procedes in one of two ways, as will be discussed below.

Once nuclear fusion begins, enough energy is generated to maintain the pressure and prevent further contraction. The pressure provides a force that pushes outward strongly enough to balance gravity's inward pull. In the center of a star, the fusion process is self-regulating. The star finds a balance between thermal pressure pushing out and gravity pushing in. It thus achieves stability on the main sequence (at a constant temperature and luminosity). When we learn how to control fusion in power-generating stations on Earth, which currently seems decades off, our energy crisis will be over.

The greater a star's mass, the hotter its core becomes before it generates enough pressure to counteract gravity. The hotter core gives off more energy, so the star becomes brighter, explaining why main-sequence stars of large mass have high luminosity (Fig. 12–9). In fact, it turns out that more massive stars use their nuclear fuel at a very *much* higher rate than less massive stars. Even though the more massive stars have more fuel to burn, they go through it relatively quickly and live shorter lives than low-mass stars, as we discussed in Chapter 11. The next two chapters examine the ultimate fates of stars, with the fates differing depending on the masses of the stars.

STELLAR ENERGY CYCLES

Several chains of reactions have been proposed to account for the fusion of four hydrogen nuclei into a single helium nucleus. Hans Bethe of Cornell University suggested some of these procedures during the 1930s. The different chain reactions prevail at different temperatures, so chains that are dominant in very hot stars may be different from the ones in cooler stars.

When the temperature of the center of a main-sequence star is less than about 20 million kelvins, the **proton-proton chain** (Fig. 12–10) dominates. This sequence uses six hydrogens, and winds up with one helium plus two hydrogens. The net transformation is four hydrogens into one helium. (Though two of the protons turn into neutrons, here this isn't the main point.) But the original six protons contained more mass than do the final single helium plus two protons.

Figure 12–9 The Pistol Nebula, gas glowing because of radiation from a massive, highly luminous, main-sequence star. The star, perhaps the most massive known, shines 10 million times more brightly than our Sun and would fill the Earth's orbit. Hidden behind the dust that lies between us and the center of our galaxy, the star and nebula are revealed here by the infrared camera aboard the Hubble Space Telescope.

Figure 12–10 The proton-proton chain; e$^+$ stands for a positron (the equivalent of an electron, but with a positive charge), ν (nu) is a neutrino, and γ (gamma) is electromagnetic radiation at a very short wavelength.

In the first stage, two nuclei of ordinary hydrogen fuse to become a deuterium (heavy hydrogen) nucleus, a positron, and a neutrino. The neutrino immediately escapes from the star, but the positron soon collides with an electron. They annihilate each other, forming gamma rays.

Next, the deuterium nucleus fuses with yet another nucleus of ordinary hydrogen to become an isotope of helium with two protons and one neutron. More gamma rays are released.

Finally, two of these helium isotopes fuse to make one nucleus of ordinary helium plus two nuclei of ordinary hydrogen. The protons are numbered to help you keep track of them.

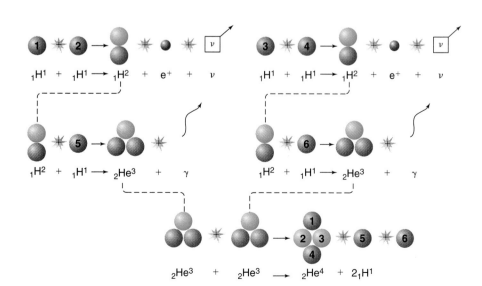

The small fraction of mass that disappears is converted into an amount of energy that we can calculate with the formula $E = mc^2$. According to Einstein's special theory of relativity, mass and energy are equivalent and interchangeable, linked by this equation.

For stellar interiors significantly hotter than that of the Sun, the **carbon-nitrogen-oxygen (CNO) cycle** dominates. This cycle begins with the fusion of a hydrogen nucleus with a carbon nucleus. After many steps, and the insertion of four hydrogen nuclei, we are left with one helium nucleus plus a carbon nucleus. Thus as much carbon remains at the end as there was at the beginning, and the carbon can start the cycle again. As in the proton-proton chain, four hydrogens have been converted into one helium, 0.007 of the mass has been transformed, and an equivalent amount of energy has been released according to $E = mc^2$. Main-sequence stars more massive than about 1.1 times the Sun are dominated by the CNO cycle.

Later in their lives, when they are no longer on the main sequence, stars can have even higher interior temperatures, above 10^8 K. They then fuse helium nuclei to make carbon nuclei. The nucleus of a helium atom is called an "alpha particle" for historical reasons. Since three helium nuclei ($_2\text{He}^4$) go into making a single carbon nucleus ($_6\text{C}^{12}$), the procedure is known as the **triple-alpha process.** A series of other processes can build still heavier elements inside very massive stars.

These processes, and other element-building methods, are called **nucleosynthesis** (new'clee-oh-sin'th-sis). The theory of nucleosynthesis in stars can account for the abundances (proportions) we observe of the elements heavier than helium. Currently, we think that the synthesis of isotopes of the lightest elements (hydrogen, helium, and lithium) took place in the first few minutes after the origin of the Universe (Chapter 19), though some of the observed helium was produced later by stars.

BROWN DWARFS

When a pre-main-sequence star has at least 8 per cent the Sun's mass (that is, it has about 80 Jupiter masses), nuclear reactions begin and continue, and it becomes a star. But if the mass is less than 8 per cent of the Sun's mass, the central temperature does not become hot enough for nuclear reactions using ordinary hydrogen to be sustained. These objects shine dimly, and grow dimmer as they grow older. They all have the same radius, about the same as that of the planet Jupiter.

These faint objects came to be called **brown dwarfs,** mainly because "brown" is a mixture of many colors and people didn't agree how such "failed stars" would look, and also because they emit very little light (Fig. 12–11). For decades, there was a debate as to whether brown dwarfs exist, but finally some were found in 1995. We now know of several dozen, because of the advances in astronomical imaging and in spectroscopy, not only in the visible but also in the infrared. The coolest ones show methane and water in their spectra, like giant planets but unlike stars.

It is difficult to tell the difference between a brown dwarf and a small, cool, ordinary star, unless the brown dwarf is exceptionally cool. One way is to see whether an object has lithium in its spectrum. Lithium is a very fragile element, and undergoes fusion in ordinary stars, converting it to other things. So if you detect lithium in the spectrum of a dim star, it is probably a brown dwarf (which hasn't begun nuclear fusion) rather than a cool ordinary dwarf star of spectral class M or L, which are the coolest stars on the main sequence (and thus have begun nuclear fusion). A complication is that very young M and L stars might not be old enough to have burned all their lithium, leading to potential confusion with brown dwarfs.

Figure 12–11 The first brown dwarf to be unambiguously discovered, Gliese 229B. A ground-based image *(top)* barely shows the brown dwarf orbiting the larger star, because the atmosphere blurs the larger star's image, but the Hubble Space Telescope *(bottom)* shows the brown dwarf clearly. (The straight streak is an artifact in the imaging.)

How do we tell between brown dwarfs and giant planets in cases where they are orbiting a more normal star? Astronomers usually distinguish between them by the way that they form: While planets form in disks of dust and gas as the central star is born, brown dwarfs form like the central star, out of the collapse of a cloud of gas. But we can't see the history of an object when we look at it, so it is hard to translate the distinction into something observable. All of the proposed tests are difficult to make. So, currently, for lack of definitive methods, the distinction is usually made on the basis of mass: Any orbiting object with a mass less than 10 times Jupiter's is called a planet, while the range 10–80 Jupiter masses corresponds to brown dwarfs.

Brown dwarfs are being increasingly studied, especially in the infrared. Hubble Space Telescope images show that one of the nearby ones is double, with the components separated by 5 A.U. By watching it over a few years, we should be able to measure its orbit and derive the masses of the components.

THE SOLAR NEUTRINO EXPERIMENT

Astronomers can apply the equations that govern matter and energy in a star, and make a model of the star's interior in a computer. Though the resulting model can look quite nice, nonetheless it would be good to confirm it observationally.

However, the interiors of stars lie under opaque layers of gas. Thus we cannot directly observe electromagnetic radiation from stellar interiors. Only neutrinos escape directly from stellar cores. Neutrinos interact so weakly with matter that they are hardly affected by the presence of the rest of the Sun's mass. Once formed, they zip right out into space, at (or almost at) the speed of light. Thus they reach us on Earth about 8 minutes after their birth. Neutrinos should be produced in large quantities by the proton-proton chain in the Sun, as a consequence of protons turning into neutrons, positrons, and neutrinos. (A positron is an "antielectron," an example of antimatter. Whenever a particle and its antiparticle meet, they annihilate each other.)

For over three decades, astrochemist Raymond Davis has carried out an experiment to search for neutrinos from the solar core, set up in consultation with the theorist John Bahcall, whose calculations continue to drive the theory. Davis set up a tank containing 400,000 liters of a chlorine-containing chemical (Fig. 12–12). One isotope of chlorine can, on rare occasions, interact with one of the passing neutrinos from the Sun. It turns into a radioactive form of argon, which Davis and his colleagues at the University of Pennsylvania can detect. He needs such a large tank because the interactions are so rare for a given chlorine atom. In fact, he detects fewer than 1 argon atom formed per day, despite the huge size of the tank.

Over the years, Davis has detected only about ⅓ the number of interactions predicted by theorists. Where is the problem? Is it that astronomers don't understand the temperature and density inside the Sun well enough to make proper predictions? Or is it that the physicists don't completely understand what happens to neutrinos after they are released?

The latest thinking is that neutrinos actually change after they are released. According to a theoretical model, the neutrinos of the specific type produced inside the Sun change, before they reach Earth, into all three types of neutrinos that are known. But Davis's experiment is sensitive only to the specific type released by the Sun. Thus only ⅓ the original prediction is expected, and that is what we detect.

The chlorine experiment, though it has run the longest by far, is no longer the only way to detect solar neutrinos. An experiment in Japan (Fig. 12–13), first

Figure 12–12 The original neutrino telescope, deep underground in the Homestake Gold Mine in South Dakota to shield it from other types of particles from space. The telescope is mainly a tank containing 400,000 liters of perchloroethylene (dry-cleaning solution).

Figure 12–13 The Super-Kamiokande experiment detects neutrinos from the Sun and is providing a test not only of how the stars shine but also of whether our basic understanding of fundamental physics is correct. Here we see thousands of light-sensitive "photomultiplier" tubes, a few of which can detect flashes of light from each interaction between an incoming neutrino and nuclear particles in the water that now fills this huge cavity.

Figure 12–14 The outside of the tank of deuterated water ("heavy water") that is the heart of the Sudbury Neutrino Observatory.

set up to study protons and whether they decay, has used a huge tank of purified water to verify that the number of neutrinos is less than expected. Other sets of experiments in Italy and in Russia that use gallium, an element that is much more sensitive to neutrinos than chlorine, also show that up to half of the expected neutrinos are missing. The chlorine was sensitive only to neutrinos of very high energy, which come out of only a small fraction of the nuclear reactions in the Sun and not out of the basic proton-proton chain. Gallium is sensitive to a much wider range of interactions, including some of the most basic ones.

Neutrinos were first thought of theoretically as a particle that has no rest mass—that is, a particle that would have no mass if it weren't moving ("at rest"). The Japanese experiment has shown that neutrinos probably have a tiny rest mass after all. Theoretically, neutrinos can oscillate from one type to another only if they have some rest mass. So this result fits with the current ideas that we are seeing neutrino oscillations from one type to another.

A U.S.-Canadian experiment in Sudbury, Ontario, Canada, began collecting data in 1999 (Fig. 12–14). It has even more sensitive detection capability than earlier experiments. It uses a large quantity of "heavy water," water whose molecules contain deuterium instead of the more normal hydrogen isotope. This experiment, like that in Japan, looks for light given off when neutrinos hit the water. The Sudbury Neutrino Observatory has the potential to resolve the remaining mysteries about the missing solar neutrinos. (www)

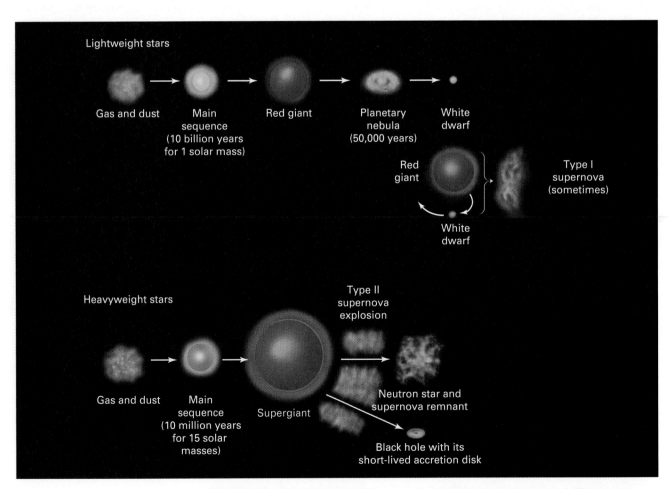

Figure 12–15 A summary of the stages of stellar evolution for stars of different masses, using the boxing terms "lightweight" and "heavyweight" to mean low mass and high mass, respectively.

The various neutrino experiments may cause a revolution in fundamental ideas of physics. Similarly, there are also important repercussions for physics from a set of astronomical observations we will discuss later, showing that measurements of distant supernovae (exploding stars) indicate that the expansion of the Universe is accelerating. The relation between physics and astronomy is close, though the point of view that physicists and astronomers have in tackling problems may be different.

● THE LIFE CYCLES OF STARS

The next two chapters discuss the various end states of stellar evolution. The mass of an isolated, single star determines its fate (Fig. 12–15). Low-mass stars, like the Sun, wind up as white dwarfs. The more massive stars are blown to smithereens in supernova explosions, becoming neutron stars or even black holes.

Concept Review

Stars are formed from regions of gas and dust that are best observed in the infrared and radio. The collapsing gas and dust becomes a **protostar** and subsequently a **pre-main-sequence star,** which is powered by gradual gravitational contraction. Eventually, the outward force from the pressure balances the inward force of gravity. The dust heats up and becomes visible in the infrared. Some planetary systems in formation may have been discovered around young stars.

Stars get their energy from **nuclear fusion.** The resulting **thermal pressure** balances gravity. The basic fusion process in the Sun and other main-sequence stars is a merger of four hydrogen nuclei **(protons)** into one helium nucleus (two protons and two **neutrons**), with the difference in mass transformed to energy according to Einstein's $E = mc^2$.

Atoms that have lost electrons are **ions.** Each **element** is defined by the number of protons in its nucleus. Forms of the same element with different numbers of neutrons are **isotopes.** Isotopes that decay spontaneously are **radioactive. Neutrinos** are light, neutral particles given off in some radioactive decays as well as during nuclear fusion in stars. They are so elusive that they escape from a star immediately.

The **proton-proton chain** is the basic fusion process in the Sun, while the **carbon-nitrogen-oxygen cycle** dominates in hotter stars. At still higher temperatures, when stars have left the main sequence, the **triple-alpha process** is dominant, where an alpha particle is a helium nucleus. Element-building processes are what is meant by **nucleosynthesis.**

Brown dwarfs are failed stars, with insufficient mass to begin nuclear fusion.

Theorists predict a certain number of neutrinos that should result from fusion in the Sun. Careful experiments are finding only about one-third to one-half that number. We may be learning new facts about neutrinos from the idea that most of them change form between the Sun and the Earth if one of the leading explanations is correct.

Questions

1. What is the source of energy in a pre-main-sequence star?
2. Arrange the following in order of development: pre-main-sequence star, nebula, protostar, Sun.
†3. Give the number of protons, the number of neutrons, and the number of electrons in ordinary hydrogen ($_1H^1$), lithium ($_3Li^7$), and iron ($_{26}Fe^{56}$) atoms.
4. (a) If you remove one neutron from helium, the remainder is what element? (b) Now remove one proton. What element is left?
5. Why is four-times ionized helium never observed?
6. Explain why nuclear fusion takes place only in the centers of stars rather than on their surfaces.
7. What is the major fusion process that takes place in the Sun?

8. What does it mean for the temperature of a gas to be higher?
†9. If 0.7% of the mass of four protons gets emitted as energy when they fuse to form a helium nucleus, how many ergs of energy are produced? (Note that an erg is a $g \cdot cm^2/sec^2$.)
10. What is a brown dwarf?
11. Why are the results of the solar neutrino experiment so important?
12. Summarize the current status of our understanding of solar neutrinos.

---

†This question requires a numerical solution.

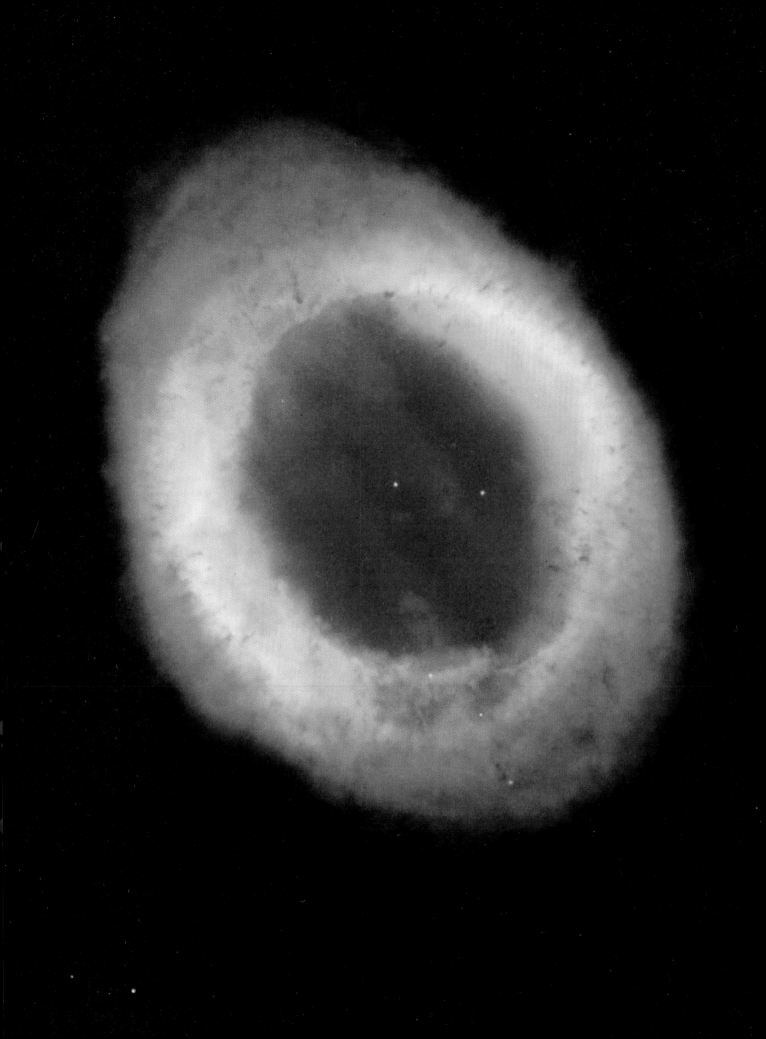

The Death of Stars: Stellar Recycling

13

ORIGINS *We discover how the heavy elements in the Earth, so necessary for life itself, came into existence. We discuss the ultimate fate of our Sun, billions of years into the future. And, we learn about the origins of interesting classes of stars, including red giants, white dwarfs, and neutron stars, and about the origins of exploding stars.*

The more massive a star is, the shorter its stay on the main sequence. The most massive stars may be there for only a few million years. A star like the Sun, on the other hand, is not especially massive and will live on the main sequence for about ten billion years. Since it has taken over four billion years for humans to evolve, it is a good thing that some stars can be stable for this long.

In this chapter, we will first discuss what will happen when the Sun dies: It will follow the same path as other single **lightweight stars,** stars born with up to about 10 times the mass of the Sun. They will go through planetary nebula (Fig. 13–1) and white-dwarf stages. Then we will discuss the death of more massive stars, greater than about 10 times the Sun's mass, which we can call **heavyweight stars.** They go through spectacular stages. Some wind up in such a strange final state—a black hole—that we devote the entire next chapter to it.

AIMS

To follow the evolution of stars of different masses after they live their main-sequence hydrogen-burning lives.
•
To describe the different ways in which stars die.

We colloquially use the boxing terms "lightweight" and "heavyweight," though we are referring to mass rather than weight.

REDSHIFT
http://www.harcourtcollege.com/astro/cosmos/rsce

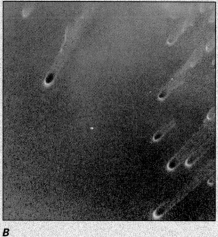

A   *B*

Figure 13–1 *(A)* A ground-based image of the Helix Nebula, a very nearby planetary nebula. *(B)* A Hubble Space Telescope close-up of part of the Helix Nebula shows comet-shaped features formed by the outward flow of stellar wind from the central star. (The full Hubble field of view is outlined in part A.) This wind pushes on nebular knots or even breaks up the shell of gas. In the Hubble image, the heads of the knots are 100 A.U. in diameter, roughly the size of our Solar System (out to the orbit of Pluto), and the tails perhaps 1000 A.U. long.

◀ The Ring Nebula, a planetary nebula, the late stage of evolution of a lightweight star like the Sun. It is imaged here in the Hubble Heritage Program.

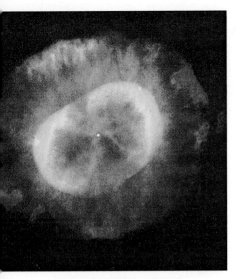

Figure 13-2 A Hubble Space Telescope view of the planetary nebula NGC 7662 shows the basic parts of elliptical planetary nebulae. We see the central cavity and shell caused by the fast wind and, around it, the material given off earlier. Colors show degrees of ionization and thus the energy of photons: singly ionized (red), doubly ionized (green), and triply or more ionized (purple).

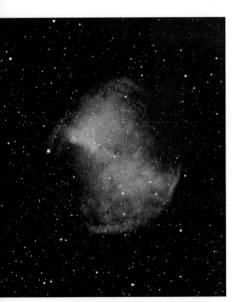

Figure 13-3 The Dumbbell Nebula, M27, a planetary nebula in the constellation Vulpecula, in a ground-based view. Its diameter in the sky is over one-fourth that of the Moon. Radiation from the hot blue central star provides the energy for the nebula to shine.

⬤ THE DEATH OF THE SUN

Red Giants

Though the details can differ, all stars containing less than about ten times the Sun's mass will have the same fate. As fusion exhausts the hydrogen in their centers, their internal pressure will diminish. Gravity will pull the core in, and the core will heat up again. Hydrogen will begin "burning" in a shell around the core. (The process is nuclear fusion, not the chemical burning we have on Earth.) The new energy will cause the outer layers of the star to swell by a factor of 10 or more. They will become very large, so large that when the Sun reaches this stage, its diameter will be 10 per cent the size of the Earth's orbit, about 20 times its current diameter. The solar surface will be relatively cool for a star, only about 3000 K, so it will appear reddish. Such a star is called a **red giant.** Red giants appear at the upper right of temperature-luminosity diagrams. The Sun will be in this stage, or on the way to it after the main sequence, for about a billion years, only 10 per cent of its lifetime on the main sequence. ✐

Red giants are so luminous that we can see them at quite a distance, and a few are among the brightest stars in the sky. Arcturus in Boötes and Aldebaran in Taurus are both red giants.

The core becomes so hot that helium will start fusing into carbon (via the triple-alpha process) and oxygen, but this stage will last only a brief time. Subsequently, the star becomes smaller and fainter. We wind up with a star whose core is carbon and oxygen, and is surrounded by shells of helium and hydrogen that are undergoing fusion. Note that stars less massive than 0.45 times the mass of the Sun don't ever produce a carbon–oxygen core; their fusion process creates only helium.

Planetary Nebulae

As the carbon–oxygen core contracts and heats up, it generates more energy. As a result, the rate at which hydrogen is fusing into helium in a shell around the core increases again. The star grows still larger. The outer layers, this time, continue to drift outward until they leave the star. Perhaps the outer layers escape as a shell of gas. Or perhaps they drift off gradually and a second round of gas comes off at a more rapid pace. This second round of gas plows into the first round, creating a visible shell (Fig. 13–2). Each of these two models has its proponents, and observations are being carried out to discover which is valid in most cases.

In any case, we know of a thousand such shells of gas in our galaxy. Each shell contains about 20 per cent of the Sun's mass. They are exceedingly beautiful. In the small telescopes of a hundred years ago, though, they appeared as faint greenish objects, similar to the planet Uranus. These objects were thus named **planetary nebulae.** We now know that planetary nebulae generally look greenish because the gas in them emits mainly a few strong spectral emission lines that include greenish ones. Uranus seems green for an entirely different reason (principally the molecule methane). But the name "planetary nebulae" remains. (Note: Planetary nebulae are almost never called just "nebulae"; always use the adjective "planetary.") ✐

The best known planetary nebula is the Ring Nebula in the constellation Lyra, which we saw in the Chapter Opener. It is visible in even a medium-sized telescope as a tiny apparent smoke ring in the sky. Only photographs reveal the vivid colors. The Dumbbell Nebula (Fig. 13-3) is another famous example. The Helix Nebula is so close to us that it covers about half the apparent diameter in the sky as the full moon, though it is much fainter.

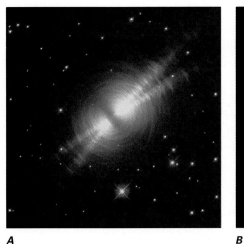
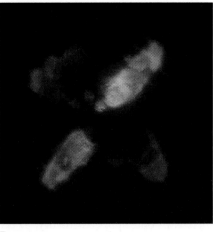

A   **B**

Figure 13-4 Hubble Space Telescope view of a planetary nebula being born, the Egg Nebula. *(A)* In this visible-light view, the central star is hidden by a dense lane of dust. We are seeing light from the central star scattered toward us by dust farther away. Some of the light escapes through relatively clear places, so we see beams coming out of the polar regions. The circular arcs are presumably the shells of gas and dust that were irregularly ejected from the central star. *(B)* The false-color view from Hubble's Near Infrared Camera/Multi-Object Spectrograph (NICMOS) shows starlight reflected by dust particles *(blue)* and radiation from hot molecular hydrogen *(red)*. The collision between material ejected rapidly along a preferred axis and the slower, outflowing shells causes the molecular hydrogen to glow.

The Hubble Space Telescope has viewed planetary nebulae with a resolution about 10 times better than from the ground, and has revealed new glories in them. (www) Its infrared camera provided views of different aspects of some of the planetary nebulae (Fig. 13–4).

The remaining part of the star in the center is the star's exposed hot core, which reaches temperatures of 100,000 K and so appears bluish. It is known as the "central star of the planetary nebula." Ultraviolet radiation from this hot star ionizes gas in the planetary nebula, causing it to glow at optical wavelengths. Each planetary-nebula stage in the life of a Sun-like star lasts only about 50,000 years, although there can be many such stages. After that time, the nebula spreads out and fades too much to be seen at a distance.

REDSHIFT
http://www.harcourtcollege.com/astro/cosmos/rsce

White Dwarfs

Somehow, through a series of winds and planetary-nebula ejections, all stars up to 8 (or perhaps even 10) times the Sun's mass manage to lose most of their mass, so the remaining stellar core is below 1.4 times the Sun's mass. (The Sun itself will have only 0.6 of its current mass at that time, in about 6 billion years.) When this contracting core reaches about the size of the Earth, 100 times smaller in diameter than it had been on the main sequence, a new type of pressure succeeds in counterbalancing gravity so that the contraction stops. This new pressure is the result of processes that can be understood only with quantum mechanics. It comes from the resistance of electrons to being packed too closely together. This resistance results in a type of star called a **white dwarf** (Fig. 13–5).

The Sun is 1.4 million km (nearly a million miles) across. When most of its mass is compressed into a volume 100 times smaller across, which is a million times smaller in volume, the density of matter goes up incredibly. A single teaspoonful of a white dwarf would weigh 5 tons! Such a high density may have been momentarily achieved in a recent terrestrial laboratory experiment.

A white dwarf's mass cannot exceed 1.4 times the Sun's mass; it would become unstable and either collapse or explode. This theoretical maximum was worked out by an Indian university student, S. Chandrasekhar (usually pronounced "chan dra sek′ har" in the United States), en route to England in 1930. It is called the "Chandrasekhar limit." In a long career in the United States, Chandrasekhar became one of the most distinguished astronomers in history, and shared the 1983 Nobel Prize in Physics with William A. Fowler for this early research. The latest major NASA spacecraft, the Chandra X-ray Observatory, is named after him. (www)

Figure 13-5 The sizes of white dwarfs are not very different from that of the Earth. A white dwarf contains about 300,000 times more mass than the Earth, however.

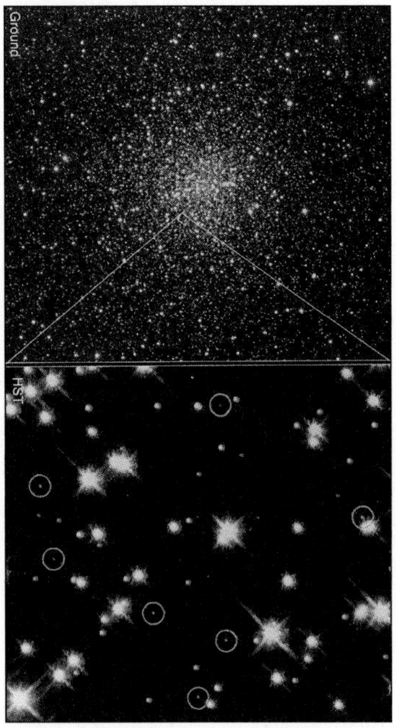

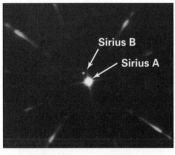

Figure 13-6 This Hubble Space Telescope view *(bottom)* of a small region only 2/3 light-year across in the M4 globular star cluster *(top)* reveals seven white dwarf stars *(inside blue circles).* The cluster may contain 40,000 white dwarfs.

Figure 13-7 Sirius A with its companion white dwarf Sirius B appearing as a faint dot nearby. Spectra are also shown. Sirius A and B have been moving apart from each other since their closest approach in 1993; they are now about 5 arc seconds apart and the separation will increase to 10 arc seconds in 2043.

Because they are so small, white dwarfs are so faint that they are hard to detect (Fig. 13–6). Only a few single ones are known. We find most of them as members of binary systems. Even the brightest star, Sirius, the Dog Star, has a white-dwarf companion, which is named Sirius B and sometimes called "The Pup" (Fig. 13–7).

White-dwarf stars have all the energy they will ever have. Over billions of years, they will gradually radiate their energy, eventually cooling off until they can no longer be seen. We can call them "retired stars," since they are spending their life savings of stored energy.

Novae

For millennia, new stars have occasionally become visible in the sky. Some of them turn out, by recent theory, to be the result of an interaction of a large star with a white dwarf. In this section, we will discuss the role of white dwarfs in some of the apparently new stars, **novae** (pronounced "no'vee"; the singular form is **nova**). In the next section, we will see how white dwarfs may also be contributors to even more luminous objects now known as supernovae.

A nova is newly visible, but is not really new. It represents a star system's brightening by a factor of 100 to a million, which corresponds to 5 to 15 magnitudes. It may remain bright for only a few days or weeks, and then fade over the years. The ejected gas may eventually become visible.

By the current theory, novae occur when one star in a binary system has evolved to the white-dwarf stage, and the other component is a red giant or almost so. Since the outer layers of the red giant are not strongly held in by the star's gravity, material from them can be pulled off. This matter surrounds the white dwarf (Fig. 13–8). When a sufficient amount of material falls down to the white dwarf's surface, it may heat up enough to begin nuclear fusion there. This process involves only $1/10,000$ or so of a solar mass, so it can happen many times. We do indeed see some novae repeat their outbursts.

● SUPERNOVAE: STELLAR RECYCLING

Though most stars with about the mass of the Sun gradually puff off faint planetary nebulae, some of them join more massive stars in finally going off with a spectacular bang. Let us consider these celestial fireworks.

Red Supergiants

Stars that are more than 10 times as massive as the Sun whip through their main-sequence lifetimes at a rapid pace. These prodigal stars use up their store of hydrogen very quickly. A star containing 15 times as much mass as the Sun may take only 10 million years from the time it reaches the main sequence until it uses up the hydrogen in its core. This time scale is 1000 times faster than that of the Sun.

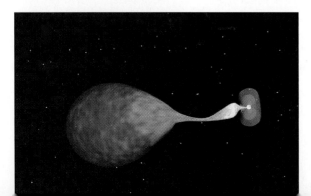

Figure 13-8 Nova Cygni 1975 came from the explosion of a white dwarf with a very high magnetic field, the first such nova discovered. This computer-graphics image shows gas flowing to the white dwarf from its companion. The white dwarf's magnetic field is shown in green.

In December 1999, Nova Aquilae 1999 No. 2 reached naked-eye brightness, the first nova to do so since 1975. The star system brightened by over 10,000 times. It was assigned variable-star number V1494 Aquilae.

For these massive stars, the helium core contracts as the outer layers expand. The star has become so bright that we call it a **red supergiant.** Betelgeuse, the star that marks the shoulder of Orion, is the best-known example (Fig. 13–9).

Eventually, the core temperature reaches 100 million degrees, and the triple-alpha process begins to transform helium into carbon. Some of the carbon nuclei then fuse with a helium nucleus (alpha particle) to form oxygen. The carbon–oxygen core of a supergiant contracts, heats up, and begins fusing into still heavier elements. The ashes of one set of nuclear reactions become the fuel for the next set. Each stage of fusion gives off energy.

Finally, even iron builds up. The iron core is surrounded by layers of elements of lower and lower mass, somewhat resembling the shells of an onion. But when iron fuses into heavier elements, it takes up energy instead of giving it off. No new energy is released to make enough pressure to hold up the star against the force of gravity pulling in. Thus the iron doesn't fuse, but instead disintegrates; the protons and electrons combine to form neutrons and neutrinos. Within a second, the star collapses. It rebounds and bursts outward, getting a large amount of energy from neutrinos, and achieves a stupendous optical luminosity rivaling the brightness of a billion normal stars. It has become a **supernova.** So much energy is available that heavy elements form in the ejected layers.

Such **supernovae,** known as **Type II,** mark the violent death of heavyweight stars. Another type of supernova, **Type Ia,** come from white dwarfs in binary systems (Fig. 13–10). ☼ (Types Ib and Ic are variations of explosions of heavy-

Figure 13–9 The red supergiant star Betelgeuse, α (alpha) Orionis, is revealed by an ultraviolet image with the Hubble Space Telescope to have an atmosphere the size of Jupiter's orbit. A huge, hot, bright spot, ten times the diameter of Earth, is visible. It is 2000 K hotter than surrounding gas.

Figure 13–10 Type Ia supernovae *(left)* come from the incineration of a white dwarf that is gaining matter from a neighboring giant and reaches the Chandrasekhar limit, 1.4 times the Sun's mass. Type II supernovae *(right)* are the explosions of massive stars, usually from the supergiant phase. When iron forms at the center of the onion-like layers of heavy elements, the star collapses. In this model of the collapse, the core overshoots its final density and rebounds. Neutrinos also push outward. The shock wave from the rebound and the neutrinos blast off the star's outer layers.

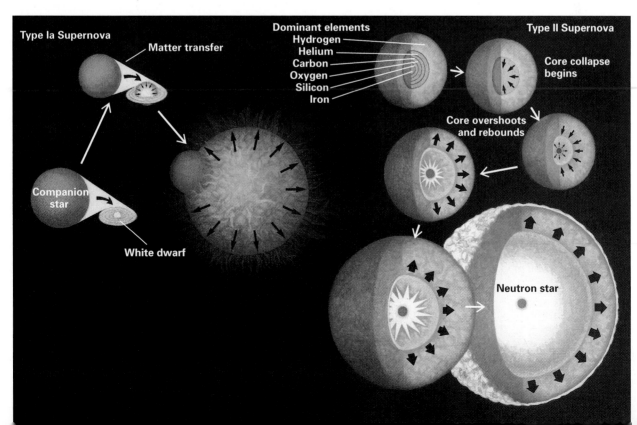

weight stars, and we won't discuss them further here.) If too much matter is added to the white dwarf by its companion, causing the white dwarf to reach the Chandrasekhar limit of 1.4 solar masses, it can no longer support itself. The white dwarf then explodes in a runaway chain of nuclear-fusion reactions. Such Type Ia supernovae become as bright as 10 billion Suns, the energy coming from the decay of radioactive heavy elements produced by the incineration. Theoretical models can now account fairly well for the spectrum and the amount of light.

When you just look at a supernova you can't tell if it is Type Ia (the explosion of a white dwarf), or Type II (that of a massive star), but the type is generally obvious from its spectrum. The basic criterion is that Type II supernovae have hydrogen spectral lines, while Type I supernovae (both Ia and others) do not. The absence of hydrogen in the spectrum of Type Ia supernovae is explained by the idea that the white dwarf could have lost its outer atmosphere before it was incinerated. However, it isn't clear why the hydrogen blown away from the atmosphere of the companion star is not visible.

Observing Supernovae

Only in the 1920s was it realized that some of the "novae"—apparently new stars—that had been seen in other galaxies were really much brighter than ordinary novae seen in our own galaxy. These supernovae are very different kinds of objects. Whereas novae are small eruptions involving only a tiny fraction of a star's mass, supernovae involve entire stars. A supernova may appear about as bright as the entire galaxy it is in (Fig. 13–11). **WWW**

Unfortunately, we have seen very few supernovae in our own galaxy, and none since the invention of the telescope. The most recent ones definitely noticed were observed by Kepler in 1604 and Tycho in 1572. A relatively nearby supernova might appear as bright as the full moon, and be visible night and day. Since studies in other large galaxies show that supernovae erupt every 30 years or so on the average, we appear to be due, although a few supernovae have probably occurred in distant, obscured parts of our galaxy. Maybe the light from a nearby supernova will reach us tonight. Meanwhile, scientists must remain content with studying supernovae in other galaxies, including one in the nearest galaxy to us, the Large Magellanic Cloud. In a following section, we will discuss how its eruption appeared to us in 1987.

Photography of the sky has revealed some two dozen regions of gas in our galaxy that are **supernova remnants,** the gas spread out by the explosion of a supernova (Fig. 13–12). The most studied supernova remnant is the Crab Nebula in the constellation Taurus (Fig. 13–13A). **WWW** The explosion was noticed widely in China, Japan, and Korea in A.D. 1054; there is still debate as to why Europeans did not see it. If we compare photographs of the Crab taken decades apart, we can measure the speed at which its filaments are expanding. Tracing them back shows that they were together at one point, at about the time the bright "guest star" was seen in the sky by the observers in Asia, confirming the identification. The rapid speed of expansion—thousands of kilometers per second—also confirms that the Crab Nebula comes from an explosive event.

The Chandra X-ray Observatory is giving us high-resolution x-ray images of supernova remnants (Fig. 13–13B).

Supernovae and Us

The heavy elements that are formed and thrown out by both Type I and Type II supernovae are necessary for life. Directly or indirectly, supernovae are the only

Figure 13–11   A supernova is visible near the right-hand edge of a spiral galaxy, 130 million light-years away from us. Though only a single star, it is comparable in brightness to the entire galaxy consisting of billions of stars. It appears as bright as foreground stars in our own Milky Way Galaxy, which are only about 1000 light-years away from us. After exploding, the supernova took only three weeks to reach its peak luminosity, but then faded to obscurity over the course of a year.

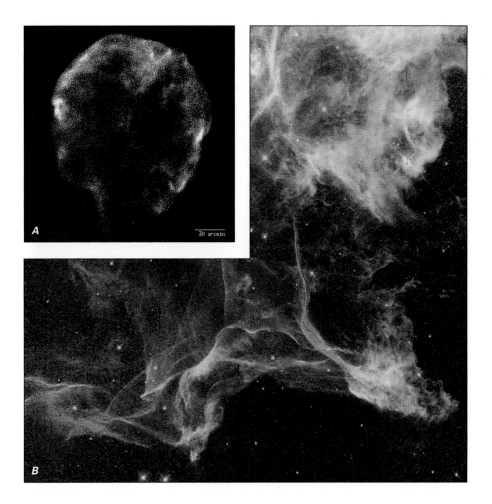

Figure 13-12   The supernova remnant known as the Cygnus Loop, itself part of the Veil Nebula. *(A)* A wide-angle ROSAT x-ray view, showing the extent of the exploded gas. *(B)* A Hubble image of a small part of the Cygnus Loop. We see the effect of the shock wave. The hydrogen radiation *(green)* comes from a thin zone, only a few A.U. across, immediately behind the shock wave. The emission from singly ionized sulfur *(red)* shows gas that has cooled after the shock wave passed. Doubly ionized oxygen *(blue)* was also formed by the shock front.

known source of most heavy elements, especially those past iron (Fe) on the Periodic Table of the Elements. They are spread through space and are incorporated in stars that form later on. The Sun is such a star, made from the debris of many previous generations. So we humans, who depend on heavy elements for our existence, are here because of supernovae and this process of recycling material!

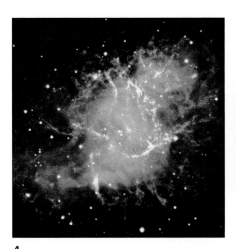

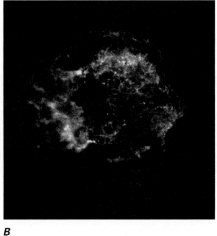

Figure 13-13   *(A)* The Crab Nebula, a prominent supernova remnant, the result of the great supernova of A.D. 1054. It is 6 light-years across. *See also* Figure 13–25B. *(B)* The supernova remnant Cassiopeia A, imaged with the Chandra X-ray Observatory. This image was the first to be released from CXO. The supernova went off in the 17th century.

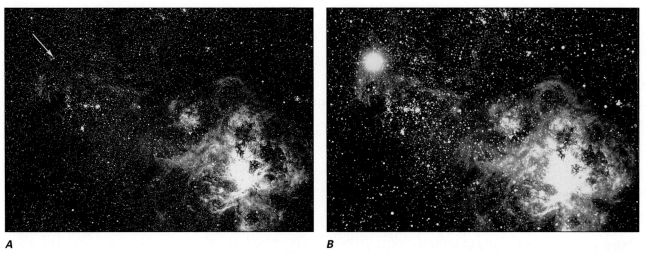

A   **B**

Figure 13–14 The region of the Tarantula Nebula *(lower right)* in the Large Magellanic Cloud. We see the region *(A)* before and *(B)* after February 24, 1987. Supernova 1987A shows clearly at upper left.

Supernova 1987A

An astronomer's delight, a supernova quite bright but at a safe distance, appeared in 1987. On February 24 of that year, Ian Shelton, then of the University of Toronto, was photographing the Large Magellanic Cloud, a small galaxy 170,000 light-years away, with a telescope in Chile. Fortunately, he chose to develop his photographic plate that night. When he looked at it, still in the darkroom, he saw a bright star where no such star belonged (Fig. 13–14). He went outside, looked up, and again saw the star in the Large Magellanic Cloud, this time with his naked eye. He had discovered the nearest supernova to Earth seen since Kepler saw one in 1604. By the next night, the news was all over the world, and all the telescopes that could see Supernova 1987A (the first supernova found in 1987) were focused on it. Some of these telescopes as well as the Hubble Space Telescope continue to observe the supernova on a regular basis to this day. Hubble's high resolution shows clear views of a ring of material expanding close to the supernova (Figs. 13–15 and 13–16). The inner debris is meeting up with the ring, and we expect the supernova to brighten substantially over the next few years.

One exciting thing about such a close supernova is that we even know which star had erupted! Pre-explosion photographs showed that a blue supergiant star

A

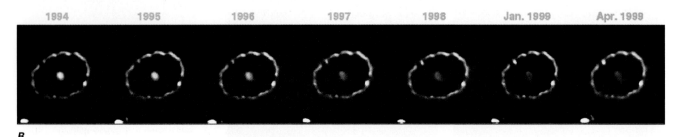

B

Figure 13–15 *(A)* The inner and outer rings of ejected gas around Supernova 1987A. *(B)* A ring of ejected gas so close to Supernova 1987A that it is clearly visible only from the Hubble Space Telescope. The ring is made of gas thrown off by the star prior to the explosion itself. Debris expanding from the explosion is reaching the ring, making spots on it brighten.

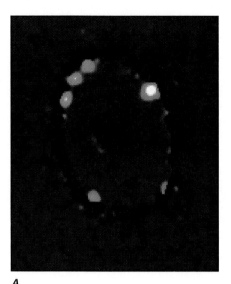

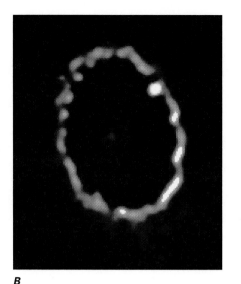

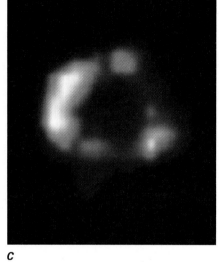

A   B   C

Figure 13–16 By 2000, new bright spots on the ring appeared, marking increased impacting of the expanding debris on the ring around Supernova 1987A. *(A)* Spots on the inner ring are brightening, and we will see if, as some predict, the whole ring brightens over the next few years as material from the supernova explosion hits it. Here we see an image from February 2, 2000. *(B)* Computer processing, subtracting earlier images from the one at left, has emphasized newly seen bright knots of superheated gas. *(C)* The Imaging Spectrometer on the Chandra X-ray Observatory shows the expanding 10-million-kelvin shell of gas heated by the impact of the shock wave caused by the supernova explosion.

had been where the supernova now is (Fig. 13–17). It had been thought that supernovae always erupt from red supergiants, not blue ones, but now we know differently. Supernova 1987A did not brighten as much as had been expected, and the fact that it was from a blue giant may explain why. Blue supergiants are smaller than red ones, so they are not able to radiate as much light in the early stages of the explosion, and more of the explosion energy is used up in expanding the star. Theoretical models coupled with the observations indicate that the star had once contained 20 times the mass of the Sun, with 6 times the mass of the Sun in the form of a helium core. Through winds and gentle ejections, it had already lost 4 solar masses by the time it went supernova.

The rate at which the supernova faded matched the rate of decay of radioactive cobalt into stable iron. Its brightness corresponded to the 0.07 solar mass of cobalt that theory predicted would be formed (initially as radioactive nickel, which decayed into cobalt). We are all awaiting the emergence of definite signs of a pulsar, the type of star we shall discuss in the next section.

When the core of a massive star collapses, producing a supernova, theorists tell us that many neutrinos are produced. In fact, 99 per cent of the energy is carried by neutrinos. The solar neutrino experiment using chlorine in liquid form (Chapter 12) was not sensitive enough to the energy range of neutrinos emitted by Supernova 1987A. Fortunately, at least two other experiments that had been set up for other purposes were operating during this event. Both experiments contained large volumes of extremely pure water surrounded by sensitive photo-tubes to measure any light given off as a result of interactions in the water.

One of the detectors, in a salt mine in Ohio, reported that 8 neutrinos had arrived and interacted within a 6-second period on February 23, 1987. Normally, an interaction of some kind of neutrino was seen only about once every five days. Another detector, in a zinc mine in Japan, detected a burst of 11 neutrinos. (The detectors were in mines to shield them from other types of particles.) These few neutrinos marked the emergence of a new observational field of astronomy: extra-solar neutrino astronomy.

The fact that the neutrinos arrived three hours before we saw the optical burst matches our theoretical ideas about how a star collapses and rebounds to make a supernova. Also, the amount of energy carried in neutrinos, taking account of the tiny fraction that we detect, matches that of theoretical predictions. Our basic ideas of how Type II supernovae occur were validated.

Figure 13–17 A composite photo in which a pre-explosion photograph of the blue supergiant star that exploded is superimposed as a negative (thus appearing black) on an image of Supernova 1987A. Several stars appear as black dots in the negative.

If neutrinos are massless, like particles of light (photons), in empty space they all travel at the same speed, the speed of light.

The observations have also given us important basic knowledge about neutrinos themselves. If neutrinos have mass, they would travel at different speeds depending on how much energy they were given. The fact that the neutrinos arrived so closely spaced in time placed a sensitive limit on how much mass they could have. So though the current observations of neutrinos show that they probably have some mass, we also know that they don't have much mass. Understanding the mass of neutrinos is important for understanding how much matter there is in the Universe, as we shall see in our discussion of galaxies and cosmology (Chapters 16 and 19).

Supernovae are so bright that they are detectable far into space. Computerized comparisons of images of whole galaxies can be used to find a few new supernovae each month. With the assurance that some new supernovae will be discovered, telescope time can be booked to follow these supernovae in some detail. The results are of great importance for finding the distance scale, expansion rate, and age of the Universe, as we shall see in Chapter 18. The results indicate that the Universe might be expanding at an accelerating rate, one of the most surprising discoveries in quite a long time.

Cosmic Rays

So far, our study of the Universe in this book has relied on information that we get by observing electromagnetic radiation—including not only visible light but also gamma rays, x-rays, ultraviolet, infrared, and radio waves. Moon rocks and meteorites have given important insights as well. But we also receive a few high-energy particles from space.

These **cosmic rays** (misnamed historically; we now know that they aren't rays at all) are nuclei of atoms moving at tremendous speeds. Some of the weaker cosmic rays come from the Sun while other cosmic rays come from farther away. Cosmic rays provide about 1/5 of the radiation environment of Earth's surface and of the people on it. (Almost all the radiation we are exposed to comes from cosmic rays, from naturally occurring radioactive elements in the Earth or in our bodies, and from medical x-rays.)

For a long time, scientists have debated the origin of the nonsolar "primary cosmic rays," the ones that actually hit the Earth's atmosphere as opposed to cosmic rays that hit the Earth's surface. (Our atmosphere filters out most of the primary cosmic rays. When they hit the Earth's atmosphere, the collisions with air molecules generate "secondary cosmic rays.")

Because cosmic rays are charged particles—we receive mostly protons and also some nuclei of atoms heavier than hydrogen, and many fewer electrons—our galaxy's magnetic field bends them. Thus we cannot trace back the paths of cosmic rays we detect to find their origin. It seems that most middle-energy cosmic rays were accelerated to their high speeds in supernova explosions.

Primary cosmic rays can be captured with high-altitude balloons or satellites. Many images taken with CCDs, whether those on the Hubble Space Telescope, on the Solar and Heliospheric Observatory, or on Earth, show streaks from cosmic rays (Fig. 13–18). When studying your CCD data, you have to eliminate the pixels that show these "cosmic-ray hits."

Stacks of suitable plastics (the observations were formerly made with thick photographic emulsions) show the damaging effects of cosmic rays passing through them. Scientists are now worried about cosmic rays damaging computer chips vital for navigation in airplanes as well as in spacecraft, and, for safety, engineers are providing chips that can work even when slightly damaged in this way.

Figure 13–18 The central part of the galaxy M77, made by combining Hubble Space Telescope images obtained through red, green, and blue filters. At *top*, many cosmic-ray hits can be seen as streaks and specks. (The shape of each one depends on the angle at which the cosmic ray hits the CCD.) At *bottom*, the cosmic-ray hits are gone. They were removed by comparing pairs of images obtained through each filter: pixels that were much brighter in one image than the other were assumed to have been hit by cosmic rays, and were replaced by the unaffected pixels.

When primary cosmic rays hit the Earth's atmosphere, they cause flashes of light that can be detected with telescopes on Earth. A project for observing secondary cosmic rays by studying light they generate as they plow through a cubic kilometer of clear natural ice is going into operation underground near the South Pole. A similar project, using clear Mediterranean water, is also progressing.

●PULSARS: STELLAR BEACONS

We have examined the fate of the outer layers of a massive star that explodes as one type of supernova. But what about the core? Let us now discuss cores that wind up as superdense stars. In the next chapter, we will see what happens when the core is too massive to ever stop collapsing.

Neutron Stars

As iron fills the core of a supergiant star, the temperature becomes so high that the iron begins to break down into smaller units like helium nuclei. This breakdown soaks up energy and reduces the pressure. The core can no longer counterbalance gravity, and it collapses.

The core's density becomes so high that electrons are squeezed into the nuclei. They react with the protons there to produce neutrons and neutrinos. The neutrinos escape, along the way helping to blow off the rest of the star. A gas composed mainly of neutrons remains. If between a few tenths the mass of the Sun and about twice the mass of the Sun are left in the remaining core, it can reach a new stable stage.

When this remaining core is sufficiently compressed, the neutrons resist being further compressed, as we can explain using laws of quantum mechanics. A pressure is created, which counterbalances the inward force of gravity. The star is now basically composed of neutrons, and is so dense that it is like a single, giant nucleus. We call it a **neutron star.** It is only about 20 kilometers across but contains a mass comparable to that of the Sun (Fig. 13–19). A teaspoonful would weigh a billion tons.

As an object contracts, its magnetic field is compressed. As the magnetic-field lines come together, the field gets stronger. A neutron star is so much smaller than the Sun that its field should be a trillion times stronger.

When neutron stars were first discussed theoretically in the 1930s, the chances of observing one seemed hopeless. But we currently can detect signs of them in several independent and surprising ways, as we now discuss.

The Discovery of Pulsars

Recall that the light from stars twinkles in the sky because the stars are point-like objects, with the Earth's atmospheric turbulence bending the light rays. Similarly, point-like radio sources (radio sources that are so small or so far away that they have no apparent length or breadth) fluctuate in brightness on time scales of a second because of variations in the density of electrons in interplanetary space. In 1967, a special radio telescope was built to study this radio twinkling; previously, radio astronomers had mostly ignored and blurred out the effect to study the objects themselves.

In 1967 Jocelyn Bell (now Jocelyn Bell Burnell) was a graduate student working on Professor Antony Hewish's special radio telescope (Fig. 13–20). As the sky swept over the telescope, which pointed in a fixed direction, she noticed that the

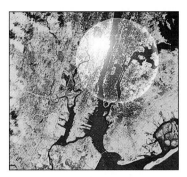

Figure 13–19 A neutron star is the size of a city, even though it may contain a solar mass or more. Here we see the ghost of a neutron star superimposed on a photograph of New York City. A neutron star might have a solid, crystalline crust about a hundred meters thick. Above these outer layers, its atmosphere probably takes up only another few centimeters. Since the crust is crystalline, there may be irregular structures like mountains, which would only poke up a few centimeters through the atmosphere.

Figure 13–20 Jocelyn Bell Burnell, the discoverer of pulsars. She did so with a radio telescope—actually a field of aerials—at Cambridge, England.

Figure 13-21 (A) The up-and-down variations on this chart led Jocelyn Bell Burnell to suspect that something interesting was going on. (B) The chart record showing the discovery of the individual pulses from CP 1919. Here downward blips are actually increases in brightness.

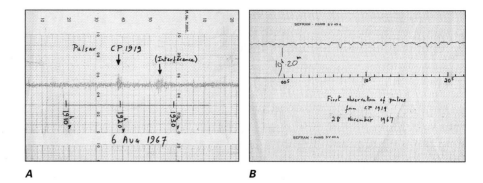

A

B

signal occasionally wavered a lot in the middle of the night, when radio twinkling was usually low.

Her observations eventually showed that the position of the source of the signals remained fixed with respect to the stars rather than constant in terrestrial time (for example, always occurring at exactly midnight). This timing implied that the phenomenon was celestial rather than terrestrial or solar.

Bell and Hewish found that the signal, when spread out, was a set of regularly spaced pulses, with one pulse every 1.3373011 seconds (Fig. 13–21). The source was briefly called LGM, for "Little Green Men," because such a signal might come from an extraterrestrial civilization. But soon Bell located three other sources, pulsing with regular periods of 0.253065, 1.187911, 1.2737635 seconds, respectively. Though they could be LGM2, LGM3, and LGM4, it seemed unlikely that extraterrestrials would have put out four such beacons at widely spaced locations in our galaxy. At the time, it also seemed unlikely that actual LGMs would have chosen to send signals that had pulsed changes in intensity or signals at such apparently randomly chosen radio frequencies (though these points no longer seem to be such major objections). The objects were named **pulsars**—to indicate that they gave out pulses of radio waves—and announced to an astonished world. It was immediately apparent that they were an important discovery, but what were they?

What Are Pulsars?

Other observatories set to work searching for pulsars, and dozens were soon found. They were all characterized by very regular periods, with the pulse itself taking up only a small fraction of a period. When the positions of many of the roughly 1400 known pulsars are plotted on a celestial map (Fig. 13–22), we easily see that they are concentrated along the relatively flat plane of our galaxy. Thus they must be in our galaxy; if they were located outside our galaxy we would see them distributed uniformly around the sky or even partly obscured near our galaxy's plane where the Milky Way might block something behind it.

The question of what a pulsar is can be divided into two parts. First, we want to know why the pulses are so regular—that is, what the "clock" is. Second, we want to know where the energy comes from.

We can get pulses from a star in two ways: if the star oscillates in size ("pulsates") or if it rotates. (The only other possibility—collapsed stars orbiting each other—would result in progressively decreasing pulse periods as the objects release energy and spiral inward.) The theory worked out for ordinary variable stars had shown that the speed with which a star oscillates depends on its average density. Ordinary stars would oscillate much too slowly to be pulsars, and even white dwarfs would oscillate somewhat too slowly. Further, neutron stars would oscillate too rapidly to be pulsars. So oscillations were excluded.

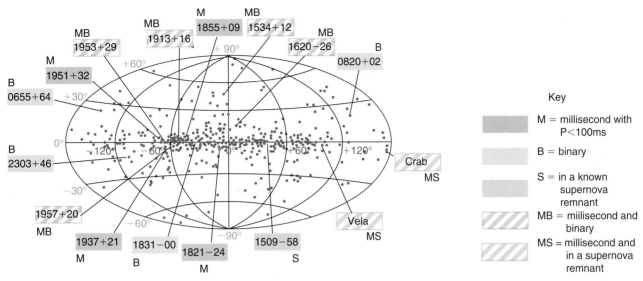

Key

█	M = millisecond with P<100ms
▒	B = binary
░	S = in a known supernova remnant
▨	MB = milisecond and binary
▧	MS = millisecond and in a supernova remnant

Figure 13–22 The distribution of over 500 of the 2000 known pulsars on a map that shows the entire sky, with the plane of the Milky Way along a horizontal line at center. From the concentration of pulsars along the plane of our galaxy, we can conclude that the pulsars are members of our galaxy. Otherwise, we would have expected to see as many near the poles of the map. The concentration of pulsars near −60° galactic longitude on this map merely represents the fact that this section of the sky has been especially carefully searched.

That left only rotation as a possibility. And it can be calculated that a white dwarf is too large to rotate fast enough to cause pulsations as rapid as those that occur in a pulsar; it would be torn apart. So, the only remaining possibility is the rotation of a neutron star. We have solved the problem by the process of elimination. There is agreement on this **lighthouse model** for pulsars (Fig. 13–23). Just as a lighthouse seems to flash light every time its beam points toward you, a pulsar is a rotating neutron star. ☼

How is the energy generated? There is much less agreement about that, and the matter remains unsettled. Remember that the magnetic field of a neutron star is extremely high. This can lead to a powerful beam of radio waves. If the magnetic axis is tilted with respect to the axis of rotation (which is also true for the Earth, whose magnetic north pole is in Hudson Bay, Canada), the beam from stars oriented in certain ways will flash by us at regular intervals. We wouldn't see other neutron stars if their beams were oriented in other directions.

In 1974, Hewish received the Nobel Prize in Physics, largely for his discovery of pulsars. Given the crucial role played by Jocelyn Bell, it is unfortunate that she was not honored in this way as well. At that time it was not the custom of the Nobel Prize committees to honor work done by a graduate student while also honoring the advisor, but largely as a result of the omission in this case the custom has changed.

The Crab, Pulsars, and Supernovae

Several months after the first pulsars had been discovered, strong bursts of radio energy were found to be coming from the direction of the Crab Nebula. Observers detected that the Crab pulsed 30 times a second, almost ten times more rapidly than the fastest other pulsar then known. This very rapid pulsation clinched the exclusion of white dwarfs from the list of possible explanations.

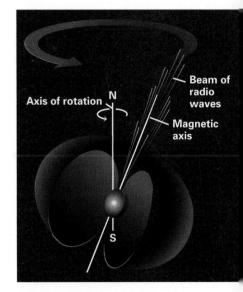

Figure 13–23 In the lighthouse model for pulsars, which is now commonly accepted, a beam of radiation flashes by us once every pulsar period. (A beam pointing in the opposite direction gives an additional pulse, if the magnetic axis is perpendicular to the rotation axis.) Similarly, a lighthouse beam appears to flash by a ship at sea.

A

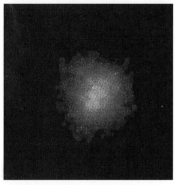

B

Figure 13–24 ROSAT views of the pulsar in the Crab Nebula. *(A)* An x-ray view of the main pulse collected only when the pulsar is on—that is, when the beam of radiation is sweeping by the Earth. *(B)* The off phase of the pulsar. The satellite-imaged x-rays, as indicated by its name, Roentgen satellite, named after Wilhelm Roentgen, the discoverer of x-rays.

The discovery of a pulsar in the Crab Nebula made the theory that pulsars were neutron stars look more plausible, since neutron stars should exist in supernova remnants like this one. And the case was clinched when it was discovered that the clock in the Crab pulsar was not precise—it was slowing down slightly. The energy given off as the pulsar slowed down was precisely the amount of energy needed to keep the Crab Nebula shining. The source of the Crab Nebula's energy had been discovered.

Astronomers soon found, to their surprise, that an optically visible star in the center of the Crab Nebula could be seen apparently to turn on and off 30 times a second. Actually the star only appears "on" when its beamed light is pointing toward us as it sweeps around. Long photographic exposures had always hidden this fact, though the star had been thought to be the remaining core because of its spectrum, which oddly doesn't show any emission or absorption lines. Later, similar observations of the star's blinking on and off in x-rays were also found (Fig. 13–24).

The high-resolution observations by the Hubble Space Telescope and Chandra X-ray Observatory of the Crab Nebula revealed interesting structure near its core (Fig. 13–25).

Slowing Pulsars and the Fast Pulsar

The Crab, when discovered, was the most rapidly pulsing pulsar, and is slowing down by the greatest amount. But most other pulsars have also been found to be slowing gradually. The theory had been that the younger the pulsar, the faster it was spinning and the faster it was slowing down. After all, the Crab came from a supernova explosion only 900 years ago.

So, in 1982, scientists were surprised to find a pulsar spinning 20 times faster—642 times per second. Even a neutron star rotating at that speed would be on the verge of being torn apart. And this pulsar is hardly slowing down at all; it may be useful as a long-term time standard to test even the atomic clocks that are now the best available to scientists. The object, which is in a binary system, is thought to be old—over a hundred million years old—because of its gradual slowdown rate. Astronomers conclude that its rotation rate has been speeded up in an interaction with its companion.

This pulsar's period is 1.56 milliseconds (0.00156 second), so it became known as "the millisecond pulsar." Dozens more millisecond pulsars have since been discovered. Each pulses rapidly enough to sound like a note in the middle of a pi-

Figure 13–25 The central region of the Crab Nebula. *(A)* An image by the Hubble Space Telescope. Wisps of gas take on an apparent whirlpool shape. *(B)* An image by the Chandra X-ray Observatory, showing an outer ring 2/3 light-year across and an inner ring.

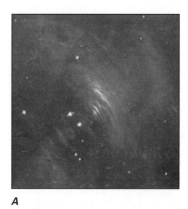

A

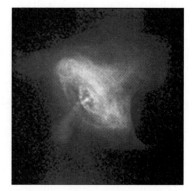

B

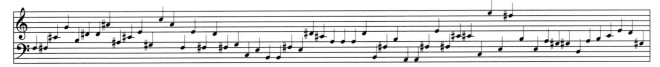

Figure 13-26 Many of the millisecond pulsars have periods fast enough to hear as musical notes when we listen to a signal at the frequency of their pulse rate. The display, of millisecond pulsars discovered through 1999, is in order of celestial longitude.

ano keyboard (Fig. 13–26) if its rotation frequency were converted to an audio signal.

Many of the millisecond pulsars we have detected are in globular clusters. So many stars are packed together in globular clusters that a companion star might have been stripped off in a few of the cases. The pace of discovery of pulsars in globular clusters is now rapid.

By contrast, the pulsar with the longest known period was identified in 1999. Its period is over 8 seconds, longer than can be accounted for by any of the current models of how the energy for the pulses is generated.

The Binary Pulsar and Gravity Waves

Joseph Taylor and Russell Hulse set out to discover lots of pulsars. One of the millisecond pulsars they found seemed less regular in its pulsing than expected. It turned out that another neutron star in close orbit with the pulsar was rapidly pulling it to and fro, shortening and lengthening the interval between the pulses that reached us on Earth. The system is called "the binary pulsar," though one of its neutron stars is not a pulsar. It has an elliptical orbit that can be traced out by studying small differences in the time of arrival of the pulses. The pulses come a little less often when the pulsar is moving away from us in its orbit, and a little more often when the pulsar is moving toward us.

Einstein's general theory of relativity explained the slight change over decades in the orientation of Mercury's orbit around the Sun. The gravity in the binary-pulsar system is much stronger, and the effect is much more pronounced. Calculations show that the orientation of the pulsar's orbit should change by 4° per year (Fig. 13–27), which is verified precisely by measurements.

Another prediction of Einstein's general theory is that **gravitational waves,** wiggles in the curvature of spacetime caused by fluctuation of the positions of masses, should travel through space. The process would be similar to the way that radiation, caused by fluctuations in electricity and magnetism, travels through space. But gravitational waves have never been detected directly. The motion of the binary pulsar in its orbit is speeding up by precisely the rate that would be expected if the system were giving off gravitational waves. (The two stars are getting closer and closer together, so according to Kepler's third law their orbital period is decreasing.) So scientists consider that the existence of gravitational waves has been verified in this way. The Nobel Prize in Physics was awarded in 1993 to Taylor and Hulse for their discovery and analysis of the binary pulsar.

The Taylor–Hulse method of detecting gravitational waves was indirect. Only the speeding up of the orbit caused by the waves, rather than the waves themselves, was measured. Efforts to sense gravitational waves directly by whether they set large metal bars vibrating have failed, because the experiments lacked the necessary sensitivity. A major pair of observational facilities has been built to try to detect gravitational waves directly in another manner. This Laser Interferometer Gravitational-wave Observatory (LIGO) uses complicated lasers, optics, and

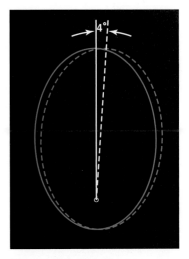

Figure 13-27 The near point of the binary pulsar's orbit to the star it is orbiting moves around by 4° per year. This measurement matches, and thus endorses, the prediction of Einstein's general theory of relativity. For convenience, the diagram shows the farthest point of the orbit rather than the nearest point; it, too, moves by 4° per year.

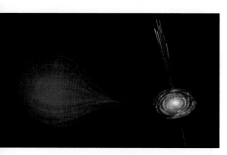

Figure 13–28 A model of SS433 in which the radiation emanates from two narrow beams of matter that are given off by the disk of matter orbiting the star.

When we discuss active galaxies in Chapter 17, we will see that high-speed jets are not unusual on the much larger galactic scale.

X-Ray Binaries

Neutron stars are now routinely studied in a way other than their existence as radio pulsars. Many neutron stars in binary systems interact with their companions. As gas from the companion is funneled toward the neutron star's poles by the strong magnetic field, the gas heats up and gives off x-rays. X-ray telescopes in orbit detect such pulses of x-rays. But in many of these binary systems, unlike the case for normal pulsars, the pulse rate usually speeds up.

One of the oddest x-ray binaries is known as SS433, based on its number in a catalogue. From measurements of Doppler shifts, we detect gas coming out of this x-ray binary at about 25 per cent of the speed of light, a huge speed for a source in our galaxy. The most widely accepted model (Fig. 13–28) considers that SS433 is a neutron star surrounded by a disk of matter it has taken up from a companion star. Our measurements of the Doppler shifts in optical light show us light coming toward us from one jet and going away from us from the other jet at the same time. The disk would wobble like a top (a precession, similar to the one we discussed for the Earth's axis in Chapter 3). As it wobbles, the apparent to-and-fro velocities decrease and increase again, as we see the jets at different angles. We have even detected the jets in radio waves and x-rays (Fig. 13–29).

X-ray binaries are obvious candidates for imaging by the Chandra X-ray Observatory.

A Pulsar with a Planet

Earlier in this book, in Chapters 2 and 7, we described the discovery of planets around other stars. But these planets were not the first to be discovered. In 1991 the first extra-solar planets were discovered by observing a pulsar.

The detections were from observations of a pulsar that pulses very rapidly—162 times each second. The arrival time of the pulsar's radio pulses varied slightly (Fig. 13–30), indicating that something is orbiting the pulsar and pulling it slightly back and forth. Alex Wolszczan, now at Penn State, has concluded that the variations in the pulse-arrival time are caused by three planets in orbit around the pulsar. (www) These planets are 0.19, 0.36, and 0.47 A.U. from the pulsar, within about the same distance that Mercury is from the Sun. They revolve in 25.3-, 66.5- and 98.2-day periods, respectively. The system is 2000 light-years from us, too faint for us to detect optically. The presence of the planets was conclusively verified when they interacted gravitationally as they passed by each other. The two most massive planets are calculated to be somewhat larger than Earth, each containing about 4 times its mass. The innermost planet is much less massive than Earth. The existence of a fourth planet farther out in the system is possible but uncertain.

Astronomers think that neutron stars are formed in supernova explosions, so any original planets almost certainly didn't survive the explosion. Most likely, the planets formed after the supernova explosion, from a disk of material in orbit around the neutron star remnant. These pulsar planets are not the ones on which we expect life will have arisen!

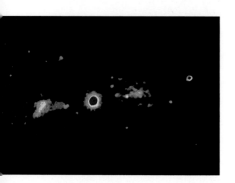

Figure 13–29 An x-ray image of SS433 from the Einstein Observatory. The jets show clearly.

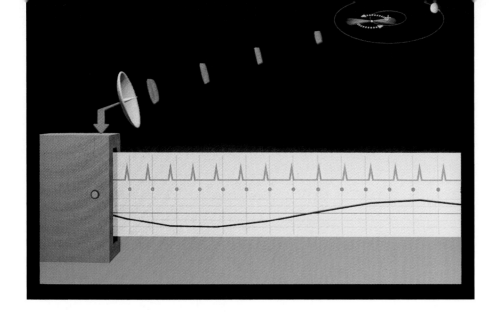

Figure 13-30 The pulses from the pulsar, shown in blue schematically on a chart, sometimes arrive slightly before or after regular pulses would *(red dots)*. The average pulse period is 6.2 milliseconds. The difference is graphed at the bottom. It reveals that there must be planets orbiting the pulsar.

Concept Review

Stars have different fates depending on the mass with which they were born. When the Sun and other **lightweight stars** (stars with mass up to 10 times the Sun's) exhaust their central hydrogen, they will swell and become **red giants.** The outer layers often drift off as **planetary nebulae.** The remaining core contracts until electrons won't be compressed further and it becomes a **white dwarf.** Matter falling onto a white dwarf from a companion star can flare up in a burst of nuclear fusion, which we see as a **nova** (plural **novae**).

Stars that have more than 10 times the Sun's mass, **heavyweight stars,** become **red supergiants.** Heavy elements build up in layers inside these stars. When the innermost layer is iron, the star collapses and the surrounding layers rebound, becoming a **Type II supernova. Type Ia** supernovae, in contrast, occur when white dwarfs in binary systems receive too much mass from their companions to remain in that state, and undergo a nuclear runaway. Usually the explosions leave detectable **supernova remnants.** The Type II supernova that was seen in the Large Magellanic Cloud starting in 1987 was the brightest supernova since the year 1604, and provided many

valuable insights. Neutrinos from the supernova show that we had the basic ideas of the theory of Type II supernovae correct. The heavier elements in our bodies came from past supernova explosions. **Cosmic rays** are particles speeded up to high energy, perhaps from supernova explosions.

The cores of massive stars, after the supernova explosions, consist of neutrons that cannot be compressed further. They are then **neutron stars.** Some give off beams of radiation as they rotate, and we detect pulses in the radio spectrum. We know thousands of these **pulsars.** They are explained by this **lighthouse model.** One pulsar is proving very useful for testing the general theory of relativity: its orbital period around another neutron star is getting shorter at a rate that agrees with the idea that **gravitational waves** are given off. Some pulsars in binary systems have been speeded up to very fast rotation rates. Strange planets have been discovered around one pulsar, but they must have formed after the supernova explosion. Other neutron stars are detected from the x-rays they give off when they are in binary systems.

Questions

1. Why does a red giant appear reddish?
2. Sketch a temperature-luminosity diagram, label the axes, and point out the location of red giants.
3. What forces balance to make a white dwarf?
4. What is the relation of novae and white dwarfs?
5. Is a nova really a new star? Explain.
6. Distinguish between what is going on in novae and supernovae.
7. In what way do we distinguish observationally between Type Ia and Type II supernovae?
8. What are the physical differences between Type Ia and Type II supernovae, in terms of the kinds of stars that explode and their explosion mechanisms?
9. From where did the heavy elements in your body come?
10. Why do we think that the Crab Nebula is a supernova remnant?
11. What are two reasons why Supernova 1987A was significant?
12. Why does the iron core of a supergiant star collapse and then produce a supernova?
13. What is the difference between cosmic rays and x-rays?
14. What keeps a neutron star from collapsing?
15. Compare the Sun, a white dwarf, and a neutron star in size. Include a sketch.
16. In what part of the spectrum do all pulsars give off the energy that we study?
17. How do we know that pulsars are in our galaxy?
18. Why do we think that the lighthouse model explains pulsars?
19. How did studies of the Crab Nebula pin down the explanation of pulsars?
20. How has the binary pulsar been especially useful?
21. Compare the discovery of gravitational waves with the method of detecting them now being worked on.
22. Why is SS433 so unusual?

251

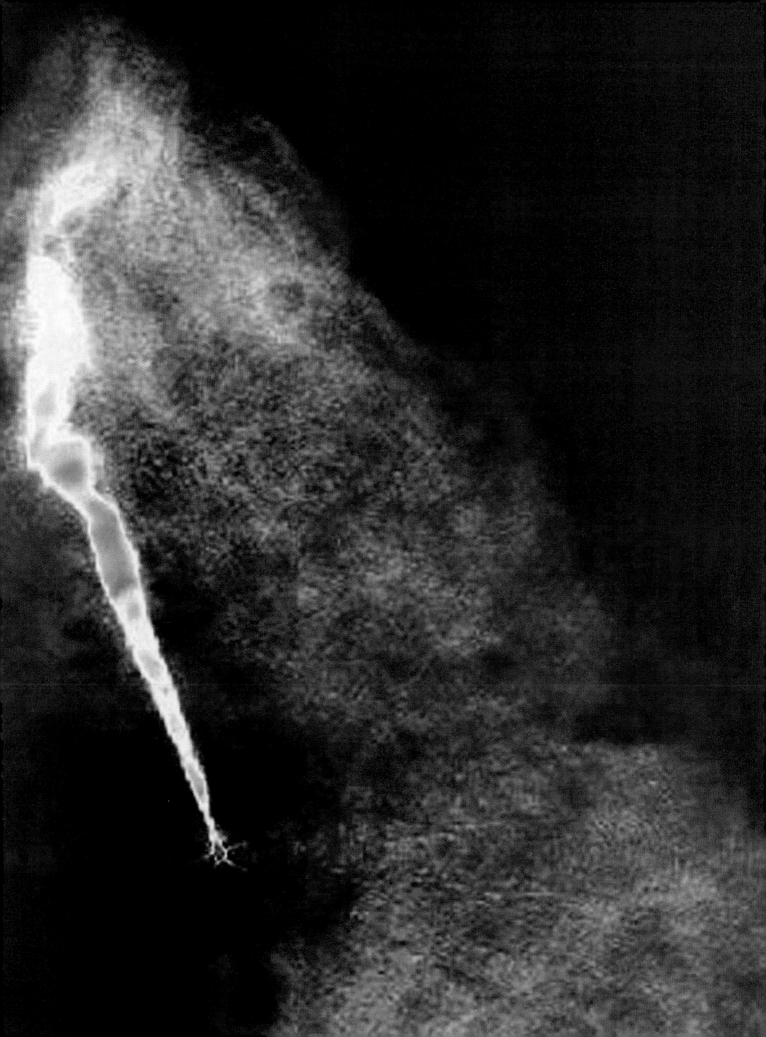

14

Black Holes: The End of Space and Time

ORIGINS *Gravity is the dominant force shaping the structure and evolution of the Universe. Black holes provide a laboratory for testing the most extreme predictions of Einstein's general theory of relativity, our best existing description of gravity.*

The strange forces of electron and neutron pressure support dying light-weight stars and some heavyweight stars against gravity. The strangest case of all occurs at the death of the most massive stars, which contained much more than 10 and up to about 100 solar masses when they were on the main sequence. After these stars undergo supernova explosions, some may retain cores of over 2 or 3 solar masses. Nothing in the Universe is strong enough to hold up the remaining mass against the force of gravity, so it collapses.

The result is a **black hole,** in which the matter disappears from contact with the rest of the Universe (Fig. 14–1). Later, we shall discuss how much more massive black holes form than those that result from the collapse of a star.

AIMS

To understand the properties of black holes, regions in which gravity is so incredibly strong that not even light can escape.

•

To discuss the observational evidence for stellar-mass black holes in binary systems, and for supermassive black holes in the centers of some galaxies.

⬤ THE FORMATION OF A STELLAR-MASS BLACK HOLE

Astronomers had long assumed that the most massive stars would somehow lose enough mass to wind up as white dwarfs. When the discovery of pulsars ended this prejudice, it seemed more reasonable that black holes could exist. If more than 2 or 3 times the mass of the Sun remains after the supernova explosion, the star collapses through the neutron star stage. We know of no force that can stop the collapse. (In some cases, the supernova explosion itself may fail, so the remaining mass can be much larger than 2 or 3 solar masses.)

We may then ask what happens to a 5- or 10- or 50-solar-mass star as it collapses, if it retains more than 2 or 3 solar masses. It must keep collapsing, getting denser and denser. We have seen that Einstein's general theory of relativity predicts that a strong gravitational field will redshift and appear to bend radiation. As the mass contracts and the star's surface gravity increases, radiation is continuously redshifted more and more. Also, radiation leaving the star other than perpendicularly to the surface is bent more and more. Eventually, when the mass has been compressed to a certain size, radiation from the star can no longer

Figure 14–1 Directions to a small, deep cove on the shore of the Bay of Fundy. It got its name (probably a century or two ago) because it appears very dark as seen from the sea.

◀ A radio image at high resolution of the center of the Virgo Cluster galaxy known as M87, with a jet emanating from a supermassive black hole near its core. Later in the chapter, we will discuss evidence that a black hole is indeed present.

Figure 14-2 As the star contracts, a light beam emitted other than straight outward, along a radius, will be bent.

Figure 14-3 Light can be bent so that it falls back onto the star.

escape into space. The star has withdrawn from our observable Universe, in that we can no longer receive radiation from it. We say that the star has become a "black hole."

Why do we call it a black hole? We think of a black surface as a surface that reflects none of the light that hits it. Similarly, any radiation that hits a black hole continues into its interior and is not reflected. In this sense, the object is perfectly black.

THE PHOTON SPHERE

Let us consider what happens to radiation emitted by the surface of a star as it contracts. Although what we will discuss applies to radiation of all wavelengths, let us simply visualize standing on the surface of the collapsing star while holding a flashlight.

On the surface of a supergiant star, if we shine the beam at any angle, it seems to go straight out into space. As the star collapses, two effects begin to occur. (We will ignore the outer layers, which are unimportant here. We also ignore the star's rotation.) Although we on the surface of the star cannot notice the effects ourselves, a friend on a planet revolving around the star could detect them and radio information back to us about them. For one thing, our friend could see that our flashlight beam is redshifted.

Second, our flashlight beam would be bent by the gravitational field of the star (Fig. 14–2). If we shine the beam straight up, it would continue to go straight up. But the further we shine it away from the vertical, the more it would be bent from the vertical. When the star reaches a certain size, a horizontal beam of light would not escape (Fig. 14–3).

From this time on, only if the flashlight is pointed within a certain angle of the vertical does the light continue outward. This angle forms a cone, with its apex at the flashlight, and is called the **exit cone** (Fig. 14–4). As the star grows smaller yet, we find that the flashlight has to be pointed more directly upward in order for its light to escape. The exit cone grows smaller as the star shrinks.

When we shine our flashlight in a direction outside the exit cone, the light is bent sufficiently that it falls back to the surface of the star. When we shine our flashlight exactly along the side of the exit cone, the light goes into orbit around the star, neither escaping nor falling onto the surface (Fig. 14–5).

The sphere around the star in which light can orbit is called the **photon sphere.** Its radius is calculated theoretically to be 4.5 km for each solar mass present. It is thus 13.5 km in radius for a star of 3 solar masses, for example.

As the star continues to contract, theory shows that the exit cone gets narrower and narrower. Light emitted within the exit cone still escapes. The photon sphere remains at the same height even though the matter inside it has contracted further, since the total amount of matter within has not changed.

THE EVENT HORIZON

We might think that the exit cone would simply continue to get narrower as the star shrinks. But Einstein's general theory of relativity predicts that the cone vanishes when the star contracts beyond a certain size. Even light travelling straight up can no longer escape into space, as was worked out by Karl Schwarzschild in solving Einstein's equations in 1916. The radius of the star at this time is called the **Schwarzschild radius.** The imaginary surface at that radius is called the **event horizon.** (A horizon on Earth, similarly, is the limit to which we can see.) Its radius is exactly $\frac{2}{3}$ times that of the photon sphere, 3 km for each solar mass.

Formally, the equation for the Schwarzschild radius is $R = 2GM/c^2$, where M is the star's mass, G is Newton's constant of gravitation, and c is the speed of light.

We can visualize the event horizon in another way, by considering a classical picture based on the Newtonian theory of gravitation. The picture is essentially that conceived in 1783 by John Michell and in 1796 by Pierre Laplace. You must have a certain minimum speed, called the **escape velocity,** to escape from the gravitational pull of another body. For example, we have to launch rockets at 11 km/sec (40,000 km/hr) or faster in order for them to escape from the Earth's gravity. For a more massive body of the same size, the escape velocity would be higher. Now imagine that this body contracts. We are drawn closer to the center of the mass. As this happens, the escape velocity rises.

When all the mass of the body is within its Schwarzschild radius, the escape velocity becomes equal to the speed of light. Thus even light cannot escape (Fig. 14–6). If we begin to apply the special theory of relativity, which deals with motion at very high speeds, we might then reason that since nothing can go faster than the speed of light, nothing can escape.

Now let us return to the picture according to the general theory of relativity, which explains gravity and the effects caused by large masses. The size of the Schwarzschild radius is directly proportional to the amount of mass that is collapsing. A star of 3 solar masses, for example, would have a Schwarzschild radius of 9 km. A star of 6 solar masses would have a Schwarzschild radius of 18 km. One can calculate the Schwarzchild radii for less massive stars as well, although the less massive stars would be held up in the white dwarf or neutron-star stages and not collapse to their Schwarzchild radii. The Sun's Schwarzschild radius is 3 km. The Schwarzschild radius for the Earth is only 9 mm; that is, the Earth would have to be compressed to a sphere only 9 mm in radius in order to form an event horizon and be a black hole.

Anyone or anything on the surface of a star as it passes its event horizon would not be able to survive. An observer would be torn apart by the tremendous difference in gravity between head and foot. (This resulting force is called a tidal force, since this kind of difference in gravity also causes tides on Earth.) If the tidal force could be ignored, though, the observer on the surface of the star would not notice anything particularly wrong as the star passed its event horizon. But the observer's flashlight signal would never get out.

Once the star passes inside its event horizon, it continues to contract. Nothing can ever stop its contraction. In fact, the classical mathematical theory predicts that it will contract to zero radius—it will reach a **singularity.** (Quantum effects, however, probably prevent it from reaching exactly zero radius.)

Even though the mass that causes the black hole has contracted further, the event horizon doesn't change. It remains at the same radius forever, as long as the amount of mass inside doesn't change.

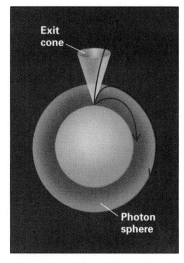

Figure 14–4 When the star (shown by the inner sphere) has contracted enough, only light emitted within the exit cone escapes. Light emitted on the edge of the exit cone goes into the photon sphere. The further the star contracts within the photon sphere, the narrower the exit cone becomes.

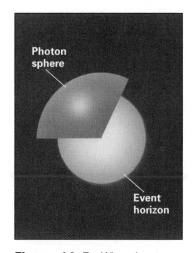

Figure 14–5 When the star becomes smaller than its gravitational radius, we can no longer observe it. It has passed its event horizon.

Figure 14–6 This drawing by Charles Addams is reprinted with permission of The New Yorker Magazine, Inc., © 1974.

Note that if the Sun were turned into a black hole (by unknown forces), Earth's orbit would not be altered: The masses of the Sun and Earth would remain constant, as would the distance between them, so the gravitational force would be unchanged. Indeed, the gravitational field would remain the same everywhere outside the current radius of the Sun. Only at smaller distances would the force be stronger. ☼

TIME DILATION

According to the general theory of relativity, if you were far from a black hole and watching a space shuttle carrying some of your friends fall into it, your friends' clock would appear to run progressively slower as they approached the event horizon. From your perspective, time would be slowing down (or "dilated") for them; indeed, it would take an infinite amount of time (as measured by your clock) for them to cross the horizon. From your friends' perspective, in contrast, it takes a finite amount of time to cross the event horizon, and shortly thereafter they reach the singularity.

On the other hand, if your friends were to approach the event horizon and subsequently *escape* from the vicinity of the black hole, they would have *aged* less than you did. This is a method for jumping into the future while aging very little! However, it doesn't increase longevity—your friends' lives would not be extended. They would not read more books or see more movies than you.

ROTATING BLACK HOLES

Once matter is inside a black hole, it loses its identity in the sense that from outside a black hole, all we can tell is the mass of the black hole, the rate at which it is spinning, and what total electric charge it has. These three quantities are sufficient to completely describe the black hole. Thus, in a sense, black holes are simple objects to describe physically, because we only have to know three numbers to characterize each one. The theorem that describes the simplicity of black holes is often colloquially stated by astronomers active in the field as "a black hole has no hair."

The theoretical calculations about black holes we have just discussed are based on the assumption that black holes do not rotate. But this assumption is only a convenience; we think, in fact, that the rotation of a black hole is one of its important properties. It took decades before Einstein's equations were solved for a black hole that is rotating. (The realization that the solution applied to a rotating black hole came after the solution itself was found.) In this more general case, an additional special boundary—the **stationary limit**—appears, with somewhat different properties from the original event horizon. Within the stationary limit, no particles can remain at rest even though they are outside the event horizon.

The equator of the stationary limit of a rotating black hole has the same diameter as the event horizon of a nonrotating black hole of the same mass. But a rotating black hole's stationary limit is squashed. The event horizon touches the stationary limit at the poles. Since the event horizon remains a sphere, it is smaller than the event horizon of a nonrotating black hole (Fig. 14–7).

The space between the stationary limit and the event horizon is the **ergosphere**. This is the region in which particles cannot be at rest. In principle, we can get energy and matter out of the ergosphere.

A black hole can rotate up to the speed at which a point on the event horizon's equator is travelling at the speed of light. The event horizon's radius is then half the Schwarzschild radius. If a black hole rotated faster than this, its event

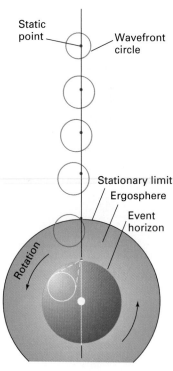

Figure 14–7 Top view of a rotating black hole. The region between the stationary limit and the event horizon of a rotating black hole is called the ergosphere (from the Greek word *ergon*, "work") because, in principle, work can be extracted from it.

horizon would vanish. Unlike the case of a nonrotating black hole, for which the singularity is always unreachably hidden within the event horizon, in this case distant observers could receive signals from the singularity. Such a point would be called a **naked singularity.** Most theorists assume the existence of a law of "cosmic censorship," which requires all singularities to be "clothed" in event horizons—that is, not naked. Since so much energy might erupt from a naked singularity, we can conclude from the fact that we do not find signs of them that there are probably none in our Universe.

DETECTING A BLACK HOLE

A star collapsing to become a black hole would blink out in a fraction of a second, so the odds are unfavorable that we would actually see a collapsing star as it went through the event horizon. And a black hole is too small to see directly. But all hope is not lost for detecting a black hole. Though the black hole disappears, it leaves its gravity behind. It is a bit like the Cheshire Cat from *Alice in Wonderland*, which fades away leaving only its grin behind (Fig. 14–8).

Like all objects in the Universe, the black hole attracts matter, and the matter accelerates toward it. Some of the matter will be pulled directly into the black hole, never to be seen again. But other matter will go into orbit around the black hole, and will orbit at a high speed. This added matter forms an **accretion disk** (Fig. 14–9); "accretion" is growth in size by the gradual addition of matter.

It seems likely that the gas in orbit will be heated to a very high temperature by friction. The gas radiates strongly in the x-ray region of the spectrum, giving off bursts of radiation sporadically as clumps develop in the disk. The inner 200 km should reach hundreds of millions of kelvins. Thus, though we cannot observe the black hole itself, we can hope to observe x-rays from the gas surrounding it. A NASA satellite known as the Rossi X-ray Timing Explorer has the capability of measuring such rapid changes in x-rays, and detects a number of interesting flickering x-ray sources. The Chandra X-ray Observatory and the XMM-Newton Mission, both launched in 1999, should find still more black-hole candidates. (www)

In fact, a large number of x-ray sources are binary-star systems. Some may contain black holes, but the majority probably have neutron stars. It is not enough to find an x-ray source that gives off sporadic pulses, for other mechanisms besides matter revolving around a black hole can lead to such pulses. We also have to show that a collapsed star of greater than 3 solar masses is present. We can determine masses only for certain binary stars. When we search the position of the x-ray sources, we look for a single-lined spectroscopic binary (that is, an apparently single star whose spectrum shows a periodically changing Doppler shift that indicates the presence of an invisible companion; see Chapter 11). Then, if we can show that the companion is too faint to be a normal, main-sequence star, it must be a collapsed star. If, further, the mass of the unobservable companion is greater than 3 solar masses, it must be a black hole.

The most discussed, though no longer the most persuasive, case is named Cygnus X-1 (where X-1 means that it was the first x-ray source to be discovered in the constellation Cygnus). A 9th-magnitude star called HDE 226868 has been found at its location (Fig. 14–10). This star has the spectrum of a blue supergiant, and should have a mass of about 33 times that of the Sun. Its radial velocity varies with a period of 5.6 days, indicating that the supergiant and the invisible companion are orbiting each other with that period. From the orbit, it is deduced that the invisible companion must certainly have a mass greater than 7 solar masses, and a probable mass of about 16 solar masses. This makes it very

REDSHIFT
http://www.harcourtcollege.com/astro/cosmos/rsce

Figure 14–8 Lewis Carroll's Cheshire Cat, from *Alice in Wonderland*, shown here in John Tenniel's drawing. The Cheshire Cat is analogous to a black hole in that it left its grin behind when it disappeared, just as a black hole leaves its gravity behind when it disappears from view. Alice thought that the Cheshire Cat's persisting grin was "the most curious thing I ever saw in my life!" We might say the same about the black hole and its persisting gravity.

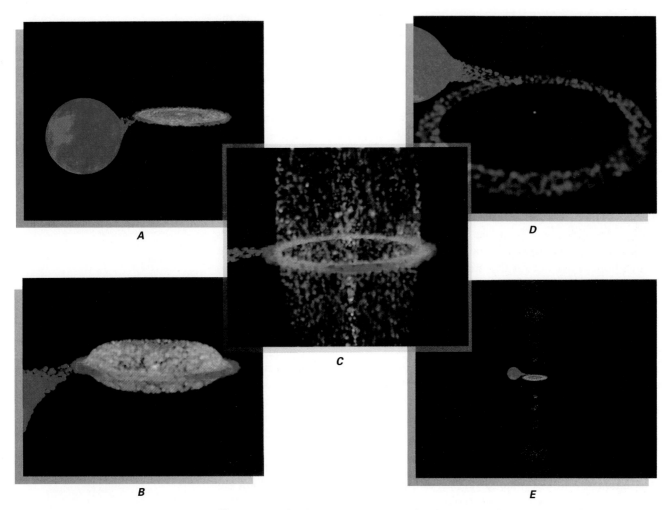

Figure 14–9 A computer animation showing the occasional disruption of the accretion disk surrounding a black hole. The black hole is orbiting a massive companion star, the red sphere on the left. The gas falling into the black hole is compressed and heated to millions of degrees. Different temperatures in this drawing are shown as different colors. *(A)* The hottest material, shown as blue/white in the center of the disk, is closest to the black hole. The Chandra X-ray Observatory should pick up the radiation from such sources. *(B)* A disruption passes through the accretion disk's gas. *(C)* The gas in the disk accelerates to close to the speed of light and is ejected in jets. *(D)* The ejection of gas empties the disk center, and the black hole (the dot in the middle) attracts more gas. *(E)* The ejection repeats every half hour or so, forming the long jets that are seen from a distance.

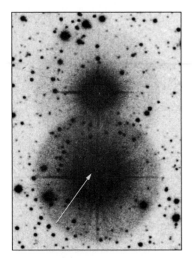

Figure 14–10 The black hole Cygnus X-1 is orbiting the star marked with the arrow. The image of this supergiant star appears large only because it is overexposed on the film. Stars appear black in this negative print.

likely that it is a black hole. But if the visible star is abnormal, and doesn't really have a mass of 33 times that of the Sun, the invisible companion could be less massive, so the case for it being a black hole is not absolutely conclusive.

Although Cygnus X-1 provided the first good evidence for the existence of stellar-mass black holes in our galaxy, the large (and uncertain) mass of the visible star complicates the measurement of the mass of the dark object. More conclusive evidence for black holes comes from binary systems in which the visible star has a very small mass (for example, a K- or M-type main sequence star); uncertainties in its mass are nearly irrelevant (see *Figure It Out: Binary Stars and Kepler's Third Law*). However, when such systems brighten suddenly at x-ray wavelengths, light from the accretion disk also dominates at other wavelengths, and the low-mass star becomes too difficult to see. Observers must wait until the outburst subsides, and the companion star once again becomes visible, before they are able to measure the orbital parameters.

Figure It Out

Binary Stars and Kepler's Third Law

Consider Kepler's third law, as generalized by Newton and discussed in Chapter 5: $(m_1 + m_2)P^2 = (4\pi^2/G)R^3$. If m_1 (the mass of the invisible object) is much larger than m_2 (the mass of the visible star), then we can safely ignore m_2 in the sum. Thus $m_1 P^2 = (4\pi^2/G)R^3$, or $m_1 = (4\pi^2/GP^2)R^3$, regardless of the exact value of m_2.

If the visible star has a circular orbit of radius R, it travels the circumference $(2\pi R)$ in one orbital period (P) at speed v, so $2\pi R = vP$ (distance = speed × time). Dividing by 2π, we have $R = vP/2\pi$. Substituting this expression for R into the previous equation for m_1, we find $m_1 = (4\pi^2/GP^2)(vP/2\pi)^3 = v^3P/2\pi G$. If v and P are respectively measured as the amplitude (height) and period of the radial-velocity curve (observed speed versus time) of the visible star, then m_1 is found from this equation. Actually, this is the *minimum* mass of the invisible object, since the measured radial velocity is less than the true orbital velocity if the orbit isn't exactly perpendicular to the line of sight.

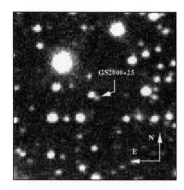 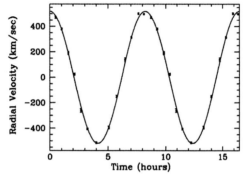

(left) An image of the visible star in the x-ray binary system known as GS 2000+25, long after it had faded from its x-ray outburst of 1988. *(right)* Radial-velocity curve of the visible star, obtained by one of us (A.F.) with the Keck-I telescope.

For example, the figure shows the radial-velocity curve for the visible star in an x-ray binary system. When the star is moving most directly toward us or away from us, we see the maximum blueshift or redshift of the absorption lines, −520 km/sec and +520 km/sec, respectively. At other times in the circular orbit, a smaller part of the velocity is radial, so intermediate values are measured, but we conclude that $v = 520$ km/sec for this system. The orbital period is measured to be only 8.3 hours. Using these values in the equation $m_1 = v^3P/2\pi G$ (and being careful with units!), we find that $m_1 = 5.0$ solar masses. This is just the minimum mass of the dark object; further considerations suggest that the true mass is 8 to 9 Suns. Hence, the visible star's companion is almost certainly a black hole.

The first well-studied system of this type is A0620−00, which is in the constellation Monoceros. During an x-ray outburst in 1975, the source was even brighter than Cygnus X-1, but it subsequently faded and the properties of the visible star were measured. The derived *minimum* mass of the invisible companion was 3.2 Suns, close to but above the limit at which a collapsed star could be a neutron star instead of a black hole. In 1992, an even more convincing case was found: The dark object in the x-ray binary star V404 Cygni has a mass of at least 6 Suns, but probably closer to 12 solar masses. One of us (A.F.) has found two additional black-hole candidates with more than 5 times the Sun's mass, as well as two somewhat weaker cases for black holes. By mid-2000, about a dozen likely (or nearly certain) stellar-mass black holes in binary systems in our galaxy had been identified and measured.

REDSHIFT
http://www.harcourtcollege.com/astro/cosmos/rsce

Figure 14–11 Stephen Hawking, who holds the Chair of Mathematics at Cambridge University once held by Isaac Newton. Among his many theoretical ideas were black-hole radiation and the evaporation of mini black holes.

Very recently (2000), a few solitary black holes (not in binary systems) may have been detected as they gravitationally bent and magnified light from more distant stars along our line of sight. This phenomenon of "gravitational lensing" is discussed in Chapter 16.

MINI BLACK HOLES

We have discussed how black holes can form by the collapse of massive stars. But theoretically a black hole should result if a mass of any amount is sufficiently compressed. No object containing less than 2 or 3 solar masses will contract sufficiently under the force of its own gravity in the course of stellar evolution. The density of matter was so high at the time of the origin of the Universe (see Chapter 19) that smaller masses may have been sufficiently compressed to form **mini black holes.**

Stephen Hawking (Fig. 14–11), an English astrophysicist, has suggested their existence. Mini black holes the size of pinheads would have masses equivalent to those of asteroids. There is no observational evidence for a mini black hole, but they are theoretically possible. Hawking has deduced that small black holes can emit energy in the form of elementary particles (neutrinos and so forth). The mini black holes would thus evaporate and eventually disappear, with the final stage being an explosion of gamma rays.

Hawking's idea that black holes radiate may seem to be a contradiction to the concept that mass can't escape from a black hole. But when we consider effects of quantum physics, the simple picture of a black hole that we have discussed up to this point is not sufficient. Physicists already know that pairs of particles and antiparticles can form simultaneously in empty space, though they disappear by destroying each other a short time later. Hawking suggests that a black hole so affects space near it that the particle or antiparticle disappears into the black hole, allowing its partner to escape. ☼ Photons, which are their own antiparticles, appear too.

As the mass of the mini black hole decreases, the evaporation rate increases, and the typical energy of emitted particles and photons increases. The final result is an explosion of gamma rays. Although "gamma-ray bursts" have indeed been detected in the sky (Chapter 15), their observed properties are not consistent with the explosions of mini black holes. Instead, they may result from neutron stars merging with black holes, or from the collapse of a rapidly rotating star to become a black hole.

SUPERMASSIVE BLACK HOLES

At the other extreme of mass, we can consider what a black hole would be like if it contained millions or billions of times the mass of the Sun. The more mass involved, the lower the density needed for a black hole to form. For a very massive black hole, one containing hundreds of millions or even billions of solar masses, the density would be so low when the event horizon formed that it would be close to the density of water.

Thus if we were travelling through the Universe in a spaceship, we couldn't count on detecting a black hole by noticing a volume of high density, or by measuring large tidal effects. We could pass through the event horizon of a high-mass black hole without being stretched tidally to oblivion, though visual effects

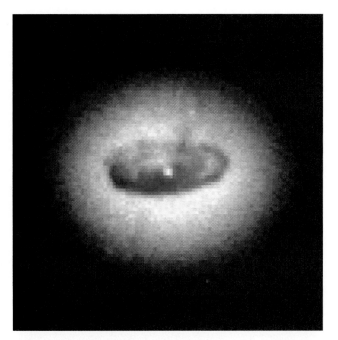

Figure 14–12 A Hubble Space Telescope image of a spiral-shaped disk of dust fueling a massive black hole in the center of the galaxy NGC 4261. The disk shown is 800 light-years wide and is 100 million light-years from us. Astronomers have measured the speed of the gas swirling around the central object. The object must have 1.2 billion times the mass of our Sun in order to have enough gravity to keep the orbiting gas in place. But it is concentrated into a region of space not much larger than our Solar System. Only a black hole easily meets both these requirements.

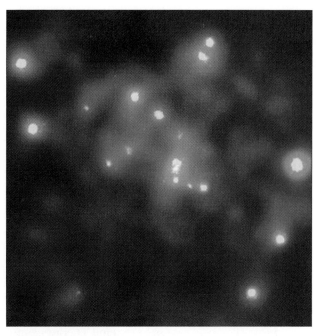

Figure 14–13 A Chandra view of a tiny region at the core of the Andromeda galaxy. The relatively cool object (blue in this color coding) is at the very center of the galaxy and is deduced to be a black hole millions of times more massive than the Sun.

would certainly be bizarre. We would never be able to get out, but it would be over an hour before we would notice that we were being drawn into the center at a rapidly accelerating rate.

Where could such a **supermassive black hole** be located? The center of our galaxy almost certainly contains a black hole of over two million solar masses. Though we do not observe radiation from the black hole itself, the gamma rays, x-rays, and infrared radiation we detect come from the gas surrounding the black hole. Other galaxies and quasars (see Chapter 17) are also probable locations for massive black holes. The Hubble Space Telescope is being used to take images of galaxies with the highest possible resolution, and is finding extremely compact, bright cores at the centers of some of them (Fig. 14–12). Many of these cores probably contain black holes. (www)

The Hubble Space Telescope is able to take spectra very close to the centers of certain galaxies. The Doppler shifts on opposite sides of the galaxies' cores show how fast these points are revolving around the centers of the galaxies. The speed, in turn, shows how much mass is present in such a small volume. Only a giant black hole can have so much mass in such a small volume.

It seems that normal galaxies may have black holes of what is now considered "moderate" size: millions of times the mass of the Sun. Some galaxies that give off exceptional amounts of radio waves or x-rays, and quasars (Chapter 17), probably contain black holes having billions of times the mass of the Sun. The Chandra X-ray Observatory, the XMM-Newton Mission, and the Hubble Space Telescope, working together, should greatly increase our knowledge of such supermassive black holes. Chandra has confirmed, for example, the presence of such a black hole at the center of the Andromeda Galaxy (Fig. 14–13).

Radio techniques that provide very high resolution can be used to study the jets of gas emitted from supermassive black holes. The jet from the galaxy M87, the central galaxy in the Virgo Cluster of galaxies, has been imaged in various ways (Fig. 14–14). The image has been traced so close to the central object that it shows a widening of the jet. This widening seems to indicate that the jet comes from the accretion disk rather than from the central black hole itself.

Most astronomers now agree that black holes are widespread and very important in our Universe.

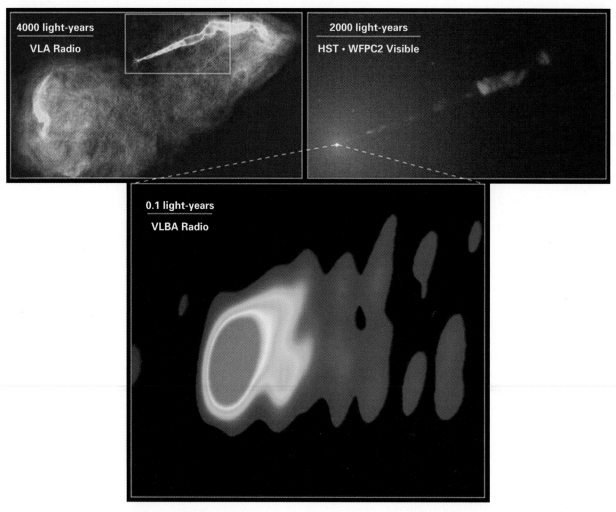

Figure 14–14 A combination of observations in different parts of the spectrum reveals the region of formation of the jet at the center of the giant elliptical galaxy M87 (see Chapter 17). A radio image made with the Very Large Array (VLA) of radio telescopes in New Mexico shows the jet, which is imaged in visible light with the Hubble Space Telescope. The region of origin of that jet can be imaged in radio waves with the Very Long Baseline Array (VLBA), a network of radio telescopes spanning the globe, giving the resolution of a single radio telescope close to the size of the Earth.

Concept Review

Black holes result when too much mass is present in a collapsed star to stop at the neutron-star stage. The strong gravitational field bends light, according to the general theory of relativity. Only light within an **exit cone** escapes. Light on the edge of the exit cone orbits in the **photon sphere.** When the radius of the star becomes equal to the **Schwarzschild radius,** it is so compact that the exit cone closes, and no light escapes; the star is within its **event horizon.** In essence, the **escape velocity** from the star's surface exceeds the speed of light. The picture is somewhat more complicated for rotating black holes, which have a **stationary limit.** Within the **ergosphere,** the region between the stationary limit and the event horizon, particles cannot remain at rest. The **singularity** at the center of a black hole, which contains all the collapsed matter, is thought to be always clothed by an event horizon, so it is never a **naked singularity.**

We look for black holes from collapsed stars in regions where flickering x-rays come from an **accretion disk.** We think a black hole is present where, as for Cygnus X-1, we find a visible object that is being yanked gravitationally to and fro by an invisible object that is too massive and too faint to be anything other than a black hole. About a dozen probable stellar-mass black holes in our galaxy have been identified and measured. **Mini black holes** may have formed in the early Universe, but there is no direct evidence for them. If they were indeed present, they should gradually evaporate by a quantum process, finally resulting in a violent burst of gamma rays. Astronomers conclude that **supermassive black holes,** with millions or billions of times the mass of the Sun, exist in the centers of quasars and many galaxies.

Questions

1. Why doesn't the pressure from electrons or neutrons prevent a star from becoming a black hole?
2. Discuss the escape of light from the surface of a star that is collapsing to form a black hole.
3. Explain the bending of light as a property of a warping of space, as discussed in Chapter 9.
†4. What is the Schwarzschild radius of a nonrotating black hole of 10 solar masses? What is the radius of its photon sphere?
†5. What would be your Schwarzschild radius, if you were a black hole?
†6. What is the relation in size of the photon sphere and the event horizon of a nonrotating black hole?
7. If you were an astronaut in space, could you escape from within the photon sphere of a nonrotating black hole?

From within the ergosphere of a rotating black hole? From within the event horizon of any black hole?
8. **(a)** How could the mass of a black hole that results from a collapsed star increase? **(b)** How could mini black holes, if they exist, lose mass?
9. Would we always notice when we reached a black hole by its high density? Explain.
10. Could we easily detect a black hole that is not part of a binary system, or is not surrounded by stars and gas in a galaxy? How?
11. Under what circumstances does the presence of an x-ray source associated with a spectroscopic binary suggest to astronomers the presence of a black hole?

---

†This question requires a numerical solution.

15

The Milky Way: Our Home in the Universe

ORIGINS *Our Sun is part of the Milky Way Galaxy, an enormous collection of billions of stars bound by gravity. New stars forming from dense clouds of gas and dust in spiral arms provide clues to the Sun's birth.*

We have already described the stars, which are important parts of any galaxy, and how they are born, live, and die. In this chapter, we describe the gas and dust that accompany the stars, and from which they are born. Clouds of this gas and dust are called **nebulae** (pronounced "neb'yu-lee"; singular: **nebula**); nebula is Latin for "fog" or "mist." We also discuss the overall structure of the Milky Way Galaxy and how, from our location inside it, we detect this structure.

AIMS

To learn about our galaxy and its parts and constituents, and to appreciate the position of our Sun and Earth, far from its center.

•

To discuss the gas and dust from which new stars form in our galaxy.

THE MILKY WAY

On the clearest moonless nights, when we are far from city lights, we can see a hazy band of light stretching across the sky (Fig. 15–1). This band is the **Milky Way**—the dust, gas, and stars that make up the galaxy in which our Sun is located. All this matter is our celestial neighborhood. The nearest star is but 4 light-years away. If we look a few thousand light-years outward, away from the direction of the Milky Way, we see out of our galaxy. Then it is much farther to the other galaxies and beyond.

Don't be confused by the terminology: The Milky Way itself is the band of light that we can see from the Earth, and the Milky Way Galaxy is the whole galaxy in which we live. Like other galaxies, our Milky Way Galaxy is composed of perhaps a few hundred billion stars plus many different types of gas, dust, planets, and so on. In the directions in which we see the Milky Way in the sky, we are looking through the relatively thin, pancake-like disk of matter that forms a major part of our Milky Way Galaxy. This disk is about 90,000 light-years across, an enormous, gravitationally bound system of stars.

The Milky Way appears very irregular when we see it stretched across the sky—there are spurs of luminous material that stick out in one direction or another, and there are dark lanes or patches in which much less can be seen (see

◀ A nebula in our Milky Way, NGC 3603, that shows various stages of the life cycle of stars. A blue supergiant star lights up a ring of gas around it. Stellar winds from a cluster of stars at the center hollows out space. Dark spots at upper right contain newly forming stars. This view is with the Hubble Space Telescope.

Figure 15–1 The Milky Way, shown in this image and the next, which shows the whole panorama in visible light. This image shows the Perseus Spiral Arm of our galaxy in types of radiation that are not visible to the eye: radio waves and infrared light, translated into false color. The observations and display are part of the Canadian Galactic Plane Survey.

To see the Milky Way, go far away from city lights on a dark moonless night and look up in the sky. Depending on the season, you should be able to see a faint band of light somewhere in the sky. It will become more and more apparent as your eyes become adapted to the darkness, a process that usually takes about 15 minutes. The Milky Way is best seen in the summer sky, especially toward the constellations Scorpius and Sagittarius, but the winter Milky Way is also quite bright. Notice how irregular some parts of the Milky Way appear, due to different amounts of gas, dust, and stars. There is an especially prominent, dark rift through Cygnus (the Swan), visible in the summer and early Fall.

After viewing with your unaided eye, try using binoculars; you will probably see many open star clusters and nebulae sprinkled throughout the Milky Way. Refer to the star charts near the beginning and end of this book for their names and positions, and also consult Chapter 4 (Observing the Constellations). The Orion Nebula, in the sword of Orion, is an impressive emission nebula. The Pleiades (the Seven Sisters, in Taurus) form an obvious open star cluster surrounded by a faint reflection nebula. If you have a small telescope, you can get reasonably detailed views of several other star clusters and nebulae that are marked on the star charts.

Starparties: Observing the Milky Way). This patchiness is due to the splotchy distribution of gas, dust, and stars. Here on Earth, we are inside our galaxy together with all of the matter we see as the Milky Way (Fig. 15.2). Because of our position, we see a lot of our own galaxy's matter when we look along the plane of our galaxy. On the other hand, when we look "upward" or "downward" out of this plane, our view is not obscured by matter, and we can see past the confines of our galaxy.

● THE ILLUSION THAT WE ARE AT THE CENTER

The gas in our galaxy is more or less transparent to visible light, but the small solid particles that we call "dust" are opaque. So the distance we can see through our galaxy depends mainly on the amount of dust that is present. This is not surprising: We can't always see across a large, smoke-filled room. Similarly, the dust between the stars in our galaxy dims the starlight by absorbing it or by scattering it in different directions.

The dust in the plane of our galaxy prevents us from seeing very far toward its center. With visible light, we can see only one-tenth of the way in, about 2000 light-years. Because of widespread dust, with the unaided eye and small telescopes we can see (in visible light) just about the same distance in any direction we look in the plane of the Milky Way. These direct optical observations fooled astronomers at the turn of the century into thinking that the Earth was near the center of the Universe (Fig. 15–3).

We shall see in this chapter how the American astronomer Harlow Shapley (pronounced to rhyme with "map'lee," as in "road map") in 1917 realized that

Figure 15-2   A visible-light view of the Milky Way, which wraps around the sky.

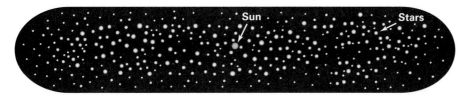

Figure 15–3 A cross-sectional view of the structure of the Universe, as perceived by astronomers near the turn of the 20th century. The effects of interstellar gas and dust had not been taken into account, so they mistakenly concluded that the Sun was near the center of the Universe.

our Sun was not in the center of the Milky Way. This fundamental idea took humanity one step further away from thinking that we were at the center of the Universe. Copernicus, in 1543, had already made the first step in removing the Earth from the center of the Universe.

In the 20th century, astronomers began to use wavelengths other than optical ones to study the Milky Way Galaxy. In the 1950s and 1960s especially, radio astronomy gave us a new picture of our galaxy. In the 1980s and 1990s, we benefited from infrared observations at wavelengths too long to pass through the Earth's atmosphere, from the Infrared Astronomical Satellite (IRAS), the Cosmic Background Explorer (COBE), and the Infrared Space Observatory (ISO). Infrared and radio radiation can pass through the galaxy's dust and allow us to see our galactic center and beyond. A new generation of telescopes on high mountains enables us to see parts of the infrared and submillimeter spectrum. Giant arrays of radio telescopes spanning not only local areas but also continents and the Earth itself enable us to get pinpoint views of what was formerly hidden from us.

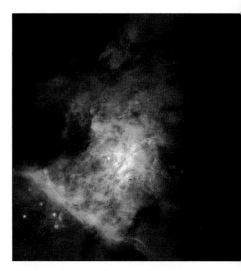

Figure 15–4 The central part of the Orion Nebula, a large emission nebula about 1500 light-years away. The gas is heated by the cluster of hot stars.

⬭ NEBULAE

The original definition of nebula was a cloud of gas and dust that we see in visible light, though we now detect nebulae in a variety of ways. When we see the gas actually glowing in the visible part of the spectrum, we call it an **emission nebula** (Fig. 15–4). Gas is ionized by ultraviolet light from very hot stars within the nebula; it then glows at optical (and other) wavelengths when electrons recombine with ions and cascade down to lower energy levels, releasing photons. Additionally, free electrons can collide with atoms (neutral or ionized) and lose some of their energy of motion, kicking the bound electrons to higher energy levels. Photons are emitted when the excited bound electrons jump down to lower energy levels, so the gas glows even more. The spectrum of an emission nebula therefore consists of emission lines.

Emission nebulae often look red (on long-exposure images; the human eye doesn't see these colors directly), because the red light of hydrogen is strongest in them. Other types of emission nebulae can photograph as green, because of green light from oxygen ions; other colors are possible as well. Don't be misled by the pretty, false-color images that you often see. In them, color is assigned to some specific type of radiation and need not correspond to colors that the eye would see when viewing the objects.

Sometimes a cloud of dust obscures our vision in some direction in the sky. When we see the dust appear as a dark silhouette (Fig. 15–5), we call it a **dark nebula** (or, often, an **absorption nebula**, since it absorbs light from stars behind it).

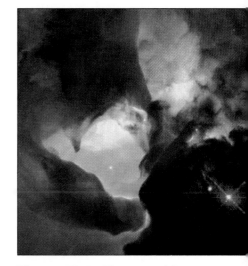

Figure 15–5 The center of the Lagoon Nebula, imaged in high detail with the Hubble Space Telescope. Emission from ionized sulfur atoms (red), from doubly ionized oxygen atoms (blue), and from electrons recombining with ionized hydrogen (green) are put together in this false-color view. The central hot star appears as bright white, with red surroundings and reddish spikes that are an artifact caused in the telescope. This star provides the ultraviolet radiation that ionizes the brightest region of the nebula.

A Closer Look

Why Is the Sky Blue?

Just as interstellar dust scatters blue light more than it scatters red light, scattering by the electrons in air molecules in the terrestrial atmosphere sends blue light preferentially around in all directions. Since we are close to the scattering electrons, our sky appears blue (see the figure). Also, as the Sun approaches the horizon, we have to look more and more diagonally through the Earth's layer of air. Our line of sight through the air is then longer than a line of sight straight up, and most of the blue light is scattered out before it reaches us, especially as the Sun gets very near the horizon. Relatively more red light reaches us, accounting for the reddish color of sunsets.

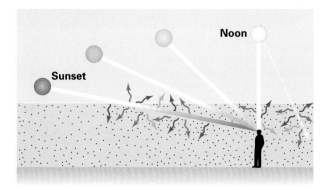

When the Sun is high in the sky, it appears white. The atmosphere scatters green, blue, and violet photons more effectively than yellow, orange, and red photons. So, away from the Sun, the sky looks blue. As the Sun approaches the horizon, the path of light through the atmosphere increases, so more of the short-wavelength photons get scattered away. Thus, the Sun itself looks progressively redder.

REDSHIFT
http://www.harcourtcollege.com/astro/cosmos/rsce

The Horsehead Nebula (Fig. 15–6) is an example of an object that is simultaneously an emission and an absorption nebula. The reddish emission from glowing hydrogen gas spreads across the sky near the leftmost (eastern) star in Orion's belt. A bit of absorbing dust intrudes onto the emitting gas, outlining the shape of a horse's head. We can see in the picture that the horsehead is a continuation of a dark area in which very few stars are visible. In this region, dust is obscuring the stars that lie beyond.

Figure 15–6   The Horsehead Nebula in Orion is dark dust superimposed on glowing gas.

A

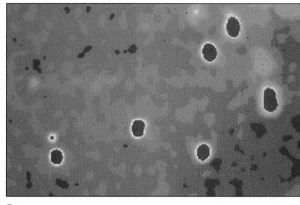

B

Figure 15-7 *(A)* The Pleiades in visible light, showing the blue reflection nebula. The spikes from bright stars in the open cluster are caused by the telescope optics. *(B)* This x-ray view of the Pleiades penetrates the dust to show the hot stars. It was made from the EXOSAT in orbit. *(C)* The Pleiades have left a wake as they moved through the interstellar medium, as we see in this infrared image.

C

Clouds of dust surrounding hot stars, like some of the stars in the star cluster known as the Pleiades (Fig. 15–7), are examples of **reflection nebulae.** They merely reflect the starlight toward us without emitting visible radiation of their own. Reflection nebulae usually look bluish for two reasons: (1) They reflect the light from hot stars, which are bluish, and (2) dust reflects blue light more efficiently than it does red light. (Similar scattering of sunlight in the Earth's atmosphere makes the sky blue; see *A Closer Look: Why Is the Sky Blue?*) Whereas an emission nebula has its own spectrum, as does a neon sign on Earth, a reflection nebula shows the spectral lines of the star or stars whose light is being reflected. Dust tends to be associated more with young, hot stars than with older stars, since the older stars would have had a chance to wander away from their dusty birthplaces.

The Great Nebula in Orion (Fig. 15–4) is an emission nebula. In the winter sky, we can readily observe it through even a small telescope or binoculars, sometimes with a tinge of color. We need long photographic exposures or large telescopes to study its structure in detail. Deep inside the Orion Nebula and the gas and dust alongside it, we think stars are being born this very minute. Some of the new, large telescopes are able to observe in the infrared, which penetrates the dust. An example in a different region of the sky is shown in Fig. 15–8.

We have already discussed some of the most beautiful emission nebulae in the sky, composed of gas thrown off in the late stages of stellar evolution. They include planetary nebulae (Fig. 15–9) and supernova remnants.

Figure 15-8 An image of the Chameleon I region made with the Antu unit telescope of the Very Large Telescope. The three colors used were visual (yellow), red, and infrared. The infrared color penetrates the dust especially well.

Figure 15-9 The Butterfly Nebula, a planetary nebula imaged with the Very Large Telescope of the European Southern Observatory.

Note that this "reddening" is not a Doppler redshift! Interstellar reddening suppresses blue light relative to red, but does not shift the wavelengths of absorption or emission lines, as the Doppler effect does.

Thus, nebulae are closely associated with both stellar birth and stellar death. The chemically enriched gas blown off by unstable or exploding stars at the ends of their lives becomes the raw material from which new stars and planets are born. As we emphasized in Chapter 13, we are made of the ashes of stars!

⬤ THE PARTS OF OUR GALAXY

It was not until 1917 that the American astronomer Harlow Shapley realized that we were not in the center of the galaxy. He was studying the distribution of globular clusters and noticed that, as seen from Earth, they were all in the same general area of the sky. They mostly appear above or below the galactic plane and thus are not heavily obscured by the dust. When he plotted their distances and directions, he noticed that they formed a spherical halo around a point thousands of light-years away from us (Fig. 15–10). Shapley's touch of genius was to realize that this point is likely to be the center of the galaxy. After all, if we are at a party and discover that everyone we see is off to our left, we soon figure out that we aren't at the party's center.

Though Shapley correctly deduced that the Sun is far from our galactic center, he actually overestimated the distance. The reason is that dust dims the starlight, making the stars look too far away, yet he didn't know about this "interstellar extinction." The amount of dimming can be determined by measuring how much the starlight has been **reddened**: blue light gets scattered and absorbed more easily than red light, so the star's color becomes redder than it should be for a star of a given spectral type. This is the same reason sunsets tend to look orange or red, not white (see *A Closer Look: Why Is the Sky Blue?*).

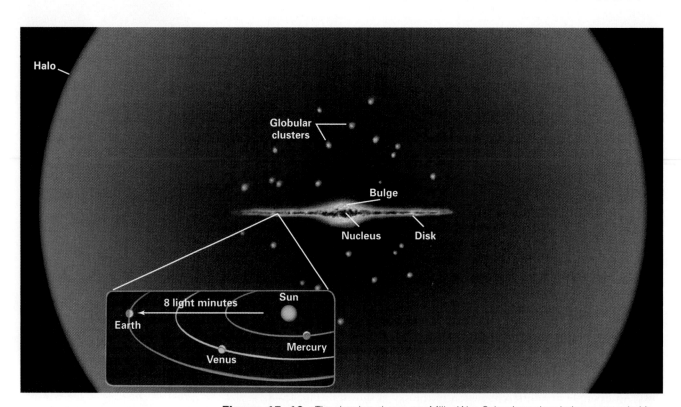

Figure 15-10 The drawing shows our Milky Way Galaxy's nuclear bulge surrounded by the disk, which contains the spiral arms. The globular clusters are part of the halo, which extends above and below the disk.

From the fact that most of the clusters appear in less than half of our sky, Shapley deduced that the galactic center is in the direction indicated.

Our galaxy has several parts:

1. *The nuclear bulge.*   Our galaxy has the general shape of a pancake with a bulge at its center that contains millions of stars, primarily old ones. This nuclear bulge has the galactic **nucleus** at its center. The nucleus itself is only about 10 light-years across.

2. *The disk.*   The part of the pancake outside the bulge is called the galactic disk. It extends 45,000 light-years or so out from the center of the galaxy. The Sun is located about one half to two thirds of the way out. The disk is very thin—2 per cent of its width—like a phonograph record. It contains all the young stars and interstellar gas and dust, as well as some old stars. The disk is slightly warped at its ends, perhaps by interaction with our satellite galaxies, the Magellanic Clouds. Our galaxy looks a bit like a hat with a turned-down brim.

It is very difficult for us to tell how the material is arranged in our galaxy's disk, just as it would be difficult to tell how the streets of a city were laid out if we could only stand on one street corner without moving. Still, other galaxies have similar properties to our own, and their disks are filled with great **spiral arms,** regions of dust, gas, and stars in the shape of a pinwheel. (We shall see examples in Chapter 16.) So, we assume the disk of our galaxy has spiral arms too. Though the direct evidence is ambiguous in the visible part of the spectrum, radio observations have better traced the spiral arms.

The disk looks different when viewed in different parts of the spectrum (Fig. 15–11). Infrared and radio waves penetrate the dust that blocks our view in visible light, while gamma rays and x-rays show the hot objects best. (www)

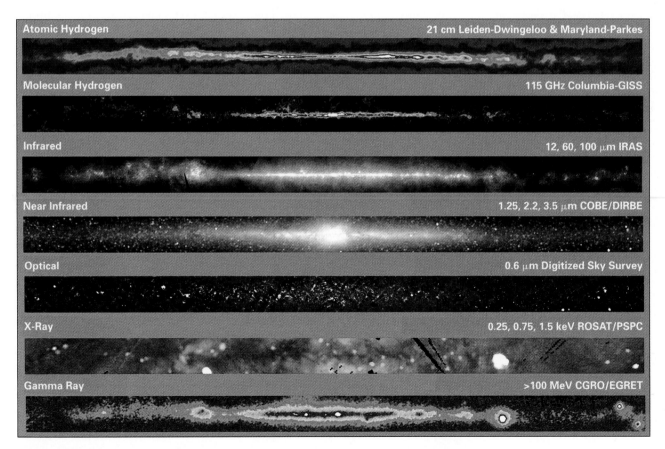

Figure 15–11   Views of the Milky Way at a wide variety of wavelengths. In all cases, we see the concentration toward the disk of the galaxy. In some wavelength bands, such as the IRAS infrared view, the radiation comes all the way from the galactic center.

3. *The halo.* Old stars (including the globular clusters) and very dilute interstellar matter form a roughly spherical galactic halo around the disk. The inner part of the halo is at least as large across as the disk, perhaps 60,000 light-years in radius. The gas in the inner halo is hot, 100,000 K, though it contains only about 2 per cent of the mass of the gas in the disk. As we discuss in Chapter 16, the outer part of the halo extends much farther, out to perhaps 200,000 or 300,000 light-years. Believe it or not, this galactic outer halo apparently contains 5 or 10 times as much mass as the nucleus, disk, and inner halo together—but we don't know what it consists of! We shall see that such "dark matter" (invisible, and detectable only through its gravitational properties) is a very important constituent of the Universe.

THE CENTER OF OUR GALAXY AND INFRARED STUDIES

We cannot see the center of our galaxy in the visible part of the spectrum because our view is blocked by interstellar dust. Radio waves and infrared, on the other hand, penetrate the dust. The Hubble Space Telescope, with its superior resolution, has seen isolated stars where before we saw just blur (Fig. 15–12).

In about 2001, NASA plans to launch an 0.85-m infrared telescope—the Space Infrared Telescope Facility (SIRTF). Its infrared detectors will be more sensitive than those on earlier infrared telescopes. SIRTF is to complete NASA's series of Great Observatories, already including the Compton Gamma-Ray Observatory, the Chandra X-ray Observatory, and the Hubble Space Telescope. (www)

One of the brightest infrared sources in our sky is the nucleus of our galaxy, only about 10 light-years across. This makes it a very small source for the prodigious amount of energy it emits: as much energy as radiated by 80 million Suns. It is also a radio source and a variable x-ray source.

High-resolution radio maps of our galactic center (Fig. 15–13) show a small bright spot that could well be gas surrounding a central giant black hole (Fig. 15–14). The appearance of a spiral is an optical illusion: the "arms" are only apparently superimposed on each other. Extending somewhat farther out, a giant Arc of parallel filaments stretches perpendicularly to the plane of the galaxy (Fig. 15–15).

As we discuss more fully in Chapter 17, the very rapid motions of stars measured near the galactic center strongly implies the presence of a supermassive black hole, 2.6 million times the Sun's mass.

ALL-SKY MAPS OF OUR GALAXY

The study of our galaxy provides us with a wide range of types of sources. Many of these have been known for decades from optical studies (Fig. 15–16). The infrared sky looks quite different (Fig. 15–17), with its appearance depending on wavelength. The radio sky provides still a different picture.

Maps of our galaxy in the x-ray and gamma-ray regions of the spectrum show the hottest sources. Different experiments on the Compton Gamma-Ray Observatory produced maps of the steady gamma-rays (Fig. 15–18) and those of bursts of gamma rays that last only minutes (Fig. 15–19). The **gamma-ray bursts,** which were seen at random places in the sky roughly once per day, are especially intriguing. Though some models suggested that they are produced within our galaxy (either very close to us, or in a very extended halo), more recent observations have conclusively shown that they are actually in galaxies billions of light-years

Figure 15–12 A cluster of stars near the center of our galaxy, imaged with the infrared camera aboard the Hubble Space Telescope.

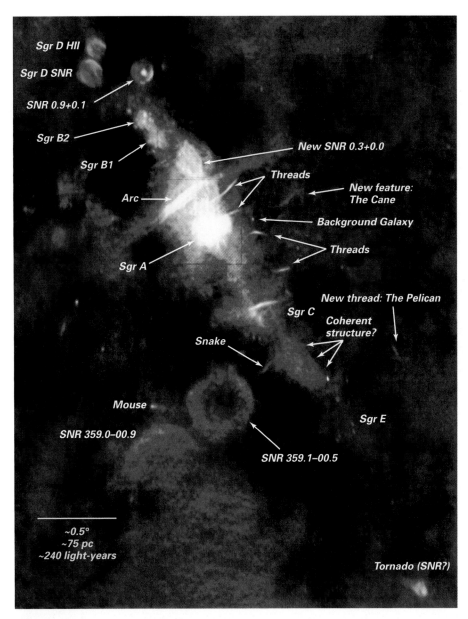

Figure 15–13 labels:
Sgr D HII
Sgr D SNR
SNR 0.9+0.1
Sgr B2
Sgr B1
New SNR 0.3+0.0
Arc
Threads
New feature: The Cane
Sgr A
Background Galaxy
Threads
New thread: The Pelican
Sgr C
Coherent structure?
Snake
Mouse
Sgr E
SNR 359.0–00.9
SNR 359.1–00.5
Tornado (SNR?)

~0.5°
~75 pc
~240 light-years

Figure 15–13 A wide-field view of the center of our galaxy, observed at the VLA at a wavelength of 90 cm.

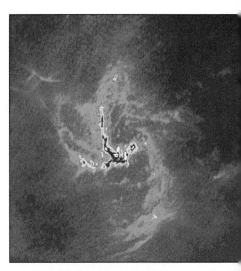

Figure 15–14 After special processing of the data, we see a spiral of gas within the central ten light-years of our galaxy. The wavelength was 6 cm, the resolution was 1 arc second, and the field of view was 3 arc minutes.

away. They may be produced when extremely massive stars collapse to form black holes, or when a neutron star merges with another neutron star or with a black hole.

The Chandra X-ray Observatory is producing more detailed images of x-ray sources than had ever before been available. Studies of the highest-energy electromagnetic radiation like x-rays and gamma rays and of rapidly moving cosmic-ray particles are part of the field of **high-energy astrophysics.**

THE SPIRAL STRUCTURE OF OUR GALAXY

It is always difficult to tell the shape of a system from a position inside it. Think, for example, of being somewhere inside a maze of tall hedges; we would find it difficult to trace out the pattern. If we could fly overhead in a helicopter, though,

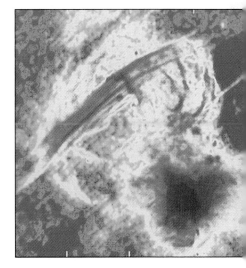

Figure 15–15 The VLA shows that a vast Arc of parallel filaments stretches over 130 light-years perpendicularly to the plane of the galaxy, which runs from upper left to lower right. The field of view is 25 arc min across; the wavelength used was in the continuous emission near 21 cm. The Arc was also visible on Figure 15–13.

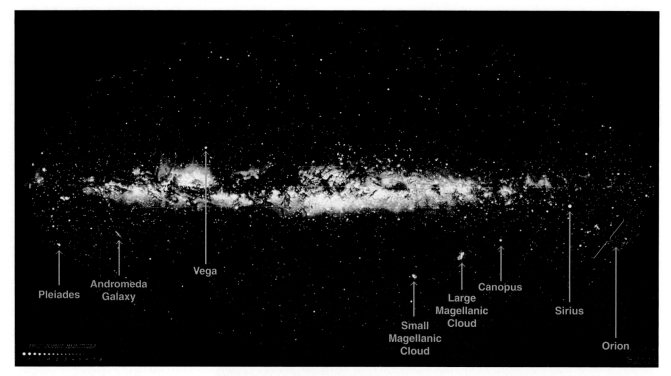

Pleiades

Andromeda
Galaxy

Vega

Small
Magellanic
Cloud

Large
Magellanic
Cloud

Canopus

Sirius

Orion

Figure 15–16 A drawing of the Milky Way as it appears to the eye, in visible light. Seven thousand stars plus the Milky Way are shown in this panorama.

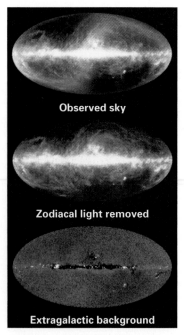

Observed sky

Zodiacal light removed

Extragalactic background

Figure 15–17 An infrared map of the sky, from the Cosmic Background Explorer (COBE) satellite.

the pattern would become very easy to see (Fig. 15–20). Similarly, we have difficulty tracing out the spiral pattern in our own galaxy, even though the pattern would presumably be apparent from outside the galaxy. Still, by noting the distances and directions to objects of various types, we can determine the Milky Way's spiral structure.

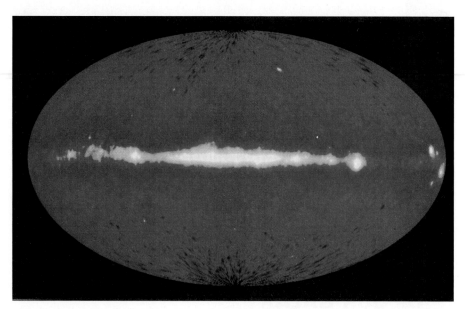

Figure 15–18 This gamma-ray map made by the EGRET (Energetic Gamma-Ray Experiment Telescope) aboard the Compton Gamma-Ray Observatory shows the plane of the Milky Way and several individual gamma-ray sources.

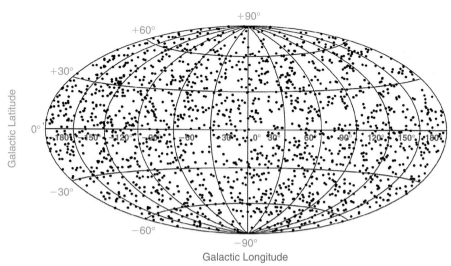

A

B

Figure 15–20 The view from the level of a maze *(A)* makes it hard to figure out the pattern, which is readily visible from above *(B)*.

Figure 15–19 Gamma-ray bursts, mapped by the Burst and Transient Source Experiment (BATSE) on the Compton Gamma-Ray Observatory. Surprisingly, the sources are randomly distributed, so they seemed either very close to us or far outside of our galaxy. In either case, they are clearly not. Recent studies with x-ray telescopes, the Hubble Space Telescope, and ground-based telescopes have revealed that the gamma-ray bursters are very distant, and hence are incredibly powerful.

Young open clusters are good objects to use for this purpose, for they are always located in spiral arms. We think that they formed there and that they have not yet had time to move away (Fig. 15–21). We know their ages from the length of their main sequences on the temperature-luminosity diagram (Chapter 10). Also useful are O and B stars, or loose groups of them; the lives of such stars are so short we know they can't be old. But since our methods of determining the distances to open clusters, as well as to O and B stars, from their optical spectra and magnitudes are uncertain to 10 per cent, they give a fuzzy picture of the distant parts of our galaxy. Parallaxes measured from the Hipparcos spacecraft do not go far enough out into space to help in mapping our galaxy, but the Full-sky Astrometric Explorer (FAME), scheduled for launch in 2004, should help make excellent maps.

Other signs of young stars are the presence of emission nebulae. We know from studies of other galaxies that emission nebulae are preferentially located in spiral arms. In mapping the locations of emission nebulae, we are really again studying the locations of the O stars and the hottest of the B stars, since it is ultraviolet radiation from these hot stars that provides the energy for the nebulae to glow.

When the positions of the open clusters, the O and B stars, and the H II regions of known distance are studied (by plotting their distances and directions as seen from Earth), they appear to trace out bits of three spiral arms, which are relatively nearby.

Interstellar dust prevents us from using this technique to study parts of our galaxy farther away from the Sun. However, another valuable method of mapping the spiral structure in our galaxy involves spectral lines of hydrogen and of carbon monoxide in the radio part of the spectrum. Radio waves penetrate the interstellar dust, allowing us to study the matter throughout our galaxy, though getting the third dimension that allows us to trace out spiral arms remains difficult. We will discuss the method later on in this chapter.

Figure 15–21 The Jewel Box, an open cluster of stars. The bright orange star is kappa Crucis, a red supergiant, while most of the other stars are bluish and therefore very hot.

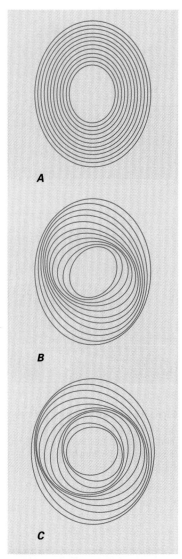

Figure 15–22 Each part of the figure includes the same set of ellipses; the only difference is the relative alignment of their axes. Consider that the axes are rotating slowly and at different rates. The places where their orbits are close together take a spiral form, even though no actual spiral of physically connected points exists. The spiral structure of a galaxy may arise from an analogous effect.

Though all hydrogen is ionized in an H II region, only some of the helium is; heavier elements have often lost 2 or 3 electrons.

WHY DOES OUR GALAXY HAVE SPIRAL ARMS?

The Sun revolves around the center of our galaxy at a speed of approximately 200 kilometers per second. At this rate, it takes the Sun about 250 million years to travel once around the center, only 2 per cent of the galaxy's current age. But stars at different distances from the center of our galaxy revolve around the galaxy's center in different lengths of time. (As we will see in Chapter 16, the galaxy does not rotate like a solid disk.) For example, stars closer to the center revolve much more quickly than does the Sun. Thus the question arises: Why haven't the arms wound up very tightly, like the cream in a cup of coffee that has been stirred?

The leading current solution to this conundrum says, in effect, that the spiral arms we now see do not consist of the same stars that would previously have been visible in those arms. The spiral-arm pattern is caused by a **spiral density wave,** a wave of increased density that moves through the gas in the galaxy. This density wave is a wave of compression, not of matter being transported. It rotates more slowly than the actual material and causes the density of passing material to build up. Stars form at those locations, and give the optical illusion of a spiral (Fig. 15–22), but the stars then move away from the compression wave.

Think of the analogy of a crew of workers fixing potholes in two lanes of a 4-lane highway. A bottleneck occurs at the location of the workers; if we were in a traffic helicopter, we would see an increase in the number of cars at that place. As the workers continued slowly down the road, fixing potholes in new sections, we would seem to see the bottleneck move slowly down the road. Cars merging from four lanes into the two open lanes need not slow down if the traffic is light, but they are compressed more than in other (fully open) sections of the highway. Thus the speed with which the bottleneck advances is much smaller than that of individual cars.

Similarly, in our galaxy, we might be viewing only some galactic bottleneck at the spiral arms. The new, massive stars heat the interstellar gas so that it becomes visible. In fact, we do see young, hot stars and glowing gas outlining the spiral arms, which are checks of this prediction of the density-wave theory. This mechanism may work especially well in galaxies with a companion that gravitationally perturbs them. ☼

MATTER BETWEEN THE STARS

The gas and dust between the stars is known as the **interstellar medium.** The nebulae represent regions of the interstellar medium in which the density of gas and dust is higher than average.

For many purposes, we may consider interstellar space as being filled with hydrogen at an average density of about 1 atom per cubic centimeter. (Individual regions may have densities departing greatly from this average.) Regions of higher density in which the atoms of hydrogen are predominantly neutral are called **H I clouds** (pronounced "H one clouds"; the Roman numeral "I" refers to the neutral, basic state). Where the density of an H I region is high enough, pairs of hydrogen atoms combine to form molecules (H_2). The densest part of the gas associated with the Orion Nebula might have a million or more hydrogen molecules per cubic centimeter. So hydrogen molecules (H_2) are often found in H I clouds.

A region of ionized hydrogen, with one electron missing, is known as an **H II region** (from "H two," the second state—neutral is the first state and once ionized is the second). Since hydrogen, which makes up the overwhelming proportion of interstellar gas, contains only one proton and one electron, a gas of

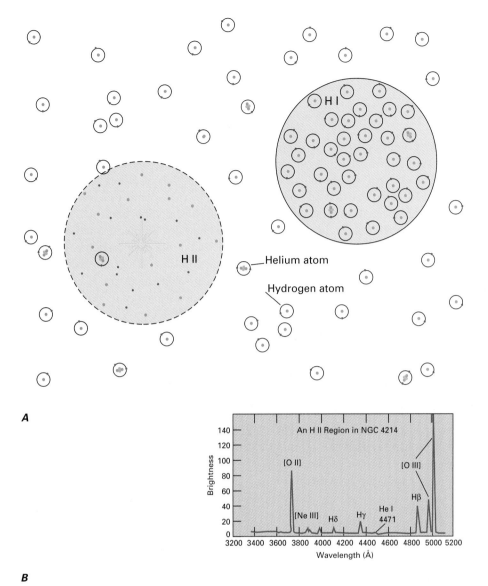

A

B

A

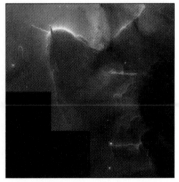

B

Figure 15-23 H I clouds are regions of neutral hydrogen of higher density than average, and H II regions are regions of ionized hydrogen. The protons and electrons that result from the ionization of hydrogen by ultraviolet radiation from a hot star, and the neutral hydrogen atoms, are shown schematically. The larger dots represent protons or neutrons, and the smaller dots represent electrons. The star that provides the energy for the H II region is shown.

Figure 15-24 (A) The Trifid Nebula, M20, in Sagittarius, is the red H II region divided into three visible parts by absorbing dust lanes. The red Hα is diluted with blue light scattered from the hot central stars. The blue reflection nebula at the top is unconnected to the Trifid. In the dark regions, the extinction is so high we cannot see stars behind the dust causing the extinction. (B) A high-resolution Hubble Space Telescope view.

ionized hydrogen contains individual protons and electrons. Wherever a hot star provides enough energy to ionize hydrogen, an H II region (emission nebula) results (Fig. 15–23 and Fig. 15–24).

Studying the optical and radio spectra of H II regions and planetary nebulae tells us the abundances of several of the chemical elements (especially helium, nitrogen, and oxygen). How these abundances vary from place to place in our galaxy and in other galaxies helps us choose between models of element formation and of galaxy evolution.

Tiny grains of solid particles are given off by the outer layers of red giants. They spread through interstellar space, and dim the light from distant stars. This "dust" never gets very hot, so most of its radiation is in the infrared. The radiation from dust scattered among the stars is faint and very difficult to detect, but the radiation coming from clouds of dust surrounding newly formed stars is easily observed from the ground and from infrared spacecraft. They found infrared radiation from so many stars in our galaxy that we think that about one star forms in our galaxy each year.

Since the interstellar gas is "invisible" in the visible part of the spectrum (except at the wavelengths of certain weak emission lines), different techniques

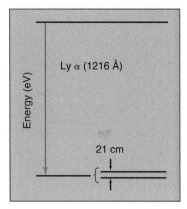

Figure 15–25 21-cm radiation results from an energy difference between two sublevels in the lowest principal energy state of hydrogen. The energy difference is much smaller than the energy difference that leads to Lyman α. So (because $E = h\nu = hc/\lambda$) the wavelength is much longer.

Figure 15–26 When the electron in a hydrogen atom flips over so that it is spinning in the opposite direction from the spin of the proton *(top)*, an emission line at a wavelength of 21 cm results. When an electron takes energy from a passing beam of radiation, causing it to flip from spinning in the opposite direction from the proton to spinning in the same direction *(bottom)*, then a 21-cm line in absorption results.

are needed to observe the gas in addition to observing the dust. Radio astronomy is the most widely used technique, so we will now discuss its use for mapping our galaxy.

⬤ RADIO OBSERVATIONS OF OUR GALAXY

The first radio astronomy observations were of continuous radiation; no spectral lines were known. If a radio spectral line is known, Doppler-shift measurements can be made, and we can tell about motions in our galaxy. What is a radio spectral line? Remember that an optical spectral line corresponds to a wavelength of the optical spectrum that is more intense (for an emission line) or less intense (for an absorption line) than neighboring wavelengths. Similarly, a radio spectral line corresponds to a wavelength at which the radio radiation is slightly more, or slightly less, intense. A radio station is an emission line on a home radio.

Since hydrogen is by far the most abundant element in the Universe, the most-used radio spectral line is a line from the lowest energy levels of interstellar hydrogen atoms. This line has a wavelength of 21 cm. A hydrogen atom is basically an electron "orbiting" a proton. Both the electron and the proton have the property of spin, as if each were spinning on its axis. The spin of the electron can be either in the same direction as the spin of the proton or in the opposite direction. The rules of quantum mechanics prohibit intermediate orientations. The energies of the two allowed conditions are slightly different.

If an atom is sitting alone in space in the upper of these two energy states, with its electron and proton spins aligned in the same direction, it has a certain small probability that the spinning electron will spontaneously flip over to the lower energy state and emit a bundle of energy—a photon (Fig. 15–25). We thus call this a **spin-flip transition** (Fig. 15–26). The photon of hydrogen's spin-flip

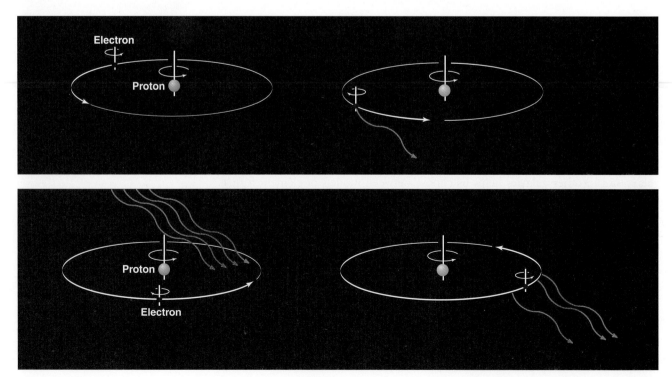

transition corresponds to radiation at a wavelength of 21 cm—the **21-cm line.** If the electron flips from the higher to the lower energy state, we have an emission line. If it absorbs energy from passing continuous radiation, it can flip to the higher energy state and we have an absorption line.

If we were to watch any particular group of hydrogen atoms in the slightly higher-energy state, we would find that it would take 11 million years before half of the electrons had undergone spin-flips; we say that the "half-life" is 11 million years for this transition. Thus, hydrogen atoms are generally quite content to sit in the upper state! But there are so many hydrogen atoms in space that enough 21-cm radiation is given off to be detected. The existence of the line was predicted in 1944 and discovered in 1951, marking the birth of spectral-line radio astronomy.

MAPPING OUR GALAXY

The 21-cm hydrogen line has proven to be a very important tool for studying our galaxy because this radiation passes unimpeded through the dust that prevents optical observations very far into the plane of the galaxy. It can even reach us from the opposite side of our galaxy, whereas light waves penetrate the dust clouds in the galactic plane only about 10 per cent of the way to the galactic center.

Astronomers have ingeniously been able to find out how far it is to the clouds of gas that emit the 21-cm radiation. They use the fact that gas closer to the center of our galaxy rotates with a shorter period than the gas farther away from the center. Though there are substantial uncertainties in interpreting the Doppler shifts in terms of distance from the galaxy's center, astronomers have succeeded in making some maps. The 21-cm maps show many narrow arms (Fig. 15–27) but no clear pattern of a few broad spiral arms like those we see in other galaxies (Chapter 16). The question emerged: Is our galaxy really a spiral at all? Only in recent years, with the additional information from studies of molecules in space that we describe in the next section, have we made further progress.

Figure 15–27 A map of our galaxy's 21-cm radiation in a Mercator projection.

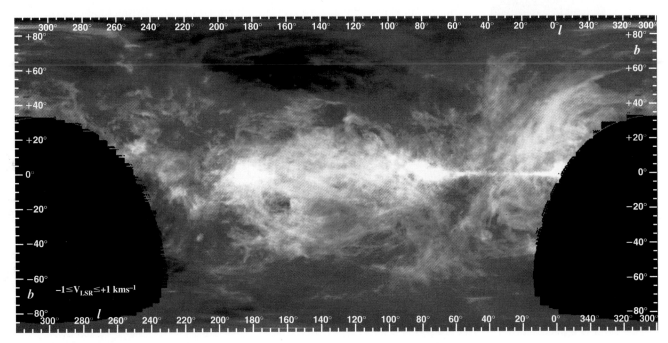

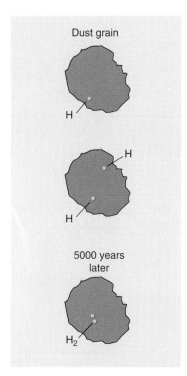

Dust grain

H

H

H

5000 years later

H₂

Figure 15–28 Hydrogen molecules are formed in space with the aid of dust grains at an intermediate stage.

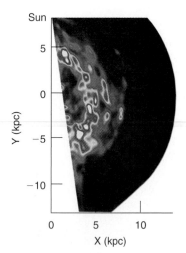

Sun

Figure 15–29 A map based on CO data does not show clear spiral arms. The point at (0, 0) on the graph marks the center of our galaxy. The regions between 4–6 kpc and near 7 kpc are more ring-like than spiral. Studies of other galaxies are showing that clear spiral arms are common in the outer regions of galaxies but not necessarily in the inner regions, which may be the case with our galaxy as well.

RADIO SPECTRAL LINES FROM MOLECULES

Radio astronomers had only the hydrogen 21-cm spectral line to study for a dozen years, and then only the addition of one other group of lines for another five years. Then radio spectral lines of water (H_2O) and ammonia (NH_3) were found. The spectral lines of these molecules proved surprisingly strong, and were easily detected once they were looked for. Over 100 additional types of molecules have since been found.

The earlier notion that it would be difficult to form molecules in space was wrong. In some cases, atoms apparently stick to interstellar dust grains, perhaps for thousands of years, and molecules build up (Fig. 15–28). Though hydrogen molecules form on dust grains, most of the other molecules may be formed in the interstellar gas, or in the atmospheres of stars, without need for grains. Studying the spectral lines provides information about physical conditions—temperature, densities, and motion, for example—in the gas clouds that emit the lines. Studies of molecular spectral lines have been used together with 21-cm observations to improve the maps of the spiral structure of our galaxy (Fig. 15–29). Observations of carbon monoxide (CO) in particular have provided better information about the parts of our galaxy farther out from the galaxy's center than our Sun. We use the carbon monoxide as a tracer of the more abundant hydrogen molecular gas, since the carbon monoxide produces a far stronger spectral line and is much easier to observe.

THE FORMATION OF STARS

Astronomers have found that **giant molecular clouds** are fundamental building blocks of our galaxy. Giant molecular clouds are 150 to 300 light-years across. There are a few thousand of them in our galaxy. The largest giant molecular clouds contain about 100,000 to 1,000,000 times the mass of the Sun. Their internal densities are about 100 times that of the interstellar medium around them. Since giant molecular clouds break up to form stars, they only last 10 million to 100 million years.

Most radio spectral lines seem to come only from the molecular clouds. (Carbon monoxide is the major exception, for it is widely distributed across the sky.) Infrared and radio observations together have provided us with an understanding of how stars are formed from these dense regions of gas and dust. Carbon-monoxide observations reveal the giant molecular clouds, but it is molecular hydrogen (H_2) rather than carbon monoxide that contains a vast majority of the mass. It is difficult to detect the molecular hydrogen directly, since it emits extremely little. So we must satisfy ourselves with observing the tracer, carbon monoxide.

Many radio spectral lines have been detected only in a particular cloud of gas, the Orion Molecular Cloud, located not very far from the main Orion Nebula. It contains about 500 times the mass of the Sun. It is relatively accessible to our study because it is only about 1500 light-years away. Even though less than 1 per cent of the Cloud's mass is dust, that is still a sufficient amount of dust to prevent ultraviolet light from nearby stars from entering and breaking the molecules apart. Thus molecules can accumulate.

We know that young stars are found in this region of the Orion Nebula (Fig. 15–30). The Trapezium (Fig. 15–31), a group of four hot stars readily visible in a small telescope, is the source of ionization and of energy for the Orion Nebula. The Trapezium stars are relatively young, about 100,000 years old. The Orion Nebula, though prominent at visible wavelengths, is but an H II region located along the near side of the much more extensive molecular cloud (Fig. 15–32).

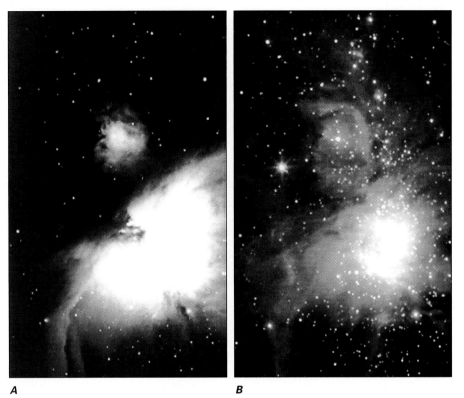

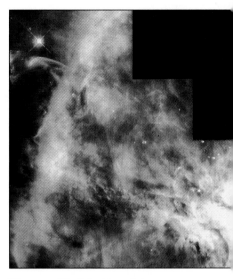

Figure 15-31 Part of the Orion Nebula only 1.6 light-years across diagonally, imaged with the Hubble Space Telescope at much higher detail than possible from the ground. One of the Trapezium stars that provides energy for the nebula to glow is at top left. The long bright bar is where we look along a long "wall" on a surface of gas. In the image, nitrogen emission is shown as red, hydrogen emission is shown as green, and oxygen emission is shown as blue.

A   *B*

Figure 15-30 *(A)* A visible-light image of the Orion Nebula, an H II region on the side of a molecular cloud. *(B)* A view of Orion taken with an infrared array. Three infrared wavelengths known as J, H, and K (1.25 μm, 1.6 μm, and 2.2 μm, respectively) were used to make this false-color view.

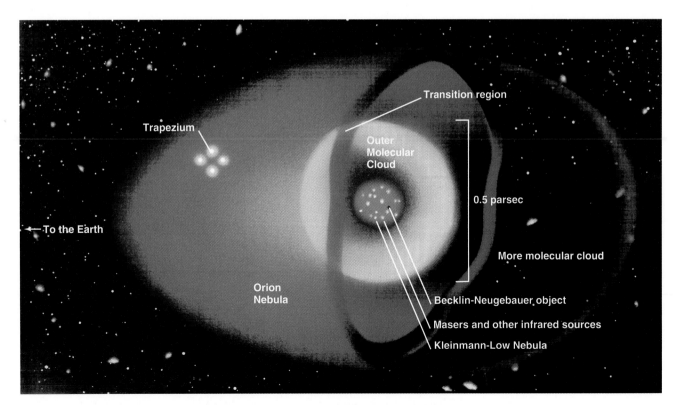

Figure 15-32 A model of the structure of the Orion Nebula and the Orion Molecular Cloud, proposed by Ben Zuckerman, now of UCLA.

Figure 15-33 Optical and infrared views of the heart of the Orion Molecular Cloud. The left image is from the visible-light camera and the right image is from the infrared NICMOS on Hubble. Most of the stars revealed by NICMOS were hidden from optical view. NICMOS reveals a chaotic, active region of star birth. Stars and dust that have been heated by the intense starlight show as yellow-orange. Emission from excited hydrogen molecules shows as blue. The brightest region is the "Becklin-Neugebauer object" (BN), a massive, young star. Outflowing gas from BN causes the bow shock above it. The view is about 0.4 light-years across the diagonal; details as small as our Solar System show.

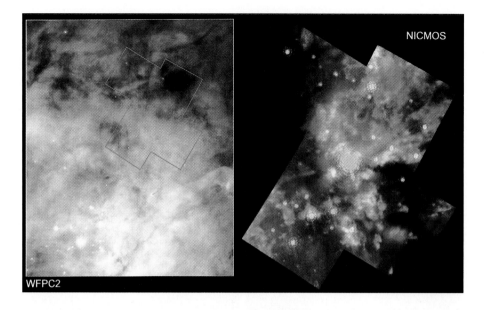

The properties of the molecular cloud can be deduced by comparing the radiation from its various molecules and by studying the radiation from each molecule individually. The density increases toward the center. The cloud may actually be as dense as a million particles per cm^3 at its center. This is still billions of times less dense than our Earth's atmosphere, though it is substantially denser than the average interstellar density of about 1 particle per cm^3.

The Near-Infrared Camera/Multi-Object Spectrograph (NICMOS) on the Hubble Space Telescope was able to record infrared light that had penetrated the dust, bringing us images of newly formed stars within the Orion Molecular Cloud (Fig. 15-33).

AT A RADIO OBSERVATORY

What is it like to go observing at a radio telescope? First, you decide just what you want to observe, and why. You have probably worked in the field before, and your reasons might tie in with other investigations under way. Then you decide at which telescope you want to observe; let us say it is the Very Large Array (VLA) of the National Radio Astronomy Observatory (Fig. 15-34). (www) You send in a proposal describing what you want to observe and why.

Your proposal is read by a panel of scientists. If the proposal is approved, it is placed in a queue waiting for observing time. You might be scheduled to observe for a five-day period to begin six months after you have submitted your proposal.

At the same time you might apply (usually to the National Science Foundation) for financial support to carry out the research. Your proposal possibly contains requests for some support for yourself, perhaps for a summer, and support for a student or students to work on the project with you. You are not charged directly for the use of the telescope itself—that cost is covered in the observatory's overall budget.

You carry out your observing at the VLA headquarters at Socorro, New Mexico. A trained telescope operator runs the mechanical aspects of the telescope. You give the telescope operator a computer program that includes the coordinates of the points in the sky that you want to observe and how long to dwell at each location. The telescopes operate around the clock—one doesn't want to waste any observing time.

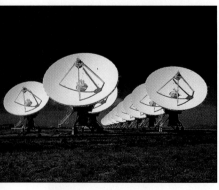

Figure 15-34 The Very Large Array (VLA), near Socorro, New Mexico, with its dishes in its most compact configuration.

The electronics that are used to treat the incoming signal collected by the radio dishes are particularly advanced. Computers combine the output from the 27 telescopes and show you a color-coded image, with each color corresponding to a different brightness level (Fig. 15–35). Standard image-processing packages of programs are available for you to use back home, with a different package being most often used in the radio community from that used in the optical community.

You are expected to publish the results as soon as possible in one of the scientific journals, often after you have given a presentation about the results at a professional meeting, such as one of those held twice yearly by the American Astronomical Society.

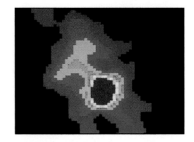

Figure 15–35 The sum of all our partial maps showed the Arc near the center of our galaxy.

Concept Review

A **nebula** is a cloud of gas and dust. **Emission nebulae** glow because of heating by hot stars, while **absorption nebulae** (**dark nebulae**) block radiation from behind. **Reflection nebulae,** like those near the stars of the Pleiades, reflect radiation. Dust dims and **reddens** starlight.

Our galaxy, the **Milky Way Galaxy,** has a **nuclear bulge** centered in a flat **disk** centered on the galaxy's **nucleus.** A spherical **halo** includes the globular clusters, and has a much greater diameter than the disk. We can detect the center of our galaxy in infrared light, radio waves, or gamma rays that penetrate the dust between us and it. The galactic nucleus is a bright infrared source and almost certainly contains a very massive black hole. Although mysterious **gamma-ray bursts** were at one time considered to be in our galaxy, we now know that they originate billions of light-years away, perhaps from the formation of black holes. The gamma rays and x-rays throughout our galaxy are studied as part of **high-energy astrophysics.**

Our galaxy's **spiral arms** may be an illusion caused by a slowly rotating **spiral density wave.** The matter between the stars, the **interstellar medium,** is mainly hydrogen gas. Emission nebulae are mostly **H II regions,** regions of ionized hydrogen. Hydrogen molecules (H_2) are very difficult to detect, but we think they are plentiful in regions where they are protected by dust from ultraviolet radiation. Radio astronomy has enabled us to map our galaxy by finding the distances to **H I regions,** clouds of neutral hydrogen gas (H I), in given directions. The observed **21-cm line** comes from the **spin-flip transition** of hydrogen atoms, when the spin of the proton and electron relative to each other changes. Observations of interstellar molecules, primarily carbon monoxide, have added to our ability to map our galaxy.

Giant molecular clouds, containing a hundred thousand to a million times the mass of the Sun, are fundamental building blocks of our galaxy. Infrared satellites and radio telescopes have permitted mapping the Orion Molecular Cloud and others.

Questions

1. Why do we think our galaxy is a spiral?
2. How would the Milky Way appear if the Sun were closer to the edge of the galaxy?
3. Compare (a) absorption (dark) nebulae, (b) reflection nebulae, and (c) emission nebulae.
4. How can something be both an emission and an absorption nebula? Explain and give an example.
5. If you see a red nebula surrounding a blue star, is it an emission or a reflection nebula? Explain.
6. Discuss the key observations that led to the discovery of the Sun's location relative to the center of our galaxy.
†7. If the Sun is 8 kpc from the center of our galaxy and it orbits with a speed of 200 km/sec, show that the Sun's orbital period is about 250 million years. (Assume the orbit is circular.)
8. Why may some infrared observations be made from mountain observatories while all x-ray observations must be made from space?
9. Describe infrared and radio results about the center of our galaxy.
10. What are three tracers that we use for the spiral structure of our galaxy? What are two reasons why we expect them to trace spiral structure?

11. Discuss how infrared observations from space have added to our knowledge of our galaxy.
12. While driving at night, you almost instinctively judge the distance of an oncoming car by looking at the apparent brightness of its headlights. Is your answer correct if you don't account for fog along the way? Discuss your answer.
13. Describe the relation of hot stars to H I and H II regions.
14. What determines whether the 21-cm lines will be observed in emission or absorption?
15. Describe how a spin-flip transition can lead to a spectral line, using hydrogen as an example.
†16. Suppose the rest wavelength of the 21-cm line of hydrogen was exactly 21.0000 cm. If this line from a particular cloud is observed at a wavelength of 21.0021 cm, is the cloud moving toward us or away from us? How fast?
17. Why are dust grains important for the formation of interstellar molecules?
18. Describe the relation of the Orion Nebula and the Orion Molecular Cloud.
19. Optical astronomers can observe only at night. In what time period can radio astronomers observe? Explain any difference.

†This question requires a numerical solution.

A Universe of Galaxies

16

ORIGINS *We find that galaxies, of which many billions are known to exist, are the fundamental units of the Universe. Essentially all stars, including our Sun, are within them. Understanding the birth and lives of galaxies is necessary for a complete picture of our existence.*

At the beginning of the 20th century, the nature of the faint, fuzzy "spiral nebulae" was unknown. In the mid-1920s, Edwin Hubble showed that they are distant galaxies like our own Milky Way Galaxy, and that the Universe is far larger than previously thought. Galaxies are the fundamental units of the Universe, just as stars are the basic units of galaxies. Like stars, some are found in clusters, and there are superclusters separated by enormous voids. By looking back in time at very distant galaxies and clusters, we can study how they formed and evolved. Surprisingly, we now know that all of these enormous structures consist largely of "dark matter" that emits little or no electromagnetic radiation.

GALAXIES: "ISLAND UNIVERSES"

The Discovery of Galaxies

In the 1770s, the French astronomer Charles Messier was interested in discovering comets. To do so, he had to be able to recognize whenever a new fuzzy object (a candidate comet) appeared in the sky. To minimize possible confusion, he thus compiled a list of about 100 diffuse objects that could always be seen, as long as the appropriate constellation was above the horizon. Some of them are nebulae, and others are star clusters, which can appear fuzzy through a small telescope such as that used by Messier. To this day, the objects in Messier's list are commonly known by their Messier numbers (Fig. 16–1). They are among the brightest and most beautiful objects in the sky visible from mid-northern latitudes.

Other astronomers subsequently compiled additional lists of nebulae and star clusters. By the early part of the 20th century, several thousand nebulae and clusters were known. The nebulae were especially intriguing: Although some of them, such as the Orion Nebula (Fig. 16–2), seemed clearly associated with bright stars in our Milky Way Galaxy, the nature of others was more controversial. When

AIMS

To study the properties and evolution of galaxies—giant collections of stars far beyond our Milky Way Galaxy—and to see how our perception of the Universe changed with their discovery.
•
To explore how pervasive dark matter is.
•
To introduce the expansion of the Universe.

REDSHIFT
http://www.harcourtcollege.com/astro/cosmos/rsce

◀ NGC 2997 in the constellation Antlia, a type Sb spiral galaxy whose structure is similar to that of our own Milky Way Galaxy. This ground-based view shows the whole galaxy. Note the bluish arms where stars are being formed and the red regions of ionized hydrogen.

A

B

Figure 16–1 *(A)* M81 and M82 in the constellation Ursa Major, two nearby galaxies in Charles Messier's catalogue of fuzzy, comet-like objects. *(B)* A closeup of M82 from the new Japanese national telescope, Subaru, on Mauna Kea. Extending outward from the galaxy's center are filaments of gas, seen in the red H-alpha light of hydrogen, tracing hydrogen that is forced outward by a superwind from numerous supernovae.

The horizontal streak extending from the bright star is the result of an overloaded pixel in the CCD camera used to make the image.

examined with the largest telescopes then available, many of them showed traces of spiral structure, like pinwheels, but no obvious stars (Fig. 16–3).

The name "island universes" had originated with the philosopher Immanuel Kant in 1755, though at that time there was no evidence for the existence of galaxies outside our own.

Some astronomers thought that these so-called **spiral nebulae** were merely in our own galaxy, while others suggested that they were very far away—"island universes" in their own right, so distant that the individual stars appeared blurred together. The distance and nature of the spiral nebulae was the subject of the well-publicized **Shapley-Curtis debate,** (www) held in 1920 between the astronomers Harlow Shapley and Heber Curtis. (www) Shapley argued that the Milky Way Galaxy is larger than had been thought, and could contain such spiral-shaped clouds of gas. Curtis, in contrast, believed that they are separate entities, far beyond the outskirts of our galaxy. This famous "Shapley-Curtis debate" is an interesting example of the scientific process at work.

Shapley's wrong conclusion was based on rather sound reasoning but erroneous measurements and assumptions. For example, one distinguished astronomer thought he had detected the slight angular rotation of a spiral nebula, and Shap-

Figure 16–2 The central region of the Orion Nebula in our Milky Way Galaxy, which is powered by the cluster of hot, massive stars. Green shows ionized oxygen.

Figure 16–3 The type Sc spiral galaxy M101 in Ursa Major, as seen with a small telescope.

Figure 16–4 The Andromeda Galaxy, also known as M31 and NGC 224, is the nearest large spiral galaxy to the Milky Way. Its Hubble type is Sb, and it is accompanied by two small elliptical galaxies, M32 *(left of middle)* and NGC 205 *(lower middle)*. These galaxies are only 2.4 million light-years from Earth. The older red and orange stars give the central regions of M31 a yellowish cast, in contrast to the blueness of the spiral arms from the younger stars there.

ley correctly argued that this change would require a preposterously high physical rotation speed if the nebula were very distant. It turns out, however, that the measurement was faulty. Shapley also argued that a bright nova that had appeared in 1885 in the Andromeda Nebula (M31, the largest spiral nebula; Fig. 16–4) would be far more powerful than any previously known nova if it were very distant. Unfortunately, the existence of supernovae (this object turned out to be one), which are indeed more powerful than any known nova, was not yet known. On the other hand, Curtis's conclusion that the spiral nebulae were external to our Galaxy was based largely on an incorrect notion of our galaxy's size; his preferred value was much too small. Moreover, he treated the nova in Andromeda as an anomaly.

The matter was dramatically settled in the mid-1920s, when observations made by **Edwin Hubble** (Fig. 16–5) at the Mount Wilson Observatory in California proved that the spiral nebulae were indeed "island universes" (now called galaxies) well outside the Milky Way Galaxy. Using the 100-inch (2.5-m) telescope, Hubble discovered very faint Cepheid variable stars in several objects, including the Andromeda Nebula.

As we saw in Chapter 11, Cepheids are named after their prototype, the variable star δ (delta) Cephei. Their light curves (brightness vs. time) have a distinctive, easily recognized shape. Cepheids are intrinsically very luminous stars, 500 to 10,000 times as powerful as the Sun, so they can be seen at large distances, out to a few million light-years with the 100-inch telescope used by Hubble.

Cepheids are very special to astronomers because measuring the period of a Cepheid's brightness variation gives you its average luminosity. And comparing its average luminosity with its average apparent brightness tells you its distance, using the inverse-square law of light. The Cepheids in the spiral nebulae observed by Hubble turned out to be exceedingly distant. The Andromeda Nebula, for example, was found to be over 1 million light-years away (the value is now known to be about 2.4 million light-years)—far beyond the measured distance of any known stars in the Milky Way Galaxy.

From the distance and the measured angular size of the Andromeda Nebula, its physical size was found to be enormous. Clearly, the "spiral nebulae" were actually huge, gravitationally bound stellar systems like our own Milky Way Galaxy, not relatively small clouds of gas like the Orion Nebula (and so the Andromeda Nebula was renamed the Andromeda Galaxy). The effective size of the Universe, as perceived by humans, increased enormously with this realization. In essence,

Figure 16–5 Edwin Hubble, who did his ground-breaking work at the Mt. Wilson Observatory above Los Angeles, California. Here he is shown at the Palomar Observatory, near San Diego, California.

Hubble brought the Copernican revolution to a new level; not only is the Earth just one planet orbiting a typical star among over 100 billion stars in the Milky Way Galaxy, but also ours is just one galaxy among myriads in the Universe! Indeed, it is humbling to consider that the Milky Way is one of roughly 100 billion galaxies within the grasp of the world's best telescopes such as the Keck twins and the Hubble Space Telescope.

Types of Galaxies

Galaxies come in a variety of shapes. In 1925, Hubble set up a system of classification of galaxies, and we still use a modified form of it.

Spiral Galaxies

There are two main "Hubble types" of galaxies. We are already familiar with the first kind—the **spiral galaxies.** The Milky Way Galaxy and its near-twin, the Andromeda Galaxy (M31; Figure 16–4), are relatively large examples containing several hundred billion stars. (Most spiral galaxies contain a billion to a trillion

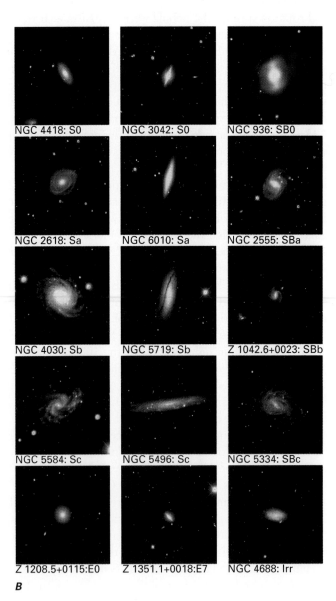

Figure 16-6 *(A)* The Hubble "tuning fork" classification of galaxies. It is *not* meant to imply an evolutionary sequence among galaxies. *(B)* Images of all the Hubble types of galaxies taken during the Sloan Digital Sky Survey.

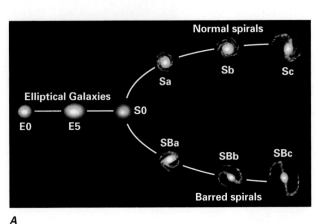

A

B

A **B**

Figure 16–7 *(A)* An edge-on view of NGC 4565, a type Sb spiral galaxy in the constellation Coma Berenices. The dust lane in the plane of the galaxy can be seen. *(B)* The Sombrero galaxy (M104) in the constellation Virgo is also seen close to edge-on, with its dust lane showing prominently.

stars.) Spiral galaxies consist of a bulge in the center, a halo around it, and a thin rotating disk with embedded spiral arms. ⟳   There are usually two main arms, with considerable structure such as smaller appendages. Doppler shifts indicate that spiral galaxies rotate in the sense that the arms trail.

Figure 16–6, which is based on a very similar "tuning fork diagram" drawn by Hubble, shows different types of spiral galaxies. The bulge tends to be large in tightly wound spiral galaxies (known as "Sa"), and smaller in more loosely wound ones ("Sb" and "Sc"; "Sd" exist but are rare). Moreover, the loosely wound spirals generally have the most gas and dust, and the largest amount of active star formation within the arms at the present time. The disk contains many young stars, especially in the arms, but old stars are also present. Old stars dominate the bulge, and especially the halo. Spiral galaxies viewed along, or nearly along, the plane of the disk (that is, "edge-on") often exhibit a dark dust lane that divides the disk into two halves (Fig. 16–7).

In nearly one-half of all spirals, the arms unwind not from the nucleus, but rather from a straight bar of stars, gas, and dust that extends to both sides of the nucleus (Fig. 16–8). These "barred spirals" are similarly classified in the Hubble scheme from "SBa" to "SBd" (the "B" stands for "barred"), in order of increasing openness of the arms and decreasing size of the bulge (Fig. 16–6). In many cases, the distinction between a barred and nonbarred spiral is subtle. Studies show that our own Milky Way Galaxy is a barred spiral, probably of type SBbc (that is, intermediate between SBb and SBc).

Since young, massive stars heat the dusty clouds from which they formed, resulting in the emission of much infrared radiation, the current rate of star formation in a galaxy can be estimated by measuring its infrared power. Space telescopes such as the Infrared Astronomical Satellite (IRAS, in the mid-1980s) and the Infrared Space Observatory (ISO, in the mid-1990s) were very useful for this kind of work, and it will be continued with new instruments such as the Space Infrared Telescope Facility (SIRTF, to be launched around 2002). About half of the energy emitted by our own Milky Way Galaxy is in the infrared, indicating that a lot of stars are being formed. But we don't know why the Andromeda Galaxy, which in optical radiation resembles the Milky Way, emits only 3 per cent of its energy in the infrared—mostly in a ring rather than in spiral arms (Fig. 16–9).

Elliptical Galaxies

Hubble recognized a second major galactic category: **elliptical galaxies** (Fig. 16–10). These objects have no disk and no arms, and generally very little gas and dust. Unlike spiral galaxies, they do not rotate very much. At the present time, nearly

Figure 16–8 A barred spiral galaxy, NGC 1365, in the constellation Eridanus.

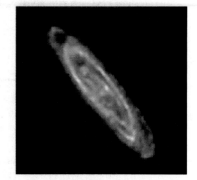

Figure 16–9 An infrared view of the Andromeda Galaxy, M31, obtained with the Infrared Space Observatory (ISO) at a wavelength of 175 μm. Considerable star formation is occurring in the main ring and a few fainter ones.

289

A

B

Figure 16–10 *(A)* M87 (NGC 4486), a giant elliptical galaxy in the constellation Virgo. Two more-distant galaxies appear to the lower right. *(B)* The central region of the small elliptical galaxy M32 (see also Figure 16–4), imaged with the Hubble Space Telescope.

REDSHIFT
http://www.harcourtcollege.com/astro/cosmos/rsce

all of them consist almost entirely of old stars. The dearth of gas and dust is consistent with this composition: there is insufficient raw material from which new stars can form.

Elliptical galaxies can be roughly circular in shape (which Hubble called type E0), but are usually elongated (from E1 to E7, in order of increasing elongation). Since the classification depends on the observed appearance, rather than on the intrinsic shape, some E0 galaxies must actually be elongated, but are seen end-on (like a cigar viewed from one end).

Although spiral galaxies constitute a majority of large galaxies, ellipticals constitute a majority of *all* galaxies. Most are dwarf ellipticals, like the two main companions of the Andromeda Galaxy (Figures 16–4 and 16–10*B*), containing only a few million solar masses—a few per cent of the mass of our Milky Way Galaxy. Some, however, are enormous, consisting of several trillion stars in a volume several hundred thousand light-years in diameter (Fig. 16–10*A*). Many ellipticals may have resulted from collisions of spiral galaxies.

Other Galaxy Types

"Lenticular" galaxies (also known as "S0" [S-zero] galaxies) have a shape resembling an optical lens; they combine some of the features of spiral and elliptical galaxies. They have a disk, like spiral galaxies. On the other hand, they lack spiral arms, and generally contain very little gas and dust, like elliptical galaxies. Hubble put them at the intersection between spiral and elliptical galaxies in his classification diagram (Figure 16–6). Though sometimes called "transition galaxies," this designation should not be taken literally: The diagram is not meant to imply that spiral galaxies evolve with time into ellipticals (or vice versa) in a simple manner, contrary to the belief of some astronomers decades ago. ✵

A few per cent of galaxies at the present time in the Universe show no clear regularity. Examples of these "irregular galaxies" include the Small and Large Magellanic Clouds, small satellite galaxies that orbit the much larger Milky Way Galaxy (Fig. 16–11). Sometimes traces of regularity—perhaps a bar—can be seen. Irregular galaxies generally have lots of gas and dust, and are rapidly forming stars. Indeed, some of them emit 10 to 100 times as much infrared as optical energy, probably because the rate of star formation is greatly elevated.

Some galaxies are called "peculiar". These often look roughly like spiral or elliptical galaxies, but have one or more abnormalities. For example, some pecu-

Figure 16–11 *(A)* The Large Magellanic Cloud and *(B)* the Small Magellanic Cloud—small, irregular galaxies that orbit around our Milky Way Galaxy. Their respective distances from us are about 170,000 light-years and 210,000 light-years. The Clouds were first reported to Europe by 16th-century Portuguese navigators who had travelled to the Cape of Good Hope at the southern tip of Africa. They were named a few decades later in honor of Magellan, who was then circumnavigating the world by that route.

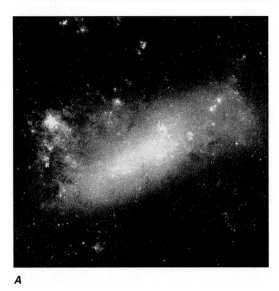

A

B

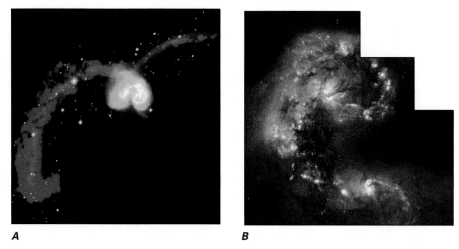

A   **B**

Figure 16–12 *(A)* The peculiar, interacting spiral galaxies NGC 4038 and NGC 4039, known as The Antennae. False color (optical: green and white; radio: blue) is used to highlight the tidal tails in this view. *(B)* A Hubble Space Telescope close-up of The Antennae that shows details of gas and dust in the two central regions.

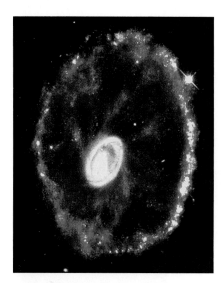

liar galaxies look like interacting spirals (Fig. 16–12), or like spirals without a nucleus (that is, like rings, Fig. 16–13), or like ellipticals with a dark lane of dust and gas.

Clusters and Superclusters of Galaxies

Most galaxies are not solitary; instead, they are generally found in gravitationally bound binary pairs, small groups, or larger **clusters of galaxies.** Binary and multiple galaxies consist of several members. An example is the Milky Way with its two companions (the Magellanic Clouds; Fig. 16–14), or Andromeda and its two main satellites (Figure 16–4).

Figure 16–13 A Hubble Space Telescope image of the Cartwheel Galaxy, in the constellation Sculptor, a ring galaxy probably caused by one of its satellite galaxies passing through it.

Figure 16–14 A view of the Milky Way *(at bottom)*, together with the Large and Small Magellanic Clouds, two companion galaxies (see also Fig. 16–11) that are easily visible to the naked eye from the southern hemisphere.

starparties

OBSERVING GALAXIES

The stars that make up the outlines of constellations, and essentially all other stars that you see in the night sky (even faint ones that require binoculars), are part of our Milky Way Galaxy. But a separate "island universe," the Andromeda Galaxy (M31; see Figure 16–4), is visible to the unaided eye if you know where to look. On a clear, dark, moonless night you can see it as a faint, fuzzy patch of light between the brightest stars of the constellations Andromeda and Cassiopeia (consult the Sky Maps inside the back cover of this text). Through binoculars, it can be traced over an arc length of several degrees. (The width of your thumb at the end of your outstretched arm covers an arc length of about 2 degrees.) Ponder that you are seeing light that left the Andromeda Galaxy about 2.4 million years ago, around the time when early hominids were developing on Earth! The smaller and fainter galaxy M33 (Figure 16–15), 2.6 million light-years away, is also sometimes visible in the constellation Triangulum near Andromeda, but it is considerably more difficult to see.

With a good pair of binoculars, it is possible to view the galaxies M81 and M82 (Figure 16–1). They are about 11 million light-years away, in the direction of the constellation Ursa Major. M81 is a type Sb spiral galaxy, while M82 is an irregular galaxy with a very large amount of gas and dust.

Try looking for galaxies in the Virgo Cluster with your college telescope, if one is available. In clear, dark skies, many galaxies will be visible in a relatively small area of the sky. The light that you see actually left those galaxies about 60 million years ago, close to the time when dinosaurs became extinct on Earth.

The Local Group is a small cluster of about 30 galaxies, some of which are binary or multiple galaxies. Its two dominant members are the Andromeda (M31) and Milky Way Galaxies. M33, the Triangulum Galaxy (Fig. 16–15), is a smaller spiral. M31 and M33, at respective distances of 2.4 and 2.6 million light-years, are the farthest objects you can see with your unaided eye (see *Starparties: Observing Galaxies*). The Local Group also contains four irregular galaxies, at least a dozen dwarf irregulars, four regular ellipticals, and the rest dwarf ellipticals or the related "dwarf spheroidals" such as Leo I (Fig. 16–16). The diameter of the Local Group is about 3 million light-years.

The Virgo Cluster (in the direction of the constellation Virgo), at a distance of about 60 million light-years, is the largest relatively nearby cluster (Fig. 16–17). It consists of at least 1000 galaxies spanning the full range of Hubble types, covering a region in the sky over 15° in diameter—about 15 million light-years. The Coma Cluster of galaxies (in the direction of the constellation Coma Berenices) is very rich, consisting of over 10,000 galaxies at a distance of about 300 million light-years (Fig. 16–18).

Figure 16–15 The Triangulum Galaxy, M33, a member of the Local Group. It is about 2.6 million light-years away, in nearly the same direction as the Andromeda Galaxy but considerably smaller. At lower right we see a nebula in M33, imaged with the Hubble Space Telescope.

Figure 16–16 The dwarf spheroidal galaxy Leo I, a member of the Local Group. The individual stars are easily resolved in this photograph.

Figure 16–17 The central part of the extensive Virgo Cluster of galaxies, the nearest large cluster to the Milky Way Galaxy. Fewer than a dozen of the roughly 1000 cluster galaxies are visible here.

Figure 16–18 The extremely rich Coma Cluster, containing over 10,000 galaxies, only a fraction of which are visible in this image. There are two dominant, large elliptical galaxies.

A majority of the galaxies in rich clusters are ellipticals, not spirals. There is often a single, very large central elliptical galaxy (sometimes two) that is cannibalizing other galaxies in its vicinity, growing bigger with time (Fig. 16–19).

X-ray observations of rich clusters reveal a hot intergalactic gas (10 to 100 million K) within them, containing as much (or more) mass as the galaxies themselves (Fig. 16–20). The gas is clumped in some clusters, while in others it is spread out more smoothly with a concentration near the center. This may be an evolutionary effect; the clumps occur in clusters that only recently formed from the gravitational attraction of their constituent galaxies and groups of galaxies. As clusters age, the gas within them becomes more smoothly distributed and partly settles toward the center.

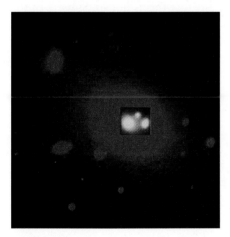

Figure 16–19 An example of galactic cannibalism. Computer processing displays the central region of the large galaxy in yellow, and we also see two smaller galaxies in the process of being swallowed. The large galaxy, NGC 6166, is the central one in the rich cluster Abell 2199.

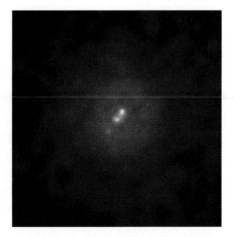

Figure 16–20 A false-color x-ray image of 3C 295, a cluster of galaxies, taken with the Chandra X-ray Observatory. We see more irregularities than had been expected from earlier observations of clusters of galaxies, which didn't show such faint detail. It appears that the cluster has been built up from smaller elements like the constituent galaxies or small groups of galaxies.

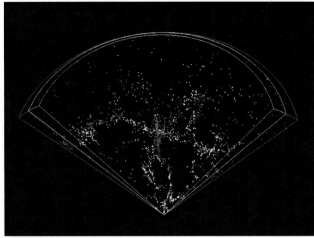

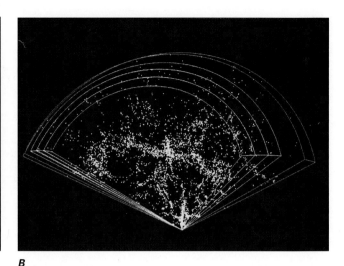

A

B

Figure 16–21 Slices of the Universe, showing the distribution of galaxies. Each slice shows distance from the Earth (going along straight lines outward from the Earth, which is at the bottom point) versus position on the sky (in right ascension) for all galaxies in a strip of declination 6° in size. (Right ascension corresponds to longitude and declination to latitude on the sky; see Chapter 4.) Each wedge extends out to a distance of about one billion light-years. *(A)* Distance versus position of galaxies in two adjacent slices centered on declinations of 29.5° *(pink dots)* and 35.5° *(white dots)*. *(B)* Positions of galaxies in four adjacent slices. Note how the structures apparent in one slice continue to the next, indicating that the galaxies define giant bubbles, like the suds in a kitchen sink, or perhaps long sponge-like structures in space. The Coma Cluster of galaxies is in the center, apparently at the intersection of several bubbles.

Clusters are seen to vast distances, in a few cases up to 6 billion light-years away. When we survey their spatial distribution, we find that they form clusters of clusters of galaxies, appropriately called **superclusters.** These vary in size, but a typical diameter is about 100 million light-years. The Local Group, dozens of similar groupings nearby, and the Virgo Cluster form the Local Supercluster.

Superclusters often appear to be elongated and flattened. The thickness of the Local Supercluster, for example, is only about 10 million light-years, or one-tenth of its diameter. They tend to form a network of bubbles, like the suds in a kitchen sink (Fig. 16–21). Large concentrations of galaxies (that is, several adjacent superclusters) surround relatively empty regions of the Universe, called **voids,** that have typical diameters of about 100 million light-years (but sometimes up to 300 million light-years). Though the voids resemble the aftermath of enormous cosmic explosions, it is more likely that they formed as a consequence of matter gravitationally accumulating into superclusters.

Does the clustering continue in scope? Are there clusters of clusters of clusters, and so on? The present evidence suggests that this is not so. Surveys of the Universe to very large distances do not reveal many obvious super-superclusters. There are, however, a few giant structures such as the "Great Wall" that crosses the center of the slice shown in Figure 16–21. We will see in Chapter 19 that the "seeds" from which these objects formed were present within 300,000 years after the birth of the Universe.

⬤ THE DARK SIDE OF MATTER

There are now strong indications that much of the matter in the Universe does not emit any detectable electromagnetic radiation, but has a gravitational influence on its surroundings. One of the first clues was provided by the flat (nearly constant) **rotation curves** of spiral galaxies.

The Rotation Curve of the Milky Way Galaxy

The rotation curve of any spinning galaxy is a plot of its orbital speed as a function of distance from its center. For example, the rotation curve of the Milky Way Galaxy has been determined through studies of the motions of stars and clouds of gas (Fig. 16–22). It rises from zero in the center, to a value of about 200 km/sec at a radial distance of about 5000 light-years. The rotation curve farther out is

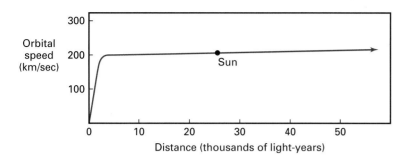

Figure 16-22 The rotation curve of our Milky Way Galaxy. The vertical axis shows the speed with which stars travel as they orbit the center, and the horizontal axis is their distance from the center. Note how constant ("flat") the speed is beyond distances of a few thousand light-years; it certainly doesn't decline, and might even rise a little.

rather "flat"—that is, the orbital speed stays constant, all the way out to distances well beyond that of the Sun.

With a distance of about 26,000 light-years from the center, and a speed of 200 km/sec, the Sun takes about 250 million years to orbit our galaxy. So, the Sun has orbited the Galaxy 18 times since its birth 4.6 billion years ago. If we think of "galactic years" being analogous to Earth's years, the Sun is a young adult. (But, of course, the Sun is about half-way through its main-sequence life, and thus is a middle-aged star in those terms.)

The speed of a star at any given distance from the center is determined by the gravitational field of matter enclosed within the orbit of that star—that is, by the matter closer to the center. (It can be shown that matter at larger distances, outside the star's orbit, does not affect the star's speed.) So, we can use the rotation curve to map out the distribution of mass within our galaxy. The speeds and distances of stars near our galaxy's edge, for example, are used to measure the mass in the entire galaxy. A similar method is used to find the amount of mass in the Sun by studying the orbits of the planets, or the amount of a planet's mass by observing the orbits of its moons. In the 17th century, Newton developed this technique as part of his derivation and elaboration of Kepler's third law of orbital motion (see Chapter 5).

Specifically, Kepler's third law can be manipulated to give an expression for the mass *(M)* enclosed within an orbit of radius *R* from the center (see *Figure It Out: The Rotation Curve*). If we insert the Sun's distance from the galactic center and the Sun's orbital speed into the formula, we find that the matter within the Sun's orbit has a mass of about 100 billion (10^{11}) solar masses! But the mass of a typical star is about half that of the Sun. Thus, if most of the matter in a galaxy is in the form of stars (rather than interstellar gas and dust, black holes, etc.), we conclude that there are about *200 billion stars* within the Sun's orbit—that is, closer to the galaxy's center than the Sun is.

The next thing to notice is that the flat rotation curve of the Milky Way Galaxy is quite different from the rotation curve of our Solar System. The orbital speeds of distant planets are slower than those of planets near the Sun (see *Figure it Out: The Rotation Curve*), instead of being roughly independent of distance. In the Solar System, the Sun's mass greatly dominates all other masses; the masses of the planets are essentially negligible. But in the Milky Way Galaxy, the flat rotation curve implies that except in the central region (where the rotation curve isn't flat), the mass grows with increasing distance from the center.

The growth in mass of our galaxy continues to large distances beyond the Sun's orbit. (The rotation curve way out from the center is usually determined from the measured speeds of clouds of hydrogen gas, which can be easily seen at radio wavelengths.) This is very puzzling because few stars are found in those regions: The number of stars falls far short of accounting for the derived mass. For example, at a distance of 130,000 light-years from the center, the enclosed mass is about 5×10^{11} solar masses, and the corresponding number of typical stars

Figure It Out

The Rotation Curve

The rotation curve, a plot of orbital speed versus distance from the orbit's center, can be used to calculate the mass enclosed within an object's orbit. We shall first apply it to our Solar System.

Recall from Chapter 5 that Newton's generalized version of Kepler's third law is

$$P^2 = \left[\frac{4\pi^2}{G(m_1 + m_2)} \right] R^3,$$

where m_1 and m_2 are the masses of the two bodies (1 and 2). Suppose m_2 (a planet) is negligible relative to m_1 (the Sun). Then we can ignore m_2 in the sum $(m_1 + m_2)$, and Kepler's third law becomes $P^2 = \left[\frac{4\pi^2}{Gm_1} \right] R^3$. Let's call this equation (1).

If the planet's orbit is circular (roughly true), then the circumference of the orbit $(2\pi R)$ must equal the planet's speed multiplied by the period: $2\pi R = vP$. This is just an application of distance = speed × time, which is correct for constant speed. Thus, $P = 2\pi R/v$. If we now substitute this into equation (1), and rearrange terms, we find that $m_1 = v^2 R/G$. So, knowing Earth's distance from the Sun ($R = 1.5 \times 10^8$ km) and its orbital speed ($v = 30$ km/sec), we can deduce that the Sun's mass is $m_1 = 2 \times 10^{33}$ grams. This is about 330,000 times the Earth's mass, thereby justifying our assumption that Earth's mass (m_2) is negligible.

The formula $m_1 = v^2 R/G$ also turns out to give the mass of a galaxy (m_1) within a radius R from its center. Let's use $R = 26,000$ light-years and $v = 200$ km/sec, as is the case for the Sun in the Milky Way Galaxy. Being careful to keep track of units, we find that the mass within the Sun's orbit is $m_1 = (200 \text{ km/sec})^2 (10^5 \text{ cm/km})^2 (26,000 \text{ ly})(9.5 \times 10^{17} \text{ cm/ly})/(6.67 \times 10^{-8} \text{ cm}^3/\text{g/sec}^2) = 1.5 \times 10^{44}$ g. But the mass of the Sun is about 2×10^{33} g, so $m_1 = (1.5 \times 10^{44} \text{ g})/(2 \times 10^{33} \text{ g/solar-mass}) \approx 7.4 \times 10^{10}$ solar masses, or roughly 10^{11} solar masses, as stated in the text.

Equivalently, we can write $v = (Gm_1/R)^{1/2}$. In the case of the Solar System, m_1 (the mass of the Sun) is constant, so we see that $v \propto (1/R)^{1/2}$. Therefore, the speed of a planet is inversely proportional to the square root of its orbital radius; distant planets move more slowly than those near the Sun.

The figure below shows this relationship graphically. Such a rotation curve is characteristic of systems in which a large central mass dominates over the masses of orbiting particles. Note how it differs from the rotation curve of the Milky Way Galaxy (as was shown in Figure 16–22), and of other rotating galaxies.

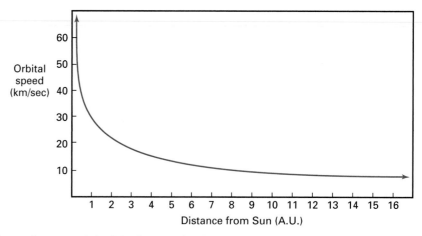

The rotation curve of the Solar System, showing the orbital speed for planets at different distances from the Sun.

(each having half a solar mass) would be about a trillion—yet there are too few stars visible, by a large margin. Indeed, studies of the outer parts of the Milky Way Galaxy throughout the electromagnetic spectrum do not reveal sufficient quantities of material to account for the derived mass.

Dark Matter Everywhere

We conclude that the Milky Way Galaxy contains large quantities of **dark matter**—material that exerts a gravitational force, but is invisible or at least very difficult to see! This material has sometimes been called the "missing mass," especially in older texts, but the term is not appropriate because the *mass* is present. Instead, it is the *light* that's missing. ☼

Many other spiral galaxies also have flat (speed roughly constant) rotation curves, as was first shown by Vera Rubin (Fig. 16–23). Estimates suggest that over 80 per cent of the mass of a typical spiral galaxy consists of dark matter. However, it has been shown that the amount of matter in the disk cannot exceed what is visible by more than a factor of 2. Instead, the dark matter is probably concentrated in an extended, spherical, outer halo of material that extends to perhaps 200,000 light-years from the galactic center.

Figure 16–23 Vera Rubin, who was the first to show that many spiral galaxies have flat rotation curves and thus contain substantial dark matter.

Similar studies show that elliptical galaxies also contain large amounts of dark matter. Gas and stars are moving so quickly that they would escape if visible matter alone produced the gravitational field.

The orbital speeds of galaxies in binary pairs, groups, and clusters can be used to determine the masses of these systems. Astronomers find that in essentially all cases, the amount of mass required to produce the observed orbital speeds is larger than that estimated from the visible light (which is assumed to come from stars and gas). A related technique is to measure the typical speeds of particles of gas bound to a cluster of galaxies—and once again, the particles could not be gravitationally bound to the cluster if its mass consisted only of that provided by the visible matter.

The fractional amount of extra material (dark matter) tends to increase with the size of the system. In other words, small groups contain relatively more dark matter than do binary galaxies, and an even larger percentage of the mass in clusters is dark. Over 90 per cent of the matter in large clusters of galaxies is dark.

Decades ago, the Caltech astronomer Fritz Zwicky was the first to point out that clusters of galaxies could not remain gravitationally bound if they contain only visible matter. He postulated the existence of some form of dark matter. However, this idea was largely ignored—it was too far ahead of its time.

What is the physical nature of the dark matter in single and binary galaxies, groups, and clusters? We just don't know—this is one of the outstanding unsolved problems in astrophysics. A tremendous number of very faint stars (or even brown dwarfs) is a possibility, though it seems unlikely, extrapolating from the numbers of the faintest stars that we can study. Perhaps there are plentiful corpses of dead stars (white dwarfs, neutron stars), but then where is the chemically enriched gas that they must have ejected near the ends of their lives? Other candidates for the dark matter are small black holes, massive planets ("Jupiters"), neutrinos, and exotic undiscovered subatomic particles (such as **WIMPs**— "weakly interacting massive particles").

If it is unsatisfactory to you that most of the mass in our galaxy (and indeed, most of the mass in the Universe!) is in some unknown form, you may feel better by knowing that astronomers also find the situation unsatisfactory. But all we can do is go out and conduct our research, and try to find out more. Clever new techniques, such as one described below, may provide the crucial clues that we seek.

Figure 16–24 *(A)* The formation of an "Einstein ring" from the gravitational lensing of a compact object by a point-like lens that is along a straight line between the observer and the compact object. The total brightness of the object is amplified by the lensing. *(B)* A slight misalignment of the lens leads instead to the formation of several discrete images. If the lensed object is a galaxy, these images often appear as arcs.

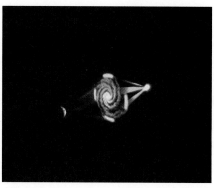

A   *B*

Gravitational Lensing

The phenomenon of **gravitational lensing** of light provides a powerful probe of the amount (and in some cases the nature) of dark matter. If the light from a distant object passes through a strong gravitational field, the light is bent—that is, it follows a curved path through the warped spacetime. This is analogous to (but differs in detail from) the effect that a simple glass lens has on light. We have already encountered it in Chapter 9: Recall that Einstein's general theory of relativity predicts that the Sun should bend the light of stars beyond it, and that this effect was first measured in 1919 by Arthur Eddington during a total solar eclipse.

If an observer, a point-like lens, and a more distant, very compact object are perfectly aligned (colinear), the object will look like a circle (known as an "Einstein ring") centered on the lens (Fig. 16–24*A*). More generally, deviations from symmetry (for example, slight misalignment of the lens) lead to the formation of several discrete, well-separated images (Fig. 16–24*B*). The images are amplified in brightness, in some cases by large amounts.

When a massive cluster of galaxies lenses many distant galaxies, a collection of arcs tends to be seen (Fig. 16–25). The number of arcs, their amplification factors, and their distorted pattern depend on the mass of the cluster, as well as on the distribution of mass within the cluster. This method measures the total mass (visible and dark) in the cluster, and gives results consistent with those obtained from other techniques. Again, we conclude that dark matter dominates most clusters of galaxies.

Observations of gravitational lensing have provided evidence that a significant fraction (perhaps up to one third) of the dark matter in the halo of the Milky Way Galaxy consists of compact objects having masses of a few tenths of a solar

> A Hubble Space Telescope image showing gravitational lensing appears on this book's cover.

Figure 16–25 *(A)* Distant galaxies gravitationally lensed by the cluster Abell 2218 appear as arcs centered on the cluster. *(B)* The cluster Cl 0024+1654 and gravitationally lensed background galaxies. Both images were obtained with the Hubble Space Telescope. The blue color of many of the lensed galaxies indicates that their rate of star formation is high.

A   *B*

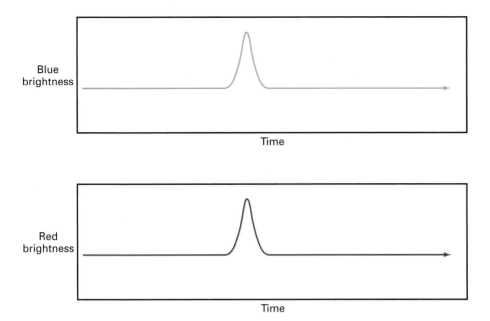

Figure 16–26 Light curves, through two different filters, of a star that has been gravitationally lensed by a dark, compact object. The star's light is temporarily focused toward us, with the same amplification at all wavelengths. Unlike the case for variable stars, which brighten and fade periodically, a star is generally lensed only once while we are watching, since the chances of a lens lining up perfectly are quite small.

mass. ☼ Projects in which the apparent brightness of millions of stars in the Large and Small Magellanic Clouds are systematically monitored have revealed that occasionally, a star brightens and fades over the course of a few weeks. The light curve has exactly the shape expected if a compact lens were to pass between the star and us, temporarily focusing the star's light toward us (Fig. 16–26).

The masses of these so-called "massive compact halo objects," or **MACHOs,** derived in a statistical manner from the observed duration of the events, are difficult to understand. They are too massive to be brown dwarfs, yet cannot be normal main-sequence stars because we would see them. They might be white dwarfs, but the population of stars that produced them must have been devoid of very low-mass stars, since we do not see enough stars in the halo.

There is, however, considerable controversy about the interpretation of the lensing events toward the Large and Small Magellanic Clouds. In two especially favorable cases, the distance of the lens has been determined—and it turns out to be in the Small Magellanic Cloud itself, rather than in the halo of the Milky Way Galaxy! If most of the lenses are not in our galaxy's halo, then the evidence for dark, compact objects greatly decreases. Indeed, the most recent estimates (mid-2000) suggest that only about 20 per cent of the halo's mass consists of MACHOs. More studies are needed to determine better the distribution of lens distances.

Regardless of the reality of the MACHOs, or of their nature if they are indeed real, much or most of the dark matter in the halo of our galaxy may consist of subatomic particles like the WIMPs mentioned earlier. Physicists studying the fundamental forces of nature suggest that many exotic particles exist, though none has been unambiguously detected in a laboratory experiment. It is exciting to think that astronomical observations may end up providing the crucial evidence for the existence of tiny, otherwise undetectable particles.

◖◗ THE BIRTH AND LIFE OF GALAXIES

It is difficult or impossible to determine what any given nearby galaxy (or our own Milky Way Galaxy) used to look like, since we can't view it as it was long ago. However, as discussed in Chapter 1, the finite speed of light effectively allows us to view the past history of the Universe: We see different objects at different times in the past, depending on how long the light has been travelling toward us. At least in a statistical manner we can explore galactic evolution by examining galaxies at progressively larger distances or **lookback times**, and hence progressively farther back in the past.

An important but probably valid assumption is that we live in a typical part of the Universe, so that nearby galaxies are representative of galaxies everywhere. Hence, galaxies several billion light-years away, viewed as they were billions of years ago, are likely to resemble what today's nearby galaxies used to look like. The refurbished Hubble Space Telescope has led to the most progress in this field, since it provides detailed images of faint, distant galaxies. Also, the Chandra X-ray Observatory has revealed objects that might be very primitive, distant galaxies; the seemingly uniform x-ray glow that previous x-ray telescopes had detected is actually produced by many individual discrete sources.

It is crucial to know the distances of the very distant galaxies, but they are so far away that no Cepheid variables or other individual stars can generally be seen and compared with nearby examples. So, an indirect technique is used: Hubble's law, as described below.

The Expansion of the Universe

Early in the 20th century, Vesto Slipher noticed that the optical spectra of "spiral nebulae" (later recognized by Edwin Hubble to be separate galaxies) almost always show a **redshift.** The absorption or emission lines seen in the spectra have the same *patterns* as in the spectra of normal stars or emission nebulae, but these patterns are displaced (that is, shifted) to longer (redder) wavelengths.

In 1929, using newly derived distances to some of these galaxies (from Cepheid variable stars), Hubble discovered that the displacement of a given line (that is, the redshift) is *proportional* to the galaxy's distance. (When the redshift we observe is greater by a certain factor, the distance is greater by the same factor.) Under the assumption that the redshift results from the Doppler effect, it is an indication of the galaxy's recession speed. He thus concluded that most galaxies are moving away from us, regardless of their direction in the sky.

Moreover, the recession speed, v, of a given galaxy is proportional to its current distance, d (Fig. 16–27A). This relation is known as **Hubble's law,** $v = H_0 d$

Figure 16–27 *(A)* The speed-distance relation, in Hubble's original diagram from 1929. Note that the units of velocity (speed) on the vertical axis should be km/sec, not km. One million parsecs equals 3.26 million light-years. Dots are individual galaxies; open circles represent groups of galaxies. The scatter to one side of the line or the other is substantial. *(B)* By 1931, Hubble and Humason had extended the measurements to greater distances, and Hubble's law was well established. All the points shown in the 1929 work appear bunched in the lower-left corner of this graph. The relation $v = H_0 d$ represents a straight line of slope H_0. These graphs use older distance measurements than we now use, and so give different values for H_0 than we now derive.

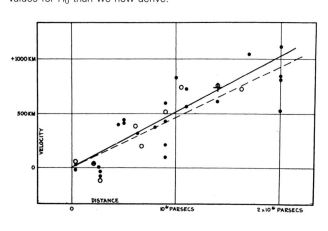

A

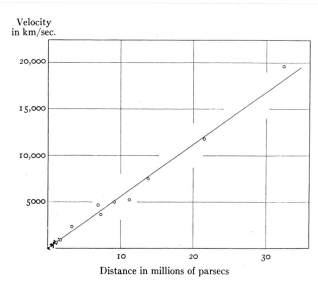

Velocity in km/sec.

Distance in millions of parsecs

B

Figure It Out

Redshifts and Hubble's Law

For a given absorption or emission line in a spectrum of a galaxy, define $\Delta\lambda$ ("delta lambda") to be the displacement (shift) in the wavelength [that is, the observed wavelength (λ) minus the wavelength measured at rest in a laboratory gas (λ_0)]. Although different lines have different displacements $\Delta\lambda$, the ratio of the displacement to the rest wavelength ($\Delta\lambda/\lambda_0$) is the same for all lines in a galaxy's spectrum. This quantity, $z = \Delta\lambda/\lambda_0$, is defined as the redshift (z) of the galaxy. When multiplied by 100 per cent, the redshift z can be considered the percentage by which lines are shifted toward longer wavelengths. For example, if $z = 0.5$, then a line with $\lambda_0 = 6000$ Å is shifted by $\Delta\lambda = 3000$ Å (50 per cent of 6000 Å) to an observed wavelength of $\lambda = 9000$ Å.

Empirically, Hubble found that the redshift is larger in the spectra of distant galaxies than in the spectra of nearby galaxies. Thus, redshift is proportional to distance (or, $z \propto d$). If this redshift z is produced by motion away from us, then we can use the Doppler formula to derive v, the speed of recession. Since $z = \Delta\lambda/\lambda_0 \approx v/c$ (the Doppler formula; see Chapter 10), we have $z \approx v/c$, and therefore $v \approx cz$, where c is the speed of light (3×10^5 km/sec). The approximation $v \approx cz$ is accurate as long as the redshift z is not larger than about 0.2 (but the definition $z = \Delta\lambda/\lambda_0$ is exact for all redshifts). For example, if $z = 0.1$, then $0.1 = \Delta\lambda/\lambda_0 \approx v/c$, and so $v \approx 0.1c = 0.1 \times (3 \times 10^5$ km/sec$) = 3 \times 10^4$ km/sec.

Since speed v is proportional to redshift z, which Hubble found to be proportional to distance d, a galaxy's speed of recession must be proportional to its distance. This relation is known as Hubble's law, $v = H_0 d$, where H_0 is the constant of proportionality, Hubble's constant. Thus, a galaxy twice as far away from us as another galaxy recedes twice as fast; if 10 times as distant, it recedes 10 times as fast.

Note that $v = H_0 d$ does *not* imply that the speed of a given galaxy increases with time as its distance increases. ☼ We have defined H_0 to be the present-day value of Hubble's constant H, but the value of H actually decreases with time. (Indeed, in most simple models of the Universe, $H \propto 1/t$.) If no mysterious long-range repulsive forces exist in the Universe, then the speed of a given galaxy is at best constant with time, and in fact should gradually decrease due to the attractive gravitational force of all other matter in the Universe. Since "Hubble's constant" changes with time, some astronomers prefer to call it the "Hubble parameter." It is a constant only in *space*, having the same value throughout the Universe at any given time.

(see *Figure It Out: Redshifts and Hubble's Law*), where H_0 (pronounced "H naught") is the *present-day* value of the constant of proportionality, H (the factor by which you multiply d to get v). H_0 is known as **Hubble's constant.** For various reasons, Hubble's original data were suggestive but not conclusive. Subsequently, Hubble's assistant and disciple Milton Humason joined Hubble in very convincingly showing the relationship (Fig. 16–27B).

This behavior is similar to that produced by an explosion: Bits of shrapnel are given a wide range of speeds, and those that are moving fastest travel the largest distance in a given amount of time. The implication of Hubble's law is that the Universe is expanding! However, as we shall see in Chapter 18, there is no unique center to the expansion. Moreover, the expansion of the Universe marks the creation of *space itself*, unlike the explosion of a bomb in a preexisting space. ☼

As we discuss in Chapter 18, one of the greatest debates in 20th century astronomy has been over the value of Hubble's constant. There we will explain in detail how it is determined. Measurements of the recession speeds of galaxies at known distances show that $H_0 = 50$ to 80 km/sec/Mpc (the exact number is still uncertain), with a most likely value of 65 ± 7 km/sec/Mpc. For example, a galaxy 10 Mpc (32.6 million light-years) away recedes from us with a speed of about

REDSHIFT
http://www.harcourtcollege.com/astro/cosmos/rsce

"Mpc" means megaparsec, a million parsecs. Parsecs are units of distance used by astronomers; 1 parsec (1 pc) is equal to 3.26 light-years.

Figure It Out

Using Hubble's Law to Determine Distances

Suppose you find a galaxy that is so faint and far away that conventional techniques for measuring distances (such as Cepheid variable stars) cannot be used, or give inaccurate results. You can instead do the following. (1) Obtain a spectrum of the galaxy. (2) Measure the redshift, $z = \Delta\lambda/\lambda_0$, of an absorption or emission line that you know has rest wavelength λ_0. (3) Compute the approximate recession speed v from $z \approx v/c$—that is, solve for the recession speed $v \approx cz$. (4) Compute the distance d from Hubble's law, $v = H_0 d$, using your preferred value for Hubble's constant H_0—in other words, solve for the distance $d = v/H_0$.

For example, suppose an absorption line with a known rest wavelength $\lambda_0 = 5000$ Å is at an observed wavelength $\lambda = 5100$ Å in the spectrum of a galaxy. Then $\Delta\lambda = \lambda - \lambda_0 = 5100$ Å $- 5000$ Å $= 100$ Å, so $z = \Delta\lambda/\lambda_0 = (100$ Å$)/(5000$ Å$) = 0.02$. Therefore the recession speed is $v \approx cz = (3 \times 10^5$ km/sec$) \times 0.02 = 6{,}000$ km/sec. If we assume that $H_0 = 65$ km/sec/Mpc, we find that the galaxy's distance is $d = v/H_0 = (6{,}000$ km/sec$)/(65$ km/sec/Mpc$) \approx 92$ Mpc, or about 300 million light-years since 1 Mpc = 3.26 million light-years.

Several large "redshift surveys" of galaxies have each produced many thousands of distances, allowing study of the large-scale, three-dimensional distribution of galaxies. These are the studies that led to the conclusion that galaxies tend to be found in superclusters separated by large voids (see Figure 16–21).

The above conversion between redshift and recession speed ($z \approx v/c$) is not accurate when z is larger than about 0.2, and it fails badly at redshifts approaching or exceeding 1. Instead, the relativistic equation

$$z = \sqrt{\frac{1 + \dfrac{v}{c}}{1 - \dfrac{v}{c}}} - 1$$

provides a better approximation at high redshifts. (Even this equation isn't exactly right, since the redshift is due to the expansion of space rather than motion through space, but here it will suffice.)

When the derived "distance" of an object is very large (say, billions of light-years), it is better to think of it as a lookback time—that is, how long the light that we are currently receiving has been travelling from the object. Since the Universe expanded during the light's long journey, it had to cover a greater distance than if the Universe were static. Thus, the derived "distance" corresponds neither to the object's distance when the light was emitted (which was smaller), nor to the object's distance now (which is larger), but rather to the distance at some intermediate time. This is much more confusing than simply saying that the light was emitted a certain time ago—the lookback time. Such lookback times can be compared with the age of the Universe for various models; see the example in Table 16–1. So, remember—if a newspaper article says that astronomers found a galaxy 8 billion light-years away, what it really means is that the light we are seeing was emitted 8 billion years ago!

650 km/sec, and a galaxy 20 Mpc away recedes with a speed of about 1300 km/sec. Hubble's constant is always quoted in the strange units of km/sec/Mpc. ✺ The value $H_0 = 65$ km/sec/Mpc simply means that for each megaparsec (3.26 million light-years) of distance, galaxies are receding 65 km/sec faster.

The expansion of the Universe will be the central theme in Chapters 18 and 19; we will discuss its implications and associated phenomena. For now, however, let us simply use Hubble's law to determine the distances of very distant galaxies and other objects. Knowing the value of H_0, a measurement of a galaxy's recession speed v then gives the distance d, since $d = v/H_0$; see *Figure It Out: Using Hubble's Law to Determine Distances*. Note that Hubble's law *cannot* be used to find the distances of stars in our own Galaxy, or of galaxies in our Local Group; these objects are gravitationally bound to us, and hence the expansion of the intervening space is overcome. ☼

The Search for Distant Galaxies

With the Hubble Space Telescope, we obtained relatively clear images of faint galaxies that are suspected to be very distant. The main imaging camera exposed on a small area of the northern sky for 10 days in December 1995. Though it covers only about one 30-millionth of the area of the sky (roughly the apparent size of a grain of sand held at arm's length), this **Hubble Deep Field** contains several thousand extremely faint galaxies (Fig. 16–28). If we could photograph the entire sky with such depth and clarity, we would see about 100 billion galaxies, each of which contains billions of stars!

Three years later, the Hubble Space Telescope got very deep images of another region, this time in the southern celestial hemisphere: the Hubble Deep Field—South (Fig. 16–29). It looks similar to the northern field, even though it is nearly in the opposite direction in the sky, providing some justification for our assumption that the Universe is reasonably uniform over large scales. Other deep surveys further strengthen our belief that we live in a rather typical place in the Universe.

Spectra obtained with large telescopes, especially the Keck telescopes in Hawaii, confirm that many galaxies in the Hubble Deep Fields and other deep surveys have large redshifts and hence are very far away. Though a few of the galaxies have relatively low redshifts, typical redshifts of the faintest objects are between 1 and 4 (Fig. 16–30). Light that we observe at visible wavelengths actually corresponds to ultraviolet radiation emitted by the galaxy, but shifted red-

Figure 16–28 Part of the original Hubble Deep Field, a tiny section of the sky in the northern celestial hemisphere, near the Big Dipper. About a thousand galaxies at a range of distances are visible in this exposure. In fact, only a few foreground stars in our own Milky Way Galaxy can be seen.

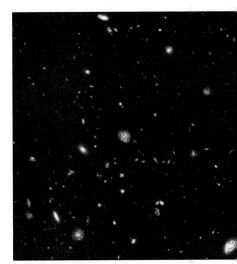

Figure 16–29 Part of the Hubble Deep Field—South. Overall, it looks similar to the original Hubble Deep Field in the northern celestial hemisphere.

Figure 16–30 Part of the original Hubble Deep Field, with measured redshifts (from the Keck telescopes) indicated for many of the galaxies.

TABLE 16–1

Redshift	Lookback time
0	0 years
0.2	2.6 billion years
0.4	4.6 billion years
0.6	6.2 billion years
0.8	7.4 billion years
1.0	8.3 billion years
1.5	10.0 billion years
2.0	11.0 billion years
2.5	11.7 billion years
3.0	12.2 billion years
4.0	12.9 billion years
5.0	13.3 billion years
6.0	13.5 billion years
∞	14.5 billion years

Note: Here we assume that the Universe's matter density is 30% of the critical density, and that the energy density associated with the cosmological constant is 70% of the critical density; see Chapter 18 for definitions and details. The Universe is 14.5 billion years old in this model. If the cosmological constant is actually 0, the qualitative results are similar, though the numbers differ in detail.

ward by 100 per cent to 400 per cent! If we convert these redshifts to "distances" (or, more precisely, lookback times—see Table 16–1), we find that the galaxies are billions of light-years away. We see them as they were billions of years in the past, when the Universe was much younger than it is now. Note that when astronomers say that light from a high-redshift galaxy comes from "the distant universe," what they really mean is "distant parts of our Universe." We do not receive light from other universes, even if they exist!

In the past few years, several galaxies with redshifts exceeding 5 (that is, >500 per cent) were discovered (Fig. 16–31), and there is evidence for some with redshifts over 6. A few objects are suspected of having even higher redshifts, though the spectra are not yet good enough to be certain. The lookback time corresponding to redshift 6 is about 13.5 billion years (Table 16–1); we are seeing denizens of an era shortly (a billion years) after the birth of the Universe. They are probably newly formed galaxies.

Out of the thousands of galaxies found in images such as the Hubble Deep Fields, how do we go about choosing those that are likely to be at high redshift? (After all, with limited time for spectroscopy using large telescopes such as Keck, it is important to improve the odds if the goal is to find the highest redshifts.) One very effective technique is to first measure the "color" of each galaxy. For two reasons, those that are likely to be very far away look very red, and have little if any of the blue or ultraviolet light normally emitted by stars. First, their redshift moves all of the light to redder (longer) wavelengths. Second, there is often another galaxy or large cloud of gas along the way, and the hydrogen gas within it completely absorbs the ultraviolet light.

A

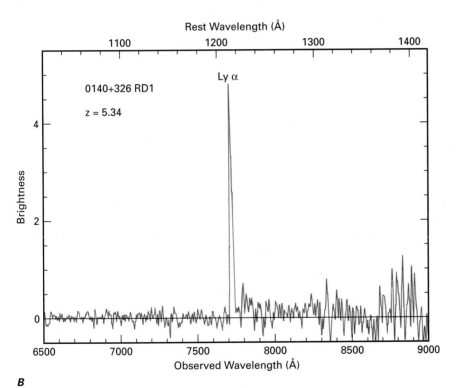

B

Figure 16-31 *(A)* The image and *(B)* the spectrum of a distant galaxy, obtained with the Keck telescopes. The measured redshift is 5.34, one of the highest known. The spectrum shows ultraviolet light that was redshifted to optical wavelengths.

The Evolution of Galaxies

Comparisons of the appearance of distant galaxies and nearby galaxies provide clues to the way in which galaxies evolve. These comparisons must be done carefully to avoid wrong conclusions. For example, a visible-light image of a high-redshift galaxy actually corresponds to ultraviolet radiation emitted by the galaxy,

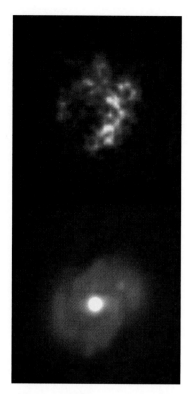

Figure 16–32 *Ultraviolet (top) and visible-light (bottom)* images of a nearby galaxy, M100 in the constellation Coma Berenices, obtained with the Hubble Space Telescope. A ring-shaped region in which intense bursts of star formation have occurred is obvious in the ultraviolet image. The nucleus consists mostly of all stars, and is faint in the ultraviolet. These images show how dramatically different galaxies can appear at different wavelengths.

Figure 16–33 An infrared image of a high-redshift galaxy ($z = 1$), obtained with the Hubble Space Telescope. This can be compared with visible-light (optical) images of nearby galaxies that have only a small redshift, since the intrinsic wavelengths viewed are then the same in both cases.

so it would not be fair to compare the image with the visible-light appearance of a nearby (and hence essentially unshifted) galaxy.

One way around this problem is to compare the visible-light images of high-redshift galaxies with ultraviolet images of nearby galaxies, as obtained with the Hubble Space Telescope. These images emphasize regions containing hot, massive, young stars that glow brightly in the ultraviolet (Fig. 16–32). Another technique is to obtain infrared images of the high-redshift galaxies, and compare them with the visible-light images of nearby galaxies. Such images tend to be dominated by light from older, less massive stars that more accurately reflect the overall shape of the galaxy rather than pockets of recent, intense star formation (Fig. 16–33). So far, the clearest infrared images have been made with the Hubble Space Telescope. (Unfortunately, its infrared camera ran out of solid nitrogen coolant in late-1998, sooner than anticipated. A new method of cooling the camera will be used in equipment to be installed during the servicing mission in 2001.)

These data, together with various types of analysis such as computer simulations (for example, of what happens when two galaxies merge), provide many interesting results, some of which are still preliminary because the studies are not finished. One spectacular conclusion is that most spiral galaxies used to look quite peculiar; there were essentially no large galaxies with distinct, well-formed spiral arms beyond redshift 2. By redshift 1 there were quite a few of them, but many took on their current, mature shapes more recently, in the past 5 billion years (that is, at redshifts below about 0.5).

Another conclusion is that there used to be a large number of small, blue, irregular galaxies that formed stars at an unusually large rate (Fig. 16–34). Their strange shape might be partly caused by an irregular distribution of star clusters. However, they appear peculiar even at infrared wavelengths, which are more sensitive to older stars, so they must be structurally disturbed. Some of them prob-

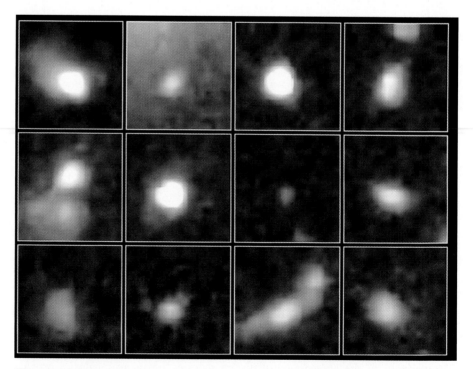

Figure 16–34 Small, generally blue, irregular galaxies that have unusually high rates of star formation, imaged with the Hubble Space Telescope. These objects have lookback times of about 11 billion years. Some of them probably merged to form larger galaxies; others may have faded away with time.

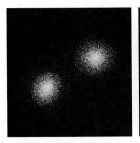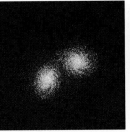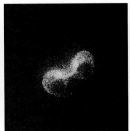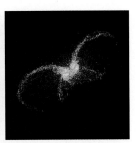

Figure 16–35 Calculations made with a supercomputer, showing a time sequence of the very close encounter of two identical model galaxies. A time interval of 250 million years separates successive frames. About 10,000 particles represent each galaxy. The formation of tidal tails is obvious. Extension of such calculations to even later times shows the formation of an elliptical galaxy from the merged pair.

The Antennae (Fig. 16–12) are thought to have taken their shape in the manner shown above.

ably later merged together to form larger galaxies, including disturbed spirals. Perhaps others faded and are now difficult to find because they are so dim.

It appears that most elliptical galaxies formed early in the Universe, beyond redshift 2 (lookback time 11 billion years); there are many old-looking, well-formed ellipticals between redshifts of 1 and 2. On the other hand, we also think that an elliptical galaxy can be produced by the **merging** of two spiral galaxies. Many new stars are created from the interstellar gas in the spirals. Computer models of the merging process also show that long "tails" of material are sometimes temporarily formed (Fig. 16–35), just as we see in nearby examples of interacting galaxies (Figure 16–13). Later these tails disappear, leaving a more normal looking elliptical galaxy, but with a population of stars younger than in the really ancient ellipticals. In fact, the Milky Way Galaxy and the Andromeda Galaxy are approaching each other and may collide (or barely miss each other) in about 5 or 6 billion years. Subsequently, they are likely to merge and become an elliptical galaxy.

The total rate at which stars currently form in the Universe is rather small compared with what it was billions of years ago. We see that the star formation rate has decreased since a redshift of 1 or 2 (8 to 11 billion years ago) to the present time. The rate may have been constant at still larger redshifts, up to about 4 or 5 (13 billion years ago), though we are unsure because many high-redshift galaxies are cloaked with dust. The dust seems to have been produced by the first few generations of stars, making it difficult to detect high-redshift galaxies at visible wavelengths. But infrared and sub-millimeter telescopes are finding them in progressively larger numbers, so we can expect a more accurate census in the near future.

As galaxies age, they evolve chemically, primarily because of supernovae that create many of the heavy elements through nuclear reactions and disperse them into the cosmos. Large, massive galaxies, whose gravitational field doesn't allow much of the gas to escape, tend to become more chemically enriched than small galaxies that are not able to retain the hot gas. So, we don't expect to find many rocky, Earth-like planets in small (dwarf) galaxies like the Magellanic Clouds. The formation of massive galaxies like the Milky Way seems to be a critical step for the existence of humans.

Large-Scale Structure in the Universe

We have also studied the evolution of **large-scale structure** in the Universe. By getting the redshifts of many relatively nearby galaxies over large portions of the sky, and of many very distant galaxies in smaller areas, the growth of clusters, superclusters, and voids can be traced.

Most clusters formed relatively recently, at redshifts below 0.5 (within the past 5 billion years), and in fact many are still forming now. Thus, galaxies generally preceded clusters. However, cluster formation did begin earlier. A few very

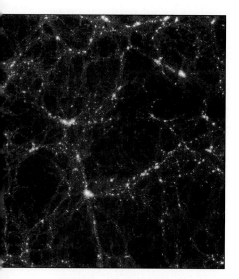

Figure 16–36 This supercomputer calculation of the formation of clusters of galaxies shows the existence of filaments and voids. The details of the results depend on whether cold dark matter, hot dark matter, or a mixture of the two fills the Universe.

large, well-formed clusters have been found at redshift 1 (8 billion light-years away), and evidence exists for substantial concentrations of matter (which later formed galaxies) at a redshift of 4, corresponding to a lookback time of about 13 billion years.

The observed distribution of superclusters and voids can be compared with the predictions of various theoretical models using computer simulations (Fig. 16–36). One important conclusion is that dark matter probably pervades the Universe; otherwise, it is difficult to produce very large structures. Galaxies and clusters seem to form at unusually dense regions ("peaks") in the dark matter distribution, like snow on the peaks of mountains.

Agreement with observations seems best when most of the dark matter used in the simulations is "cold"—that is, moving relatively slowly. Work is in progress to determine what specific type of **cold dark matter** is likely to account for most of the material. Simulations that use primarily **hot dark matter** (such as neutrinos, with speeds close to that of light) do not produce galaxy distributions that resemble those observed. Specifically, hot dark matter has a hard time clustering on small scales, like those of individual galaxies and small clusters.

As we mentioned previously, there seems to be a large amount of dark matter in the halos of galaxies, and especially in clusters of galaxies. Indeed, dark matter may exceed the mass of luminous matter by a factor of 10. It is not yet known how much dark matter exists between clusters of galaxies and in voids, but it could be substantial. If so, close to 99 per cent of the Universe on the largest scales may consist of invisible material!

Concept Review

Around the turn of the 20th century, some astronomers believed that the **spiral nebulae** were clouds of gas within our Milky Way star system, while others thought they might be huge, completely separate systems of stars bound together by gravity. Their distance and nature was the subject of the famous **Shapley-Curtis debate.** In the mid-1920s, **Edwin Hubble** recognized Cepheid variable stars within a few of these objects, and their observed faintness implied an enormous distance. Thus, the true nature of the spiral nebulae was finally unveiled: They are distant galaxies like our own Milky Way Galaxy, and the Universe is far larger than previously thought. Galaxies are the fundamental units of the Universe; each one contains billions of stars spread over a diameter exceeding 10,000 light-years.

There are various types of galaxies. **Spiral galaxies** contain a thin, rotating disk of stars and gas; new stars tend to form in the spiral arms. They also have a spherical bulge of old stars. **Elliptical galaxies** have no disk or arms, and generally very little gas and dust; they consist almost entirely of old stars. They are roughly spherical or elongated in shape, and don't rotate much. Lenticular (S0) galaxies are a cross between spirals and ellipticals. The shapes of irregular and peculiar galaxies are strange. Galaxies frequently occur in **clusters of galaxies,** and these in turn congregate in **superclusters** separated by large **voids.**

We now have strong indications that much of the matter in the Universe emits little if any detectable electromagnetic radiation, but it has a gravitational influence on its surroundings. One of the first clues was provided by the flat (nearly constant)

rotation curves of spiral galaxies. These are in stark contrast with the orbital speeds of planets in the Solar System, which decrease with increasing distance from the Sun. There is also evidence for such **dark matter** in pairs or groups of galaxies, and in large clusters. The composition of the dark matter is unknown; possibilities include white dwarfs, neutron stars, black holes, brown dwarfs, large "planets," neutrinos, and various subatomic particles such as **WIMPs** (weakly interacting massive particles). Studies of **gravitational lensing** suggest that part of the dark matter in our Milky Way Galaxy consists of **MACHOs** (massive compact halo objects), though most of it may be subatomic particles.

Clues to the formation and evolution of galaxies are found from observations of very distant galaxies, at large **lookback times;** we are viewing them as they were when the Universe was far younger than it is today. Their distances are obtained from the **redshifts** measured in their spectra and interpreted according to the Doppler formula and **Hubble's law.** The Universe is expanding, with the speeds of different galaxies being proportional to their separations from us, and the constant of proportionality is called **Hubble's constant.** Galaxies at a variety of lookback times are seen in the **Hubble Deep Fields** and other long-exposure photographs of the Universe. Recently, several galaxies with redshifts exceeding 5 or even 6 have been found; we see them as they were only one billion years after the birth of the Universe.

Though most elliptical galaxies formed very long ago, some were produced relatively recently by the **merging** of two spiral galaxies. Early in the history of the Universe, there were many small, irregular galaxies that may have coalesced to form large spirals. Massive galaxies can retain the heavy elements synthesized and blown out by supernovae, while small galaxies tend not to become as chemically enriched. Computer models of the **large-scale structure** of the Universe suggest that clusters and superclusters of galaxies formed at the peaks of a more uniform distribution of dark matter. **Cold dark matter,** which moves slowly compared with light, provides a better match to the data than does rapidly moving **hot dark matter.**

Questions

1. Summarize the topic (and its significance) of the Shapley-Curtis debate.
2. What was the physical principle used by Edwin Hubble to demonstrate that the "spiral nebulae" are very distant and large?
3. Explain why Cepheid variable stars are so useful for determining the distances of galaxies.
4. What are the main morphological types of galaxies, and their most important physical characteristics?
5. In which types of structures are galaxies found?
6. Why can't we determine the distances to galaxies using the geometric method of trigonometric parallax (triangulation), as we do for stars?
7. How is the discovery of other galaxies an extension of the Copernican revolution?
8. Describe the rotation curve of the Milky Way Galaxy and contrast its shape with that of the Solar System.
†9. Show that the mass of the Milky Way Galaxy within a radius of 10,000 light-years from the center is about 3×10^{10} solar masses. (Assume a rotation speed of 200 km/sec at that radius.)
10. How does the shape of the Milky Way Galaxy's rotation curve imply the presence of dark matter, given the distribution of visible light in the galaxy?
11. What is the evidence for dark matter in larger structures such as clusters of galaxies?
12. Describe the phenomenon of gravitational lensing.
13. Define MACHOs, state the tentative evidence for their presence, and discuss their possible nature.
14. How can we statistically study the evolution of galaxies many billions of years ago?
15. What if the speed of light were infinite? Would there still be a way for astronomers to look back at galaxies as they appeared long ago?
16. What is meant by the redshift of a galaxy?
17. State Hubble's law, explain its meaning, and argue that the Universe is expanding.
†18. What is the recession speed of a galaxy having a distance of 50 Mpc, if Hubble's constant is 65 km/sec/Mpc?
†19. What is the redshift z of a galaxy whose Hα absorption line (rest wavelength 6563 Å) is measured to be at 6707 Å?
†20. Calculate the distance of a galaxy with a measured redshift of $z = 0.03$, using 65 km/sec/Mpc for Hubble's constant.
21. Of what importance are spectra and clear images of distant galaxies, in studies of galactic evolution?
22. Summarize some of the tentative conclusions drawn about the formation and evolution of galaxies.

---

†This question requires a numerical solution.

People in Astronomy

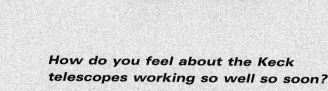

SANDRA FABER

"An observer with a telescope is like an explorer in the New World."

Sandra Faber is Professor of Astrophysics and Astronomy, and at the Lick Observatory, University of California at Santa Cruz. She was born in Boston, grew up in Cleveland and Pittsburgh, and attended Swarthmore College, Swarthmore, Pennsylvania. After graduate school at Harvard's Department of Astronomy, she accepted a position at the Lick Observatory, "the only job I have ever had."

Professor Faber has always wondered about the large-scale features in the Universe: why there are galaxies, why they look as they do, and how the Universe began. Her work takes her regularly to the telescope, and she was for a time Co-Chair of the Scientific Steering Committee of the Keck Observatory, as they planned and built their 10-meter telescopes.

Among the prizes she has received are the Bart J. Bok Prize from Harvard University and the Dannie Heineman Prize of the American Astronomical Society. She is a Trustee of the Carnegie Institution of Washington and a member of the National Academy of Sciences. In 1996 she was elevated to University Professor at the University of California, one of only four such faculty in the system.

In 1980, she joined six other scientists in a study that eventually showed a large-scale flow of galaxies at a million miles per hour toward the direction in the sky where the constellation Centaurus is located. Our Milky Way is part of this flow. The flow is caused by the gravitational attraction of a large supercluster of galaxies, one of the largest structures yet seen in the Universe. They nicknamed it the Great Attractor. Its existence implies, yet again, that most of the matter in the Universe is dark and invisible to telescopes, in this case the dark halos that surround the visible galaxies that compose the Great Attractor.

Professor Faber has two daughters in their 20s. Her husband, Andy, is an attorney in San Jose.

How do you feel about the Keck telescopes working so well so soon?

It is one of the most wonderful things that has happened in my professional career. Although I didn't have that much to do with the mechanical concept of the Keck, I think that I did lend critical support for the scientific case, without which the telescope never would have gotten off the ground. This was at a time when the idea of big, new optical telescopes was unpopular; yet we now see sprouting up around the world a whole generation of large optical telescopes of which Keck was the forerunner. We were pioneers.

How about the Hubble Space Telescope?

The Space Telescope has been the biggest roller coaster of my scientific career. First there was euphoria just after launch, when all seemed to be going well. Then we discovered the flawed primary mirror (I was part of a three-person group that diagnosed the error and reported to NASA). The whole project hastily regrouped and replanned to limp along and do some science with the flawed mirror. And then there were the heroic efforts of hundreds of people who conceived of a strategy to fix the telescope and carried it out brilliantly. The telescope has performed beautifully after the repair mission and has delivered more important data than any other telescope in history, and I now find myself with the rewarding assignment of entertaining audiences with slides of gorgeous Hubble images.

The main lesson is never give up. Pull victory from defeat. The second lesson is, a team of dedicated people can accomplish amazing things.

What things have you learned with the Hubble Space Telescope?

I've been part of a team searching for giant black holes at the centers of galaxies. Hubble can find them by spotting stars very close to the hole that are orbiting superfast.

Some of these black holes are many billions of solar masses in size. Our team has shown that a big black hole lurks at the center of nearly every big galaxy.

I've also been part of another team using the Hubble Space Telescope along with the Keck Telescope to study the most distant galaxies in the Universe. We are literally looking back in time over billions of years to the epoch of galaxy formation. It has been a great thrill to actually see changes in the infant galaxies and to try to piece together the story of galaxy evolution by comparing the infants to local, mature galaxies nearby.

Tell us about your study of the Great Attractor.

Like most of the important things I have done scientifically, the motivation for the project was all wrong. We started out to survey the properties of nearby elliptical galaxies, such things as their brightness, radii, and so on. And we wound up finding a method that could estimate the absolute size of each galaxy and hence tell you how far away each object is. Knowing that, from the Hubble law you could predict the redshift/velocity of every galaxy. When we compared these predictions with the measured velocities, we found a big discrepancy, and this could be interpreted simply as a streaming motion of all the nearby galaxies toward the center of a hitherto unidentified mass concentration. This came as a total surprise. We couldn't believe there was such a large supercluster of galaxies so close-by that nobody had noticed before. But fortunately there was a graduate student in Cambridge, England—Ofer Lahav—who had just stored complete galaxy catalogues in the computer and he was able to make a gigantic map of all the galaxies in that direction in the sky. In this new picture, the Great Attractor appeared for the first time.

Were you surprised at all the interest your results generated?

No. I think we generated some of the interest because we were so surprised ourselves. We were stunned. In graduate school, I was taught that the Hubble expansion was very uniform. The typical streaming motion of galaxies was only supposed to be about 100 km/sec, so it was a total surprise to find peculiar motions 6 times larger than that.

When you were in grad school at Harvard years ago, would you have been surprised at such a discovery?

Totally. I have never had long-range goals as a professional astronomer. I've always been a short-range opportunist, so it always comes as a surprise when something interesting turns up.

What in your view is the relation of observation with theory?

Deep down I feel a little sorry for theoreticians, because they see the Universe only through the eyes of the observers. An observer at a telescope with a good project is like an explorer in the New World. The view over each new ridge is new.

How did you get interested in astronomy?

I was one of those children who was deeply interested in science. It really didn't matter too much what kind. I had star charts, I had a rock collection, I had books on spiders. It was only later, graduating from high school, that I began to focus on astronomy. The reason was simple: I wanted to know where the Universe came from and why it is how it is.

And are you making progress toward understanding the Universe?

I think so, in broad outline—very broad. I believe there are many universes—an infinite number, perhaps. I'd love to know how different they all are from one another, but I don't know the answer to that. However, I do think that ours is roughly the way it is because we are in it. It takes certain restrictions to create intelligent life. Within those restrictions, it is a matter of chance, but the basic restrictions are set by our existence in this Universe.

Perhaps there is an analogy here. Ancient peoples might have wondered why the Earth is as it is. We now know that there are probably millions of planets, but most of them are like the other eight planets in our Solar System, that is, not hospitable to our kind of life. However, out of those millions, there are probably very many planets rather similar to our Earth that would do quite nicely. And within that broad restriction, our existence on this particular Earth is probably just a matter of chance.

In the same way, our Universe is probably just one of many hospitable universes we could inhabit. Our being in this particular one is of no special interest. The really interesting implication is that there exist "out there" many more universes of vastly different types, most of them so bizarre that intelligent life would find them quite hostile. Recent breakthroughs in quantum cosmology have even found a plausible way to generate all those universes in a never-ending, multiply infinite cascade of big bangs. This idea is mostly speculation right now, but chasing it down is going to provide a lot of excitement in the years ahead.

17

Quasars and Active Galaxies

ORIGINS *We explore the significance of quasars as an early phase in the origin and evolution of large galaxies, such as our Milky Way Galaxy. We find that quasars allow us to study the distribution of matter, both visible and dark, in the vast space between us and them.*

AIMS

To describe the discovery of quasars, the source of their stupendous power, and their relationship to galaxies with unusually energetic centers.

Quasars, and the way in which they became understood, have been one of the most exciting stories of the last forty years of astronomy. First noticed as seemingly inconsequential stars, quasars turned out to be some of the most powerful objects in the Universe, and represent violent forces at work. We think that giant black holes, millions or even billions of times the Sun's mass, lurk at their centers. A quasar shines so brightly because its black hole is pulling in the surrounding gas, causing the gas to glow vividly before being swallowed. Our interest in quasars is further piqued because many of them are among the most distant objects we have ever detected in the Universe. Since, as we look out, we are seeing light that was emitted farther and farther back in time, observing quasars is like using a time machine that enables us to see the Universe when it was very young. We find that quasars were an early stage in the evolution of large galaxies. As time passed, gas in the central regions was used up, and the quasars faded, becoming less active. Indeed, we see examples of active galaxies relatively near us, and in some of these the presence of a massive black hole has been all but proven.

ACTIVE GALACTIC NUCLEI

The central regions of normal galaxies tend to have large concentrations of stars. For example, at infrared wavelengths we can see through our Milky Way Galaxy's dust and penetrate to the center. When we do so, we see that the bulge of our galaxy becomes more densely packed with stars as we look closer to the nucleus. With so many stars confined there in a small volume, the nucleus itself is relatively bright. This concentrated brightness appears to be a natural consequence of galaxy formation; gas settles in the central region due to gravity, and subsequently forms stars.

◄ A Hubble Space Telescope image of quasar PKS 2349-014, whose spectral lines are redshifted by 17.3 per cent—that is, $z = 0.173$. The loops of gas around this quasar suggest that it is being fueled by gas from the merging of two galaxies. The galaxies above the quasar are probably interacting with it.

Figure 17–1 Hubble Space Telescope images of the central regions of the normal galaxy NGC 7626 *(left)* and the active galaxy NGC 5548 *(right)*. Both galaxies have the same general appearance (Hubble type), but note the bright, star-like nucleus in the active galaxy. Rays of light are a telescope artifact, and indicate that the active nucleus is very compact. The normal galaxy does not show such rays because its nucleus is much less concentrated.

In a minority of galaxies, however, the nucleus is far brighter than usual at optical and infrared wavelengths, when compared with other galaxies at the same distance (Fig. 17–1). Indeed, when we compute the optical luminosity (power) of the nucleus from its apparent brightness and distance, we have trouble explaining the result in terms of normal stars: it is difficult to cram so many stars into so small a volume.

Such nuclei are also often very powerful at other wavelengths, such as x-rays, ultraviolet, and radio. These galaxies are called "active" to distinguish them from normal galaxies, and their luminous centers are known as **active galactic nuclei.** Clusters of ordinary stars rarely, if ever, produce so much x-ray and radio radiation.

Active galaxies that are extraordinarily bright at radio wavelengths often exhibit two enormous regions (known as "lobes") of radio emission far from the nucleus, up to a million light-years away. The first "radio galaxy" of this type to be detected, Cygnus A (Fig. 17–2), emits about a million times more energy in the radio region of the spectrum than does the Milky Way Galaxy. Close scrutiny

Figure 17–2 A radio map of Cygnus A, with shading indicating the intensity of the radio emission in the two giant lobes. An electronic image of the faint central object seen at optical wavelengths is superposed at the proper scale. *(lower inset)* This optical ground-based view of the faint central object, taken under very good conditions, reveals some structure. *(upper inset)* A Hubble Space Telescope optical image shows yet more structure. Particles feeding the large radio lobes come from the very central part.

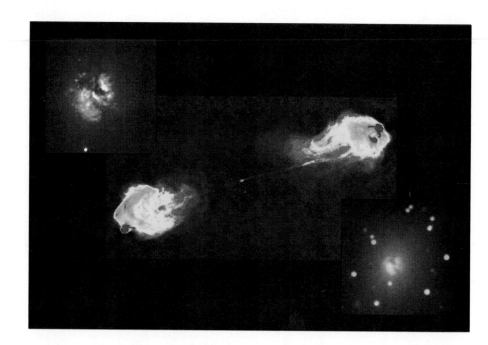

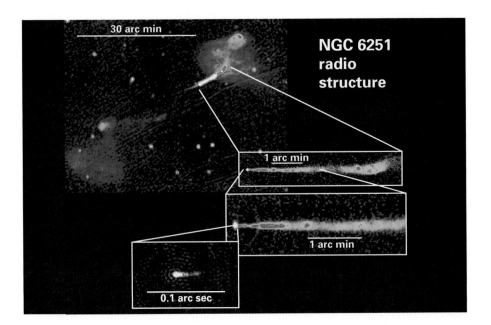

Figure 17-3 Radio maps of the active galaxy NGC 6251 shown on four scales. (At the distance of NGC 6251, 30 arc minutes corresponds to about 3 million light years, 1 arc minute is about 100,000 light years, and 0.1 arc second is about 170 light years.) One of the radio jets is pointing roughly in our direction, and thus appears much brighter than the other jet, which points away from us.

of such galaxies sometimes reveals two long, narrow, oppositely directed "jets" joining the nuclei and the lobes of radio galaxies (Fig. 17–3). The jets are thought to consist of charged particles moving close to the speed of light and emitting radio waves. Sometimes radio galaxies appear rather peculiar when we look at visible wavelengths, and the jet is visible in x-rays, as in the case of Centaurus A (Fig. 17–4).

Optical spectra of the active nuclei often show the presence of gas moving with speeds in excess of 10,000 km/sec, far higher than in normal galactic nuclei. We measure these speeds from the spectra, which have broad emission lines (Fig. 17–5). Atoms that are moving toward us emit photons that are then blueshifted, while those that are moving away from us emit photons that are then redshifted, thereby broadening the line by the Doppler effect. Early in the 20th century, Carl Seyfert was the first to systematically study galaxies with unusually bright optical nuclei and peculiar spectra, and in his honor they are often called "Seyfert galaxies."

Although spectra show that gas has very high speeds in supernovae as well, the overall observed properties of active galactic nuclei generally differ a lot from

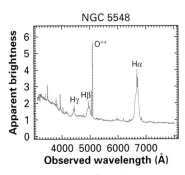

Figure 17-5 Optical spectrum of the nucleus of the Seyfert (active) galaxy NGC 5548, an image of which is shown in Figure 17–1. Note the broad emission lines, along with narrower ones. The broad lines are formed by gas that is moving very rapidly close to the center.

A   *B*

Figure 17-4 (A) Centaurus A (NGC 5128) looks like an elliptical or S0 galaxy seen through an extensive dust lane. (B) A Chandra Observatory x-ray view of the jet in Centaurus A.

those of supernovae, making it unlikely that stellar explosions are responsible for such nuclei. Indeed, it is difficult to see how stars of any kind could produce the unusual activity. However, for many years active galaxies were largely ignored, and the nature of their central powerhouse was unknown.

QUASARS

Interest in active galactic nuclei was renewed with the discovery of **quasars** (shortened form of "quasi-stellar radio sources"), the recognition that quasars are similar to active galactic nuclei, and the realization that both kinds of objects must be powered by a strange process that is unrelated to stars. ☼

The Discovery of Quasars

In the late 1950s, as radio astronomy developed, astronomers found that some celestial objects emit strongly at radio wavelengths. Catalogs of them were compiled, largely at Cambridge University in England, where the method of pinpointing radio sources was developed. For example, the third such Cambridge catalog is known as "3C," and objects in it are given numerical designations like 3C 48.

Although the precise locations of these objects were difficult to determine with single-dish radio telescopes (since they had poor angular resolution), sometimes within the fuzzy radio image there was an obvious probable optical counterpart such as a supernova remnant or a very peculiar galaxy. More often, there seemed to be only a bunch of stars in the field—yet which of them might be special could not be identified, and in any case there was no known mechanism by which stars could produce so much radio radiation.

Special techniques were developed to pinpoint the source of the radio waves in a few instances. Specifically, the occultation (hiding) of 3C 273 by the Moon provided an unambiguous identification with an optical star-like object. When the radio source winked out, we knew that the Moon had just covered it while moving slowly across the background of stars. Thus, we knew that 3C 273 was somewhere on a curved line marking the front edge of the Moon. When the radio source reappeared, we knew that the Moon had just uncovered it, so it was somewhere on a curved line marking the Moon's trailing edge at that time. These two curves intersected at two points, and hence 3C 273 must be at one of those points. Though one point seemed to show nothing at all, the other point was coincident with a bluish, star-like object of 13th magnitude—600 times fainter than the naked-eye limit.

REDSHIFT
http://www.harcourtcollege.com/astro/cosmos/rsce

Figure 17–6 Optical photographs of four of the first known quasars. They appear star-like, but measurements with radio telescopes show that they are much brighter than normal stars at radio wavelengths.

3C 48

3C 147

3C 273

3C 196

ESO New Technology Telescope

Palomar Sky Survey

Figure 17–7 An optical photo-graph of 3C 273, the first quasar to be unambiguously identified. The object, near the center, looks like a normal star, especially in the wide-angle view shown at the lower right. However, there is also a faint "jet" *(boxed)* pointing away from the quasar. A Hubble Space Tele-scope close-up image of the jet is shown at the upper right.

Figure 17–7 An optical photo-graph of 3C 273, the first quasar to be unambiguously identified. The object, near the center, looks like a normal star, especially in the wide-angle view shown at the lower right. However, there is also a faint "jet" *(boxed)* pointing away from the quasar. A Hubble Space Tele-scope close-up image of the jet is shown at the upper right.

When the positions of other radio sources were determined accurately enough, it was found that they, too, often coincided with faint, bluish-looking stars (Fig. 17–6). These objects were dubbed "quasi-stellar radio sources," or "quasars" for short. Optically they looked like stars, but stars were known to be faint at radio wavelengths, so they had to be something else. 3C 273 seemed to be especially interesting: a jet-like feature stuck out from it, visible at optical wavelengths (Fig. 17–7) and radio wavelengths (Fig. 17–8).

Several astronomers, including Maarten Schmidt of Caltech, photographed the optical spectra of some quasars with the 5-m Hale telescope at the Palomar Observatory. These spectra turned out to be bizarre, unlike the spectra of nor-mal stars. They showed bright, broad emission lines, at wavelengths that did not correspond to lines emitted by laboratory gases at rest. Moreover, different quasars had emission lines at different wavelengths.

Schmidt made a breakthrough in 1963, when he noticed that several of the emission lines visible in the spectrum of 3C 273 had the pattern of hydrogen—a series of lines with spacing getting closer together toward shorter wavelengths though not at the normal hydrogen wavelengths (Fig. 17–9). He realized that he could simply be observing hot hydrogen gas (with some contaminants to produce the other lines) that was Doppler shifted. The required redshift would be huge, about 16% (that is, $z = \Delta\lambda/\lambda_0 = 0.16$), corresponding to 16% of the speed of light (since $z \approx v/c$, or $v \approx cz$, valid for z less than about 0.2).

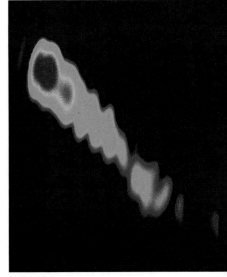

Figure 17–8 The radio jet of the quasar 3C 273 stems from the nucleus.

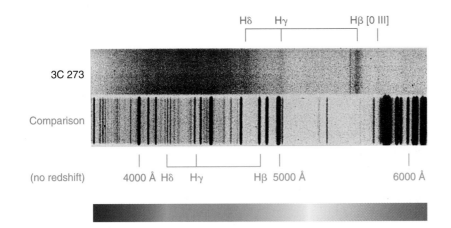

Hδ   Hγ   Hβ [O III]

3C 273

Comparison

(no redshift)   4000 Å Hδ   Hγ   Hβ 5000 Å   6000 Å

Figure 17–9 The spectrum of the quasar 3C 273. The lower spec-trum is of a hot lamp in the tele-scope dome; it consists of hydrogen, helium, neon, and other elements that emit lines at known wavelengths. This "comparison spectrum" establishes the scale of wavelength. A color bar shows the colors of the different wavelengths of the comparison spectrum. The upper spectrum is the spectrum of the quasar. The hydrogen Balmer lines Hβ, Hγ, and Hδ in the quasar spectrum are at longer wavelengths *(labels in red)* than in the compari-son spectrum *(labels in green)*. The redshift of 16% corresponds, ac-cording to Hubble's law (with $H_0 = 65$ km/sec/Mpc), to a distance of 2.4 billion light-years.

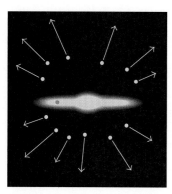

Figure 17–10 The idea that quasars were local, and were ejected from our galaxy, could explain why they all have high redshifts. (The blue dot shows the position of the Sun.) But this hypothesis has a number of problems and was quickly rejected by most astronomers.

This possibility had not been recognized because nobody expected stars to have such large redshifts. Also, the spectral range then available to astronomers, who took spectra on photographic film, did not include the bright Balmer-α line of hydrogen (that is, Hα), which is normally found at 6563 Å but was shifted over to 7600 Å in 3C 273. As soon as Schmidt announced his insight, the spectra of other quasars were interpreted in the same manner. Indeed, one of Schmidt's Caltech colleagues immediately realized that the spectrum of quasar 3C 48 looked like that of hydrogen redshifted by an even more astounding amount: 37%.

Subsequent searches for blue stars revealed a class of "radio-quiet" quasars—their optical spectra are similar to those of quasars, yet their radio emission is weak or absent. These are often called QSOs ("*quasi-stellar objects*"), and they are about ten times more numerous than "radio-loud" quasars. Consistent with the common practice of using the terms interchangeably, here we will simply use "quasar" to mean either the radio-loud or radio-quiet variety, unless we explicitly mention the radio properties.

The Nature of the Redshift

How were the high redshifts produced? The Doppler effect is the most obvious possibility. But it seemed implausible that quasars were discrete objects ejected like cannonballs from the center of the Milky Way Galaxy (Fig. 17–10); their speeds were very high, and no good ejection mechanism was known. Also, we would then expect some quasars to move slightly across the sky relative to the stars, since the Sun is not at the center of the galaxy, but such motions were not seen. Even if these problems could be overcome, we would then have to conclude that only the Milky Way Galaxy (and not other galaxies) ejects quasars—otherwise, we would have seen "quasars" with blueshifted spectra, corresponding to those objects emitted toward us from other galaxies.

Similarly, there were solid arguments against a "gravitational redshift" interpretation, one in which a very strong gravitational field causes the emitted light to lose energy on its way out. This possibility was completely ruled out later, as we shall see.

If, instead, the redshifts of quasars are due to the expansion of the Universe (as is the case for normal galaxies), then quasars are receding with enormous speeds and hence must be very distant. 3C 273, for example, has $z = 0.16$, so $v \approx 0.16c \approx 48,000$ km/sec. According to Hubble's law, $v = H_0 d$, so if $H_0 = 65$ km/sec/Mpc, then $d = v/H_0 \approx (48,000$ km/sec$)/(65$ km/sec/Mpc$) \approx 740$ Mpc ≈ 2.4 billion light-years, a sixth of the way back to the origin of the Universe! A few galaxies with comparably high redshifts (and therefore distances) had previously been found, but they were fainter than 3C 273 by a factor of 10 to 1000, and they looked fuzzy (extended) rather than star-like.

3C 273 turns out to be one of the closest quasars. Other quasars found during the 1960s had redshifts of 0.2 to 1, and hence are billions of light-years away. Note that redshifts greater than 1 do *not* imply speeds larger than the speed of light, because the approximation $z \approx v/c$ is reasonably accurate only when v/c is less than about 0.2. For higher speeds we must instead use the relativistic Doppler formula to calculate the nominal speed. ☼ Even calling it a Doppler effect is a bit misleading and not entirely correct: the redshift is produced by the *expansion* of space, not by motion *through* space, and the concept of "speed" then takes on a somewhat different meaning.

Similarly, as discussed in Chapter 16 for galaxies, it makes more sense to refer to the "lookback time" of a given quasar (the time it has taken for light to reach us) than to its distance: $v = H_0 d$ is inaccurate at large redshifts for a num-

"Relativistic," to an astronomer or physicist, means that such high speeds are involved that formulas from Einstein's special theory of relativity must be used instead of the simpler formulas that are valid only at low speeds.

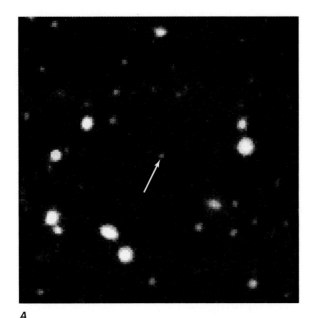

A

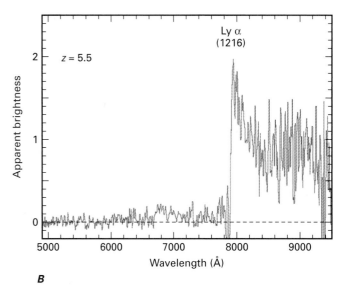

B

Figure 17-11 *(A)* An image of one of the farthest known quasars, RD J030117+002025. *(B)* Its spectrum shows that all the emission lines are shifted by 5.5 times their original wavelengths. The high redshift puts the ultraviolet line Lyman-α, normally at 1216 Å, into the near-infrared region of the spectrum. (The original wavelength is shown in parentheses.) At these redshifts, we are looking back to about 1 billion years after the birth of the Universe. In April 2000, a quasar was found with an even higher redshift, 5.8. Its name is SDSS 1044-0125.

ber of reasons. The lookback time formula is complicated, but some representative values are given in Table 16–1. The highest redshift known for a quasar as of mid 2000 is $z = 5.8$ (Fig. 17–11), which means that a feature whose laboratory (rest) wavelength is 1000 Å is observed to be at a wavelength 580 per cent larger, or 1000 Å + 5800 Å = 6800 Å. (Recall that $z = \Delta\lambda/\lambda_0$.) The corresponding nominal speed of recession is $0.96c$, and the quasar's lookback time is about 13.5 billion years (in a model where the Universe is 14.5 billion years old). We see the quasar as it was when the Universe was about 7 per cent of its current age!

How do we detect quasars? Many of them are found by looking for faint objects with unusual colors—that is, the relative amounts of blue, green, and red light differ from those of normal stars. Low-redshift quasars tend to look bluish, because they emit more blue light than typical stars. But the light from high-redshift quasars is shifted so much toward longer wavelengths that these objects appear very red, especially since intergalactic clouds of gas absorb much of the blue light. Quasars have also been found in maps of the sky made with x-ray satellites, and of course with ground-based radio surveys. After finding a quasar candidate with any technique, however, it is necessary to take a spectrum in order to verify that it is really a quasar and to measure its redshift. As we have seen, the spectra of quasars are quite distinctive, and are rarely confused with other types of objects. Thousands of quasars are now known, and more are being discovered very rapidly.

Figure It Out
The Relativistic Doppler Effect

The relativistic Doppler formula is $z = [(1 + v/c)/(1 - v/c)]^{1/2} - 1$, where z is the redshift and v is the speed of recession. Note that for $v = 0$, $z = [(1 + 0)/(1 - 0)]^{1/2} - 1 = 0$, the same answer as for the nonrelativitistic approximation. But for $v = 0.9c$, $z = [(1 + 0.9)/(1 - 0.9)]^{1/2} - 1 = [1.9/0.1]^{1/2} - 1 = \sqrt{19} - 1 = 3.4$, far greater than the nonrelativistic approximation would have given. So even this $z > 1$ corresponds to a speed less than the speed of light.

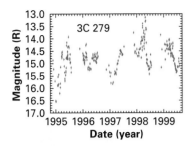

Figure 17–12 The apparent brightness of optical radiation from 3C 279 varies on a timescale of months.

The Energy Problem

Astronomers who conducted early studies of quasars (mid-1960s) recognized that quasars are very powerful, 10 to 1000 times brighter than a galaxy at the same redshift. Yet their diameters appeared relatively small (even through Earth's blurry atmosphere) compared with galaxies, so their energy must be efficiently produced from within a small volume. Already this made them unusual and intriguing.

However, these astronomers were in for a big surprise when they figured out just how small quasars really are. They noticed that some quasars vary in brightness over short time scales—days, weeks, months, or years (Fig. 17–12). This implies that the emitting region is probably smaller than a few light-days, light-weeks, light-months, or light-years in diameter, in all cases a far cry from the tens of thousands of light-years for a typical galaxy.

The argument goes as follows: suppose we have a glowing, spherical, opaque object that is 1 light-month in radius (Fig. 17–13). Even if all parts of the object brightened instantaneously by an intrinsic factor of two, an outside observer would see the object brighten gradually over a time scale of 1 month, because light from the near side of the object would reach the observer 1 month earlier than light from the edge. Thus, the time scale of an observed variation sets an upper limit (that is, a maximum value) to the size of the emitting region: the actual size must be smaller than this upper limit.

Although this conclusion can be violated under certain conditions (such as when different regions of the object brighten in response to light reaching them from other regions, creating a "domino effect"), such models generally seem unnatural. Proper use of Einstein's theory of relativity (in case the light-emitting material is moving very fast) can also change the derived upper limit to some extent, but the basic conclusion still holds: quasars are very small, yet they release tremendous amounts of energy. For example, a quasar only 1 light-month across can be 100 times more powerful than an entire galaxy of stars 100,000 light-years in diameter!

The nature of the prodigious (yet physically small) power source of quasars was initially a mystery. How does such a small region give off so much energy? After all, we don't expect huge explosions from tiny firecrackers. There was some indication that these objects might be related to active galactic nuclei: they have similar optical spectra and are bright at radio wavelengths. So, perhaps the same mechanism might be used to explain the unusual properties of both kinds of objects. In fact, maybe active galactic nuclei are just low-power versions of quasars! If so, quasars should be located in the centers of galaxies. Later we will see that this is indeed the case.

Figure 17–13 Why a large object can't appear to fluctuate in brightness as rapidly as a smaller object. Say that each object abruptly brightens at one instant. The wave emitted from the edge of the object takes longer to reach the observer than light from the near side of the object, because it has to travel farther. We don't see the full variation until waves from all parts of the object reach us. The same is true for both opaque and transparent objects. For the latter, light from the far side takes longer to reach us than light from the near side.

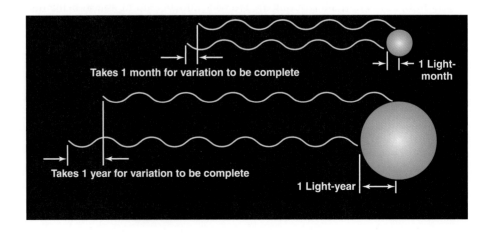

The fact that the incredible power source of quasars is very small immediately rules out some possibilities. Such a process of elimination is often useful in astronomy; recall, for instance, how we deduced that pulsars are rapidly spinning neutron stars. It turns out that for quasars, chemical energy is woefully inadequate: they cannot be wood on fire, or even chemical explosives, because even the most powerful of these is insufficient to produce so much energy within such a small volume.

Even nuclear energy, which works well for stars, is not possible for the most powerful quasars. They cannot be radiation from otherwise-unknown supermassive stars or chains of supernovae going off almost all the time, or other more exotic stellar processes, because once again the efficiency of nuclear energy production is not high enough. To produce that much nuclear energy, a larger volume of material would be needed.

The annihilation of matter and antimatter is energetically feasible, since it is 100% efficient. That is, *all* of the mass in a matter–antimatter collision gets turned into photons (radiation), and in principle a very small volume can therefore be tremendously powerful. However, the observed properties of quasars do not support this hypothesis. Specifically, matter-antimatter collisions tend to emit excess amounts of radiation at certain wavelengths, and this is not the case for quasars.

The release of gravitational energy, on the other hand, can in some cases be very efficient, and seemed most promising to several theorists studying quasars in the mid-1960s. We have already discussed how the gravitational contraction of a ball of gas (a protostar), for example, both heats the gas and radiates energy. But to produce the prodigious power of quasars, a very strong gravitational field is needed. The conclusion was that a quasar is a **supermassive black hole**, perhaps 10 million to a billion times the mass of the Sun, taking up ("accreting") gas. The black hole is in the center of a galaxy. The rate at which matter can be swallowed, and hence the power of the quasar, is proportional to the mass of the black hole, but it is typically a few solar masses per year.

The matter generally swirls around the black hole, forming a rotating disk called an **accretion disk** (Fig. 17–14). As the matter falls toward the black hole, it gains speed (kinetic energy) at the expense of its gravitational energy, just as a ball falling toward the ground accelerates. Friction between the gas particles in the accretion disk causes them to heat up, and they emit electromagnetic radiation, thereby converting part of their kinetic energy into light. Energy is radiated *before* the matter is swallowed by the black hole; nothing escapes from within the black hole itself. This process can convert the equivalent of about 10% of the rest-mass energy of matter into radiation, more than 10 times more efficiently than nuclear energy. (Recall that the fusion of hydrogen to helium converts only 0.7% of the mass into energy.)

A spinning, very massive black hole is also consistent with the well-focused "jets" that emerge from some quasars. No material actually comes from within the black hole; instead, its origin is the accretion disk. The charged particles in the jets are believed to shoot out perpendicularly to the accretion disk, along the black hole's axis of rotation (Figure 17–14). They emit radiation as they are accelerated. In addition to the radio radiation, high-energy photons such as x-rays can also be produced (Fig. 17–15). The impressive focusing might be provided by a magnetic field, as in the case of pulsars, or by the central cavity in the disk. Recall that jets are also seen in radio galaxies, which appear to be closely related to quasars (Figure 17–3). As discussed in more detail later in this chapter, we know that the particles move with very high speeds because a jet can sometimes *appear* to travel faster than the speed of light—an effect that occurs only when an object travels nearly along our line of sight, nearly at the speed of light.

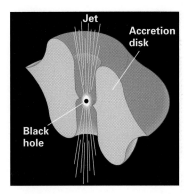

Figure 17–14 Cross-sectional view of an accretion disk (doughnut) surrounding a black hole, with high-speed jets of particles emerging from the disk's nozzle. The nozzle points along the black hole's rotation axis.

Figure 17–15 An image of the quasar PKS 0637-752 ($z = 0.65$) obtained with the Chandra X-ray Observatory. It is so distant that we see it as it was about 6 billion years ago. Note the "jet" that extends about 200,000 light-years out from the quasar; its power at x-ray energies exceeds that at radio energies, an important constraint that a successful theory of jet formation will have to explain.

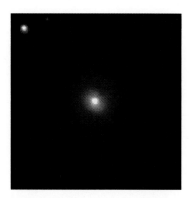

Figure 17-16 An active galaxy whose nucleus is much brighter than the surrounding galactic disk, making the latter hard to detect. This image from a ground-based telescope has been enhanced to emphasize the faint outer parts. Two stars are shown for comparison *(upper left)*.

What Are Quasars?

The idea that quasars are energetic phenomena at the centers of galaxies is now strongly supported by observational evidence. First of all, the observed properties of quasars and active galactic nuclei are strikingly similar. In some cases, the active nucleus of a galaxy is so bright that the rest of the galaxy is difficult to detect because of contrast problems, making the object look almost like a quasar (Fig. 17-16). This is especially true if the galaxy is very distant: we see the bright nucleus as a point-like object, while the spatially extended outer parts (known as "fuzz" in this context) are hard to detect because of their faintness and because of blending with the nucleus.

In the 1970s, a statistical test was carried out with quasars. A selection of quasars, sorted by redshift, was carefully examined. Faint fuzz (presumably a galaxy) was discovered around most of the quasars with the smallest redshifts (the nearest ones), a few of the quasars with intermediate redshifts, and none of the quasars with the largest redshifts (the most distant ones). Astronomers concluded that the extended light was too faint and too close to the nucleus in the distant quasars, as expected. In the 1980s, optical spectra of the fuzz in a few nearby quasars revealed absorption lines due to stars, but the vast majority of objects were too faint for such observations. In any case, the data strongly suggested that quasars could indeed be extreme examples of galaxies with bright nuclei.

More recently, images obtained with the Hubble Space Telescope demonstrate conclusively that quasars live in galaxies, almost always at their centers. (www) With a clear view of the skies above the Earth's atmosphere, and equipped with CCDs, the Hubble Space Telescope easily separates the extended galaxy light from the point-like quasar itself at low redshifts. In some cases the galaxy is obvious (Fig. 17-17), but in others it is barely visible, and special techniques are used to reveal it (Fig. 17-18). Further solidifying the association of quasars with galaxies, ground-based optical spectra of some relatively nearby quasars ($z = 0.2$–0.3) show unambiguous stellar absorption lines at the same redshift as that given by the quasar emission lines (Fig. 17-19).

Quasars exist almost exclusively at high redshifts and hence large distances. The peak of the distribution is at $z \approx 2$ (Fig. 17-20), though new studies at x-ray wavelengths suggest that it might have been at an even higher redshift. With lookback times of about 10 billion years, quasars must be denizens of the *young* Universe. What happened to them? Quasars probably faded with time, as the central black hole gobbled up most of the surrounding gas; the quasar shines only while it is pulling in material.

Figure 17-17 *(A)* The quasar PG 0052+251, at redshift $z = 0.155$, is in the center of a normal spiral galaxy. *(B)* The quasar PHL 909, at redshift $z = 0.171$, is at the core of a normal elliptical galaxy. Both of these images were obtained with the Hubble Space Telescope.

A

B

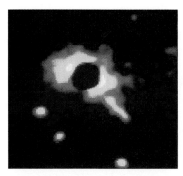

Figure 17–18 Hiding the bright quasar 3C 273 with a dark disk, as was done here in pre-Hubble days, has revealed extended light ("fuzz") that is almost certainly a galaxy. Quasars are now thought to be very powerful events in the centers of galaxies.

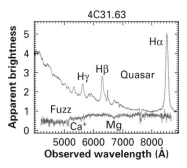

Figure 17–19 Keck telescope spectra of the quasar 4C 31.63 ($z = 0.296$) and its "fuzz," the latter revealing absorption lines produced by the relatively cool outer parts of normal stars. This shows that the "fuzz" seen around nearby quasars is really galaxy light. The quasar is actually much brighter than the "fuzz," but was scaled here by an arbitrary amount for clarity.

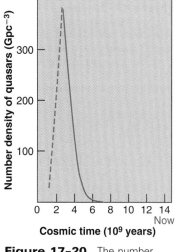

Figure 17–20 The number density of quasars (number per billion cubic parsecs) is plotted versus cosmic time, for an assumed Universe age of 14 billion years. There was a bright, spectacular era of quasars billions of years ago, and essentially none now remain.

Thus, some of the nearby active and normal galaxies may have been luminous quasars in the distant past, but now exhibit much less activity because of a slower accretion rate. Perhaps even the nucleus of the Milky Way Galaxy, which is only slightly active, was more powerful in the past, when the putative black hole had plenty of material to accrete. Of course, many of the weakly active galaxies we see nearby were probably never luminous enough to be genuine quasars. Either their central black hole wasn't sufficiently massive to pull in much material, or there was little gas available to be swallowed.

Though most quasars are very far away, some have relatively low redshifts (like 0.1). If quasars were formed early in the Universe, how can these quasars still be shining? Why hasn't all of the gas in the central region been used up? Hubble Space Telescope images show that in many cases, the galaxy containing the quasar is interacting or merging with another galaxy (Fig. 17–21). This result suggests that gravitational tugs end up directing a fresh supply of gas from the outer part of the galaxy (or from the intruder galaxy) toward its central black hole, thereby fueling the quasar and allowing it to continue radiating so strongly. Some quasars may have even faded for a while, and then the interaction with another galaxy rejuvenated the activity in the nucleus.

A few astronomers have disputed the conclusion that the redshifts of quasars indicate large distances, partly because of the implied enormously high luminos-

Figure 17–21 Quasars in interacting or merging galaxies, in each case imaged with the Hubble Space Telescope. (A) Debris from a collision between two galaxies fueling a quasar. A ring galaxy left by the collision is at bottom; a foreground star in our galaxy is at top. (B) A tidal tail above a quasar, perhaps drawn out by a galaxy that is no longer there. (C) A quasar merging with the bright galaxy that appears just below it. The swirling wisps of dust and gas surrounding them indicate that an interaction is taking place. (D) A pair of merged galaxies have left loops of gas around this quasar.

A　　　　　　　*B*　　　　　　　*C*　　　　　　　*D*

Figure 17–22 Two galaxies with relatively low redshifts are seen close in the sky to high-redshift quasars. Most astronomers think these are just chance projections along nearly the same lines of sight; the quasars and nearby galaxies are not physically associated with each other.

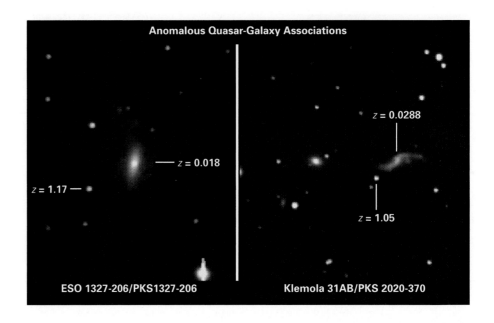

ity produced in a small volume. If Hubble's law doesn't apply to quasars, maybe they are actually quite nearby. Specifically, Halton Arp has found some cases where a quasar seems associated with an object of a different, lower redshift (Fig. 17–22). However, most astronomers blame the association on chance superposition. We now have little reason to doubt the conventional interpretation of quasar redshifts. Quasars clearly reside in the centers of galaxies having the same redshift. They are simply the more luminous cousins of active galactic nuclei, and a plausible energy source has been found. In addition, gravitational lensing (see below) shows that quasars are indeed very distant.

Detection of Supermassive Black Holes

We argued above, essentially by the process of elimination, that the central engine of a quasar or active galaxy consists of a supermassive black hole swallowing material from its surroundings. Is there any more direct evidence for this? Well, the high speed of gas in quasars and active galactic nuclei, as measured from the widths of emission lines, suggests the presence of a supermassive black hole. A strong gravitational field causes the gas particles to move very quickly, and the different emitted photons are Doppler shifted by different amounts, resulting in a broad line. On the other hand, alternative explanations such as supernovae might conceivably be possible; they, too, produce high-speed gas, but without having to use a supermassive black hole.

Recently, however, very rapidly rotating disks of gas have been found in the centers of several mildly active galaxies. Their motion is almost certainly produced by the gravitational attraction of a compact central object, because we see the expected decrease of orbital speed with increasing distance from the center, as in Kepler's laws for our Solar System. The galaxy NGC 4258 (Fig. 17–23) presents the most convincing case, one in which radio observations were used to obtain very accurate measurements. The typical speed is $v = 1120$ km/sec at a distance of only 0.4 light-year from the center. This implies a mass of about 3.6×10^7 Suns in the nucleus.

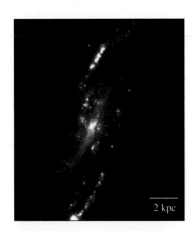

2 kpc

Figure 17–23 An optical image of the spiral galaxy NGC 4258 that emphasizes star-forming regions. Radio observations of the nucleus have revealed gas orbiting a central, dark, very massive object, almost certainly a black hole.

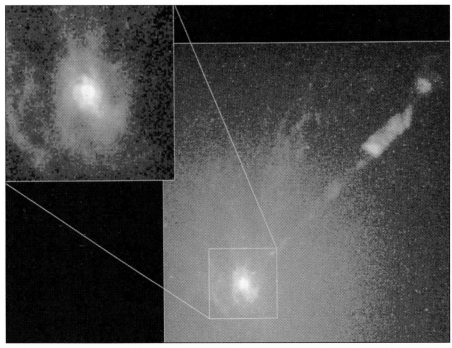

A

B

The corresponding density is over 100 million solar masses per cubic light-year, a truly astonishing number. If the mass consisted of stars, there would be no way to pack them into such a small volume, at least not for a reasonable amount of time: they would rapidly collide and destroy themselves, or undergo catastrophic collapse. The natural conclusion is that a supermassive black hole lurks in the center. Indeed, this is now regarded as the most conservative explanation for the data: if it's not a black hole, it's something even stranger!

One of the most massive black holes ever found is that of M87, an active galaxy in the Virgo cluster that sports a bright radio and optical jet (Fig. 17–24). Spectra of the gas disk surrounding the nucleus were obtained with the Hubble Space Telescope (Fig. 17–25), and the derived mass in the nucleus is about 3 billion Suns.

If some nearby, relatively normal-looking galaxies were luminous quasars in the past, and a significant fraction even show some activity now, we suspect that supermassive black holes are likely to exist in the centers of many large galaxies today. Sure enough, when detailed spectra of the nuclear regions of a few galaxies were obtained (especially with the Hubble Space Telescope), strong evidence was found for rapidly moving stars. The masses derived from Kepler's third law were once again in the range of a million to a billion Suns.

Figure 17-24 *(A)* The nucleus of the active elliptical galaxy M87, showing a jet of high-speed charged particles and, enlarged, an unusual spiral disk in the center. *(B)* The nucleus (bright point at top) and jet of M87, in a computer-processed view from the Faint Object Camera aboard the Hubble Space Telescope. In the jet, which extends 8000 light-years, the image reveals detail as small as 10 light-years across.

Figure It Out
The Central Mass in a Galaxy

We saw in Chapter 16 that the mass enclosed within a circular orbit of radius R is $M = v^2 R/G$, where v is the orbital speed and G is Newton's constant of gravitation. For NGC 4258, we measure $v = 1120$ km/sec at a distance of 0.4 light-year from the center. Properly converting units, we find that $M = 7.1 \times 10^{40}$ g $= 3.6 \times 10^7$ solar masses.

Figure 17–25 Spectra of the regions shown on the image of the center of M87, taken with the Faint Object Spectrograph aboard the Hubble Space Telescope, reveal Doppler shifts of the gas. (The single emission line appears at different wavelengths.) The speed of revolution of 550 km/sec at this distance of 60 light-years from the nucleus allows astronomers to calculate how much mass must be inside those locations to keep the gas in orbit. The result is about 3 billion solar masses, after various effects like the inclination of the disk are taken into account.

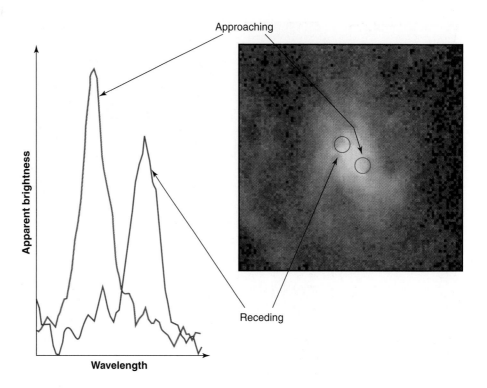

Probably the most impressive and compelling case is our own Milky Way Galaxy. At infrared wavelengths, stars in the highly obscured nucleus were seen from Earth, and their motions were measured over the course of a few years (Fig. 17–26). The implied mass of the central dark object is 2.6 million Suns, and it is confined to a volume < 0.03 light-year in diameter! Our galaxy could certainly have been more active in the past, though never as powerful as the most luminous quasars, which require a black hole of 10^8 to 10^9 solar masses.

The Effects of Beaming

Radio observations with extremely high angular resolution, generally obtained with the technique of very-long-baseline interferometry, have shown that some quasars consist of a few small components. In many cases, observations over a few years reveal that the components are apparently separating very fast (Fig.

Figure 17–26 (left) Stars in the central region of the Milky Way Galaxy, photographed several times with the Keck-I telescope using a special technique that enhances angular resolution. Some of the stars clearly change position relative to each other over the course of a few years. (right) The observed speeds of stars versus their distance from the center of our galaxy. The curve shows the predicted speeds if the center contains a black hole 2.6 million times as massive as the Sun. Note that 1 pc = 3.26 light-years.

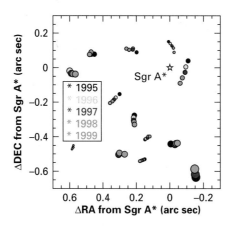

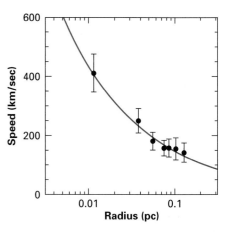

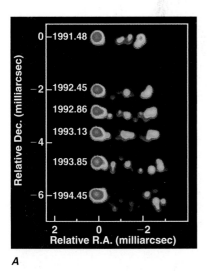
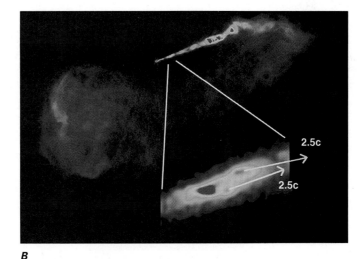

A   **B**

Figure 17-27 *(A)* A series of views of the quasar 3C 279 with radio interferometry at a wavelength of 1.3 cm. The apparent speed translates (for $H_0 = 65$ km/sec/Mpc) to a speed of $9.5c$, though such "superluminal speeds" can be explained in conventional terms. *(B)* Superluminal speeds also appear in the active galaxy M87. The inset shows features that seem to move at $2.5c$. Red shows the brightest and blue shows the faintest emission.

17–27), given the conversion from the angular change in position we measure across the sky to the actual physical speed in km/sec at the distance of the quasar. Indeed, some of the components appear to be separating at **superluminal speeds**—that is, at speeds greater than that of light. But Einstein's special theory of relativity says that no objects can travel through space faster than light.

Astronomers can explain how the components only *appear* to be separating at greater than the speed of light even though they are actually moving at allowable speeds (less than that of light). If one of the components is a jet approaching us almost along our line of sight, and nearly at the speed of light, then according to our perspective the jet is nearly keeping up with the radiation it emits. If the jet moves a certain distance in our direction in 1 year, the radiation it emits at the end of that period gets to us sooner than it would have if the jet were not moving toward us. So in less than 1 year, we see the jet's motion over 1 full year. In the interval between our observations, the jet had several times longer to move than we would naively think it had. So it could, without exceeding the speed of light, appear to move several times as far. 🜨

Whether a given object looks like a quasar or a less active galaxy with broad emission lines probably depends on the orientation of the jet relative to our line of sight: jets pointing at us appear far brighter than those that are misaligned. Thus, quasars are probably often beamed roughly toward us, a conclusion supported by the fact that many radio-loud quasars show superluminal motion. However, if the jet is pointing straight at us, it can greatly outshine the emission lines, and the object's optical spectrum looks rather featureless, unlike that of a normal quasar. It is then called a "BL Lac object," after the prototype in the constellation Lacerta, the Lizard. At the other extreme, if the jet is close to the plane of the sky, dust and gas in a torus (doughnut) surrounding the central region may hide the active nucleus from us (Figure 17–14). The galaxy nucleus itself may then appear relatively normal, although the active nature of the galaxy could still be deduced from the presence of extended radio emission from the jet.

This general idea of beamed, or directed, radiation probably accounts for many of the differences seen among active galactic nuclei. For example, in one type of

Figure 17-28 The appearance from Earth of an active galaxy can depend on the orientation relative to our line of sight of the torus (doughnut) that surrounds it. If we see the nucleus from a direction within the indicated cones, the nucleus appears bright, and we can see the broad emission lines produced by gas in rapid motion near the supermassive black hole. If instead we view from outside the cones, then the active nucleus and central, rapidly moving clouds can be hidden at optical wavelengths by gas and dust in the torus. Narrow emission lines from clouds farther from the nucleus (and within the cones), however, should still be visible.

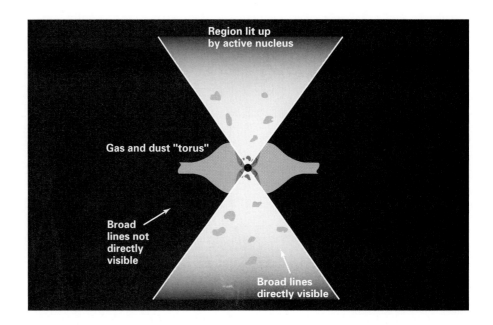

Seyfert galaxy, the very broad emission lines are not easily visible, despite other evidence that indicates considerable activity in the nucleus. (For example, bright narrow emission lines can be seen.) We think that in some cases, the broad emission lines are present, but simply can't be directly seen because they are being blocked by an obscuring torus of material (Fig. 17–28). But light from the broad lines can still escape along the axis of this torus and reflect off of clouds of gas elsewhere in the galaxy. Observations of these clouds then reveal the broad lines, but faintly.

Similarly, some galaxies hardly show any sort of active nucleus directly—it is too heavily blocked from view by gas and dust along our line of sight, in the central torus. However, radiation escaping along the axis of this torus can still light up exposed parts of the galaxy, indirectly revealing the active nucleus (Fig. 17–29).

Figure 17-29 Radiation from the core of this active galaxy, NGC 5252, illuminates matter only within two oppositely directed cones. Presumably the cones are aligned along the axis of a rotating black hole surrounded by a torus of gas and dust that blocks our direct view of the active nucleus, as shown in Figure 17–28.

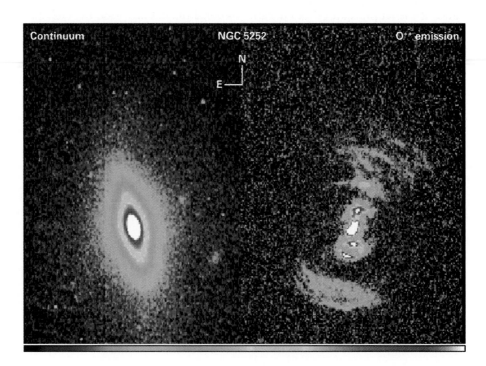

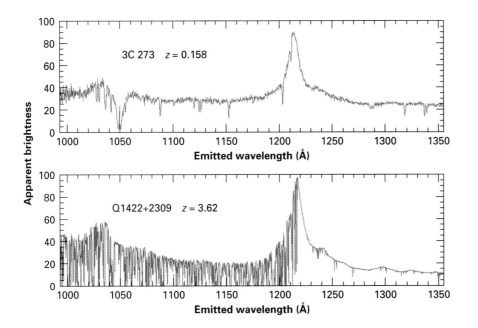

Figure 17–30 Spectrum of a high-redshift quasar *(bottom)*, showing a large number of absorption lines produced by clouds of gas and galaxies along our line of sight to the quasar. Most of these lines, especially the ones in the blue *(left)* part of the spectrum, correspond to the hydrogen Lyman-α transition produced by intergalactic clouds at many different (and generally large) distances from us. We see the lines at many different wavelengths because of the different redshifts of the clouds. The spectrum of 3C 273 *(top)*, a low-redshift quasar, has far fewer absorption lines, indicating that intergalactic clouds of gas are scarce at the present time in the universe.

Quasars as Probes of the Universe

Quasars are powerful beacons, allowing us to probe the amount and nature of intervening material. For example, numerous narrow absorption lines are seen in the spectra of high-redshift quasars (Fig. 17–30). These spectral lines are produced by clouds of gas at different redshifts between the quasar and us. The lines can be identified with hydrogen, carbon, magnesium, and other elements.

Analysis of the line strengths and redshifts allows us to explore the chemical evolution of galaxies, the distribution and physical properties of intergalactic clouds of gas, and other interesting problems. The lines are produced by objects that are generally too faint to be detected in other ways. One surprising conclusion is that all of the clouds have at least a small quantity of elements heavier than helium. Since stars and supernovae produced these heavy elements, the implication is that an early episode of star formation preceded the formation of galaxies.

Another way in which quasars are probes of the Universe is the phenomenon of **gravitational lensing** of light (Chapter 16). In fact, such lensing was first confirmed through studies of quasars. In 1979, two quasars were discovered close together in the sky, only a few seconds of arc apart (Fig. 17–31*A*). They had the same redshift, yet their spectra were essentially identical, arguing against a possible binary quasar. A cluster of galaxies with one main galaxy (Figure 17–31*B,C*)

> The quasar absorption lines are narrow because they are formed by cold gas in intergalactic space between us and the quasar. Spectral lines formed in the quasar itself are broad; they are emission lines.

Figure 17–31 *(A)* Hubble Space Telescope wide-angle view of two quasars, 0957+561 A and B, that are only about 6 seconds of arc apart in the sky. They have the same redshifts (z = 1.4136) and essentially identical spectra. *(B)* Narrow-angle view of the two quasars, obtained with a ground-based telescope; they appear slightly oblong because of imperfect tracking by the telescope during the exposure. An intervening galaxy can be seen only 1 second of arc from the quasar on the right, in the direction of the quasar on the left. *(C)* The image of the quasar on the left has been subtracted from that on the right to reveal just the intervening galaxy.

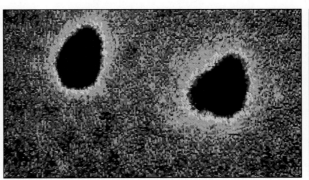

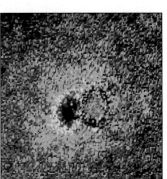

A   *B*   *C*

Figure 17–32 The geometry of gravitational lensing. Light from a distant quasar can take two different paths on its way to Earth; each is bent because of the gravitational field of the intervening galaxy or cluster of galaxies. The two paths produce two distinct images of the same quasar close together in the sky. Note that the lengths of the two paths generally differ.

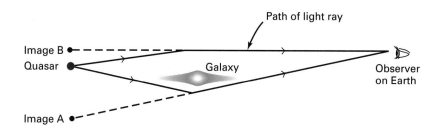

was subsequently found along the same line of sight, but at a smaller redshift. The most probable explanation is that light from the quasar is bent by the gravity of the cluster (warped spacetime), leading to the formation of two distinct images (Fig. 17–32). The cluster is acting like a gravitational lens.

Since then, over two dozen gravitationally lensed quasars have been found. For a point lens and an exactly aligned object, we can get an image that is a ring centered on the lensing object. Such a case is called an "Einstein ring," and a few are known (Fig. 17–33A). Some gravitationally lensed quasars even have quadruple quasar images that resemble a cross (Fig. 17–33B,C). Only gravitational lensing seems to be a reasonable explanation of these objects, the redshifts of whose components are identical. Moreover, in some cases continual monitoring of the brightness of each quasar image has revealed the same pattern of light variability, but with a time delay between the different quasar images. This delay occurs because the light travels along two different paths of unequal length to form the two quasar images (Figure 17–32). The variability pattern is not expected to be identical in two entirely different quasars that happen to be bound in a physical pair.

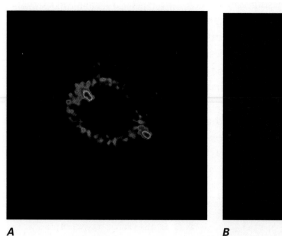

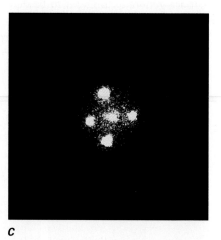

A   *B*   *C*

Figure 17–33 (A) This particular quasar is so exactly aligned with the intervening gravitational lens that its radio radiation is spread out into an "Einstein ring." A second object is less well aligned; its image appears slightly inside the ring on one side and slightly outside the ring on the other side. (B) The "clover leaf" (H 1413+117), a quadruply imaged quasar at redshift 2.55. The four images of comparable brightness are only 1 second of arc apart, and barely resolved in this ground-based photograph. (C) Hubble Space Telescope image of the "Einstein cross" (Q2203+0305), a quasar ($z = 1.70$) that is gravitationally lensed into four images by a relatively nearby galaxy ($z = 0.039$). The nucleus of the galaxy is visible in the center of the image. The four images of comparable brightness are separated by less than 2 seconds of arc and have essentially identical spectra. The rare configuration and identical spectra show that we are indeed seeing gravitational lensing rather than a cluster of quasars.

Multiply imaged (pronounced "mul´te-plee") quasars are an exciting verification of a prediction of Einstein's general theory of relativity. The lensing details are sensitive to the total amount and distribution of matter (both visible and dark) in the intervening cluster. In some cases, the lens itself seems to consist almost entirely of dark matter!

Concept Review

The central regions of some galaxies are unusually luminous—they are very powerful emitters of all kinds of electromagnetic radiation, and they sometimes spew out enormous jets of particles. It is unlikely that normal stellar processes can account for these objects, called **active galactic nuclei.** The mysterious **quasars** are apparently related: they are relatively bright, star-like objects that have bizarre optical spectra and in some cases shine very brightly at radio wavelengths. The discovery that quasars have high redshifts implied that they are at vast distances and have astonishing luminosities. Rapid variations in their brightness were used to deduce that quasars are small, in some cases less than one light-week across—yet they are more powerful than entire galaxies 100,000 light-years in diameter. The energy of a quasar probably comes from matter falling into a **supermassive black hole,** 10 million to a billion times the mass of the Sun, in the central region of a galaxy. The matter forms a rotating **accretion disk** around the black hole.

We now know that quasars are indeed phenomena in the centers of a minority of galaxies. Hubble Space Telescope images of low-redshift quasars reveal nebulous structures that closely resemble galaxies, and spectra confirm that the extended light comes from stars. Although a few quasars appear close in the sky to galaxies with much lower redshifts, these are thought to be chance projections. Quasars lived in the distant past; as the material surrounding the supermassive black holes was used up, the quasars faded, becoming less-active galaxies and finally normal-looking galaxies. There is strong evidence for the existence of gigantic black holes in the centers of some galaxies, and even our own Milky Way Galaxy harbors such a beast. Some quasars appear to be revived through gravitational interactions and collisions with galaxies, which direct gas toward the central black hole.

Apparent **superluminal speeds** of components seen in some quasars are produced by material moving near (but below) the speed of light, close in direction to our line of sight; they are not violations of Einstein's special theory of relativity. The spectra of quasars often exhibit absorption lines due to material at intermediate distances, allowing astronomers to study galaxies and clouds of gas that are otherwise too dim to see. Some quasars look multiple because of **gravitational lensing** by intervening galaxies or clusters.

Questions

1. Summarize the main observed characteristics of active galactic nuclei.
2. Describe the historical development of the study of quasars, and list their peculiar properties.
3. Why is it useful to find the optical objects that correspond in position with radio sources?
4. What was the key breakthrough in the interpretation of quasar spectra? What was its significance?
5. Why do you think it was initially difficult for astronomers to entertain the possibility that the spectra of quasars are highly redshifted?
†6. We observe a quasar with a spectral line at 5000 Å that we know is normally emitted at 4000 Å. **(a)** By what percentage is the line redshifted? **(b)** By approximately what percentage of the speed of light is the quasar receding? (Nonrelativistic formula acceptable.) **(c)** At what speed is the quasar receding in km/sec? **(d)** Using Hubble's law, to what distance does this speed correspond?
7. Explain what is meant by the "lookback time" of a quasar.
8. Outline the argument used to infer that the physical size of quasars is small.
9. What is the most probable physical mechanism that produces a quasar's energy? Does it seem paradoxical that black holes, from which nothing can escape, have something to do with it?
10. What are three differences between quasars and pulsars?
11. What is the evidence suggesting that quasars live in the centers of galaxies?
12. Summarize what generally happens to a quasar as it ages.
13. Suppose *no* nearby galaxies exhibited evidence for the presence of supermassive black holes in their centers. Would this be a problem for the hypothesis of what powers quasars?
14. Why do we think that some nearby quasars are rejuvenated?
15. Explain how we deduce the presence and measure the mass of a supermassive black hole in the center of a galaxy.
16. Describe how quasars can be used as probes of matter between them and us.
17. What do we mean by a "gravitationally lensed quasar"?

———————————————

†This question requires a numerical solution.

18

Cosmology: The Birth and Life of the Cosmos

ORIGINS *Knowledge of the age, geometry, structure, and evolution of the Universe is fundamental to our overall understanding of our place in the Cosmos. The ultimate fate of the Universe is closely linked to its past history.*

Cosmology is the study of the structure and evolution of the Universe on its grandest scales. Some of the major issues studied by cosmologists include its birth, age, size, geometry, and ultimate fate. As in the rest of astronomy, we are trying to discover the fundamental laws of physics, and use them to understand how the Universe works. We do not claim to determine the purpose of the Universe or why humans, in particular, exist; these questions are more in the domain of theology, philosophy, and metaphysics. However, in the end we will see that some of our conclusions, which are solidly based on the methods of science, nonetheless seem to be untestable with our present knowledge.

⬤ OLBERS'S PARADOX

We begin our exploration of cosmology by considering a deceptively simple question: "Why is the sky dark at night?" This seems simple or even absurd ("The sky is dark because the Sun is down, dummy!"). Actually, though, it is very profound. If the Universe is *infinite* in size and age, with stars spread throughout it, every line of sight should intersect a shining star (Fig. 18–1)— just as in a hypothetical infinite forest, every line of sight eventually intersects

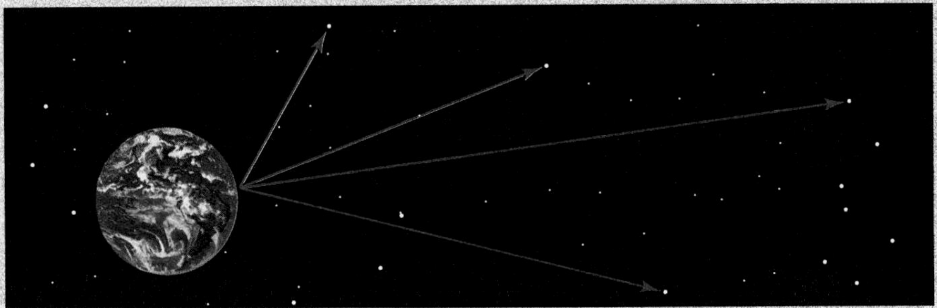

AIMS

To see how the simple observation that the night sky is dark has profound implications for the origin or extent of the Universe.

•

To learn how astronomers determine the age of the Universe.

•

To explore various possibilities for the overall geometry, expansion history, and future of the Universe—including the recent suggestion that a long-range cosmic "antigravity" exists.

Figure 18–1 If we look far enough in any direction in an infinite universe, our line of sight should eventually hit the surface of a star. This leads to Olbers's paradox.

◀ The dark night sky with comet Hale-Bopp above the twin Keck telescope domes on Mauna Kea volcano in Hawaii. The puzzle of why the sky isn't always bright is known as Olbers's paradox.

Figure 18–2 The lower halves of trees are sometimes painted white, and mimic seeing a uniform expanse of starlight. Note that the surface brightness of a tree is the same whether the tree is close to us or far away.

The name "Olbers" has an "s" on the end before we make it possessive, so we write of "Olbers's paradox" (or, alternatively, "Olbers' paradox"). It is incorrect to write "Olber's paradox," with an apostrophe before a sole "s," since his name wasn't "Olber."

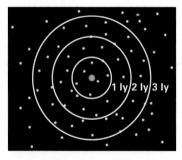

Figure 18–3 A model non-expanding universe in which the stars were suddenly created everywhere at the same instant. In the first year the observer would see only those stars within 1 light-year of him or her (that is, all stars within the innermost sphere), and most of the sky would appear not to have stars. With each additional year, he or she would see all stars within another light-year farther away (progressively larger spheres). The size of gaps between the stars would steadily decrease as time passes and more stars became visible.

a tree. So, the sky should everywhere be bright, even at night. But it obviously isn't, thereby making a paradox—a conflict of a reasonable deduction with our common experience.

One might argue that distant stars appear dim according to the inverse-square law of light, so they won't contribute much to the brightness of the night sky. But the apparent size of a star also decreases with increasing distance. (This can be difficult to comprehend: Stars are so far away that they are usually approximated as points, or as the blur circle produced by turbulence in Earth's atmosphere, but intrinsically they really *do* have a nonzero angular area.) So, a star's brightness per unit area remains the same regardless of distance (Fig. 18–2). If there is indeed an infinite number of stars, and if we can see all of them, then every point on the sky will be covered with a star, making the entire sky blazingly bright!

This dark-sky paradox has been debated for hundreds of years. It is known as **Olbers's paradox,** though the 19th-century astronomer Wilhelm Olbers wasn't the first to realize the problem. Kepler and others considered it, but not until the 20th century was it solved.

Actually, there are several conceivable resolutions of Olbers's paradox, each of which has profound implications. For example, the Universe might have finite size. It is as though the whole Universe were a forest, but the forest has an edge—and if there are sufficiently few trees, one can see to the edge along some lines of sight. Or, the Universe might have infinite size, but with few or no stars far away. This is like a forest that stops at some point, or thins out quickly, with an open field (the rest of the Universe) beyond it.

Another possible solution is that the Universe has a finite age, so that light from most of the stars has not yet had time to reach us. If the forest suddenly came into existence, an observer would initially see only the most nearby trees (due to the finite speed of light), and there would be gaps between them along some lines of sight.

There are other possibilities as well. Most of them are easily ruled out by observations or violate the Copernican principle (that we are at a typical, non-special place in the Universe), and some are fundamentally similar to the three main suggestions discussed above. One idea is that dust blocks the light of distant stars, but this doesn't work because if the Universe were infinitely old, the dust would have time to heat up and either glow brightly or be destroyed.

It turns out that the primary true solution to Olbers's paradox is that the Universe has a *finite age*, about 14 billion years. There has been far too little time for the light from enough stars to reach us to make the sky bright. Effectively, we see "gaps" in the sky where there are no stars, because light from the distant stars still hasn't been detected.

For example, consider the static (non-expanding) universe in Fig. 18–3. If stars were suddenly created as shown, then in the first year the observer would see only those stars within 1 light-year of him or her, because the light from more distant stars would still be on its way. After two years, the observer would see more stars—all those within 2 light-years—and the gaps between stars would be smaller. After 10 years, the observer would see all stars within 10 light-years, so the sky would be even brighter. But it would be a long time before enough very distant stars became visible and filled the gaps.

Our own Universe, of course, is not so simple—it is expanding. To some degree, the expansion of the Universe also helps solve Olbers's paradox: As a galaxy moves away from us, its light is redshifted from visible wavelengths (which we can see) to longer wavelengths (which we can't see). Indeed, the energy of each photon actually decreases because of the expansion. But the effects of expansion are minor in resolving Olbers's paradox, compared with the finite, relatively short age of the Universe.

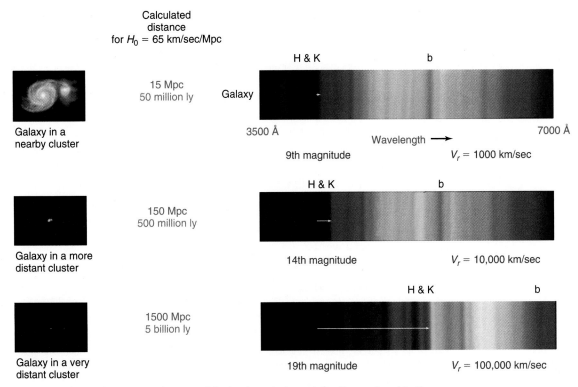

Figure 18-4 Spectra are shown at right for the galaxies at left, all reproduced to the same scale. Distances are based on Hubble's constant = 65 km/sec/Mpc. Notice how the farther away a galaxy is, the smaller it looks. The arrow on each horizontal spectrum shows how far the "H and K" lines of singly ionized calcium are redshifted.

Regardless of the actual resolution of Olbers's paradox, the main point is that such a simple observation and such a silly sounding question leads to incredibly interesting possible conclusions regarding the nature of the Universe. So the next time your friends are in awe of the beauty of the stars, point out the profound implications of the darkness, too!

AN EXPANDING UNIVERSE

Hubble's Law

To see that the Universe has a finite age, we must consider the expansion of the Universe. In Chapter 16, we described how spectra of galaxies studied by Edwin Hubble led to this amazing concept, one of the central themes of cosmology. He found that in every direction, all but the closest galaxies have spectra that are shifted to longer wavelengths. Moreover, the measured **redshift**, z, of a galaxy is *proportional* to its distance from us, d (Fig. 18–4). If this redshift is produced by motion away from us, then we can use the Doppler formula to derive the speed of recession, v. The recession speed can be plotted against distance for many galaxies in a Hubble diagram (Fig. 18–5), and a straight line nicely represents the data. The final result is **Hubble's law**, $v = H_0 d$, where H_0 is the *present-day value* of **Hubble's constant**, H. (Note that H is constant throughout the Universe at a given time, but its value decreases with time.)

An explosion can give rise to Hubble's law. If we kick a pile of balls, for example, then some of them are hit directly and given a large speed, while others fly off more slowly. After a while, we see that the most distant ones are moving

> Recall from Chapter 16 that the redshift z of an object is defined by $z = \Delta\lambda/\lambda_0$, where $\Delta\lambda = \lambda - \lambda_0$ is the observed wavelength of an absorption or emission line in the object's spectrum minus its laboratory (rest) wavelength. If $z \lesssim 0.2$, a reasonable approximation is the Doppler formula, $z = \Delta\lambda/\lambda_0 \approx v/c$, where v is the object's speed of recession.

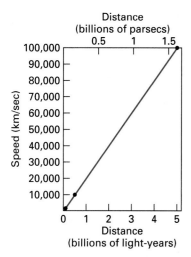

Figure 18–5 The Hubble diagram for the galaxies shown in Figure 18–4.

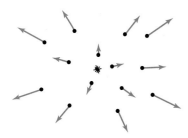

Figure 18–6 After receiving a kick, balls go flying away from their common point of origin at different speeds, depending on how directly each one was hit. After any given amount of time, it is clear that the most distant balls are those with the highest speeds; to have travelled farthest, they must be moving fastest.

Figure 18–7 From every raisin in a rising loaf of raisin bread, every other raisin seems to be moving away from you at a speed that depends on its distance from you. This leads to a relation like Hubble's law between the speed and the distance. Note also that each raisin would be at the center of the expansion measured from its own position, yet the bread is expanding uniformly. The raisins themselves do not expand, just as galaxies do not expand. For a better analogy with the Universe, consider an infinite loaf; unlike the finite analogy pictured here, there is then no overall center to its expansion.

fastest, while the ones closest to the original pile are moving slowly (Fig. 18–6). The reason is obvious: To have reached a particular distance in a given amount of time, the distant balls must have been moving quickly, while the nearby ones must have been moving slowly. Speed is indeed proportional to distance, which is the same formula as Hubble's law.

Based on Hubble's law, we conclude that the Universe is expanding. Like an expanding gas, its density and temperature are decreasing with time. Extrapolating the expansion backward in time, we can reason that the Universe began at a specific instant when all of the material was in a "singularity" at essentially infinite density and temperature. (This is not the only possible conclusion, but other evidence, to be discussed later, strongly supports it.) We call this instant the **big bang**—the initial event that set the Universe in motion—though we will see below that it is not like a conventional explosion. "Big bang" is both the technical and popular name for the current class of theories that deal with the birth and evolution of the Universe.

Expansion Without a Center

Does the observed motion of galaxies away from us imply that we are the center of expansion, and hence in a very special position? Such a conclusion would be highly anti-Copernican: looking at the billions of other galaxies, we see no scientifically based reason for considering our galaxy to be special, in terms of the expansion of the Universe. Historically, too, we have encountered this several times. The Earth is not the only planet, and it isn't the center of the Solar System, just as Copernicus found. The Sun is not the only star, and it isn't the center of the Milky Way Galaxy. The Milky Way Galaxy is not the only galaxy . . . and it probably isn't the center of the Universe.

A different conclusion that is consistent with the data, and also satisfies the Copernican principle, is that there is *no* center—or, alternatively, that all places can claim to be the center. Consider a loaf of raisin bread about to go into the oven. The raisins are spaced at various distances from each other. Then, as the yeast causes the dough to expand, the raisins start spreading apart from each other (Fig. 18–7). If we were able to sit on any one of those raisins, we would see our neighboring raisins move away from us at a certain speed. Note that raisins far away from us recede faster than nearby raisins because there is more dough between them and us, and all of the dough is expanding.

For example, suppose that after one second, each original centimeter of dough occupies 2 cm (Figure 18–7). From our raisin, you will see that another raisin initially 1 cm away has a distance of 2 cm after 1 second. It therefore moved with a speed of 1 cm/sec. A different raisin initially 2 cm away has a distance of 4 cm after 1 second, and moved with a speed of 2 cm/sec. Yet another raisin initially

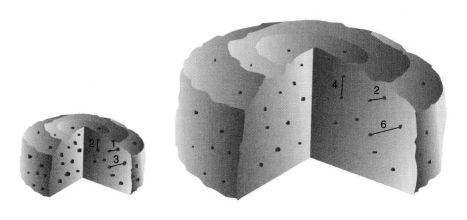

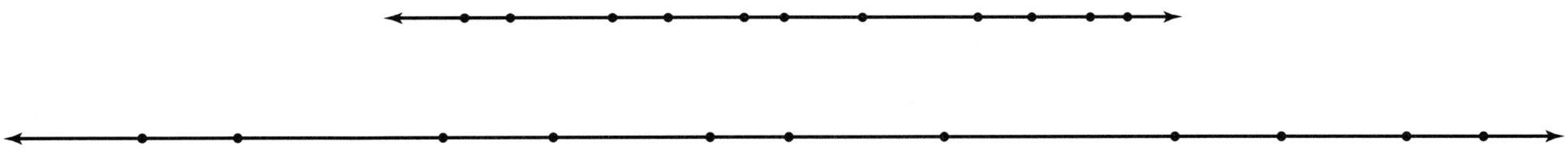

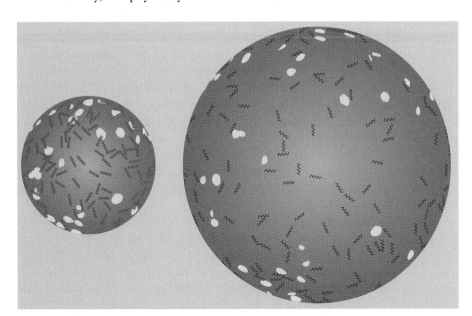

Figure 18-8 A one-dimensional representation of the expanding Universe, in which balls are attached to an infinite, expanding rubber band. No matter which ball you choose to sit on, all others move away from it. Distant balls recede faster than nearby ones, since each bit of rubber expands uniformly and there is more rubber between you and distant balls than between you and nearby ones.

3 cm away has a distance of 6 cm after 1 second, so its speed was 3 cm/sec. We see that speed is proportional to distance ($v \propto d$), as in Hubble's law.

It is important to realize that it doesn't make any difference which raisin we sit on; all of the other raisins would seem to be receding, regardless of which one was chosen. (Of course, any real loaf of raisin bread is finite in size, while the Universe may have no limit so that we would never see an edge.) The fact that all the galaxies are receding from us does not put us in a unique spot in the Universe; there is no center to the Universe. Each observer at each location would observe the same effect.

A convenient one-dimensional analogue is an infinitely long rubber band with balls attached to it (Fig. 18–8). Imagine that we are on one of the balls. As the rubber band is stretched, we would see all other balls moving away from us, with a speed of recession proportional to distance. But again, it doesn't matter which ball we chose as our home.

Another very useful analogy is an expanding spherical balloon (Fig. 18–9). Suppose we define this hypothetical universe to be only the *surface* of the balloon. It has just two spatial dimensions, not three like our real Universe. We can travel forward or backward, left or right, or any combination of these directions— but "up" and "down" (that is, out of, or into, the balloon) are not allowed. All of the laws of physics are constrained to operate along these two directions; even light travels only along the surface of the balloon.

If we put flat stickers on the balloon, they recede from each other as the balloon expands, and flat creatures on them would deduce Hubble's law. A clever observer could also reason that the surface is curved: by walking in one direction, for example, the starting point would eventually be reached. With enough data, the observer might even derive an equation for the surface of the balloon, but it would reveal an *unreachable third spatial dimension*. The observer could conclude that the center of expansion is in this "extra" dimension, which exists only mathematically, not physically!

Figure 18-9 The surface of a spherical balloon provides a good example of a finite, expanding universe. The space inside and outside the balloon is forbidden in this hypothetical universe. Stickers on the balloon represent galaxies, and their distances from each other increase according to Hubble's law as the balloon expands. The center of expansion is the center of the balloon, but this is not part of the surface itself, and hence is outside this hypothetical universe. Note also that while the stickers don't expand, light waves do expand, since they stretch with the rubber—and this fundamentally gives rise to the redshift. (In this simple drawing, the wavelength is longer at the right, though fewer waves are shown in each set of waves displaced.)

It is possible that we live in an analogue of such a spherical universe, but with three spatial dimensions that are physically accessible, and an additional, inaccessible spatial dimension around which space mathematically curves. We will discuss this idea in more detail later.

Besides illustrating Hubble's law and the absence of a unique center, the above analogies accurately reproduce two additional aspects of the Universe. First, according to Einstein's general theory of relativity, which is used to quantitatively study the expansion and geometry of the Universe, it is *space itself* that expands. The dough or the rubber expands, making the raisins, balls, and stickers recede from each other. They do *not* travel *through* the dough or rubber. Similarly, in our Universe, galaxies do not travel through a preexisting space; instead, space itself is expanding. (We sometimes say that the "fabric of space-time is expanding.") In this way, the expansion of the Universe differs from a conventional explosion, which propels material through a preexisting space.

Second, note that the raisins, balls, and stickers themselves don't expand—only the space *between* them expands. (We intentionally didn't draw ink dots on the balloon, because they would expand, unlike stickers.) Similarly, galaxies and other gravitationally bound objects such as stars and planets do not expand: The gravitational force is strong enough to overcome the tendency of space within them to expand. Nor do people expand, because electrical forces (between atoms and molecules) strongly hold us together. Strictly speaking, even most clusters of galaxies are sufficiently well bound to resist the expansion. Only the space between the *clusters* expands, and even in these cases the expansion is sometimes diminished by the gravitational pull between clusters (as in superclusters).

However, electromagnetic waves or photons *do* expand with space; they are not tightly bound objects. Thus, for example, blue photons turn into red photons (Figure 18–9). In fact, this is what actually produces the observed redshift of galaxies. Strictly speaking, the redshift is not a Doppler effect, since nothing is moving *through* the Universe, and the Doppler effect was defined in terms of the motion of an object relative to the waves it emits. The Doppler formula remains valid at low speeds, though, and it is convenient to think about the redshift as a Doppler effect, so we will continue to do so—but be aware of the deeper meaning of redshifts.

⬤ THE AGE OF THE UNIVERSE

The discovery that our Universe had a definite beginning in time, the big bang, is of fundamental importance. The Universe isn't infinitely old. But humans generally have a fascination with the ages of things, from the Dead Sea scrolls to movie stars. Naturally, then, we would like to know how old the Universe itself is!

There are at least two ways in which to determine the age of the Universe. First, the Universe must be at least as old as the oldest objects within it. Thus, we can set a minimum value to the age of the Universe by measuring the ages of progressively older objects within it. For example, the Universe must be at least as old as you—admittedly, not a very meaningful lower limit! More interestingly, it must be at least 200 million years old, since there are dinosaur fossils of that age. Indeed, it must be at least 4.4 billion years old, since Moon rocks and meteorites of that age exist.

The oldest discrete objects whose ages have reliably been determined are globular star clusters in our Milky Way Galaxy (Fig. 18–10). The oldest ones are now thought to be 13–14 billion years old, though the exact values are still controversial. (Through 1996, globular clusters were thought to be about 14–17 billion years old, but the preferred values are now 11–14 billion years.) Also significant is the fact that no discrete objects have been found that are likely to

Figure 18–10 A Hubble image of a globular cluster in the Large Magellanic Cloud, seen as clearly as Earth-bound telescopes image clusters within our own galaxy. Two bright foreground stars also appear.

be much older than the oldest globular clusters. So, a reasonable conclusion is that the age of the Universe is at least 14 billion years, but not much older.

A different method for finding the age of the Universe is to determine the time elapsed since its birth, the big bang. At that instant, all the material of which any observed galaxies consist was essentially at the same location. Thus, by measuring the distance between our Milky Way Galaxy and any other galaxy, we can calculate how long the two have been separating from each other if we know the current recession speed of that galaxy.

At this stage, we have made the simplifying assumption that the recession speed has always been *constant*. So, the relevant expression is distance equals speed multiplied by time: $d = vt$. For example, if we measure a friend's car to be approaching us with a speed of 60 miles/hour, and the distance from his home to ours is 180 miles, we calculate that the journey took 3 hours if the speed was always constant and there were no rest breaks. In the case of the Universe, the amount of time since the big bang assuming a constant speed for any given galaxy is called the **Hubble time** (see *Figure It Out: The Hubble Time*).

If, however, the recession speed was *faster* in the past, then the true age is *less* than the Hubble time. Not as much time had to elapse for a galaxy to reach a given distance from us, compared to the time needed with a constant recession speed. Using the previous example, if our friend started his journey with a speed of 90 miles/hour, and gradually slowed down to 60 miles/hour by the end of the trip, then the average speed was clearly larger than 60 miles/hour, and the trip took fewer than 3 hours. (This is why people often break the posted speed limit!)

Astronomers have generally expected such a decrease in speed because all galaxies are gravitationally pulling on all others, thereby presumably slowing down the expansion rate. In fact, many cosmologists have believed that the deceleration in the expansion rate is such that it gives a true age of exactly two thirds of the Hubble time. This is, in part, a theoretical bias; it rests on an especially pleasing cosmological model. Later in this chapter we will see how attempts have been made to actually measure the expected deceleration.

Figure It Out

The Hubble Time

There is a simple relationship between the value of Hubble's constant and the time elapsed since the big bang, assuming that the recession speed of any given galaxy has always been constant. For galaxies, we know that Hubble's law holds: $v = H_0 d$. Rearranging this relation, we have $d = v/H_0 = vT_0$, where we define T_0 (the "Hubble time") to be $1/H_0$, the reciprocal or inverse of Hubble's constant. Note that this equation looks just like the relation $d = vt$ for the distance travelled by an object at constant speed.

Suppose Hubble's constant is 50 km/sec/Mpc. What is the corresponding Hubble time? This is mostly an exercise in keeping track of units. First, from Appendix 2 we find that 1 pc is approximately equal to 3.09×10^{18} cm, so 1 Mpc $\approx 3.09 \times 10^{24}$ cm $= 3.09 \times 10^{19}$ km. We then convert $H_0 = 50$ km/sec/Mpc $= 50$ km/(Mpc sec) to units of inverse seconds: $H_0 = (50$ km/(Mpc sec))/$(3.09 \times 10^{19}$ km/Mpc) $= 1.62 \times 10^{-18}$/sec. Thus, $T_0 = 1/H_0 = 6.2 \times 10^{17}$ sec. But one year (that is, 365.25 days at 86,400 seconds per day) is about 3.15×10^7 sec, so $T_0 \approx (6.2 \times 10^{17}$ sec)/$(3.15 \times 10^7$ sec/year) $= 19.7 \times 10^{10}$ years. This is roughly 20 billion years! But if H_0 were actually twice as large, 100 km/sec/Mpc, then the corresponding Hubble time would be half as large, about 10 billion years. This illustrates why astronomers want to determine the value of Hubble's constant accurately.

Note that as the Universe ages, T_0 increases, and so H_0 decreases—the value of Hubble's constant decreases with time. However, at a given time, its value is thought to be the same everywhere in the Universe—hence the name "Hubble's constant."

Figure 18-11 The period-luminosity relation of Cepheid variable stars. Long-period Cepheids have higher average luminosities than short-period Cepheids.

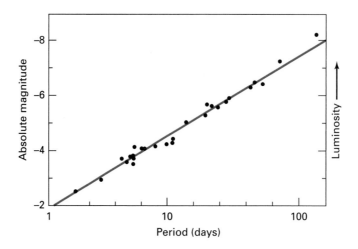

The Quest for Hubble's Constant

To determine the Hubble time, we must measure Hubble's constant, H_0. This can be done if we know the distance (d) and recession speed (v) of another galaxy, since rearrangement of Hubble's law tells us that $H_0 = v/d$. Many galaxies at different distances should be used, so that an average can be taken.

The recession speed is easy to measure from a spectrum of the galaxy and the Doppler formula. We can't use galaxies within our own Local Group (like the Andromeda galaxy, M31), however, since they are bound to the group by gravity and are not expanding away.

Galaxy distances are notoriously hard to determine, and this leads to large uncertainties in the derived value of Hubble's constant. We can't use triangulation because galaxies are much too far away. In principle, the distance of a galaxy can be determined by measuring the apparent angular size of an object (such as a nebula) within it, and comparing it with an assumed physical size. But this generally gives only a crude estimate of distance, because the physical sizes of different objects in a given class are not uniform enough.

More frequently, we measure the apparent brightness of a star, and compare this with its intrinsic brightness (luminosity) to determine the distance. This is similar to how we judge the distance of an oncoming car at night: We intuitively use the inverse-square law of light (discussed in Chapter 10) when we see how bright a headlight appears to be. We must be able to recognize that particular type of star in the galaxy, and we assume that all stars of that type are "standard candles"—that is, they all have the same luminosity.

Historically, the best such candidates have been the Cepheid variables, at least in relatively nearby galaxies. Though not all of uniform luminosity, they do obey a period-luminosity relation (Fig. 18–11), as shown by Henrietta Leavitt in 1912. Thus, if the variability period of a Cepheid is measured, its average luminosity can be read directly off the graph. But individual stars are difficult to see in distant galaxies: They merge with other stars when viewed with ground-based telescopes (Fig. 18–12). Other objects that have been used include luminous nebulae, globular star clusters, and novae—but all of them have uncertainties. They aren't excellent standard candles, or are difficult to see in ground-based images, or depend on the assumed distances of some other galaxies.

Even entire *galaxies* can be used, if we determine their luminosity from other properties. The luminosity of a spiral galaxy is correlated with how rapidly it rotates, for example. Also, the brightest galaxy in a large cluster has a roughly standard luminosity. Again, however, significant uncertainties are associated with these techniques, or they depend on the proper calibration of a few key galaxies.

Figure 18-12 M83, a spiral galaxy in the constellation Centaurus, is relatively near Earth (only about 12 million light-years away), yet individual stars are difficult to discern in ground-based photographs because atmospheric turbulence appears to blur images and merge stars.

Throughout the 20th century, many astronomers have attempted to measure the value of Hubble's constant. Edwin Hubble himself initially came up with 550 km/sec/Mpc, but several effects that were at that time unknown to him conspired to make this much too large. From the 1960s to the early 1990s, the most frequently quoted values were between 70 and 50 km/sec/Mpc, due largely to the painstaking work of Allan Sandage (Fig. 18–13), a disciple of Edwin Hubble himself. This yields a Hubble time of about 14–20 billion years, the probable maximum age of the Universe. The true expansion age could therefore be 9–13 billion years, if it is only two-thirds of the Hubble time, as many cosmologists have been prone to believe. These lower numbers may give rise to a discrepancy if the globular clusters are 11–14 billion years old. There is certainly an age crisis if the clusters are 14–17 billion years old, as was thought until the mid-1990s.

But some astronomers obtained considerably larger values for H_0, up to 100 km/sec/Mpc. This value gives a Hubble time of about 10 billion years, and two-thirds of it is only 6.7 billion years. Even the recently revised ages of globular star clusters are substantially greater (11–14 billion years), leading to a sharp age crisis. The various teams of astronomers who get different answers all claim to be doing careful work, but there are many potential hidden sources of error, and the assumptions might not be completely accurate. The debate over the value of Hubble's constant is often heated, and sessions of scientific meetings at which the subject is discussed are well attended.

Note that the value of Hubble's constant also has a broad effect on the perceived size of the observable Universe, not just its age. For example, if Hubble's constant is 50 km/sec/Mpc, then a galaxy whose recession speed is measured to be 5000 km/sec would be at a distance $d = v/H_0 = (5000 \text{ km/sec})/(50 \text{ km/sec/Mpc}) = 100$ Mpc. On the other hand, if Hubble's constant is actually 100 km/sec/Mpc, then the same galaxy is only half as distant: $d = (5000 \text{ km/sec})/(100 \text{ km/sec/Mpc}) = 50$ Mpc.

The aptly named Hubble Space Telescope was expected to provide a major breakthrough in the field. It would obtain distances to many important galaxies, mostly through the use of Cepheid variable stars (Fig. 18–14). Indeed, very large amounts of telescope time were to be dedicated to this "Key Project" of measuring galaxy distances and deriving Hubble's constant. But astronomers had to wait a long time, even after the launch of the Hubble Space Telescope in April 1990, because the primary mirror's spherical aberration made it too difficult to detect and reliably measure Cepheids in the chosen galaxies.

Finally, in 1994, the Hubble "Key Project" team announced their first results, based on Cepheids in only one galaxy (Fig. 18–15). Their value of H_0 was about 80 km/sec/Mpc, higher than many astronomers had previously thought. This implied that the Hubble time was 12 billion years; the Universe could be no older, but perhaps significantly younger (down to 8 billion years) if the expansion decelerates with time. Being less than 14–17 billion years (the ages pre-

> Recall that 1 Mpc = 1 megaparsec, a million parsecs, which is 3.26 million light-years.

Figure 18–13 Allan Sandage *(right)* of the Carnegie Observatories in Pasadena, California, who has devoted much of his career to the measurement of Hubble's constant. He is seen here with Maarten Schmidt of Caltech, the first person to realize that the spectra of quasars are highly redshifted.

| April 23 | May 4 | May 9 | May 16 | May 20 | May 31 |

Figure 18–14 Cepheid variable stars in the galaxy M100, as observed with the Hubble Space Telescope. They have been used to calibrate the cosmic distance scale through the measurement of Hubble's constant.

Figure 18–15 A Hubble Space Telescope image of the galaxy M100 in the Virgo Cluster, the first galaxy whose distance was determined by the Key Project team that used Cepheid variable stars.

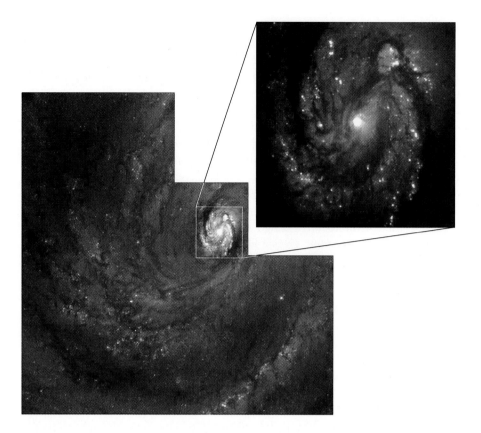

ferred for globular star clusters at that time), this brought the age crisis to great prominence among astronomers, who shared it with the public. How could the Universe, as measured with the mighty Hubble Space Telescope, be younger than its oldest contents? There were several dramatic headlines in the news (Fig. 18–16).

Admittedly, the Hubble team's quoted value of H_0 had an uncertainty of 17 km/sec/Mpc, meaning that the actual value could be between about 63 and 97 km/sec/Mpc. Thus, the Universe could be as old as 15–16 billion years, especially if there has been little deceleration. The ages of globular clusters were uncertain as well, so it was not entirely clear that the age crisis was severe. But, as is often the case with newspaper and popular magazine articles, these subtleties are ignored or barely mentioned; only the "bottom line" gets reported, especially if it's exciting.

In mid-1999, the Hubble team announced a final answer, which was based on Cepheid variables in 18 galaxies. Their preferred value of H_0 decreased from 80 to 70 km/sec/Mpc, with an uncertainty of about 7 km/sec/Mpc. Also, by this time additional techniques for measuring H_0 had been developed or perfected by other astronomers, and many of them were giving values of around 65 km/sec/Mpc, perhaps even as low as 58 km/sec/Mpc. A "best bet" estimate of $H_0 = 65$ km/sec/Mpc seems reasonable, and we will use it in the remainder of this book.

A value of 65 km/sec/Mpc for Hubble's constant means that the Universe has been expanding for 15 billion years, if there is no deceleration. By assuming only a small amount of deceleration (not as much as many theorists would have preferred), the Hubble team announced a "final" expansion age of 12 billion years for the Universe. Moreover, the preferred ages of globular clusters had recently shifted from 14–17 billion years to only 11–14 billion years, based on accurate new parallaxes of stars from the Hipparcos satellite and on some new theoretical work. This meant that the age discrepancy had subsided to some extent, but did not fully disappear if the globular clusters are actually as old as 13–14 billion years.

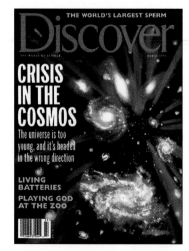

Figure 18–16 The headline of *Discover* magazine after the 1994 announcement of an expansion age of only 8 to 12 billion years for the Universe. At that time, globular star clusters were thought to be 14 to 17 billion years old.

But on what basis was the amount of deceleration estimated? We will discuss this more fully later in this chapter, with the surprising result that the assumed deceleration may have been erroneous. Instead, the expansion rate of the Universe might actually be *increasing* with time! This exciting discovery, announced in early 1998 and known as the "accelerating universe," is still somewhat controversial. It implies some very intriguing, but also troubling, new aspects to the nature and evolution of the Universe. If correct, however, it fully resolves the age crisis: We find that the expansion age of the Universe is 14 billion years, entirely consistent with the 11–14 billion year estimated ages of globular clusters.

Deviations from Uniform Expansion

A major problem with using relatively nearby galaxies for measurements of Hubble's constant is that proper corrections must be made for deviations from the **Hubble flow** (the assumed uniform expansion of the Universe). As we discussed in Chapter 16, there are concentrations of mass (clusters and superclusters) in certain regions, and large voids in others, so a specific galaxy may feel a greater pull in one direction than in another direction. It will therefore be pulled *through* space (relative to the Hubble flow), and its apparent recession speed may be affected. Though the galaxy's recession speed is easy to measure from a spectrum, it might not represent the true expansion of space!

For example, the Virgo Cluster of galaxies (Fig. 18–17) is receding from us more slowly than it would if it had no mass: The Milky Way Galaxy is "falling" toward the Virgo Cluster, thereby counteracting part of the expansion of space. Such gravitationally induced **peculiar motions** are typically a few hundred kilometers per second, but can reach as high as 1000 km/sec. Their exact size is difficult to determine without detailed knowledge of the distribution of matter in the Universe. In the case of the Virgo Cluster, the average observed recession speed is about 1100 km/sec, and the peculiar motion is thought to be about 300 km/sec, but this is uncertain. Errors in the adopted "true" recession speed directly affect the derived value of Hubble's constant.

A surprising discovery was that even the Virgo Cluster is moving with respect to the average expansion of the Universe. Some otherwise unseen "Great Attractor" is pulling the Local Group, the Virgo Cluster, and even the much larger "Hydra-Centaurus Supercluster" toward it. Redshift measurements by a team of astronomers informally known as the "Seven Samurai" showed the location of the giant mass that must be involved. It is about three times farther from us than the Virgo Cluster, and includes tens of thousands of galaxies or their equivalent mass.

Measurements of still more-distant galaxies avoid the problem of peculiar motions when trying to determine Hubble's constant. For example, compared with galaxies having recession speeds of 15,000–30,000 km/sec, the peculiar motions are negligible. So, measurements of their distances, when combined with their recession speeds, can yield an accurate value of H_0. The trick is to find their distances—and this can't be done directly with Cepheid variable stars because they aren't intrinsically bright enough.

In the 1990s, a remarkably reliable method was developed for measuring the distances of very distant galaxies. It is based on Type Ia supernovae, which are exploding stars that result from a nuclear runaway in a white dwarf (see Chapter 13). When they reach their peak power, these objects shine with the luminosity (intrinsic brightness) of about 10 billion Suns, or about a million times more than Cepheid variables. So, they can be seen at very large distances, 1000 times greater than Cepheid variables (Fig. 18–18).

Most observed Type Ia supernovae are found to have nearly the same peak luminosity, as expected since the exploding white dwarf is thought to always have

Figure 18–17 The central part of the Virgo Cluster of galaxies. This cluster is not receding from the Milky Way Galaxy as quickly as it would if galaxies had no mass and didn't gravitationally attract each other.

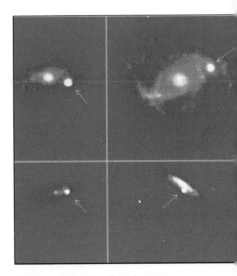

Figure 18–18 Type Ia supernovae, marked with arrows, in four different galaxies, are shown in this false-color image. Near its peak, such a supernova can be comparable in brightness to the entire galaxy in which it occurred.

the same mass (the Chandrasekhar limit). Type Ia supernovae are therefore very good "standard candles" for measuring distances. (They do show small variations in peak luminosity, but we have ways of taking this into account—essentially like reading the wattage label on a light bulb.) By comparing the apparent brightness of a faint Type Ia supernova in a distant galaxy with the supernova's known luminosity, and by using the inverse-square law of light, we obtain the distance of the supernova, and hence of the galaxy in which it exploded (see Figure 18–18).

Of course, to successfully apply this method, we need to know the peak luminosity of a Type Ia supernova. But this can be found by measuring the peak apparent brightness of a supernova in a relatively nearby galaxy—one whose distance can be measured by other techniques, such as Cepheid variable stars. So, an important part of the Hubble Key Project was to find the distances of galaxies in which Type Ia supernovae had previously been seen, and therefore to calibrate the peak luminosity of Type Ia supernovae. By mid-1999, reliable distances to half a dozen such galaxies had been measured. Indeed, our preferred value of $H_0 = 65$ km/sec/Mpc is dependent on this work.

● THE GEOMETRY AND FATE OF THE UNIVERSE

We have seen that to determine the age of the Universe, its expansion history (in addition to Hubble's constant) must be known. It turns out that, under certain assumptions, the expansion history is closely linked to the eventual fate of the Universe as well as to its overall (large-scale) geometry.

Mathematically, we use Einstein's general theory of relativity to study the expansion and overall geometry of the Universe. Since matter produces space-time curvature (as we have seen when studying black holes), we expect the average density to affect the overall geometry of the Universe. The average density should also affect the way in which the expansion changes with time: High densities are able to slow down the expansion more than low densities, due to the gravitational pull of matter. Thus, the average density appears to be the most important parameter governing the Universe as a whole.

To simplify the equations and achieve reasonable progress, we assume the **cosmological principle:** On the largest size scales, the Universe is **homogeneous** and **isotropic.** Homogeneous means that it has the same average density everywhere at a given time (though the density *can* change with time). Isotropic means that it looks the same in all directions—there is no preferred axis along which most of the galaxies are lined up, for example (Fig. 18–19). Note that we can check for isotropy only from our own position in space. However, for even greater simplicity we could suppose that the Universe looks isotropic from *all* points. (In this case of isotropy everywhere, the Universe is also necessarily homogeneous.)

The cosmological principle is basic to most big-bang theories. It is clearly incorrect on small scales: A human, Earth, the Solar System, the Milky Way Galaxy, and our Local Group of galaxies have a far higher density than average. Even the supercluster of galaxies to which the Milky Way belongs is somewhat denser than average. However, averaged over volumes about a billion light-years in diameter, the cosmological principle does appear to hold. The largest structures in the Universe seem to be superclusters and huge voids, but these are only a few hundred million light-years in diameter. Moreover, as we will see in Chapter 19, the strongest evidence comes from the "cosmic background radiation" that pervades the Universe: It looks the same in all directions, and it comes to us from a distance of about 14 billion light-years.

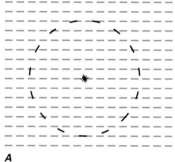

A

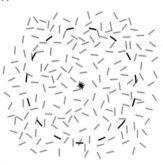

B

Figure 18–19 Illustration of a homogeneous universe that is not isotropic *(A)*, and a universe that is both homogeneous and isotropic *(B)*. The average density of galaxies in a sufficiently large volume is the same in both universes, but in *(A)* the galaxies are all lined up in the same direction. The dashed circle is the observable part of the universe.

Another assumption we will make, at least temporarily, is that there are no long-range forces other than gravity, and that only "normal" matter and energy (with an attractive gravitational force) play a significant role. Prior to Edwin Hubble's discovery that the Universe is expanding, most people thought the Universe is static (neither expanding nor contracting), which in some ways is aesthetically pleasing. Einstein knew that gravity should make the Universe contract, so in 1917 he postulated a long-range repulsive force, sort of a cosmic "antigravity," with a specific value that made the Universe static (Fig. 18–20). This "fudge factor" became known as the **cosmological constant,** denoted by the Greek capital letter Λ (lambda).

In 1929, when Hubble discovered the expansion of the Universe, the entire physical and philosophical motivation for the cosmological constant vanished. The Universe wasn't static, and no forces are needed to make it expand. After all, the Universe could have simply begun its existence in an expanding state, and is still coasting. Einstein renounced the cosmological constant, and called it the "biggest blunder of his life": He could have predicted that the Universe is dynamic, rather than static, had he not insisted on introducing the cosmological constant.

However, the concept of the cosmological constant itself should perhaps not be considered a blunder. In a sense, it is just a generalization of Einstein's relativistic equations for the Universe. The blunder was in supposing that the cosmological constant has the precise value needed to achieve a static universe—especially since this turns out to be an unstable mathematical solution (slightly perturbing the Universe leads to expansion or collapse). Nevertheless, it isn't clear what could physically produce a nonzero cosmological constant, and the simplest possibility is that the cosmological constant is zero ($\Lambda = 0$). Since there has generally been no strong observational evidence for a nonzero cosmological constant, astronomers have long assumed that its value is indeed zero. This is what we will initially assume here, too—but later in this chapter we will discuss exciting new evidence that the cosmological constant isn't zero after all.

Three Kinds of Possible Universes

Given the assumptions of the cosmological principle and no long-range antigravity, and also that no new matter or energy are created after the birth of the Universe, the general theory of relativity allows only three possibilities. These are known as "Friedmann universes" in honor of Alexander Friedmann, who was the first to derive them mathematically. In each one the expansion decelerates with time, but the ultimate fate (that is, whether the expansion ever stops and reverses) depends on the overall *average density of matter* relative to a specific **critical density.** If we define the average matter density divided by the critical density to be Ω_M, where "Ω" is the Greek capital letter Omega and the subscript "M" stands for "matter," then the three possible universes correspond to the cases where this ratio is greater than one, equal to one, and less than one. (Details are explored in *Figure It Out: The Critical Density and Ω_M.*)

The separation between any two galaxies versus time is shown in Fig. 18–21 for the three types of universes. It is best to choose galaxies in different clusters (or even different superclusters, to be absolutely safe), since we don't want them to be bound together by gravity. This galaxy separation is often called the "scale factor" of the Universe; it tells us about the expansion of the Universe itself.

If $\Omega_M > 1$ (that is, the average density is above the critical density), galaxies separate progressively more slowly with time, but they eventually turn around and approach each other (that is, the recession speed becomes negative), ending

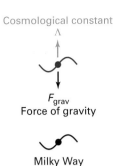

Figure 18–20 Albert Einstein introduced the cosmological constant, Λ (Lambda), in order to produce a static universe. He knew that gravity always pulls, so there had to be a counteracting repulsive force to prevent collapse. In this diagram, another galaxy is pulled toward the Milky Way Galaxy by their mutual gravitational attraction, but pushed apart with equal force provided by the cosmological constant.

REDSHIFT
http://www.harcourtcollege.com/astro/cosmos/rsce

Figure It Out

The Critical Density and Ω_M

The "critical density," ρ_{crit}, is defined to be that density of matter that would allow the Universe to expand forever, but only just barely. Its formula is $\rho_{crit} = 3H_0^2/(8\pi G)$, where H_0 is Hubble's constant and G is Newton's constant of universal gravitation. If the average matter density of the Universe (ρ_{ave}) is greater than, equal to, or less than this value, then Ω_M (Omega of the matter) is greater than, equal to, or less than 1, respectively, since we have defined $\Omega_M = \rho_{ave}/\rho_{crit}$.

For $H_0 = 65$ km/sec/Mpc, we find that $\rho_{crit} = 8 \times 10^{-30}$ g/cm³, which is the equivalent of only about 5 hydrogen atoms per cubic meter of space! Clearly, our local surroundings are much denser, but this isn't relevant. Only the average density of the Universe as a whole should be compared with ρ_{crit} to determine the value of Ω_M. In fact, we find that Ω_M is almost certainly less than 1. This means that on large scales the Universe is remarkably empty.

Note that as the Universe ages, the average density of matter in the Universe decreases. However, if the value of Ω_M initially exceeds 1, it does not later drop to 1 or less. This is because the value of H_0 decreases with time, and hence the critical density also decreases. The ratio of the average density to the critical density, Ω_M, remains either above 1 or below 1 (or is exactly equal to 1) forever.

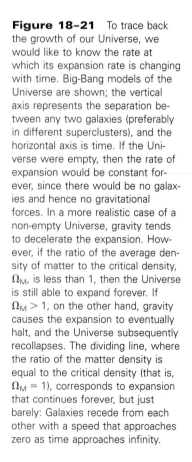

Figure 18–21 To trace back the growth of our Universe, we would like to know the rate at which its expansion rate is changing with time. Big-Bang models of the Universe are shown; the vertical axis represents the separation between any two galaxies (preferably in different superclusters), and the horizontal axis is time. If the Universe were empty, then the rate of expansion would be constant forever, since there would be no galaxies and hence no gravitational forces. In a more realistic case of a non-empty Universe, gravity tends to decelerate the expansion. However, if the ratio of the average density of matter to the critical density, Ω_M, is less than 1, then the Universe is still able to expand forever. If $\Omega_M > 1$, on the other hand, gravity causes the expansion to eventually halt, and the Universe subsequently recollapses. The dividing line, where the ratio of the matter density is equal to the critical density (that is, $\Omega_M = 1$), corresponds to expansion that continues forever, but just barely: Galaxies recede from each other with a speed that approaches zero as time approaches infinity.

in a hot "big crunch." (Some astronomers also jokingly call it a "gnab gib," which is "big bang" backwards!) A good analogy is a ball thrown up with a speed less than Earth's escape speed; it eventually falls back down. It is conceivable that another big bang subsequently occurs, resulting in an "oscillating universe," but we have little confidence in this hypothesis since the laws of physics as currently stated cannot be traced through the big crunch.

If $\Omega_M = 1$ (that is, the average density is exactly equal to the critical density), galaxies separate more and more slowly with time, but as time approaches infinity the recession speed approaches zero. Thus, the Universe will expand forever, though just barely. The relevant analogy is a ball thrown up with a speed equal to Earth's escape speed; it continues to recede from Earth ever more slowly, and it stops when time reaches infinity. This turns out to be the type of universe predicted by most "inflation theories" (which we will study in Chapter 19).

If $\Omega_M < 1$ (that is, the average density is below the critical density), galaxies separate more and more slowly with time, but as time approaches infinity the re-

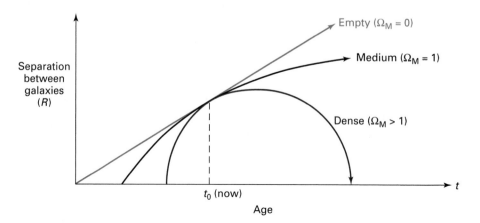

cession speed (for a given pair of galaxies) approaches a constant, nonzero value. Thus, the Universe will easily expand forever. Once again using our ball analogy, it is like a ball thrown up with a speed greater than Earth's escape speed; it continues to recede from Earth ever more slowly, but it never stops receding.

These three kinds of universes have different overall geometry. The $\Omega_M = 1$ case is known as a **flat universe** or a **critical universe.** It is described by "Euclidean geometry"—that is, the geometry worked out first by the Greek mathematician Euclid in the third century B.C. In particular, Euclid's "fifth postulate" is satisfied: Given a line and a point not on that line, only one unique parallel line can be drawn through the point (Fig. 18–22*A*). Such a universe is spatially flat, formally infinite in volume, and barely expands forever. Its age is exactly two thirds of the Hubble time, $(2/3)/H_0 = (2/3)T_0$.

In the $\Omega_M > 1$ universe, Euclid's fifth postulate fails in the following way: Given a line and a point not on that line, *no* parallel lines can be drawn through the point (Fig. 18–22*B*). Such a universe has positive spatial curvature, is finite ("closed") in volume, but has no boundaries (edges) like those of a box. Its fate is a hot "big crunch." Generally known as a **closed universe,** it is also sometimes called a "spherical" or "positively curved" universe. Its age is less than two thirds of the Hubble time.

Finally, in the $\Omega_M < 1$ universe, Euclid's fifth postulate fails in the following way: Given a line and a point not on that line, many (indeed, infinitely many) parallel lines can be drawn through the point (Fig. 18–22*C*). Such a universe has negative spatial curvature, is infinite ("open") in volume, and easily expands forever. Generally known as an **open universe,** it is also sometimes called a "hyperbolic" or "negatively curved" universe. Its age is between $(2/3)T_0$ and T_0 (the latter extreme only if $\Omega_M = 0$).

Note that in some texts and magazine articles, the $\Omega_M = 1$ universe is called "closed," but only because it is *almost* closed. It actually represents the *dividing line* between "open" and "closed."

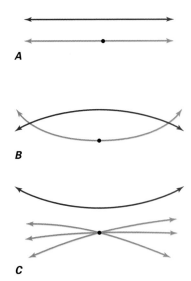

Figure 18-22 *(A)* Euclid's "fifth postulate" for spatially flat geometry states that given a line and a point not on that line, only one unique parallel line can be drawn through the point. *(B)* In a positively curved geometry, no parallel lines can be drawn through the point. *(C)* In a negatively curved geometry, many parallel lines can be drawn through the point. (Of course, we cannot accurately draw curved geometries on flat sheets of paper, so these diagrams are only meant to convey the general idea.)

Two-Dimensional Analogues

It is useful to consider analogues to the above universes, but with only two spatial dimensions (Fig. 18–23). The flat universe is like an infinite sheet of paper. One property is that the sum of the interior angles of a triangle is always 180°, regardless of the shape and size of the triangle. Moreover, the area A of a circle of radius R is proportional to R^2 (that is, $A = \pi R^2$). This relation can be measured by scattering dots uniformly (homogeneously) across a sheet of paper, and seeing that the number of dots enclosed by a circle grows in proportion to R^2.

The positively curved universe is like the surface of a sphere. The sum of the interior angles of a triangle is always greater than 180°. For example, a triangle consisting of a segment along the equator of the Earth, and two segments going up to the north pole at right angles from the ends of the equatorial segment, clearly has a sum greater than 180°. Moreover, the area of a circle of radius R falls short of being proportional to R^2. If the sphere is uniformly covered with dots, the number of dots enclosed by a circle grows more slowly with R than in flat space because a flattened version of the sphere contains missing slices.

The negatively curved universe is somewhat like the surface of an infinite horse's saddle or potato chip. These analogies are not perfect because a horse's saddle (or potato chip) embedded in a universe with three spatial dimensions is not isotropic; the saddle point, for example, can be distinguished from other points. The sum of the interior angles of a triangle is always less than 180°. The area of a circle of radius R is more than proportional to R^2. If the saddle is ho-

Figure 18-23 *(Top row)* Two-dimensional analogues to three-dimensional space. *(A)* A flat universe is like an infinite sheet of paper; the laws of Euclidean geometry are obeyed. For example, the sum of the interior angles of a triangle is equal to 180°. *(B)* A positively curved universe is like the surface of a sphere; the sum of the interior angles of a triangle is larger than 180°. *(C)* A negatively curved universe is like the surface of a saddle or potato chip in some respects; the sum of the interior angles of a triangle is smaller than 180°. *(Second row)* Consider dots uniformly distributed on each two-dimensional surface. If we count the total number of dots within circles of larger and larger radius R in flat space, we find that the total number of dots grows exactly in proportion to the area, $A = \pi R^2$. But in positively curved space there are fewer dots within a given radius, and in a negatively curved space there are more dots within a given radius. *(Third row)* It is easiest to visualize these relationships by trying to flatten out the curvature in each case. The positively curved surface, when flattened, gives rise to empty sections, while the flattened negatively curved surface has folds with extra dots. *(Bottom row)* The resulting distribution of dots in the flattened versions shows a relative deficit of dots at large radii in the positively curved surface and a relative excess of dots in the negatively curved surface.

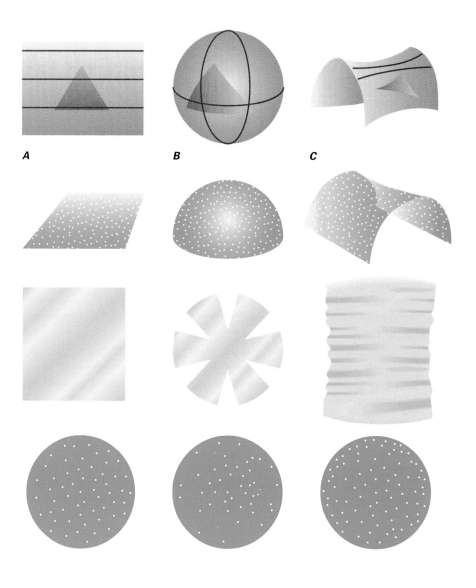

mogeneously covered with dots, the number of dots enclosed by a circle grows more quickly with R than in flat space because a flattened version of the saddle contains extra wrinkles.

With three spatial dimensions, we can generalize to the growth of volumes V with radius R. In a flat universe, the volume of a sphere is proportional to R^3 [that is, $V = (4/3)\pi R^3$]. In a positively curved universe, the volume of a sphere is not quite proportional to R^3. In a negatively curved universe, the volume of a sphere is more than proportional to R^3.

Mathematicians have pointed out that relaxing the assumption of perfect isotropy allows one to avoid the conclusion that a flat or negatively curved universe is necessarily infinite in size. The shape might be like that of a torus (doughnut) embedded in four spatial dimensions. Such a universe still looks quite isotropic; no existing measurements would be able to discern the slight deviations from isotropy. However, current data suggest that the scale over which such a universe wraps around is probably larger than the distance to which we can see, even in principle. Thus, it is not clear how we can test the hypothesis that the Universe (if flat or negatively curved) is finite. Moreover, a flat or negatively curved (but finite) universe still expands forever, just as in the case of an infinite universe. Though quite intriguing, in this book we will not further consider the possibility of a flat or negatively curved, but finite, universe.

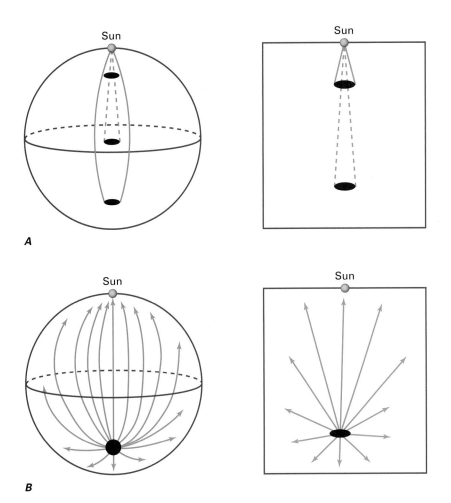

Figure 18-24 *(A)* In flat space *(right),* distant galaxies look much smaller than a nearby galaxy; in fact, the angular size is inversely proportional to distance. In positively curved space *(left),* on the other hand, the angular size is somewhat larger than expected when compared with flat space. Extremely distant galaxies can even start looking larger again, due to the curved trajectory of light rays. *(B)* In flat space *(right),* distant galaxies look very faint due to the inverse-square law of light. But in positively curved space *(right),* the light rays don't diverge as fast, and even begin to converge again at sufficiently large distances; thus, distant objects look brighter than expected.

What Kind of Universe Do We Live In?

How do we go about determining to which of the above possibilities our Universe corresponds? There are a number of different methods. Perhaps most obvious, we can measure the average density of matter, and compare it with the critical density. The value of Ω_M (again, the ratio of the average matter density to the critical density) is greater than 1 if the Universe is closed, equal to 1 if the Universe is flat (critical), and less than 1 if the Universe is open. Or, we can measure the expansion rate in the distant past (preferably at several different epochs), compare it with the current expansion rate, and calculate how fast the Universe is decelerating. This can be done by looking at very distant galaxies, which are seen as they were long ago, when the Universe was younger.

We can also examine geometrical properties of the Universe to determine its overall curvature. For example, in principle we can see whether the sum of the interior angles of an enormous triangle is greater than, equal to, or less than 180°. This is not very practical, however, since we cannot draw a sufficiently large triangle. Or, we can see whether "parallel lines" ever meet—but again, this is not practical, since we cannot reach sufficiently large distances.

A better geometrical method is to measure the angular sizes of galaxies as a function of distance. High-redshift galaxies of fixed physical size will appear larger in angular size if space has positive curvature than if it has zero or negative curvature, because light rays diverge more slowly in a closed universe than in a flat universe or in an open universe (Fig. 18–24*A*). Or, we could instead look at the apparent brightness of objects as a function of distance. High-redshift objects of fixed luminosity (intrinsic brightness) will appear brighter if space has positive curvature than if it has zero or negative curvature; again, light rays diverge more slowly in a closed universe (Fig. 18–24B).

We might also count the number of galaxies as a function of distance to see how volume grows with radius. This is analogous to the measurement of area in two-dimensional universes, as explained in the preceding section. If space is flat, volume is exactly proportional to R^3; thus, doubling the surveyed distance (R) should increase the number of galaxies by a factor of 8. On the other hand, if space has positive curvature, the factor will be smaller than 8, while if space has negative curvature, the factor will be larger than 8.

A completely different technique is to measure the relative abundances (proportions) of the lightest elements and their isotopes, which were produced shortly after the big bang. As discussed in Chapter 19, these depend on the value of Ω_M.

Some astronomers measure the motions of galaxies and clusters of galaxies through the Universe (that is, relative to the smooth Hubble flow). These are produced by the gravitational tug of other clusters, and hence provide a measure of the mass and distribution of clumped matter (both visible and dark).

There are many other, related techniques. For a number of reasons, all of them (including those listed here) are difficult and uncertain. One problem is that local, dense objects (planets, stars, galaxies, etc.) produce spatial curvature larger than the gradual, global effect that we seek. Moreover, we know that space is nearly flat, so to detect the slight average curvature one needs to look very far, and this is difficult. Another problem is that when counting galaxies or determining the average density, how does one know that a representative volume was chosen? After all, there are inhomogeneities on large scales, such as superclusters of galaxies.

A major difficulty is that galaxies evolve with time. So, we cannot assume that a known type of galaxy has a constant physical size (when measuring angular sizes as a function of distance), and the quantitative evolution of physical size is difficult to predict. We also do not know how the luminosity of a typical galaxy evolves with time, yet galaxies certainly do evolve (Fig. 18–25). Counts of galaxies to various distances might be dominated by intrinsic luminosity differences, rather than by the volume of space surveyed.

Clusters of galaxies also evolve with time, and are therefore subject to similar uncertainties. Determining the total amount of matter in a cluster is difficult; much of the matter is dark, and can be detected only through its gravitational effects. In addition, we do not know how much dark matter is spread somewhat uniformly, rather than clumped in clusters and superclusters, and this affects the calculated average density.

Of course, an observational problem is that distant objects appear small and faint, and are therefore subject to considerable measurement uncertainties.

At the present time, in mid-2000, there is consensus that $\Omega_M \approx 0.3$ (almost certainly larger than 0.1 but smaller than 0.5). If true, the Universe will expand forever and is spatially infinite (or perhaps has the shape of a finite torus in four-dimensional hyperspace). However, these conclusions are based primarily on studies of clusters of galaxies—their motions, masses, and so on. Uniformly distributed matter and other possible effects, such as the cosmological constant, are not taken into account. There are few tests of deceleration, or of overall geometry.

Yet there are theoretical reasons (see Chapter 19) for believing that Ω_M is exactly 1, if it is known to be at least relatively close to 1, such as the value of 0.3 currently favored. If it were not *exactly* 1 initially, it should have deviated *very* far from 1 (for example, 10^{-7} or 10^{15}) by the present time. This behavior is like nudging a pencil balanced on its tip: It quickly falls to the surface.

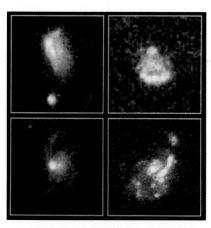

Figure 18–25 Typical galaxies photographed by the Hubble Space Telescope, at redshifts between 0.3 and 0.7 (that is, roughly 3.5 to 7 billion light-years away). They look peculiar relative to nearby galaxies, suggesting that galaxies evolve over time in shape, size, and luminosity.

◖▬ MEASURING THE EXPECTED DECELERATION

Perhaps the most direct way of determining the expected deceleration of the Universe is to measure the expansion rate as a function of time by looking at very distant objects. As discussed earlier, the separation between two clusters of galaxies varies with time in different ways, depending on the deceleration rate of the Universe (Figure 18–21). For any of the curves, the slope at a given time is the expansion rate at that time. (Hubble's constant itself is the slope divided by the separation between galaxies at that time.)

Observationally, we plot the measured distance (from the inverse-square law) versus the recession speed (from the redshift) for a set of objects; see Fig. 18–26. At small distances, Hubble's law holds: Speed is directly proportional to distance ($v = H_0 d$). At very large distances, however, there should be deviations from this. For a given distance, the speed should be higher if Ω_M is greater than 1 than if Ω_M is less than 1, because the Universe used to be expanding much more quickly if Ω_M is greater than 1. Or, for a given speed, the distance should be larger if Ω_M is less than 1 than if Ω_M is greater than 1, because the expansion didn't slow down as much if Ω_M is less than 1. The distance should be even larger if "Ω_M is less than 0" (which is physically impossible, since matter exists), because the expansion of the Universe will have accelerated, pushing objects even farther away.

What kinds of objects can be seen at sufficiently large distances to accomplish this? We could try galaxies—but they evolve with time due to mergers, bursts of star formation, and other processes that vary from one galaxy to another and are not well understood. Clusters of galaxies also evolve with time in ways that are difficult to predict accurately. Some astronomers hoped that quasars would serve well—but they exhibit a tremendous range in luminosity (intrinsic brightness), and they evolve quickly with time. Gamma-ray bursts, being so luminous, are obvious candidates, but they are rare, have faint optical counterparts, and show a wide range in luminosity. Also, their redshifts are difficult to measure.

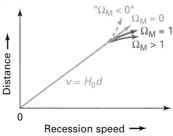

Figure 18–26 A plot of the measured distances of objects against their recession speeds (as given by their redshifts). Relatively nearby galaxies all satisfy the same relationship, Hubble's law. But at large recession speeds there are differences that depend on how rapidly the Universe's expansion rate is decelerating, and hence on Ω_M, the ratio of average density of matter to the critical density. For a given redshift, a galaxy is closer if the Universe is denser than if it is less dense. If the density of matter could be negative (that is, $\Omega_M < 0$), distances larger than those in an empty (non-decelerating) universe would be measured.

Type Ia Supernovae

In the mid to late-1990s, the most progress on this front was made with Type Ia supernovae. Type Ia supernovae are nearly ideal objects for such studies. They are very luminous, and although not exactly standard candles, accurate corrections can be made for the differences in luminosity by measuring their light curves. Their intrinsic properties should not depend very much on redshift: Long ago (that is, at high redshift), white dwarfs should have exploded in essentially the way they do now.

Starting in the mid-1990s, two teams have found and measured high-redshift ($z = 0.2$ to $z = 1$) Type Ia supernovae. The first is led by Saul Perlmutter of the Lawrence Berkeley Laboratory, and the second by Brian Schmidt of the Mt. Stromlo and Siding Spring Observatories in Australia. One of us (A.F.) has been fortunate to work with both groups, although his primary association is now with Schmidt's team.

A Type Ia supernova occurs somewhere in the observable universe every few tens of seconds. Thus, if very deep photographs are made of thousands of distant galaxies (Fig. 18–27), the chances of catching a supernova are reasonably good. The same regions of the sky are photographed about a month apart with large telescopes. Comparison of the photographs with sophisticated computer software reveals the faint supernova candidates (Figure. 18–27). A spectrum of each candidate is obtained, usually with the Keck telescopes in Hawaii. This re-

Figure 18-27 Discovery of SN 1997cj (redshift 0.5). The top two images are small subsets of much larger CCD frames obtained on April 7 and 28, 1997, with a 4-m telescope in Chile. The difference between the two images, at bottom, reveals a supernova candidate that was subsequently confirmed to be a Type Ia supernova with a spectrum obtained by one of the Keck 10-m telescopes. At left is an image obtained on May 12, 1997, with the Hubble Space Telescope, showing the supernova well-separated from the center of the host galaxy.

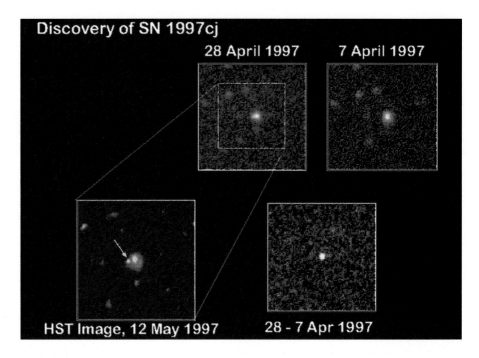

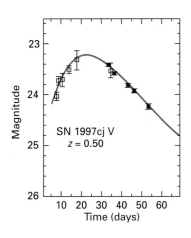

SN 1997cj V
z = 0.50

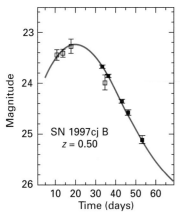

SN 1997cj B
z = 0.50

Figure 18-28 Light curves of SN 1997cj (redshift 0.5), obtained with ground-based telescopes and the Hubble Space Telescope. The data were obtained through filters that pass light emitted by the supernova at blue *(bottom)* and visual *(top)* wavelengths. These light curves are broader than those of low-redshift supernovae because the Universe is expanding, producing a kind of "time dilation."

veals whether the object is a Type Ia supernova, and provides its redshift. Follow-up observations of the Type Ia supernovae with many telescopes, including the Hubble Space Telescope (Fig. 18–28), provide their light curves. The peak apparent brightness of each supernova, together with the known (appropriately corrected) luminosity, gives the distance using the inverse-square law.

Incidentally, the light curves of high-redshift supernovae appear broader than those of low-redshift examples; that is, the former take longer to brighten and fade than the latter. This results from the expansion of the Universe: Each successive photon has farther to travel than the previous one. Indeed, this observed "time dilation" effect (Figure 18–28) currently provides the best evidence that redshifts really are produced by the expansion of the Universe, rather than by some other mechanism (such as light becoming "tired," losing energy during its long journey). The time dilation factor by which the light curves are broader is $1 + z$.

Einstein's Biggest Blunder?

Several dozen supernovae were measured in this manner by early 1998. The results are astonishing: Both teams find that the high-redshift supernovae are fainter, and hence more distant, than expected. The data agree best with the "$\Omega_M < 0$" curve in Figure 18–26, yet we know that the matter density is greater than 0 since we exist! Note that the possible presence of antimatter does not resolve this problem: Both matter and antimatter have positive energy, and exert an *attractive* gravitational force. We conclude that there is a previously unknown component of the Universe with a long-range *repulsive* effect—essentially a cosmic "antigravity," but only over very large distances. (One cannot use this to make "antigravity boots," for example.)

The extra stretching of space makes the supernovae more distant than they would have been had the Universe's expansion decelerated throughout its history. Thus, if these results are correct, we now live in an **accelerating universe,** one whose expansion rate is increasing rather than decreasing with time (Fig. 18–29)! Unless the repulsive effect turns off in the future, the Universe will expand for-

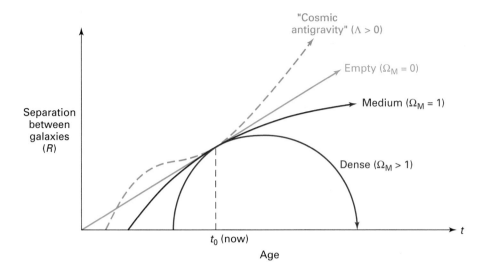

Figure 18-29 Similar to Figure 18-21, but with an additional (dashed) curve corresponding to what our Universe appears to be doing. The ratio of the average density of matter to the critical density, Ω_M, is about 0.3—but recent data also suggest that the cosmological constant (Λ, Lambda) is positive. If correct, the expansion of the Universe is now accelerating, and will continue to do so forever. The derived current age of the Universe turns out to be about 14 billion years.

ever, even if it is closed (finite in volume). Since distant galaxies are receding from each other faster and faster, eventually they will fade away, as they get so distant that their light cannot reach us.

The simplest explanation is that Einstein's cosmological constant "Lambda" (Λ), the "fudge factor" he introduced after formulating the general theory of relativity, is nonzero. In essence, space has a repulsive aspect to it, a cosmic "antigravity" over very large distances. Expressed in the same units as Ω_M, the value of Lambda that the two teams measure is $\Omega_\Lambda \approx 0.7$ (making use of the cosmic background radiation as well—see Chapter 19).

Many theorists find a positive cosmological constant to be rather disconcerting. In the context of quantum theories, it suggests that the vacuum has a nonzero energy density (a "vacuum energy" or "dark energy") due to **quantum fluctuations**—the spontaneous formation (and then rapid destruction) of virtual pairs of particles and antiparticles. (They are called "virtual" because they form out of nothing and last for only a very short time, unlike "real matter.") Theorists generally expected that $\Omega_\Lambda = 0$ due to an exact cancellation of all the quantum fluctuations. Otherwise, they predicted the large value $\Omega_\Lambda \approx 10^{120}$, which is clearly not observed. (We would definitely not exist if the cosmological constant were so large; the Universe would have expanded much too quickly for galaxies and stars to form.) If a positive cosmological constant is confirmed, it will be ironic that Einstein introduced the idea for the wrong reason (a static universe) and later completely rejected it as his "biggest blunder."

It turns out that the contribution to the density from matter and from the cosmological constant adds up to the critical density. This means that the Universe is *spatially flat* on large scales. Some theorists predicted that the Universe is flat (see Chapter 19), but without a cosmological constant. Perhaps the most natural model is one in which the Universe is formally closed (like the three-dimensional version of a sphere), but so incredibly large that it appears flat; this way, we don't need to deal with a formally infinite universe. On the other hand, some theorists have shown that an infinite universe is also a reasonable possibility, strange as it may sound.

The two teams calculate the age of the Universe to be about 14 billion years, if $H_0 = 65$ km/sec/Mpc. This is a bit smaller than the value of 15 billion years expected if the Universe were empty ($\Omega_M = 0$), but larger than the value of 10 billion years expected with $\Omega_M = 1$ (Figure 18-29). An expansion age of 14 billion years is quite consistent with the recently revised ages of globular clusters: 11–14 billion years. The pesky age crisis might finally be over!

As we briefly mentioned in Chapters 12 and 14, for every subatomic particle, there is a corresponding antiparticle. This antiparticle has the same mass as the particle, but is opposite in all other properties such as charge and the direction they are spinning. Antiparticles together make up antimatter. If a particle and its antiparticle meet, they annihilate each other; their total mass is 100 per cent transformed into energy in the amount $E = mc^2$.

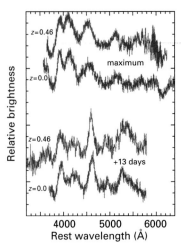

Figure 18-30 Spectrum of a Type Ia supernova at redshift 0.46, compared with that of a nearby (redshift about 0) example, at two different times: *(top)* near peak brightness, and *(bottom)* 13 days after peak brightness. There are no clear differences, suggesting that the two objects are physically very similar.

Alternatives to Acceleration?

Is it possible that the high-redshift supernovae appear fainter than expected not because of their excessively large distances, but for a different reason? For example, maybe long ago they were intrinsically less luminous than now. Keck spectra of high-redshift supernovae, however, look very similar to those of nearby supernovae (Fig. 18–30); there is no clear observational evidence for an intrinsic difference.

Or, perhaps there is dust between us and the high-redshift supernovae, making them look too faint. If this is normal dust, it would redden the light (that is, preferentially absorb and scatter blue photons)—but our procedure already accounts for this kind of dust by measurements of the reddening. If the dust grains are unusually large, on the other hand, then blue light is *not* preferentially extinguished, making such dust very difficult to detect. However, there are theoretical reasons against the formation of large dust grains, and they would have adverse observable effects on other aspects of cosmology.

Through mid-2000, no known effect other than distance has been found to account for the observed faintness of the high-redshift supernovae in a satisfactory way. Though we should remain vigilant for subtle problems that could invalidate the conclusions, for the time being they seem reliable.

Completely independent techniques that give the same results will increase our confidence in the conclusion that the Universe is currently accelerating its expansion, and that it will expand forever (unless the repulsive effect turns off in the future). Existing measurements of Ω_M combined with observations of the cosmic background radiation (Chapter 19) already strongly support the implications of the Type Ia supernovae. Satellite measurements of the cosmic background radiation, to be made by 2007, should provide definitive independent evidence. (www) There are several other promising techniques as well, so stay tuned.

THE FUTURE OF THE UNIVERSE

It seems that no matter what kind of universe we live in, the end is somewhat depressing. If the Universe is closed and the "big crunch" occurs in the reasonably near future (say, within 100 billion or a trillion years), some stars will be present throughout much of the time. During the final collapse, however, all matter and energy will be squeezed into a tiny volume, at blazingly high temperatures (since compressed gases heat up). If a rebirth occurs (but there is no good reason to suggest this), it will not contain any traces of complexity from the former universe.

If, on the other hand, the Universe is open (or critical) and expands forever, or if the cosmological constant is positive (as seems to be the case), a number of interesting physical stages will be encountered—though the end result will still be dreary. Freeman Dyson (Institute for Advanced Study) was among the first to discuss these stages, which we will now consider, in turn. The specific timescales mentioned here were calculated by Fred Adams (University of Michigan) and Greg McLaughlin (University of California at Berkeley), who assumed a universe with low matter density and zero cosmological constant (since a positive cosmological constant is still controversial). However, the overall picture remains similar even if the Universe is accelerating and continues to do so.

We now live in the **stelliferous era**—there are lots of stars! The Sun will become a white dwarf in about 6 or 7 billion years. Current M-type main-sequence stars will die as white dwarfs by the time the Universe is about 10^{13} years old. Gas supplies will be exhausted by $t \approx 10^{14}$ years, and the last M-type stars will become white dwarfs shortly thereafter (that is, within another 10^{13} years). L-type main-sequence stars won't last too much longer. Certainly by $t \approx 10^{15}$ years, normal stars will be gone.

The Universe will then enter the **degenerate era** filled with cold brown dwarfs, white dwarfs (black dwarfs, since they will be so cold and dim), and neutron stars. There will also be black holes and planets. Most stars and planets will be ejected from galaxies by $t \approx 10^{20}$ years. Black holes will swallow most of the remaining objects in galaxies by $t \approx 10^{30}$ years. All remaining objects (that is, outside of galaxies) except black holes will disintegrate by $t \approx 10^{38}$ years, due to proton decay. The time scale for this is very uncertain since the lifetime of a proton has not yet been measured.

Next, the Universe will enter the **black hole era:** The only discrete objects will be black holes. Stellar-mass black holes will evaporate by $t \approx 10^{65}$ years, due to the Hawking quantum process (see Chapter 14). Supermassive black holes having a mass of a million Suns (such as those commonly thought to be in the centers of today's galaxies) will evaporate by $t \approx 10^{83}$ years. The largest galaxy-mass black holes will evaporate by $t \approx 10^{100}$ years.

Finally, the Universe will enter the **dark era:** there will be only low-energy photons (with a characteristic temperature of nearly absolute zero), neutrinos, and some elementary particles that did not find a partner to annihilate. The electron-positron pairs may form slightly bound "positronium" atoms millions of light-years in size. (They will be disturbed very little, due to the very low space density of objects.) The positronium atoms will decay (combine to form photons) by $t \approx 10^{110}$ years.

If the Universe is open or critical it seems unlikely that "life" of any sort will be possible beyond $t \approx 10^{14}$ to 10^{15} years, except perhaps in rare cases (such as when brown dwarfs collide and temporarily form a normal star). Certainly beyond $t \approx 10^{20}$ years, the prospects for life appear grim. The Universe will be very cold, and all physical processes (interactions) will be very slow.

On the other hand, Freeman Dyson and others have pointed out that the available time scale for interactions will become very long. Hence, perhaps something akin to life will be possible. This is the "Copernican time principle"—we do not live at a special time, the only one that permits life. Life of unknown forms might be possible far into the future. However, the situation is much bleaker if the Universe is accelerating and continues to do so: It will become exceedingly cold and nearly empty much faster. Within 10^{15} years there won't even be any galaxies visible in the sky, because they will have accelerated away to invisible regions of the Universe (too far away to be traversed by light).

In Chapter 19 we will see that other universes might be born from within our Universe, or in other regions of a "super space." Thus, in some form the "universe" ("multiverse"?) might continue to exist, and to support life, essentially forever. Such ideas are intriguing and fun, but perhaps they don't currently belong to the realm of science, for we don't know any observational or experimental ways to test them.

Concept Review

Cosmology is the study of the Universe on its grandest scale—its birth, evolution, and ultimate fate. **Olbers's paradox**—the darkness of the night sky—has profound implications, and is resolved primarily by the finite age of the Universe. Historically, the first piece of evidence for this definite beginning, the **big bang,** was Edwin Hubble's discovery of the expansion of

the Universe: Most galaxies are moving away from us, and the speed of any galaxy, determined from the **redshift** of its spectrum, is proportional to its distance. This relation is known as **Hubble's law,** and the constant of proportionality is called **Hubble's constant,** H_0. The expansion of the Universe has no definite center within any physically accessible dimensions.

Note that *space itself* expands; galaxies are not flying apart from each other through a preexisting space.

There are at least two ways in which to determine the age of the Universe. First, the Universe must be at least as old as the oldest objects within it. Second, the age of the Universe is the elapsed time since the big bang, when all the material that formed our galaxy and any other was essentially in the same place. The **Hubble time** is the expansion age of the Universe under the assumption that gravity has not been slowing down the expansion; its value depends on Hubble's constant. If the true expansion age of the Universe is only two-thirds of the Hubble time, as expected by some theorists, then its value is smaller than the ages of the oldest globular star clusters, leading to an age crisis. One difficulty with measurements of Hubble's constant is that the recession speeds of relatively nearby galaxies can deviate from the **Hubble flow,** the assumed smooth expansion of the Universe. Such **peculiar motions** are produced by the gravitational tug of the Local Group and other mass concentrations. Nevertheless, several techniques were combined in the late 1990s to achieve a reasonably convincing value for Hubble's constant, $H_0 = 65$ km/sec/Mpc.

Mathematically, studies of the expansion history and overall geometry of the Universe are conducted with Einstein's general theory of relativity. To simplify the equations and achieve reasonable progress, we assume the **cosmological principle:** On the largest size scales, the Universe is **homogeneous** (has the same average density everywhere) and **isotropic** (looks the same in all directions). Also, we initially assume that Einstein's **cosmological constant** (Λ, Lambda) is zero—in other words, that there are no long-range forces other than gravity. The result is that the ultimate fate of the Universe depends on the ratio of the average matter density of the Universe to a specific **critical density.** If the average density exceeds the critical density, then this ratio, known as Ω_M, is greater than 1 and gravity will eventually cause the Universe to collapse to a fiery death informally called the "big crunch." On the other hand, if the average density is low, such that Ω_M is less than or equal to 1, the Universe will expand forever, though more and more

slowly with time; indeed, in the case $\Omega_M = 1$, the expansion will halt as time approaches infinity. The value of Ω_M also determines the overall geometry of the Universe. If the average density equals the critical density, we live in a **flat (critical) universe** whose spatial extent is probably infinite, and whose true age is two-thirds of the Hubble time. A **closed universe** has $\Omega_M > 1$; it resembles a sphere of finite volume. The last alternative, an **open universe,** occurs if $\Omega_M < 1$; its spatial extent is probably infinite, and its shape somewhat resembles that of a horse's saddle or a potato chip.

We can determine the ultimate fate of the Universe from a number of methods, all of which are difficult and have substantial uncertainties. The most successful recent technique has been to measure the apparent brightness of Type Ia supernovae at high redshifts. The result is that the Universe appears to be expanding faster now than in the past: We might live in an **accelerating universe!** This means that a long-range "antigravity" force exists, like the cosmological constant that Einstein had previously postulated with a specific value to obtain a static universe. The value actually measured for the cosmological constant resolves the age crisis of the Universe, implies that the Universe is spatially flat on large scales, and suggests that the Universe will expand forever. Physically, the cosmological constant may correspond to a vacuum energy consisting of **quantum fluctuations**, the spontaneous formation (followed by rapid destruction) of virtual pairs of particles and antiparticles.

If the Universe lasts forever, many interesting physical processes will occur. We now live in the **stelliferous era,** filled with stars, but eventually all normal stars will burn out, and the Universe will enter the **degenerate era,** dominated by objects like white dwarfs and neutron stars. After a very long time, these degenerate objects will disintegrate, leaving the Universe in the **black hole era.** But even black holes decay due to the Hawking evaporation process after exceedingly long times; the Universe will enter its last stage, the **dark era,** as a cold, nearly empty space consisting of extremely low-energy photons and a few elementary particles.

Questions

1. Describe Olbers's paradox—the darkness of the night sky—and several possible resolutions.
2. Summarize the first observational evidence found for the expansion of the Universe and a possible beginning of time.
3. Why does the recession of galaxies not necessarily imply that the Milky Way Galaxy is at the center of the Universe?
4. Explain how the effective center of expansion can be in an unobservable spatial dimension.
5. What do we mean when we say that the Universe is expanding? Are galaxies moving through some preexisting space? Do humans, planets, stars, and galaxies themselves expand?
6. What are two ways of estimating the age of the Universe?
7. Describe how the current value of Hubble's constant is measured.
8. Why can relatively nearby galaxies give erroneous values for Hubble's constant?

†9. Calculate the Hubble time (that is, the expansion age of the Universe assuming no deceleration) if Hubble's constant is 75 km/sec/Mpc.
10. Suppose we assume that the recession speed of a given galaxy never changes with time. Why is the derived age of the Universe likely to be an overestimate of its true age?
11. If the expansion rate of the Universe were increasing (rather than decreasing) with time, would the true expansion age be less than or greater than that derived by assuming a constant expansion rate?
12. What is the "age crisis" that rocked cosmology until the late-1990s? Discuss two developments that helped alleviate this crisis.
13. Why are exploding white dwarf stars so useful for measuring the distances of galaxies?
14. State the cosmological principle. Why is it a reasonable assumption?
15. Define what we mean by the cosmological constant, Λ.

16. Summarize the possible types of overall geometry for a homogeneous, isotropic universe with cosmological constant equal to zero.
17. Describe how the geometry of the Universe is intimately connected to the ultimate fate of the Universe, assuming the cosmological constant is zero.
18. Explain what we mean by the critical density of the Universe.
19. Discuss why Ω_M, the ratio of the average matter density to the critical density of the Universe, is such an important parameter.
20. Summarize different methods for determining the value of Ω_M. What are the difficulties in actually implementing these methods?

21. Does the average matter density appear to be large enough to close the Universe?
22. Discuss the main conclusion of the distant supernova studies, and its implications.
23. Explain how Einstein's cosmological constant is relevant to the conclusion in question 22.
24. Why might high-redshift supernovae appear fainter than expected, other than an acceleration of the expansion rate of the Universe?
25. What are some of the processes that will eventually occur if the Universe expands forever?

†This question requires a numerical solution.

In the Beginning

19

ORIGINS *The birth of the Universe itself is probably the most fundamental of all questions about origins. The behavior of the Universe at very early times greatly affected its contents, producing the matter of which we are made and the relative proportions of the light elements that dominate all others. It is possible that many universes exist, yet only a small fraction might have conditions suitable for life as we know it.*

By itself, the expansion of the Universe does not prove that there was a big bang; indeed, one could postulate that the Universe had no beginning in time and will have no end. The fatal blow to this "steady-state theory" was the discovery of a faint radio glow that pervades all of space and was produced when the Universe was very young. The existence of a nearly uniform amount of helium and deuterium (heavy hydrogen) throughout the Universe provides additional evidence for a very hot, dense phase in its early history. But how did the Universe achieve a uniform temperature, and why is space so nearly flat? These troubling issues led to a magnificent hypothesis: The early Universe went through a stage of phenomenally rapid expansion, doubling its size many times in just a tiny fraction of a second. Such a process may have created the Universe from essentially nothing! And if it happened once, it might happen again; perhaps there exist multiple "universes" physically disconnected from ours.

AIMS

To learn about radio radiation that now uniformly permeates the Universe and provides compelling evidence for the big bang.

•

To discuss what happened during the first few minutes after the birth of the Universe.

•

To see how a modification of the original big-bang theory in the first wink of the Universe's existence explains some otherwise puzzling observations.

•

To consider the possibility of multiple universes.

◼ THE STEADY-STATE THEORY

In the previous chapter we considered the big-bang theory of the Universe, which is motivated in part by the observed recession of galaxies away from our own Milky Way Galaxy (Fig. 19–1). We saw that the matter density of the Universe dictates the past history, as well as the future, of its expansion rate. Moreover, there is tantalizing evidence for a cosmic "antigravity" effect that may be accelerating the expansion of the Universe, causing it to grow larger and less dense at an ever-increasing rate.

Although the expansion of the Universe suggests that there was a definite beginning of time when matter was in a very compressed and hot state, this is not the only logically consistent possibility. Here we consider a reasonable alternative

Fig. 19–1 Hubble Space Telescope image of galaxies in the direction of the Coma Cluster of galaxies. The smaller, fainter galaxies are generally more distant than the brighter, larger galaxies. Essentially all galaxies are receding from us, with the distant galaxies moving away faster than the nearby ones.

◀ One quadrant of the Hubble Deep Field, the image that resulted when the Hubble Space Telescope observed one "normal," unremarkable region of space for a whole week. Almost every object in this image is a distant galaxy. An estimated 100 billion galaxies, across the whole sky, can in principle be detected with the Hubble—but this is probably just a minuscule fraction of all galaxies in the Universe, which extends far beyond the distance to which we can see.

Fig. 19–2 Fred Hoyle, one of the main advocates of the steady-state theory of the Universe. Though now known to be incorrect, this theory helped accelerate progress in cosmology because it stimulated new observational and theoretical advances.

REDSHIFT
http://www.harcourtcollege.com/astro/cosmos/rsce

that turned out to be incorrect, but that nevertheless served science well because it forced cosmologists to question critically their assumptions and conclusions.

In 1948, Fred Hoyle of Cambridge University (Fig. 19–2) and two of his associates (Hermann Bondi and Thomas Gold) proposed the **steady-state theory** as an alternative to the hot big bang. It is based on a modification of the cosmological principle called the **perfect cosmological principle**: The Universe is homogeneous and isotropic on large scales, and its *average properties never change with time.* Therefore, there was no beginning, and there will be no end; the Universe is simply expanding throughout all time. To keep the average density constant as the Universe expands, new matter must be created continuously, and new galaxies form out of this material.

We might argue against the steady-state theory on the grounds that it requires something to be produced from nothing, which is thought to be impossible over long time scales. That is, we believe in the law of "conservation of energy"—the combination of mass plus energy is neither created nor destroyed. But this objection turns out to be observationally weak. Only one hydrogen atom must be created per cubic meter, per billion years, to satisfy the requirements of the steady-state theory. This rate is below the current detection limit of laboratory experiments. In other words, we have not yet verified the law of conservation of energy to this degree of accuracy. ☼ Moreover, the original big-bang theory postulates that the Universe was created out of nothing in a single instant ($t = 0$), so it appears to violate conservation of energy as well (although we will later see that no significant violation might occur).

Despite its aesthetic appeal to some astronomers, the steady-state theory gradually lost support, as observational evidence showed that the Universe does indeed change with time. One of the first arguments was that radio galaxies seem to be more numerous and luminous at large distances. An even more compelling case was made with quasars (see Chapter 17): They are clearly denizens of the distant past; all have high redshifts, so we see them at large lookback times. There are no nearby quasars, seen as they are in the present-day Universe, but only lower-luminosity active galaxies. In recent years, Hubble Space Telescope images of distant galaxies show that they differ from nearby galaxies, providing additional evidence for the evolution of the Universe.

The fatal blow to the steady-state theory, however, was the discovery of the **cosmic background radiation**—a faint afterglow from the Universe's hot past. The steady-state theory has no explanation for it, but big-bang models require it, as we will now discuss.

● THE COSMIC BACKGROUND RADIATION

In 1964 through 1965, two young researchers (Arno A. Penzias and Robert W. Wilson) were observing with a 20-foot horn-shaped radio antenna owned by Bell Labs near Holmdel, New Jersey (Fig. 19–3). The antenna had previously been used to detect radio signals from communications satellites, but Penzias and Wilson were allowed to use it for astronomical observations. They initially wanted to accurately measure the radio brightness of Cas A, a prominent supernova remnant that is used to calibrate radio data. Then they wanted to map the 21-cm radio emission in the Milky Way Galaxy.

For the most accurate measurements, they needed to make their microwave antenna as sensitive as possible. After removing all possible sources of noise, they found a very faint but persistent hiss that seemed to defy explanation. It was independent of location in the sky, and also independent of the time of day. If pro-

Fig. 19–3 Arno Penzias *(left)* and Robert W. Wilson *(right)* with their horn-shaped antenna in the background. After they removed all possible sources of noise (by fixing faulty connections and loose antenna joints, and by removing "sticky white contributions" from pigeons), a certain amount of radiation remained. It was the 3° background radiation.

duced by a "black body" (Chapter 2), it corresponded to a temperature of about 3 K, which is just 3°C above absolute zero and −270°C. They were not aware that it might be the afterglow of the big bang!

Meanwhile, at nearby Princeton University in New Jersey, a team led by Robert Dicke was building a radio telescope with which to detect this afterglow. They had predicted that it should exist, if the Universe began in a very hot and compressed state. Actually, this prediction had already been made in the 1940s, by the Russian astrophysicist George Gamow (who was working in the United States), but the Princeton group was unaware of it. Gamow and his students, Robert Herman and especially Ralph Alpher, had based their conclusion on the idea that all of the chemical elements were created shortly after the big bang. Although incorrect in detail, this is correct in spirit; only the lightest elements were produced early in the Universe, as we shall see later. The Universe must have been very hot to do this, but expansion would have cooled it, so the radiation should now be that of a cold black body.

Another astronomer alerted the Princeton and Bell Labs teams to each other, and that's how Penzias and Wilson found out what they had discovered. The Princeton group made a measurement at another wavelength, and it agreed with the prediction of a black-body spectrum. Other, subsequent measurements were also consistent with black-body radiation from a gas at $T \approx 3$ K, a very low temperature indeed. In January 1990, astronomers announced results from NASA's Cosmic Background Explorer (COBE, pronounced "koh′bee"; Fig. 19–4), which had been launched in November 1989: The spectrum was that of a perfect, cold black body (Fig. 19–5).

From where, exactly, does a photon of the cosmic background radiation come? The early Universe was very hot and ionized: There were no bound atoms, but rather free electrons and atomic nuclei. Because photons easily scatter (bounce) off of free electrons, they cannot travel in a straight line from one object to another, and instead jump randomly from one direction to another. So the Universe looked opaque, like thick fog. Even if there were discrete objects at this time, one would not be able to see them. The Universe was filled with photons, however, since it was hot—and they simply bounced around in a sea of hot particles.

As the Universe expanded, it cooled. Eventually, about 300,000 years after the big bang, it reached a temperature of about 3000 K, and electrons were able

Fig. 19–4 The COBE spacecraft (Cosmic Background Explorer), remade smaller to fit on a Delta rocket after the explosion of the space shuttle Challenger in 1986. COBE was launched in 1989 and sent back data for four years. It fantastically improved the accuracy of measurements of the cosmic background radiation.

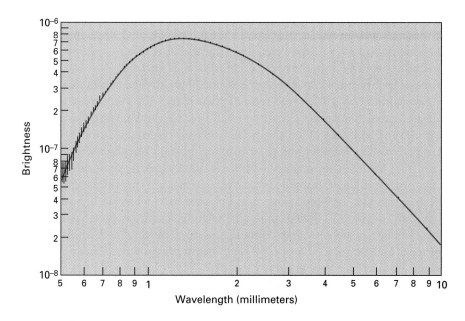

Fig. 19–5 This first spectrum based on COBE observations drew a standing ovation from astronomers when it was initially shown at an American Astronomical Society meeting. It represents astonishingly good evidence that the cosmic background radiation comes from a black body. The data points and their error bars *(black)* were fit precisely by a black-body curve *(red)* for 2.735 ± 0.06 K, a value that has since been slightly improved, based on the final data set, to 2.728 ± 0.004 K. Current error bars are thinner than the thickness of the curve drawn!

Fig. 19–6 The cosmic background radiation was set free at a temperature of 3000 K *(blue)* when the Universe became transparent at the time of recombination, about 300,000 years after the big bang. As the Universe expanded, the spectrum was transformed into that of a lower temperature gas *(red curve at lower right).* The rainbow shows the wavelengths of visible light; the cosmic background radiation is detected near its peak at microwave wavelengths and not in visible light or other wavelengths far from its peak.

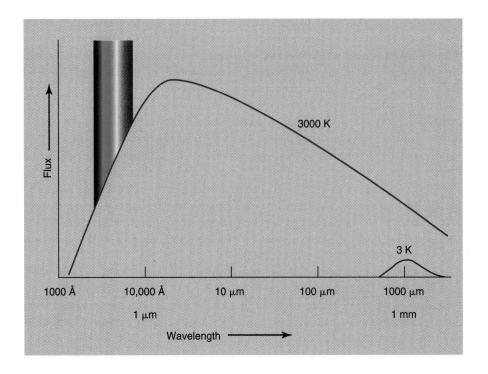

to combine with protons to form neutral hydrogen atoms. The process is called "recombination"—even though in this case, electrons were combining with protons for the *first* time in the history of the Universe. Unlike free electrons, electrons bound in atoms are able to interact with photons of only certain specific energies, absorbing them (Chapter 2). They are not effective at scattering most photons. Thus, at this time most photons stopped bouncing around, and became free to travel unhindered. The Universe therefore became transparent to almost all electromagnetic waves. Astronomers say that matter and radiation "decoupled" from each other at $t \approx 300{,}000$ years.

As space expanded, the wavelengths of photons increased, just like that of a wavy line drawn on an expanding balloon, so they lost energy. They maintained the spectrum of a black body, but of lower and lower temperature (Fig. 19–6). Typical photons went from being optical/infrared to radio (microwave). The cosmic background radiation now corresponds to such a low temperature, about 3 K, because the Universe has expanded a great deal over its 14-billion-year life. Since it started off as radiation filling the Universe, it is easy to understand why it now comes equally from every direction (that is, is isotropic).

In essence, the cosmic microwave background photons come from an opaque "wall" at redshift about 1000 (the ratio of 3000 K to 3 K). We cannot see electromagnetic radiation from beyond this wall because the Universe was opaque. This marks the boundary of the observable Universe. That is, we cannot see electromagnetic radiation from times prior to about 300,000 years after the big bang. In principle, we could see neutrinos from $t < 300{,}000$ years, since the Universe was not opaque to them. But unfortunately, neutrinos are very difficult to detect.

The cosmic background radiation is left over from when the Universe was very hot and dense. ☼ Being almost perfectly isotropic, it provides the best known evidence in support of the cosmological principle that the Universe is homogeneous and isotropic. The steady-state theory has no way to account for its presence. Penzias and Wilson shared the 1978 Nobel Prize in physics for their important discovery: They had been very careful, and had not ignored the faint (almost nonexistent) hiss. Gamow could not receive the Nobel Prize, as he had already died, but it seems unfair that Alpher was omitted. (It can be shared among a maximum of three people.)

🌑 DEVIATIONS FROM ISOTROPY

The cosmic background radiation shows a slight deviation from isotropy of about two parts per thousand. It looks slightly hotter in one direction of the sky, and slightly cooler in the opposite direction. Such an "anisotropy" is caused by a combination of motions: the Sun's motion around the center of the Milky Way Galaxy, the Milky Way's motion within the Local Group, and, most important, the Local Group's motion relative to the "Hubble flow" (that is, relative to the smoothly expanding Universe).

These gravitational perturbations are all known. The sum of the motions produces a net motion in a particular direction of space, and the slight temperature anisotropy is attributed to the Doppler effect. The radiation coming from the direction of motion is slightly blueshifted ("hotter"), while that coming from the opposite direction is slightly redshifted ("cooler"). Maps made with COBE very clearly show this anisotropy (Fig. 19–7).

A major puzzle for many years was the apparent *absence* of small ripples in the background radiation corresponding to the size scales of clusters and superclusters of galaxies. If these structures formed by the gravitational contraction of matter, there should have been slight ripples in the density of matter early in the Universe. Photons escaping from higher density clumps would have a slightly different redshift than those coming from regions of average or below-average density. There are several relevant processes, but perhaps the easiest to understand is that photons lose more energy when they come out of stronger gravitational fields. The different redshifts would translate into slight variations in the associated temperature of the radiation.

No such ripples were found until the early 1990s, despite extensive searches. Some specific models of cluster formation were consequently eliminated, because they predicted irregularities larger than the observed upper limits. In 1992, a breakthrough was made with the COBE satellite: Tiny temperature variations (about one part per hundred thousand, or 30 microkelvins) were finally found, at least in a statistical sense. (Ripples were present, but no great significance could be given to any particular one.) After a few more years, the data became quite convincing (Fig. 19–8). The angular resolution of COBE was low, so even the smallest detected ripples correspond to very large superclusters of galaxies.

Subsequently, balloon-based experiments found similar variations, but on smaller angular scales (Figure 19–8), corresponding to large clusters of galaxies. These are the "seeds" from which large-scale structure grew in the Universe. They are the imprints of minuscule ripples in the distribution of matter established shortly after the big bang.

The most recent results, published in 2000, indicate that the angular size of temperature variations in the cosmic background radiation is typically about one

A   3.3 mm

B   5.7 mm

C   9.6 mm

Fig. 19–7 Maps of the sky from all four years of COBE data at radio wavelengths of *(A)* 3.3 mm, *(B)* 5.7 mm, and *(C)* 9.6 mm. At the longest wavelength, we see a discrete source in Cygnus *(green dot at middle left)* and a sign of the galactic plane *(horizontal red at right)*. The rest of the signal is from the cosmic background radiation. The asymmetry from bottom left to top right is produced by the Sun's motion relative to the smoothly expanding Universe. The range shown is ±3 millikelvins.

Fig. 19–8 A full-sky map of the variations in the 3 K cosmic background radiation, as analyzed from all four years of COBE data, after removal of the Sun's motion relative to the Universe. The temperature variations are typically only 30 microkelvins (0.01 per cent). The red band across the center of this map results from the microwave emission of the Milky Way Galaxy. The variations above and below the Milky Way are real fluctuations in the background radiation.

The results are also shown for a ground-based experiment carried out in Saskatoon, Saskatchewan, Canada. It observes only a small area around the north celestial pole but obtains higher angular resolution within that region.

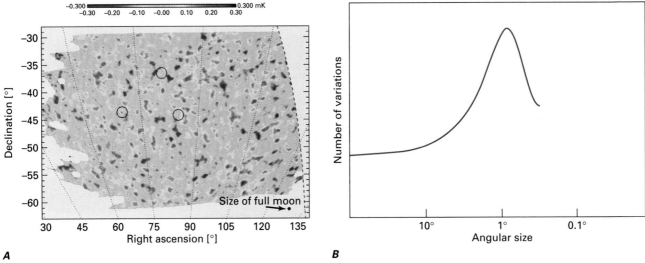

Fig. 19–9 (A) A map of the tiny variations in the 3 K cosmic background radiation over part of the sky. The full moon is shown for comparison. (B) The relative number of variations in the sky versus their angular size. Though variations appear over a wide range of scales, the most common ones appear to be about 1° in diameter, or twice the full moon.

degree (1°; Fig. 19–9). This means that the typical angular size of a ripple is around 1°, and it tells us something about the overall geometry of the Universe. In fact, the data are consistent with the presence of considerable amounts of dark matter and other energy in the Universe—enough to make the Universe spatially flat! A spatially flat geometry means that Ω, the ratio of average total density (of matter, normal energy, and other energy) to the critical density, is 1 (see Chapter 18). This "other energy" may well be the "vacuum energy" or "dark energy" associated with the cosmological constant, whose nonzero value is suggested by observations of distant supernovae (see Chapter 18). Galaxies and clusters of galaxies seem to have formed at peaks in the overall distribution of dark matter.

THE EARLY UNIVERSE

The evolution of the Universe can be studied nearly back to the moment of creation, "$t = 0$." Just how close to $t = 0$ we can get with reasonable accuracy is debated. Almost all astronomers agree that we can go quite far back in time because the early Universe was relatively simple: it consisted of particles and photons in thermal equilibrium (that is, at the same temperature), and we seem to know the few fundamental laws that governed their behavior. Compare this with a complex problem like Earth's weather, which is affected by factors such as local heating, ocean currents, cloud formation, continents, mountain ranges, and so on. It is easier to predict the properties of the early Universe than it is to predict the weather!

Certainly we can reach something like $t \approx 1$ sec, because conditions were similar to those in the central regions of stars, and the nuclear physics is well understood. We also have substantial confidence in our conclusions down to $t \approx 10^{-12}$ sec, since the corresponding temperatures and energy densities have been reproduced in high-energy particle accelerators.

It can also be argued that we can extrapolate all the way to about 10^{-35} sec, or perhaps somewhat earlier; once again, hot gases in equilibrium are relatively easy to understand. However, we do not yet know everything down to such small time scales; our physical theories are incomplete, and no conceivable particle accelerators can achieve temperatures that are high enough. Thus, some of our current conclusions about this era are rather speculative. Turning this problem around, we can use celestial observations to test the physical theories: The Universe is our great "accelerator in the sky."

Although times such as 10^{-35} sec may seem ridiculously short, with no real "action," it is important to understand that *lots* of things could have happened. This is because the interaction time scales and distances were far smaller than they are today.

A Brief History of the Early Universe

At times before 10^{-43} sec, we know essentially nothing, except that the temperature exceeded about 10^{32} K. This is the **Planck time:** It is thought that time itself might be packaged in small units ("quantized") at about this interval, or at least it becomes unpredictable. Some physicists have the notion of "space-time foam," where packets of time and space themselves flit into and out of existence. We must develop a self-consistent quantum theory of gravity to understand this era.

When the Universe was between 10^{-35} and 10^{-6} sec old, there was equilibrium between particles, antiparticles, and photons. Particle–antiparticle pairs annihilated each other and produced photons. Photons spontaneously formed particle–antiparticle pairs. **Quarks,** particles that are normally bound together by "gluons" to form protons and neutrons, were plentiful and unbound during most of this era. In 2000, physicists at CERN (the European Laboratory for Particle Physics near Geneva, Switzerland) announced that they may have reproduced a tiny amount of the "primordial soup" of free quarks and gluons by smashing nuclei of lead at very high energies, though this result is still controversial.

Quarks come in six types or "flavors": up, down, strange, charmed, top (or truth), and bottom (or beauty). Each flavor also comes in three "colors": blue, green, and red. (These names are whimsical; they don't reflect the real "character" of each quark.) Normal matter consists of the "up" and "down" quarks, together with electrons and electron neutrinos (a type of neutrino associated with electrons). For example, a proton is two "up" quarks (each with a charge of $+2/3e$, where e is the unit of electric charge) and one "down" quark (with a charge of $-1/3e$). A neutron is one "up" quark and two "down" quarks (Fig. 19–10). Each quark has a corresponding antiquark.

At some stage, probably early in this era of the quark-gluon-photon mixture, a slight imbalance of matter (quarks) over antimatter (antiquarks) was formed. It amounts to about one part per billion. It is not known which specific reactions took place, but laboratory experiments have demonstrated the existence of a matter–antimatter asymmetry for some processes (Fig. 19–11). We owe our existence to this tiny asymmetry.

The annihilation of protons, neutrons, and their antiparticles occurred at $t \approx 10^{-6}$ sec, one microsecond, when the temperature was about 10^{13} K, leaving a sea of photons. The annihilation was not complete, due to the slight imbalance of matter over antimatter. ☼

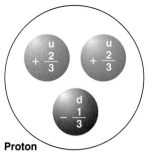

Proton
+1 Charge unit

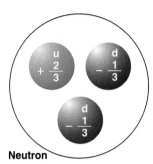

Neutron
0 Charge unit

Fig. 19–10 The most common quarks are *up (u)* and *down (d);* ordinary matter is made of them. The up quark has an electric charge of +2/3e, and the down quark has a charge of −1/3e, where e is the unit of electric charge. This fractional charge is one of the unusual things about quarks; prior to their invention, it had been thought that all electric charges come in whole numbers. Note how the charge of the proton and of the neutron is the sum of the charges of their respective quarks.

Great discoveries sometimes have humble beginnings, especially when the discoverers are not aware that they are making a great discovery. Because of this fact we do not have an attractive picture of the neutral kaon apparatus. I am sorry about this.

Sincerely,

James W. Cronin

James W. Cronin

Fig. 19–11 James Cronin and Val Fitch received the 1980 Nobel Prize in Physics for their experiment that showed an asymmetry in the decay of a certain kind of elementary particle. This result may point the way to the explanation of why there is now more matter than antimatter in the Universe.

Fig. 19–12 Methods by which protons and neutrons are converted back and forth into each other, early in the history of the Universe when temperatures were very high.

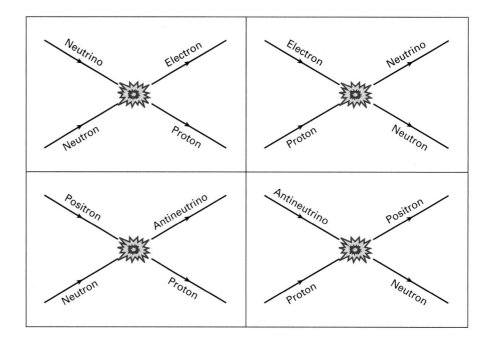

Then there was rapid conversion of protons and neutrons back and forth into each other (Fig. 19–12). However, since neutrons are slightly more massive than protons, they are harder to make, and the number of neutrons settled to only about one-quarter that of protons.

By the time the Universe was about 1 sec old, its temperature had dropped to about 10^{10} K (ten billion degrees), and the typical photon energies became too low to spontaneously produce electron–positron pairs. Electrons and positrons therefore annihilated for the last time, producing more photons. The remaining (excess) electrons balanced the positive charge of the protons. At this time, neutrons began to decay into protons, electrons, and antineutrinos; it was not possible to rapidly replenish the supply of neutrons. This further increased the imbalance between protons and neutrons.

The temperature had dropped to only a billion degrees by an age of about 100 sec, allowing the formation of heavy isotopes of hydrogen, two isotopes of helium, and a little bit of lithium. Before this time, collisions between particles would sometimes produce bound states, but subsequent collisions destroyed them. At the temperature and density of matter when $t \approx 100$ sec, collisions tended to produce a surplus of bound particles; they were not immediately destroyed (Fig. 19–13). Elements heavier than lithium were not formed at this time: the temperature and density dropped too rapidly. By an age of about ten minutes, the Universe had completed its **primordial nucleosynthesis**—the formation of the light elements shortly after the big bang.

Primordial nucleosynthesis is predicted to have happened when the proton to neutron ratio was about 7/1. For every 16 nuclear particles, 2 were neutrons, and these generally combined with 2 protons to form helium, thereby using up 4 of the 16 particles. The other 12 particles remained as protons (simple hydrogen). So, about 25 per cent of the matter (by mass) should have turned into helium, while most of the rest was hydrogen. (A little bit of the hydrogen was in the form of deuterium and tritium, but the latter is unstable and quickly decays.) The fact that helium is observed to be about 25 per cent by mass throughout the Universe (as far as we can tell) strongly supports the big-bang theory: The

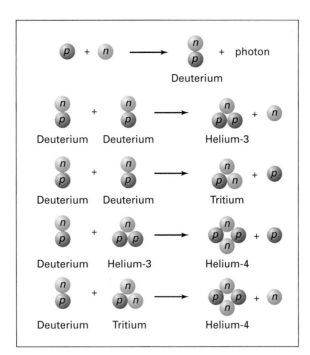

Fig. 19–13 Reactions during the first few minutes of the Universe's existence that produced various isotopes of the lightest elements, the process of primordial nucleosynthesis. In addition, a very small amount of lithium-7 was produced by other reactions.

Universe had to be hot and dense at early times. The rather uniform relative proportion (abundance) of helium suggests a primordial origin, before stars were born. Heavier elements, on the other hand, exhibit rather large variations in their relative proportions: in some stars they are rare, and in others they are common, implying an origin that is not primordial or uniform throughout the Universe.

The verification of the predictions of primordial nucleosynthesis is one of the three major pillars on which the big-bang theory rests. (The other two are the observed expansion of the Universe and the existence of the cosmic background radiation.) It is more compelling than the mere expansion of the Universe, which is consistent with the competing steady-state theory. George Gamow and his students had the right idea when they concluded that the Universe must have been hot at an early stage. Their mistake was in thinking that *all* of the elements (instead of just the lightest ones) were synthesized shortly after the big bang. They didn't know about nucleosynthesis in the cores of stars.

The approximate observed abundances of the light elements show that the general ideas of primordial nucleosynthesis are correct. Detailed measurements provide some additional constraints; the exact current proportions reflect conditions in the early Universe. For example, deuterium is rapidly destroyed, forming helium. Thus, if the density remained high for a long time (the expansion of the Universe was relatively slow), the current deuterium abundance would be very low. The fact that deuterium is reasonably abundant right now (about 10^{-5} of the amount of ordinary hydrogen) tells us that Ω_M, the ratio of the average matter density to the critical density, is less than 0.1 (10 per cent), and perhaps as low as 0.04 (4 per cent) (Fig. 19–14). However, this limit on Ω_M only refers to normal matter (protons, neutrons, etc.). There might be additional contributions to Ω_M, but they must be "exotic" matter such as neutrinos and "weakly interacting massive particles" (WIMPs). If we believe that $\Omega_M \approx 0.3$, as suggested by studies of clusters of galaxies and the peculiar motions of galaxies (Chapter 18), then a large fraction of the matter in the Universe is not only dark, but must also be composed of some kind of exotic matter!

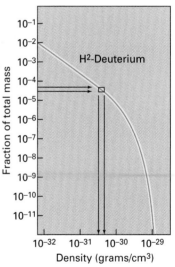

Fig. 19–14 The horizontal axis shows the current cosmic density of matter. From our knowledge of the approximate rate of expansion of the Universe, we can deduce what the density was long ago. The abundance of deuterium is very sensitive to the density; thus present-day observations of the relative amount of deuterium tell us the cosmic density, by following the arrow on the graph.

◖◗ THE INFLATIONARY UNIVERSE

Problems with the Original Big-Bang Model

The original, "standard" hot-big-bang theory is remarkably successful. Its three main observational foundations are the existence of the cosmic background radiation, the relative proportions of various isotopes of the lightest elements (hydrogen, helium, and lithium), and the expansion of the Universe. However, there are some puzzling aspects of the Universe that the original big-bang theory cannot explain, at least without imposing "initial conditions" that seem unlikely and unjustified. Here we will discuss two of the main problems.

The first of these is the great homogeneity and isotropy of the Universe. This is known as the **horizon problem.** In all directions, the temperature indicated by the cosmic background radiation is identical (2.726 K), excluding the two known effects that produce deviations from isotropy (our motion through the Universe, and early variations in density from which clusters of galaxies formed).

How can this be? Widely separated regions of the Universe could never have been in thermal equilibrium with each other (that is, at a uniform temperature): No signals, not even those travelling at the speed of light, could have traversed such vast distances over the age of the Universe. Thus, they had no way of "telling" each other which temperature to choose—they are beyond each other's horizon. For example, assume the Universe is 14 billion years old, and consider two points on opposite sides of the sky. Cosmic background photons are just now reaching us, after travelling almost the entire age of the Universe (minus about 300,000 years when the Universe was still opaque). Clearly, photons from one side have not yet reached the other side; these regions should therefore have different temperatures.

An objection might be that the big bang started as a "point," and hence could have achieved a uniform temperature. But actually, the big bang was never a true "point"; we need a quantum theory of gravity to figure out its initial size. Running the original big-bang equations backward in time, we find that the Universe was *always* larger than the distance light (or any other signal) could have travelled by that age, even if the Universe was always transparent. Indeed, the number of "causally disconnected" regions was larger in the past, making the observed homogeneity and isotropy even more perplexing. For many years, cosmologists simply *assumed* that the Universe was created with a uniform temperature. This was an "initial condition" that was adopted with no physical justification.

The second major problem with the original big-bang theory is that it has no natural way of explaining why the overall geometry of the Universe is so close to being flat. This is known as the **flatness problem.** Observations tell us that the ratio of matter density to the critical density, Ω_M, is certainly larger than 0.01, just from the luminous matter and a small correction for dark matter in galaxies. But in fact $\Omega_M \approx 0.3$ if the dark matter in clusters of galaxies is included. Moreover, as discussed in Chapter 18, observations of distant supernovae suggest that there is a significant contribution to the total density from the "dark energy" ("vacuum energy") associated with the cosmological constant, so the total Ω might be 1. And, as we saw above, recent measurements of the cosmic background radiation strongly suggest that the overall geometry of the Universe is flat, meaning that the total Ω is indeed 1. ☼

Why should we be surprised that $\Omega \approx 1$ right now? Well, according to the equations describing the evolution of the Universe in the original big-bang theory, Ω must have been *exceedingly* close to 1 in the distant past (Fig. 19–15). Other-

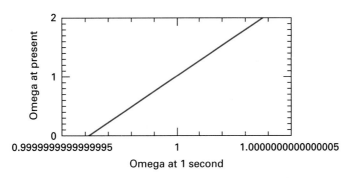

wise, as the Universe aged, Ω would have rapidly deviated from 1 by a very large extent, so we should not be measuring $\Omega \approx 1$ at present. It turns out that $\Omega = 1$ is an equilibrium value, but it is unstable in the sense that the slightest deviation from 1 quickly grows into a huge deviation. It is like a pencil standing on its point: the slightest push makes it fall over. You are unlikely to walk into a room to find a pencil standing on its point, unless the pencil was balanced *very* precisely from the start—and this is unlikely to happen by chance. On the other hand, if there is some physical reason why the pencil is *forced* to stand on its point, then you might expect to find it in that position. The original big-bang theory provides no suitable explanation for why the Universe started out with $\Omega = 1$, so for many years cosmologists simply *assumed* this initial condition; they had no reasonable explanation for it.

Inflation to the Rescue

In 1979, Alan Guth (Fig. 19–16) (now at the Massachusetts Institute of Technology) came up with a brilliant solution to these problems. Known as **inflation,** this idea is a modification to the original big-bang theory. Although his proposal had an important flaw, the basic hypothesis was very attractive. The flaw was resolved in 1981 by Andrei Linde (now at Stanford University), as well as by Paul Steinhardt (now at Princeton University) working together with Andreas Albrecht (now at the University of California, Davis). Most physicists and astronomers accept inflation, although the details and specific mechanisms currently have greater variety than they did 20 years ago. Here we discuss the historically earliest example, as formulated by Guth, Linde, Albrecht, and Steinhardt, since it serves well to illuminate the essential principles.

Suppose the Universe started out *much smaller* than implied by an extrapolation of the equations of the original big-bang theory. It might have subsequently "inflated"—expanded *extremely* fast and by an immense factor (Fig. 19–17), perhaps 10^{50} or even 10^{100} (a "googol"). This huge expansion was achieved by doubling in size every tiny fraction of a second for a significant amount of time. An intuitive feel for such "exponential" growth can be obtained by considering a chessboard. If there is one grain of wheat on the first square, and the number of grains in each successive square is doubled, the wheat corresponding to the 64th square (2^{63} grains) would be the size of a mountain that covers Manhattan Island!

Notice that before inflating, the tiny initial Universe could have achieved a uniform temperature (thermal equilibrium); it was sufficiently small that signals were able to travel back and forth across it. Specifically, collisions between particles and photons would have distributed the heat uniformly. But the Universe would have retained its homogeneity and isotropy as it inflated, because there would have been no way in which to lose these properties. Thus, the currently

Fig. 19–16 Alan Guth, inventor of the inflationary model of the Universe, a modification to the original big-bang theory during the first tiny fraction of a second of existence.

Fig. 19-17 The rate at which the Universe expands, for both inflationary and non-inflationary models. The vertical axis shows the radius of the region that expands to become the presently observed Universe, and the horizontal axis shows time. The inflationary period is marked on the graph as a vertical gray band. Before inflation, the size of the Universe in the inflationary model is much smaller than in the original big-bang model, allowing the observed Universe to come to a uniform temperature in the time available.

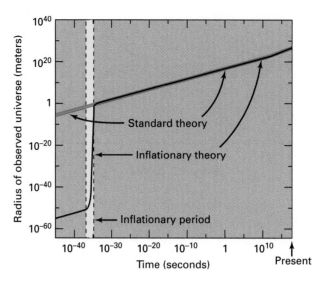

observable Universe is expected to be homogeneous and isotropic, eliminating the horizon problem. (For numerical examples, see *Figure It Out: Inflation*.)

Moreover, any initial curvature in the Universe would have been "flattened out" by the inflation, like a balloon that expands to an immense size. In effect, the Universe became so large that the entire region we can see (only a small fraction of what really exists) appears flat or nearly flat, just as the Earth's surface appears flat when viewed over small distances. The flatness problem is therefore resolved.

After the end of inflation, the Universe would have continued expanding at a rate consistent with the original big-bang theory. So, the idea of inflation affects only the first brief moments in the life of the Universe; it does not invalidate the big-bang theory.

In the inflationary universe, space itself expands far faster than the speed of light. After all, the Universe was initially much smaller than a subatomic particle, yet it quickly became an almost uncountable number of light-years across! Einstein's theory of relativity does not forbid this. Instead, it restricts the transmission of a signal (able to carry information) to a speed no greater than that of light. But the expansion of space cannot be used to transmit information from one object to another. Put another way, relativity states that two objects cannot *pass* each other at speeds exceeding that of light. But the expansion of space itself does not allow particles to pass each other.

Forces in the Universe

Why should the Universe have gone through a period of inflationary expansion? Physicists are now considering a variety of possible reasons, some quite far removed from those advocated by the early pioneers of inflation. But the originally favored reason is still a candidate and is instructive to study. It stems from theories that attempt to unify the known fundamental forces of nature.

For much of the 20th century, physicists thought that there are four fundamental forces in nature. The first force, **gravity,** is extremely weak, but it acts over large distances, and all types of matter (and energy) are affected in the same way, regardless of charge or other properties. We are already very familiar with gravity, since it has played a prominent role in many topics discussed in this book.

Electromagnetism, the second force, is about 10^{39} times stronger than gravity, and it acts over large distances, but only charged or magnetized objects are

Figure It Out

Inflation

Let us consider a numerical example to clarify the concept of inflation in the early Universe. At $t = 10^{-37}$ sec, the radius of the *currently* observable part of the Universe (at $t \approx 14$ billion years) was about 0.01 m, according to the original big-bang theory. Even if the early Universe were transparent, light could have travelled only $d = ct = (3 \times 10^8 \text{ m/sec})(10^{-37} \text{ sec}) = 3 \times 10^{-29}$ m in that amount of time. (The actual distance is somewhat greater, due to the expansion of space itself. Typically it might be a factor of 3 larger—but in this case still only about 10^{-28} m.) This distance is far too small, compared with the radius of the Universe at that time. Thermal equilibrium could not have been established, even with light signals.

Now suppose the radius of the currently observable Universe was actually only 10^{-52} m at $t = 10^{-37}$ sec. This was sufficiently small that the transmission of signals (even slow ones such as particle collisions) *could* have easily produced thermal equilibrium. If the Universe subsequently inflated by a factor of 10^{51} by $t = 10^{-35}$ sec, then its new radius would be about 0.1 m, in agreement with the original big-bang theory; see Figure 19–17. Thereafter, the radius of the currently observable part of the Universe grew at a rate consistent with the original big-bang theory. Inflation affects *only* the first roughly 10^{-35} sec of the Universe's existence.

Here is another example, this time focusing on the *entire* Universe (not just the currently observable part). Suppose the Universe had a radius of 10^{-40} m at $t = 10^{-37}$ sec. If it inflated by a factor of 10^{100} by $t = 10^{-35}$ sec, its new radius at the end of inflation would be 10^{60} m (about 10^{44} light-years)! But the currently observable part of the Universe had a radius of about 0.1 m at $t = 10^{-35}$ sec, and the radius is now about 10^{26} m (or 10^{10} light-years). Both of these radii are much smaller than the total size of the Universe after inflation.

The times discussed here, and especially the initial sizes and the inflation factors, are somewhat arbitrary; they depend on the details of the specific inflationary model under consideration. (In some versions of inflation, the Universe inflates by a factor of 10 to the *trillionth* power!) The important point is that inflation produces a truly *enormous* Universe, far larger than we can see. This means that the geometry of the Universe should appear flat ($\Omega = 1$), as observed.

---

affected. (This particular strength comparison uses an electron and a proton in a hydrogen atom. The exact numbers are different if two protons or two electrons are used.) Since most large objects are electrically neutral and not magnetized, electromagnetic forces are generally not felt over large distances. Electromagnetism is critical, however, to the structure of liquids and solids such as your body, and of course to atoms and molecules themselves.

The third force, known as the **strong nuclear force,** is about 100 times stronger than electromagnetism, but it operates only over an extremely short range. The protons and neutrons in an atomic nucleus are held together by the strong force. Actually, we now know that the "strong force" is just the residue of a more fundamental (and even stronger) "color force" that binds quarks together. However, for historical purposes we will still refer to it as the strong force.

Finally, the **weak nuclear force** is about a million times weaker than electromagnetism, and it operates only over an extremely short range. It is the force that governs certain types of radioactive decay, such as the neutron becoming a proton, an electron, and an antineutrino. Any time neutrinos or antineutrinos interact with matter, the weak force is involved.

In the late 1970s, it was suggested theoretically that the electromagnetic and weak forces are different manifestations of a single more fundamental force, **electroweak.** The electromagnetic and weak forces seem different at low energies characteristic of the Universe today, but at higher energies and temperatures (such as when the Universe was younger than about 10^{-12} sec, and hotter than about 10^{16} K), they behave in a similar manner. We say that the "symmetry is broken" at low energies. In high-energy interactions, the two forces cannot be distinguished and hence appear **symmetric** (unified). A useful analogy is a coin flicked along the surface of a table: When it is spinning rapidly at high energies, it appears symmetric, but we see either "heads" or "tails" when it comes to rest, a "broken symmetry."

High-energy experiments at CERN and elsewhere subsequently confirmed the predictions of the electroweak theory. Specifically, in 1983 the expected "W" and "Z" particles were found (Fig. 19–18). Thus, we now often say that there are only *three* fundamental forces. James Clerk Maxwell achieved a similar "unification" in the 1860s, when he showed that electricity and magnetism are different manifestations of a more fundamental interaction, electromagnetism.

Physicists are now trying to see whether the three known forces are different low-energy manifestations of only one or two fundamental forces. **Grand unified theories (GUTs)** attempt to unify the electroweak and strong nuclear forces. In principle, these forces behave in a similar (symmetric) manner at temperatures above about 10^{29} K (which characterized the Universe at times less than 10^{-37} sec). GUTs can account for the observed excess of matter over antimatter. GUTs also predict that protons are unstable, though proton decay has not yet been observed (Fig. 19–19). The experimental minimum value of about 10^{32} years for the proton lifetime already rules out some candidate GUTs, but many still remain, and we don't know which (if any) is correct.

"Theories of everything" (TOEs) attempt to unify the GUT force and gravity in a single superunified force. It is speculated that these forces are symmetric at temperatures above about 10^{32} K (which characterized the Universe at times less than about 10^{-43} sec, the Planck time). A unification of quantum mechanics and general relativity is needed—that is, a quantum theory of gravity. No fully

A Brief History of Gravity (A Limerick) by Bruce Elliot

It filled Galileo with mirth
To watch his two rocks fall
to Earth.
   He gladly proclaimed,
   "Their rates are the same,
And quite independent
of girth!"

Then Newton announced in
due course
His own law of gravity's force:
   "It goes, I declare,
   As the inverted square
Of the distance from object
to source."

But remarkably, Einstein's
equation
Succeeds to describe
gravitation
   As spacetime that's curved,
   And it's this that will serve
As the planets' unique
motivation.

Yet the end of the story's not
written;
By a new way of thinking
we're smitten.
   We twist and we turn,
   Attempting to learn
The Superstring Theory
of Witten!

Fig. 19–18 Carlo Rubbia and Simon van der Meer with the apparatus at CERN with which they and their colleagues discovered the W^+, W^-, and Z^0 particles. These discoveries verified predictions of the electroweak theory. Rubbia and van der Meer received the 1984 Nobel Prize in Physics as a result.

Fig. 19–19 This cavity, deep underground in a salt mine near Cleveland, was filled with 10,000 tons of water to make the Irvine-Michigan-Brookhaven (IMB) detector. To detect flashes of light that result as neutrinos or other elementary particles pass through it, 2,400 instruments were installed around the sides. Its detection of a few neutrinos from Supernova 1987A confirmed that massive stars can undergo a core collapse to form neutron stars.

self-consistent theory has yet emerged, but many theorists are focusing their attention on **superstring theories** (often called simply **string theories**). Modern string theory was born with the work of John Schwarz (Caltech) and Michael Green (then of Queen Mary College), and one of the current leaders in this quest is Edward Witten (Institute for Advanced Study, Princeton).

In superstring theories, each fundamental particle (quark, electron, neutrino) is thought to be a different vibrating form of a tiny one-dimensional "string" (or perhaps a tiny "membrane," having two or more dimensions). A string is an elongated, one-dimensional analog to the zero-dimensional points that have often been used in theories of matter. Strings are so small (about 10^{-33} cm) that they seem almost like points. Superstring theories thus far appear to be a promising way to get around theoretical difficulties on the extremely small scale where a quantum theory of gravity is needed.

Recent superstring theories tend to favor an 11-dimensional Universe (revised up from 10 dimensions), but 7 of the dimensions are too small to see; they are curled up on scales of about 10^{-33} cm, the Planck length (Fig. 19–20). A useful analogy is to view a sheet of sandpaper from far above: it looks only two dimensional, but is really three dimensional due to the granularity. Another example is a garden hose: from far away, it appears one dimensional (like a string) instead of three dimensional. Nobody knows why the Universe has four macroscopically large dimensions.

One prediction of superstring theories is that the Universe should contain "shadow matter." This has a gravitational effect on normal matter, but does not emit any electromagnetic radiation; it is therefore a candidate for dark matter.

> **Most recently, it has been proposed that at least one of the unseen dimensions is macroscopically large, but very difficult to detect.**

Supercooling the Universe

Right after the big bang, when temperatures were exceedingly high, there should have been just one symmetric, superunified force according to superstring theo-

Fig. 19-20 Theorists calculate that superstrings require an 11-dimensional space. We are familiar with only a four-dimensional space, with three spatial dimensions and one time dimension.

ries. Given the uncertainty over what happens before the Planck time, however, we will instead consider the situation when the Universe was about 10^{-37} sec old, focusing instead on GUTs. The grand unified (GUT) force and gravity would have existed as distinguishable forces.

At t $\approx 10^{-37}$ sec, the temperature would have dropped to 10^{29} K, below which the GUT force should have started to break symmetry, becoming the strong nuclear force and the electroweak force. This **phase transition** is similar to what happens when liquid water is cooled below a temperature of 0°C: it generally turns to ice (Fig. 19–21). When water turns to ice, excess energy is liberated so that the water molecules can line up to become crystals.

Suppose, however, that the Universe cooled below 10^{29} K *without* breaking symmetry—that is, the strong and electroweak forces remained unified as the GUT force. The **supercooled** Universe would then have contained too much energy relative to what it should have had at that temperature. For example, water can be supercooled below 0°C without turning into ice, if one is careful not to agitate the water. Its excess energy keeps it in the liquid state. The excess energy is released when the water freezes.

The energy associated with the supercooled, symmetric (GUT) Universe would have caused space to double its size every tiny fraction of a second—that is, the Universe would have entered an epoch of accelerated, inflationary expansion. (This is where the analogy with water breaks down: the excess energy associated with the supercooled phase does *not* make the water expand!) For reasons that are not intuitively obvious, the Universe's excess energy has a stretching effect on space, like a cosmic "antigravity." Called the "energy of the false vacuum" because a conventional vacuum should have zero energy, it is essentially Einstein's cosmological constant Lambda (Λ) (see Chapter 18), but in this case very early in the history of the Universe. The inflationary expansion made the Universe truly *enormous*; it is much, much larger than the regions that we can actually see.

Eventually, perhaps at an age of 10^{-35} sec, the Universe went through a phase transition, with the electroweak and strong forces becoming separate (that is, the symmetry was broken). At this time all of the excess energy of the false vacuum was released, thereby forming matter, antimatter, and photons. The energy of the false vacuum is thus the origin of all energy in the Universe. With no more "antigravity" force, the inflationary expansion of the Universe subsided, and the regular Hubble expansion began. (The Universe started "coasting," but decelerated by gravity.) In essence, inflation is what gave the Universe its "big bang"—its initial kick up to a time of about 10^{-35} sec, after which all of the events discussed previously in this chapter unfolded (Fig. 19–22).

The achievements of the inflationary theory are magnificent. First, it explains the horizon problem—the homogeneity and isotropy of the Universe. The Universe was sufficiently small to have achieved thermal equilibrium (a uniform temperature) at very early times, and it maintained this as it inflated. Second, the Universe is very close to having a spatially flat geometry. After inflation, it was so large that any observable region would appear flat (that is, $\Omega = 1$; the average total density is equal to the critical density). Third, the observable Universe has few, if any, "magnetic monopoles" (isolated north or south magnetic poles, without the opposite pole being present as well); none has been found despite intensive searches for them. There are also no apparent "cosmic strings"—long, skinny regions that contain huge amounts of mass. Many such "defects" should have been produced by phase transitions soon after the big bang, but the rapid expansion of the Universe during the inflationary era would have whisked them away, thereby accounting for their scarcity. Fourth, the predictions of inflation are consistent with the observed distribution of matter over various size scales (such as clusters of galaxies and superclusters).

A sphere of water in a vacuum, floating in space, looks the same when viewed from different directions (that is, "rotationally symmetric"). In contrast, an ice crystal has a preferred axis (the axis of symmetry), and hence does not appear the same when viewed from different directions. Physicists say that ice is less symmetric than water when they use the word "symmetry" in this manner.

Fig. 19–21 Ice (solid water) and liquid water are different states of matter. The phase change between them is, in some ways, analogous to the phase change that must have taken place early in the history of the Universe.

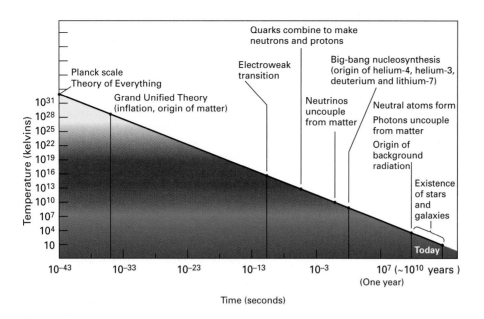

The Ultimate Free Lunch?

In the inflationary theory of the Universe, matter, antimatter, and photons were produced by the energy of the false vacuum, which was released following the phase transition. All of these consist of positive energy. ☼ This energy, however, is exactly balanced by the *negative* gravitational energy of everything pulling on everything else. The *total* energy of the Universe is zero! It is remarkable that the Universe might consist of essentially nothing, but (fortunately for us!) in positive and negative parts. That gravity is associated with negative energy is easily seen: when one drops a ball from rest (defined to be a state of zero energy), it gains energy of motion (kinetic energy) as it falls. But this gain is exactly balanced by a larger negative gravitational energy and the sum of the two remains zero.

The idea of a zero-energy Universe, together with inflation, suggests that all one needs is just a tiny bit of energy to get the whole thing started (that is, a tiny volume of energy in which inflation can begin). The Universe then goes through inflationary expansion, but without creating net energy.

What produced the energy before inflation? This is perhaps the ultimate question. It may have come out of *nothing*! The meaning of "nothing" is somewhat ambiguous here. It might be the vacuum in some preexisting space and time, or it could be nothing at all—that is, all concepts of space and time were created with the Universe itself.

According to quantum theory, there is a natural way that the energy may have come out of nothing (see *Figure It Out: Heisenberg's Uncertainty Principle*). As mentioned in Chapter 18, it turns out that "virtual" pairs of particles and antiparticles can spontaneously form and quickly disappear without measurably violating the law of conservation of energy (Fig. 19–23). These spontaneous births and deaths are called "quantum fluctuations." Indeed, we know that quantum fluctuations are occurring all the time; virtual pairs of particles (such as electrons and positrons) affect the energy levels of atoms, and the predicted energy levels disagree with the experimentally measured levels unless this process is taken into account.

Thus, perhaps there were many quantum fluctuations before the birth of our Universe. Most of them quickly disappeared—but one lived sufficiently long and had the right conditions for inflation to have been initiated. Thereafter, the initially tiny volume inflated by a huge factor, and the macroscopic Universe was born. The original particle–antiparticle pair (or pairs) may have subsequently

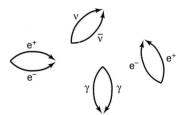

Fig. 19–23 Formation of virtual pairs of particles and antiparticles, quickly followed by their mutual destruction. According to quantum theory, such spontaneous births and deaths ("quantum fluctuations") can and do exist, even in a vacuum. Their effects have been experimentally verified.

Figure It Out

Heisenberg's Uncertainty Principle

A key feature of quantum theory is the Heisenberg Uncertainty Principle. In one of its forms, this principle states that "virtual" particles of energy ΔE can spontaneously form and live for a time Δt without measurably violating the law of conservation of energy, if the product $\Delta E \Delta t$ is no larger than about $h/2\pi$, where h is Planck's constant. These "quantum fluctuations" can and do occur, even in a complete vacuum. In a sense, space itself is uncertain of its energy by an amount up to ΔE, if the time scale over which it is measured is less than or equal to Δt, where the product of the two does not exceed $h/2\pi$.

Thus, the smaller the energy (ΔE) of the virtual particles that are spontaneously created, the larger the time interval (Δt) over which they can exist. For example, a virtual proton-antiproton pair, which has a mass of about 3.3×10^{-24} g and a corresponding energy given by $E = mc^2$, can live up to about 3×10^{-25} sec. But a virtual electron-positron pair can exist much longer, since its mass is nearly 2 thousand times less than that of a virtual proton-antiproton pair.

annihilated—but even if not, the violation of conservation of energy would be minuscule.

If this admittedly speculative hypothesis is correct, then the answer to the ultimate question is that the Universe is the "ultimate free lunch"! It came from nothing, and its total energy is zero, but it nevertheless has incredible structure and complexity.

⬤ A UNIVERSE OF UNIVERSES

Could it be that ours is not the only Universe, but rather one of many? The entire history of our Universe might be just one episode in the much grander, perhaps infinite **multiverse**. This idea is the most extreme extension of the Copernican Principle. Some scientists, notably the celebrated British cosmologist Sir Martin Rees, find it quite attractive. Others, however, are skeptical of the idea; it is certainly quite controversial at this time, and more speculative than even the "free lunch" hypothesis discussed above.

The concept of another "universe" is difficult to define. One possibility is a distinct region that occupies dimensions completely separate from those of our Universe. This other universe might, however, be connected with ours via a wormhole—a passage between two black holes.

Another possibility is a region in our physical Universe, or arising from our Universe, in which inflation occurred or ended in a different way. In some or all cases, the physical constants (or, much more speculatively, perhaps even the laws of physics) may differ from those in our Universe. By "physical constants" we mean the relative strengths of the various forces, the values of quantities such as the speed of light, and the masses of fundamental particles such as electrons and neutrinos.

As summarized below, there are scientifically valid reasons to think that other universes might exist. This interesting conclusion, based on science, by its very nature removes itself from the realm of science, at least temporarily: Generally, there is no known way to contact other universes and perform experiments on them, even in principle. One exception would be a different region in our Universe that eventually becomes observable when photons reach us, far in the future. Also, consider sending light signals through a wormhole: It is conceivable that a future generation will discover matter with antigravity properties (on small scales) and will transport it into the wormhole, perhaps thereby keeping it open.

Fig. 19–24 The cover of *Science* magazine from December 18, 1998. The caricature of Einstein is surprised because the "bubble universes" that he is blowing appear to be accelerating in their expansion, a discovery made by two teams studying distant supernovae (Chapter 18) and hailed as the top scientific breakthrough of 1998.

Some other method of communication might, in principle, eventually be found. But for now, at least, we can only speculate about the "multiverse."

There seem to be at least three or four ways to produce other universes (Fig. 19–24). For example, distinct quantum fluctuations could arise out of "nothing." This "nothing" might be the vacuum of our Universe, or the "nothing" outside our Universe, whatever that means. If our Universe was created as a quantum fluctuation out of "nothing," it seems reasonable that such a process may have occurred many times, perhaps even infinitely many. We do not yet know whether this is a viable process: a fully self-consistent quantum theory of gravity is not yet available, although theorists are quite excited by the potential of superstring theories.

Other universes may also be regions of different broken symmetry in an inflating volume. As the false vacuum inflates, the phase transition (symmetry breaking) may occur in different ways in different regions. This is analogous to water freezing along different axes in a jar: a jumble of ice crystals forms, with various orientations. The values of the physical constants could differ in these regions of different broken symmetry. (More speculatively, the physical laws themselves might differ in them.) These regions could stay separated from each other due to the presence of a still inflating false vacuum between them.

A related scenario, pursued largely by Andrei Linde (Fig. 19–25), is that an inflating volume produces new, distinct, inflating regions within it, due to quantum effects. Although most of the volume will have undergone a phase transition (symmetry breaking) and stopped inflating, some regions will continue to inflate. These inflating parts will subsequently spawn regions that continue to inflate, even as symmetry is broken throughout most of their volume (Fig. 19–26). The process never ends, and is called eternal inflation. Again, the values of the physical constants may differ in the different regions of inflation.

Yet another idea is that a new universe can sprout from within any black hole in our Universe. Even more speculatively, the main characteristics of our Universe might even be transferred to its progeny. Lee Smolin of the Pennsylvania State University promotes this notion of "universe heredity."

Fig. 19–25 Andrei Linde, who has proposed many different ways in which inflation can occur, including the idea of eternal inflation.

● A UNIVERSE FINELY TUNED FOR LIFE?

We will end this chapter on an intriguing, but very controversial note, one with which some scientists vigorously disagree. It is at the extreme boundary of what can currently be called science, and may be more suitable to theology, philosophy, or metaphysics. But it is worth bringing up for completeness, since it addresses our most basic origins.

Our existence may allow us to draw some interesting conclusions about the Universe—for example, that it contains elements heavier than hydrogen and helium. Such deductions utilize the **anthropic principle:** We exist, hence the Universe must have certain properties or we wouldn't be here to see it. In a few cases, apparently "mysterious" relationships between numbers were explained with anthropic reasoning. A previously unknown property of the energy levels of the carbon nucleus was actually *predicted* by Fred Hoyle in this manner. He argued that it must exist—otherwise there would not be enough carbon in the Universe.

The values of the physical constants (not to mention the laws of physics) seem to be spectacularly "fine tuned" for life as we know it—indeed, almost "tailor made" for humans. In many cases, if things were altered just a tiny amount, the results would be disastrous for life, and even for the production of heavy elements or molecules. The Universe would be "stillborn" in terms of its complexity. Here are a few examples.

Fig. 19–26 A series of inflating universes, each giving rise to new regions of inflation. The different colors represent changes in the values of physical constants (or, more speculatively, perhaps even different physical laws) among the universes.

If the weak nuclear force were weaker, most neutrons would not have turned into protons shortly after the big bang. The protons and neutrons would have then fused to form helium, leaving little hydrogen. A similar conclusion is reached if the weak nuclear force were much stronger: Protons would quickly combine with each other to form helium nuclei, converting two of the four protons to neutrons.

Now suppose the strong nuclear force were weaker. It turns out that the elements heavier than hydrogen would be unstable. The "periodic table of the elements" would be very boring, and stars would not produce nuclear energy. Conversely, if the strong nuclear force were stronger, protons would bind together so easily that hydrogen would not exist. Or, if the mass of a neutron and a proton were more nearly equal (they are the same to within about 0.14 per cent), most protons would be converted to neutrons. Without hydrogen, stars would live only a short time, and there would be no hydrogen-rich compounds such as water and hydrocarbons.

Additional examples abound. If gravity were considerably stronger relative to electromagnetism, stars would be more massive, and they would burn out too quickly for very complex life (as we know it) to develop. Also, if certain relationships between the energy levels of different kinds of atomic nuclei were not present, the heavy elements would not have formed; the case of carbon, predicted by Hoyle, was mentioned above. If electrons were not much less massive than protons, complex chemistry would not be possible. Large molecules such as DNA can have well-defined shapes only if the nuclei of atoms are much heavier than electrons.

Some people use these and other "cosmic coincidences" to argue for a divine Creator. But this theological conclusion is not testable by the methods of science. Therefore, here we will not consider it further, though perhaps it is true.

Or, one could say that there is only a single Universe, and it has these properties by chance. This conclusion is difficult to accept; the odds are astronomically low (pun intended), if most other possible combinations of physical constants lead to conditions unsuitable for life. However, it may be true.

Finally, one could say that there is an ensemble of universes, perhaps even infinitely many, which span a very wide range of properties (values of the physical constants, particle masses, etc.). We, of course, were dealt a "winning hand"— indeed, a royal flush: We live in a universe that allowed life and sentient beings to develop. Such universes might be exceedingly rare; most random hands are not winners. Though an excellent example of the Copernican Principle applied to the grandest imaginable scale, the multiverse idea is not directly testable, and is therefore more in the realm of philosophy and theology than of science. It seems that we are still left with the ultimate question: How and why did the Universe (or the superspace that contains it) occur? Perhaps there is no way, even in principle, for humans to answer this question.

Concept Review

The expansion of the Universe alone suggests, but does not demand, that there was a big bang. According to the alternative **steady-state theory,** the Universe had no beginning in time and will have no end. This idea is based on the **perfect cosmological principle:** The average properties of the Universe do not change at all with time. Historically, there were several observational arguments against the steady-state theory, but the fatal blow was the discovery of the **cosmic background radiation,** a faint radio glow that uniformly pervades the Universe. It was produced when the Universe was hot, be-

fore an age of about 300,000 years, and now corresponds to a very low temperature because of the cooling produced by expansion. In the 1990s, tiny ripples were detected in the cosmic background radiation, corresponding to the seeds from which superclusters and clusters of galaxies were formed. The observed angular size of typical variations indicates that the Universe is spatially flat on large scales.

We considered the first several minutes after the big bang, starting from an almost inconceivable age of 10^{-43} sec, the **Planck time.** Up to an age of about 1 microsecond, the Universe consisted of a primordial soup of **quarks,** gluons (which normally bind quarks together), and photons. A slight dominance of matter over antimatter was established, making possible our existence. The lightest elements were formed in the first 10 minutes through the process of **primordial nucleo-synthesis,** and their currently observed proportions show striking agreement with the predictions of the big-bang theory. The relative amounts of different isotopes, combined with other observations, tell us that much of the dark matter in the Universe must consist of exotic particles such as neutrinos.

Some observed aspects of the Universe, such as its incredible homogeneity (the **horizon problem**) and nearly flat spatial geometry (the **flatness problem**), are difficult to understand in the context of the original big-bang theory. Perhaps the Universe began much smaller than we had thought, was able to achieve a uniform temperature, and subsequently went through an era of extremely rapid expansion known as **inflation.** The resulting Universe would then be truly enormous, much larger than what we can see—and our relatively small observed region would naturally appear to be homogeneous and flat. Inflation affects just the first tiny fraction of a second of the Universe's existence, yet it removes some of the arbitrary assumptions of the original big-bang theory.

A clue to why inflation occurred may be provided by theories that postulate a high-energy merging of at least three of the four fundamental forces of nature: **gravity, electromagnetism, strong nuclear force,** and **weak nuclear force.** Already physicists have shown that electromagnetism and the weak nuclear force are unified at high energies to form the **electroweak** force; the two forces cannot be distinguished and hence appear **symmetric** (unified). At still higher energies, the electroweak and strong nuclear forces might merge, according to **grand unified theories (GUTs).** It is possible that even gravity can be unified with the grand unified force. Perhaps the best candidates for such "theories of everything" (TOEs) are **superstring (string) theories,** which postulate that all fundamental particles are different forms of vibration of a tiny string or membrane.

As the Universe expanded and cooled, the various forces should have become apparent as different manifestations of the same force—that is, **phase transitions** should have occurred, like water turning to ice. But if the Universe **supercooled** without undergoing a phase transition, it would have contained excess energy that caused inflation for a short time. Moreover, the Universe could have begun as a quantum fluctuation out of *nothing*, the "ultimate free lunch."

We end with the very speculative possibility that there are many, perhaps even infinitely many, universes in a **multiverse.** The values of the physical constants might be different in universes other than our own. According to the **anthropic principle,** our existence allows us to deduce some properties of the Universe. Many calculations suggest that the consequences of modified constants would be disastrous for life as we know it, or even for the production of heavy elements or molecules. Perhaps our Universe is simply one of few in the multiverse whose conditions are suitable for the formation of life.

Questions

1. Explain how, in the absence of other information, the expansion of the Universe could be consistent with no beginning and no end in time: a "steady-state universe."

2. Summarize some observational arguments against the steady-state theory.

3. Describe the origin of the cosmic background radiation and the reason it now corresponds to a very low temperature.

4. Would you believe the announcement of the discovery of a galaxy at redshift 10,000?

5. Explain why the cosmic background radiation looks slightly hotter in one direction of the sky, and slightly cooler in the opposite direction.

6. Discuss the significance of tiny inhomogeneities in the cosmic background radiation.

7. Explain why we can consider with some confidence the behavior of the Universe early in its history, even before it was 1 second old.

8. Discuss the significance of the matter–antimatter asymmetry in the Universe.

9. Summarize the main physical events that occurred in the Universe's first 10 minutes of existence.

10. Describe how the currently observed proportions of the lightest chemical elements reflect conditions in the Universe shortly after its birth.

11. What are two observational problems with the original big-bang theory of the Universe?

12. Explain how the main problems with the original big-bang theory can be resolved if, at very early times, the Universe was much smaller than we had thought and subsequently inflated by an enormous factor.

13. Explain how space itself can expand faster than light without violating Einstein's theory of relativity.

14. Describe the fundamental forces of nature and attempts to unify them.

15. Discuss what should have happened to the "grand unified force" as the Universe expanded and cooled at an age of about 10^{-37} sec.

16. Outline how the Universe may have supercooled, leading to inflationary expansion by an almost arbitrarily large factor when the Universe was only 10^{-35} sec old.

†17. Suppose that during inflation, the size-doubling time scale was only 10^{-38} sec. If inflation started at $t = 10^{-37}$ sec and continued until $t = 10^{-35}$ sec, show that about 1000 doublings occurred.

18. Explain how the *total* energy of the Universe might be zero, or nearly zero.

19. Why has it been suggested that the Universe may be the "ultimate free lunch"?

20. Define what is meant by the term "universe."

21. Describe several ways in which there might be other universes besides our own.

22. Outline the basic idea behind superstring theories and the possible existences of dimensions far too small to detect.

23. Discuss how the properties of the Universe depend on the values of the physical constants.

24. Summarize various possibilities for why our Universe seems to be so finely tuned for life.

25. State the anthropic principle and its implications.

---

†This question requires a numerical solution.

Life in the Universe

20

ORIGINS *One of the most important questions we can pursue in our quest to determine our origin is whether life, and especially intelligent, communicating life, has arisen independently elsewhere in the Universe. Are humans a common phenomenon, or are we alone?*

We have discussed the nine planets and some of the moons in our Solar System, and have found most of them to be places that seem hostile to terrestrial life forms. Yet a few locations besides the Earth—most notably Mars, with its signs of ancient running water, and Europa, with liquid water below its icy crust—have characteristics that suggest life may have existed there in the past, or might even be present now or develop in the future. **Exobiology** is the study of life elsewhere than Earth.

In our first real attempt to search for life on another planet, in the 1970s, the Viking landers carried out biological and chemical experiments with martian soil. The results seemed to show that there is probably no life on Mars (Fig. 20–1). Although studies of a martian meteorite in 1996 gave some indications of ancient, primitive life on Mars (see Chapter 6), this idea has not been generally accepted, though it is still causing much discussion. Jupiter's moon Europa and Saturn's moon Titan (see Chapter 7) are also intriguing places where scientists think it is possible that life has begun.

Since it seems reasonable that life as we know it would be on planetary bodies, we first discuss the chances of life arising elsewhere in our Solar System, as well as the kinds of stars most likely to have planets suitable for the emergence and development of intelligent life. Next we explore attempts to receive communication signals from intelligent extraterrestrials. We also consider a way to estimate the number of communicating civilizations elsewhere in our galaxy, or at least to see which factors most seriously limit our ability to do so. It is possible that humans are the only technologically advanced civilization in our galaxy, or one of very few. Finally, we explain why most scientists do not consider reported sightings of UFOs to be good evidence for extraterrestrial visitations to Earth.

NASA has formed an institute of "astrobiology" and is making a major push to investigate matters of biology that can be important to understanding the origin of life or to space exploration. The institute is "virtual," in that it has no actual buildings but rather is made of individuals who communicate by e-mail and by occasional meetings.

◀ Yoda. From *Star Wars: Episode I—The Phantom Menace*, © Lucasfilm, Ltd. & courtesy of Lucasfilm Ltd. All rights reserved. Used under authorization.

AIMS

To discuss the nature of life and conditions necessary for its emergence.
•
To assess the probability that intelligent communicating life exists elsewhere in our galaxy besides the Earth.
•
To explore ways of searching for extraterrestrial life.

Figure 20–1 The surface of Mars from a Viking Lander, which carried a small biology laboratory to search for signs of life. It probably didn't find any.

Figure 20-2 *"I'll tell you something else I think. I think there are other bowls somewhere out there with intelligent life just like ours." (Drawing by Frank Modell; copyright 1987 The New Yorker Magazine, Inc.)*

🔵 THE ORIGIN OF LIFE

It would be very helpful if we could state a clear, concise definition of life, but unfortunately that is not yet possible. Biologists state several criteria that are ordinarily satisfied by life forms—reproduction, for example. Still, there exist forms on the fringes of life, such as viruses, that need an organism in order to reproduce. Scientists cannot always agree whether some of these things are "alive" or not.

In science fiction, authors sometimes conceive of beings that show such signs of life as the capability for intelligent thought (Fig. 20–2), even though the being may share few of the other criteria that we ordinarily recognize. In Fred Hoyle's novel *The Black Cloud*, for instance, an interstellar cloud of gas and dust is as alive as (and smarter than!) human beings. But we can make no concrete deductions if we allow such wild possibilities, so exobiologists prefer to limit the definition of life to forms that are more like "life as we know it." (www)

Life on Earth is based on **amino acids**—chains of carbon, in which each carbon atom is bonded to hydrogen and sometimes to oxygen and nitrogen. Chemically, carbon is essentially unique in its ability to form such long chains; indeed, we speak of compounds that contain carbon as **organic.** But life is selective, incorporating only about 20 of all the possible amino acids. Similarly, long chains of amino acids form **proteins,** but life utilizes only a small fraction of the multitude of possible combinations of amino acids. The genetic code of any living creature is contained in one extremely long and complex structure: DNA (deoxyribonucleic acid), the famous "double helix."

How hard is it to build up long organic chains? To the surprise of many, an experiment performed in the 1950s showed that making organic molecules is easier than had been supposed. Stanley Miller and Harold Urey, at the University of Chicago, filled a glass jar with simple molecules like water vapor (H_2O), methane (CH_4), and ammonia (NH_3), along with hydrogen gas (H). They exposed it to electric sparks, simulating the vigorous lightning that may have existed in the early stages soon after Earth's formation. After a few days, long chains of atoms formed in the jar, in some cases complex enough to include simple amino acids, the building blocks of life. Later versions of these experiments created even more complex organic molecules from a wide variety of simple actions on sim-

Figure 20-3 *(A)* Bishun Khare and Carl Sagan, of Cornell University, in their laboratory several decades ago. The apparatus was used to simulate conditions thought to be present on the primitive Earth, to see whether complex compounds such as amino acids form easily from the gases of which Earth's atmosphere consisted at that time. *(B)* Louis J. Allamandola with equipment used for similar purposes in his Astrochemistry Laboratory at the NASA/Ames Research Center.

A

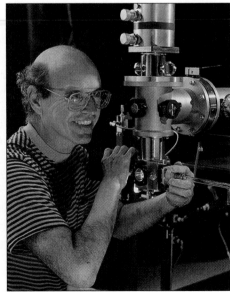

B

ple molecules (Fig. 20–3). Such molecules may have mixed in the oceans to become a "primordial soup" of organic molecules (Fig. 20–4).

Since the original experiment of Miller and Urey, most scientists have come to think that the Earth's primitive atmosphere was not made of methane and ammonia, which would have disappeared soon after the Earth's formation. Instead, it may have consisted mostly of carbon dioxide, carbon monoxide, and water, and such a mixture does not generally lead to a large abundance of amino acids. On the other hand, complex molecules may have formed near geothermal sources of energy under the oceans or on Earth's surface, or perhaps in vents deep underground, where the right raw materials existed. Also, extraterrestrial amino acids have been found in several meteorites that had been long frozen in Antarctic ice, as well as in some other meteorites (Fig. 20–5). In any case, some planets or moons in our galaxy may have had primitive atmospheres similar to the mixture used by Miller and Urey, so the results of their experiments are interesting.

However, mere amino acids or even DNA molecules are not life itself. A jar containing a mixture of all the atoms that are in a human being is not the same as a human being. This is a vital gap in the chain; astronomers certainly are not yet qualified to say what supplies the "spark" of life.

Still, many astronomers think that since it is not difficult to form complex molecules, life may well have arisen not only on the Earth but also in other locations. The appearance of very simple organisms in Earth rocks that are 3.5 billion years old, and indirect evidence for life as far back as 3.8 billion years (not long after the end of the bombardment suffered by the newly formed Earth), suggests that primitive life arises quite easily. The discovery of indigenous life on at least one other planet or moon in our Solar System would provide much support for this hypothesis. But even if life is not found elsewhere in the Solar System, there are so many other stars in space that it would seem that life might have arisen at some location.

LIFE IN THE SOLAR SYSTEM

Life elsewhere in our Solar System, if present at all, is primitive at best (single cells, or perhaps very simple multi-cellular organisms). Among all of the planets and moons, only a few have nonzero odds for life.

Mars has provided the best evidence thus far, but it is still very controversial (Fig. 20–6). Moreover, even if real, life on Mars might not have been independent of Earth—a meteorite from Mars containing simple life may have contaminated the young Earth, "seeding" it with life, though this idea is of course quite speculative. Europa, one of Jupiter's large moons, looks like a promising environment: Below its icy surface there is almost certainly water slush or even an ocean. A Europa orbiter to study this moon in more detail is being planned by NASA. (www) Titan, Saturn's largest moon, has a thick atmosphere of nitrogen molecules; in 2004, the Cassini spacecraft will send a probe into this atmosphere to study it in more detail. (www) There is evidence for substantial methane and ethane on Titan, perhaps in the form of lakes. The absence of liquid water, however, makes it more difficult for life to form. Io, another Galilean satellite, has conditions that may be suitable for life resembling that found near volcanic vents on Earth. Again, however, the apparent absence of water is a major problem.

Note that an intelligent alien who obtains and correctly interprets a spectrum of Earth could deduce the presence of life here. The large amount of free oxygen suggests the continuous production by a process like photosynthesis; otherwise, oxygen would rapidly be depleted because it is so reactive. In addition, methane quickly reacts with oxygen, so the significant amount of methane (largely

Figure 20–4 The simple solution of organic material in the oceans, from which life may have arisen, is informally known as "primordial soup."

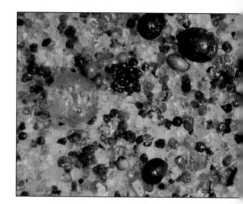

Figure 20–5 Close-up of a part of the Murchison meteorite, in which simple amino acids have been found. Analysis shows that they are truly extraterrestrial and the sample had not simply been contaminated on Earth.

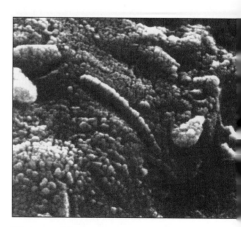

Figure 20–6 Structures in the martian meteorite ALH 84001, seen with a powerful microscope, and thought by some scientists to be remnants of primitive life forms.

from cows—"bovine flatulence," to quote the late Carl Sagan) in our atmosphere implies a steady production mechanism—the decay of organic compounds.

⬤ SUITABLE STARS FOR INTELLIGENT LIFE

If we seek indigenous (that is, originating locally, not from elsewhere), intelligent life on planets orbiting other stars, what kinds of stars have the best odds? Stars that are either near the beginning or end of their lives are not very good bets. For example, intelligent life may take a long time to form, so very young stars are not suitable. Stars heading to the red giant or white dwarf stages go through rapid changes, making it difficult for complex life to survive. White dwarfs and neutron stars certainly have passed through stages that would have destroyed life.

Some types of main-sequence stars are also not especially suitable. The lives of O-type and B-type stars are probably too short for the development of life of any kind. Planets around A-type stars might have life, but probably not intelligent life. Type M and L stars have a small **ecosphere**—the range of distances in which conditions suitable for life might be found. There is unlikely to be a planet in this narrow region around a low-mass star. Also, such a planet would be in "synchronous rotation": The same hemisphere would always face the star, so one side would be very hot, and the other very cold. ☼

Main-sequence stars of types F, G, and K are the most likely candidates. They live a long time, and models of their ecosphere lead to reasonably large sizes. (In more detail, such models must explain the "Goldilocks effect" in the case of our Solar System: why Venus is too hot, Mars is too cold, but the Earth is just right.) Single stars, stars in very wide binary systems (with planets orbiting close to one star), or closely spaced binary stars (which planets could orbit at large distances) are most suitable. Planets that move from one star to another in a binary system (for example, like a figure eight) tend to have unstable orbits and are ejected.

About three dozen extrasolar planets have already been found (Chapter 7). In most cases they orbit F, G, and K stars because the searches specifically targeted those stars. Generally these planets are gas giants, in many cases very close to the star or on highly eccentric orbits and hence probably inhospitable to intelligent life. A few of them, however, appear to have orbital properties potentially suitable for the emergence of life; perhaps at least simple life exists within their atmosphere or on moons orbiting them.

⬤ THE SEARCH FOR EXTRATERRESTRIAL INTELLIGENCE

How should we look for intelligent extraterrestrial life? Perhaps we can find evidence for such life here on Earth, in the form of alien spacecraft that have landed here. After all, Pioneers 10 and 11 and Voyagers 1 and 2 are even now carrying messages out of the Solar System in case an alien interstellar traveller should happen to encounter these spacecraft (Fig. 20–7). The odds, however, seem very small. Although spaceships can, in principle, travel between the stars, even relatively quickly according to Einstein's special theory of relativity (see *Figure It Out: Interstellar Travel*), such journeys are difficult, expensive, dangerous, and probably rare compared to communication with electromagnetic waves.

A potentially more fruitful approach is to search for electromagnetic signals, but some waves are more suitable than others. For example, x-ray and gamma-ray photons have high energy and are therefore expensive to produce, and typical atmospheric gases block them. Ultraviolet photons are absorbed by interstellar

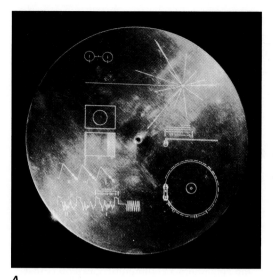

A

B

Figure 20–7 *(A)* The gold-plated copper record bearing two hours' worth of Earth sounds and a coded form of photographs, carried by Voyagers 1 and 2. The sounds include a car, a steamboat, a train, a rainstorm, a rocket blastoff, a baby crying, animals in the jungle, and greetings in various languages. Musical selections include Bach, Beethoven, rock, jazz, and folk music. *(B)* The record includes 116 photographs. Among them is this view, which one of us (J.M.P.) took when in Australia for a solar eclipse. It shows Heron Island on the Great Barrier Reef in Australia, in order to illustrate an island, an ocean, waves, beach, and signs of life.

dust in the plane of our Milky Way Galaxy. At optical wavelengths the signal from a planet orbiting a star is very difficult to distinguish from the bright light of the star itself. And at infrared wavelengths, Earth's warm atmosphere makes the sky bright.

For many years, radio waves have seemed to be the best choice: They are easy and cheap to produce, and are not generally absorbed by interstellar matter. Also, there are few sources of contamination—although the increasing number and strength of radio and television stations is a threat to radio astronomers in the same manner that city lights brighten the sky at optical wavelengths.

At least initially, it seems too overwhelming a task to listen for signals at all radio frequencies in all directions and at all times. One must make some reasonable guesses on how to proceed. A few frequencies in the radio spectrum seem especially fundamental, such as the 21-cm line of neutral hydrogen (as described in Chapter 15). This wavelength corresponds to 1420 MHz, a frequency over ten times higher than stations at the high end of the normal FM band. We might conclude that creatures on a far-off planet would decide that we would be most likely to listen near this frequency because it is so fundamental. The "water hole," the wavelength range between the radio spectral lines of H and OH, has a minimum of radio noise from background celestial sources, the telescope's receiver, and the Earth's atmosphere, and so is another favored possibility.

Humans have conducted several searches for extraterrestrial radio signals. No unambiguous evidence has been found, but the quest is worthwhile and continues. A telltale signal might be an "unnatural" set of repeating digits, such as the first 100 digits of π (see Appendix 1). We must verify that the signal could not be produced by a natural, non-intelligent source (such as a pulsar), and also that it is not of human origin (either unintentionally or intentionally).

In 1960, Frank Drake conducted the first serious search for extraterrestrial signals. He used a telescope at the National Radio Astronomy Observatory (in Virginia) for a few months to listen for signals from two of the nearest stars—tau Ceti and epsilon Eridani. He was searching for any abnormal kind of signal—a sharp burst of energy, for example. He called this investigation Project Ozma after the queen of the land of Oz (Fig. 20–8]) in L. Frank Baum's stories.

A NASA-sponsored group based at the Search for Extraterrestrial Intelligence (SETI) Institute began an ambitious effort on October 12, 1992, the 500th anniversary of Columbus's landing in the New World. It made use of sophisti-

Figure 20–8 Dorothy and Ozma climb the magic stairway. *(From Glinda of Oz, by L. Frank Baum, illustrated by Roy Neill, copyright 1920.)*

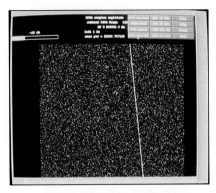

Figure 20–9 This "waterfall" display from Project Phoenix shows the instantaneous power in 462 narrow channels, each only 2 Hz wide. Each dot represents a channel, and the size of the dot is proportional to the strength of the signal. As a new spectrum is plotted in the top row across the screen, the older spectra move downward. Here, the signal changes steadily in frequency, so it slants on the screen. It is readily distinguishable from the background noise.

cated signal-processing capabilities of powerful computers to search millions of radio channels simultaneously in the microwave region of the spectrum (Fig. 20–9). Consisting of both a sky survey and a targeted study of individual stars, in its first fraction of a second it surpassed the entire Project Ozma. But Congress cut off funds anyway, after about a year. The targeted search of the project, known as Project Phoenix, is continuing, backed by funds contributed by private individ-

Figure It Out

Interstellar Travel

Interstellar travel is, in principle, possible for humans to achieve. If the speed is constant, distance is equal to speed multiplied by time ($d = vt$), and we can calculate that a rocket moving with the Earth's escape velocity (11 km/sec) would reach the bright star Sirius (8.7 light-years away) in 240,000 years. (Recall from Chapter 14 that the escape velocity is the speed needed at liftoff to escape from an object's surface.) One can imagine a huge spaceship containing a community of explorers that has permanently left Earth. They would have to be self-sufficient (for example, grow their own food), and the journey to Sirius would span roughly 10,000 generations.

But what if you want to complete the journey within a single lifetime? It is possible, according to Einstein's special theory of relativity. The key fact is that time slows down if you move rapidly: A given time interval t (according to a clock on Earth) is reduced to $t \times [1 - (v/c)^2]^{1/2}$ for the traveller, where c is the speed of light. For example, at a speed of $v = 0.995c$, the round-trip travel time to Sirius according to an observer on Earth is 17.5 years, but only 1.75 years according to the traveller. (Here we ignore the additional time it takes to accelerate up to that speed, but the general idea still holds.) At higher speeds, the traveller ages even less: If $v = 0.9999c$, 17.4 years will have passed on Earth, but the journey is only 3 months long according to the traveller's clock! Thus, by moving fast one ages less and effectively "jumps" into the future. However, the actual life span according to the traveller's clock is unchanged: In 3 months of elapsed time, for example, the traveller reads far fewer books than the Earthling does in the corresponding 17.4 years.

How can this be, according to the traveller's frame? He sees the Earth zooming away from him, and Sirius is rapidly approaching. Well, special relativity tells us that lengths along the direction of motion contract. The distance between Earth and Sirius is reduced to $L \times [1 - (v/c)^2]^{1/2}$ for the traveller, where L is the length measured by a person on Earth. At $v = 0.995c$, the round-trip distance is only $(17.4\,\text{ly})(0.1) = 1.74$ ly according to the traveller, and this is traversed in only 1.75 years at $v = 0.995c$. At $v = 0.9999c$, the round-trip distance is only 3 light-months, which is traversed in just 3 months. Thus, at high speeds, times dilate and lengths contract. Everything is self-consistent, and rapid interstellar travel is possible in principle.

The overwhelming *practical* problem is that travel close to the speed of light requires astronomically large amounts of energy, if the rocket's mass is non-negligible. According to special relativity, the total energy E of the rocket is given by $m_0 c^2 / [1 - (v/c)^2]^{1/2}$, where m_0 is the rest mass. For $v = 0.995c$, $E \approx 10 m_0 c^2$—that is, 10 times its rest mass; for $v = 0.9999c$, $E \approx 71 m_0 c^2$. These are the energies that must be expended to accelerate to such speeds, assuming 100% efficiency! (To give you some feeling for this, the *most powerful* bombs ever built by humans release the energy equivalent of only about 2 kilograms of material—that is, $E = m_0 c^2$, where $m_0 = 2$ kg.) If the efficiency of energy conversion is only 1% (the maximum possible for nuclear energy), then $E \approx 1000 m_0 c^2$ or $7100 m_0 c^2$ for the above two examples. These are truly staggering amounts of energy!

When one considers that the rockets sent to the Moon had about 1000 times as much mass as their payload, the energy requirements become even more prohibitive, since the rest mass m_0 is so large. One way to decrease the rest mass is to gather fuel for the trip along the way, perhaps from interstellar gas. Maybe some highly advanced aliens have overcome these and other formidable problems, such as reaching very high speeds in relatively small amounts of time (which requires large accelerations—and these can be harmful to life).

uals. ⓦⓌⓌ Now led by astrophysicist Jill Tarter (Fig. 20–10), a real-life model for actress Jodie Foster's character Ellie Arroway in the movie *Contact*, it examines about 1000 stars. No unexplained signals have been found so far, though there have been some exciting false alarms, as in 1997 when an intriguing signal turned out to be from the SOHO spacecraft (see Chapter 9)!

A well-known project is SERENDIP, better known as "seti@home," whose operation is based at the University of California, Berkeley (see **http://seti.berkeley.edu**). It has established a way for the general public to contribute: During otherwise unused time on your home computer, a special program can automatically analyze data from the giant radio telescope at Arecibo, Puerto Rico (Fig. 20–11) for signs of extraterrestrial signals (Fig. 20–12). Thus, there is a very small but nonzero chance that the first unambiguous evidence of intelligent life elsewhere in the Universe would be found by *your* computer, should you choose to participate! The effort already has about 2 million participants in over 200 countries, effectively forming the Earth's largest supercomputer, and the amount of computing time contributed from May 1999 through the present (mid-2000) was over 200,000 years. ⓦⓌⓌ

Although most searches for extraterrestrials have been conducted at radio wavelengths, recently some optical and near-infrared searches have begun. Because of absorption of visual light by dust in the plane of the Milky Way Galaxy, we can't expect to survey as many stars as at radio wavelengths, but there are still plenty of them. If we search for short, very intense pulses of light emitted by lasers, we can actually see quite far in the plane of our galaxy, increasing the number of available stars. More importantly, the laser pulses outshine the light from the star that a planet is orbiting. Finally, such laser pulses are very difficult or impossible to produce by any natural phenomena other than life; detection of them could provide strong evidence for the existence of extraterrestrials. ☼

⬤ COMMUNICATING WITH EXTRATERRESTRIALS

All of the searches described above are passive—astronomers are simply looking for signals from intelligent, communicating extraterrestrials. If such a signal is ever found and confirmed by several cross-checks, we might choose to "reply"— but not until some global consensus has been reached about who will speak for Earth and what they will say. Nevertheless, we have already intentionally sent our own signals toward very distant stars that may or may not be orbited by planets containing intelligent life.

Figure 20–10   Jill Tarter, Director of Project Phoenix at the SETI Institute.

Figure 20–11   The Arecibo radio telescope in Puerto Rico, used once to send a message into space. The telescope is 305 meters across, the largest telescope on Earth; you can see it in the movie *Contact*. Since its 1996 upgrade, it is sensitive enough to pick up a hypothetical cellular telephone on Venus.

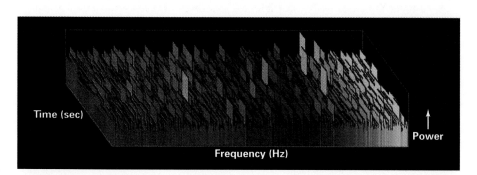

Figure 20–12   Output on the screen of a computer that has been used to analyze some of the radio data collected by the "seti@home" project. There have been no clear signs of intelligent life among the data processed so far.

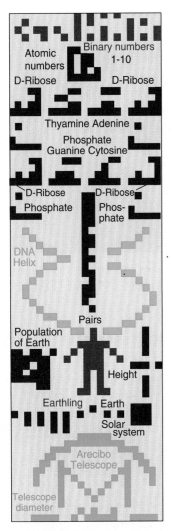

For example, on November 16, 1974, astronomers used the Arecibo radio telescope to send a coded message about people on Earth (Fig. 20–13) toward the globular star cluster M13 (Fig. 20–14) in the constellation Hercules. The idea was that the presence of 200,000 closely packed stars in that location would increase the chances of our signal being received by a civilization on one of them. But the travel time of the message (at the speed of light) is 24,000 years to M13, so we certainly cannot expect to have an answer before twice 24,000, or 48,000 years, have passed. If any creature is observing the Sun at the right frequency (2380 MHz) when the signal arrives, the radio brightness of the Sun will increase by 10 million times over a 3-minute period. A similar signal, if received from a distant star, could be the giveaway that there is intelligent life there.

In retrospect, M13 may not have been the best choice as a target, despite its large number of stars. The problem is that the stars in globular clusters, being very old, were formed from gas that did not have a large proportion of heavy elements; it had not gone through many stages of nuclear processing by massive stars and supernovae. Rocky, Earth-like planets are thus not as likely to have formed there as they would have around younger stars. A quarter century later, in 1999, a new message was sent by astronomers in Canada toward several relatively nearby (50 to 70 light-years away) Sun-like stars, including 16 Cygni B, which is known through Doppler measurements to have at least one planet orbiting it (Chapter 7). The complete message, which is much larger in size, duration, and scope than the one sent in 1974, was transmitted three times over a 3-hour period in the direction of each star. (www)

Even had we not sent these few directed messages, during much of the 20th century we have been unintentionally transmitting radio signals into space on the normal broadcast channels. Aliens within a few tens of light-years could listen to our radio and television programs. A wave bearing the voice of Winston Churchill is expanding into space, and at present is about 60 light-years from Earth. And once a week a new episode of *Friends* is carried into the depths of the Universe. Do you think aliens would get a favorable impression of us from most of what they hear? ☼

Figure 20–14 The globular star cluster M13, toward which a 3-minute message was sent with the Arecibo radio telescope in 1974.

⬤ THE STATISTICS OF INTELLIGENT EXTRATERRESTRIAL LIFE

Is it reasonable to expect that there are any signals out there that we can hope to detect with projects such as that at the SETI Institute? What if *no* intelligent creatures exist elsewhere in our Milky Way Galaxy or even in the observable Universe? Instead of phrasing an all-or-nothing question about extraterrestrial life, we can use a procedure developed by Frank Drake, then of Cornell University, and extended by Carl Sagan and Joseph Shklovskii, among others. The overall problem is broken down into a chain of simpler questions, the results of which are multiplied together in what is called the **Drake equation** to give the final answer. ☼

First, we consider the probability that stars at the centers of planetary systems are suitable to allow intelligent life to evolve. For example, as we have already discussed, the most massive stars evolve rather quickly, remaining stable for too short a time to allow intelligent life to evolve.

Second, we ask what the chances are that a suitable star has planets. With the new detection of planets orbiting other stars, most astronomers think that the chances are probably pretty high.

Third, we need planets with suitable conditions for the origin of life. A planet like Jupiter might be ruled out, for example, because it lacks a solid surface and because its surface gravity is high. (Alternatively, though, one could consider a liquid region, if it were at a suitable temperature, to be as advantageous as the oceans on Earth may have been to the development of life here. Or a moon of the planet could provide the solid surface.) Also, planets probably must be in orbits in which the temperature does not vary too much.

Fourth, we have to consider the fraction of the suitable planets on which life actually begins. This is a very large uncertainty, for if this fraction is zero (with the Earth being the only exception), then we get nowhere with this entire line of reasoning. Still, the discovery that amino acids can be formed in laboratory simulations of at least some kinds of potential primitive atmosphere, and the discovery of complex molecules (such as ammonia, formic acid, and vinegar) in interstellar space, indicate to many astronomers that it is easy to form complicated molecules. Amino acids, much less complex than DNA but also basic to life as we know it, have even been found in some meteorites.

Moreover, life is found in a wide range of extremes on Earth, including oxygen-free environments near geothermal sulfur vents on the ocean bottom or surface hot springs, under rocks in the Antarctic, and deep underground in some other parts of the Earth (Fig. 20–15). These bacteria do not survive in the presence of oxygen; their discovery supports the idea that life evolved before oxygen appeared in Earth's atmosphere. So environments on other planets may not be as hostile to life as we had thought, even though they couldn't support the types of animal and plant life with which we are most familiar. It is these hardy examples of life on Earth that give us hope that the discovery of primitive life on Mars will someday be confirmed.

If we want to hear meaningful signals from aliens, we must have a situation where not just life but intelligent life has evolved. We cannot converse with algae or paramecia, and certainly not with the organic compounds reported on Mars. Furthermore, the life must have developed a technological civilization capable of interstellar communication.

Such a civilization must also live for a fairly long time; otherwise, its existence would be just like a flashbulb going off in an otherwise perpetually dark room. Humans now have the capability of destroying themselves either dramatically in a flurry of hydrogen bombs or more slowly by, for example, altering our

A

B

Figure 20–15   *(A)* We see the inside of an Antarctic rock, with a lichen growing safely insulated from the external cold. The image shows the outer 1 cm of the rock. *(B)* Bacteria that live over 1 km below the Earth's surface, which fluoresce as red patches on this microscope image. The bacteria, discovered in deep wells, survive on hydrogen generated by a chemical reaction between water and iron silicates in the surrounding volcanic rock. Thus they do not need sunlight (or oxygen). Do similar types of microbes exist on Mars?

climate, lessening our ozone shield (which keeps harmful ultraviolet radiation out), or increasing the level of atmospheric pollution. Natural disasters must also be avoided. That an Earth-crossing asteroid or comet will eventually impact the Earth with major consequences seems statistically guaranteed on a time scale of a few hundred million years, unless we take preventive action. It is a sobering question to ask whether the typical lifetime of a technological civilization is measured in decades, or whether all the problems that we have—political, environmental, and otherwise—can be overcome, leaving our civilization to last for millions or billions of years.

We can try to estimate answers for each of these simpler questions within our chain of reasoning, though some of them are actually more like guesses. We can then use these answers together to figure out the larger question of the probability of communicating extraterrestrial life (see *Figure It Out: The Drake Equation*). Fairly reasonable assumptions can lead to the conclusion that there may be thousands, or even tens of thousands, of planets in our galaxy on which technologically advanced life has evolved. Perhaps overly optimistically, Carl Sagan estimated that there might be a million such planets.

On the other hand, adopting more pessimistic (but possibly more realistic) values for the probabilities of intelligent and technologically advanced life, and for the typical lifetime of such a civilization, gives a much bleaker picture. Indeed, humans may be at the pinnacle of intelligence and technological capability in our galaxy, with few, if any comparably advanced civilizations.

One interesting argument is as follows: If there are many communicating civilizations (say, a million), why haven't we detected any of them? Where are they all? Surely most must be far more advanced than we are, and have sent signals if not spacecraft that have reached Earth, yet there is no evidence for them. This reasoning, promoted among others by Enrico Fermi of the University of Chicago, suggests that advanced creatures such as us might be very rare. (Note that various surface features claimed by some people to be built by extraterrestrials are either nonexistent, such as the "face" on Mars, (www) or have more conventional explanations, like the huge drawings on the Peruvian desert known as the Nazca lines.)

Indeed, a more extreme version of this argument points out that there could be self-sustaining colonies voyaging through space for generations. They need not travel close to the speed of light if families are aboard. Even if colonization of space took place at a rate of only 1 light-year per century, our entire galaxy would still have been colonized during a span of just a hundred million years, less than one per cent of its lifetime. The fact that we do not find extraterrestrial life here indicates that our Solar System has not been colonized, which in turn may imply that technologically advanced life has not arisen elsewhere in our galaxy.

Another possibly relevant fact suggesting that we are very rare in our galaxy is that out of the roughly 50 billion species that have lived over Earth's history, we are the only species to have developed space communication or even acquired technology. Similarly, it is sobering to realize that a communicating civilization developed on Earth only in the last century, despite evidence for primitive life on Earth the past 3.5 to 3.8 billion years.

Recently, astronomers and other scientists have carefully considered the many factors that affect complex life on Earth. For example, the stability of the Earth's axis of rotation depends on the presence of our rather massive Moon. If we had no moon, or only a small one, then Earth's axis would undergo rather rapid, random changes in its orientation, causing large variations in the seasons and presumably making it more difficult for complex life to develop. Similarly, the presence of the very massive planet Jupiter in our Solar System, yet fairly far from Earth, is a blessing: Jupiter's gravitational tugs have cleared out our Solar System, making collisions between Earth and large "killer meteoroids" infrequent.

Figure It Out

The Drake Equation

We can make an educated guess regarding the present number of communicating civilizations (those that should be able to contact each other) in our Milky Way Galaxy by using the "Drake equation," first written by Frank Drake in 1961: $N = R f_s f_p n_e f_l f_i f_c L$. Let us examine each of the terms in this equation.

R is the star-formation rate in our galaxy. It is roughly equal to the total number of stars divided by the age of our galaxy. A reasonable estimate is $R = 10$ stars per year. The true rate is lower now, but was higher in the past.

f_s is the fraction of stars that are suitable. Various criteria can be applied, such as F, G, and K stars; and a sufficiently high abundance of heavy elements to make rocky planets possible. Perhaps 0.1 is a good guess.

f_p is the fraction of stars with planetary systems. This fraction could be roughly 0.1—many binary systems might not have planets.

n_e is the number of Earth-like planets or moons per planetary system. We could guess that this number is typically between 0.1 and 1. It might be a little larger than 1, if a planetary system has several suitable moons. For example, in our Solar System, it is not unreasonable to suppose Mars, Europa, and Titan (besides Earth) had or will have conditions suitable for the formation of life at some time in their existence.

f_l is the fraction of Earth-like planets on which life (even if primitive) arises. This fraction, though quite speculative, might be in the range 10^{-3} to 1.

f_i is the fraction of life with intelligence (that is, the ratio of intelligent species to all species). Though perhaps even more speculative, it might be 10^{-6} to 1.

f_c is the fraction of intelligent life that has the ability and desire to communicate with aliens. Once again, we can only speculate; perhaps 10^{-3} to 1 is reasonable.

L is the lifetime of a communicating civilization, or the cumulative lifetime of such civilizations on a given planet (if there are more than one at any given time, or if the annihilation of one is followed at some later time by another). Humans have sent radio signals for only about 100 years—and we hope that we have many years ahead of us! However, perhaps advanced civilizations tend to destroy themselves after a short time. The possible range for L is enormous: Let's say 100 to 10^9 years.

The overall result is the product of all these factors, most of which are highly uncertain. The pessimistic estimates give a value of 10^{-12}, suggesting that creatures like humans are exceedingly rare—only one case per 10^{12} galaxies. There are about 10^{11} galaxies in the observable parts of the Universe, so we might be the only communicating civilization.

On the other hand, the most optimistic estimates give a value of 10^9—essentially the lifetime L, since all of the other factors are either about 1 or balance each other out (note that $R \times f_s \approx 1$). This means that one star out of several hundred might now have a communicating civilization. The fraction is even larger if some civilizations colonize more than one star.

Although use of the Drake equation doesn't answer our question very precisely at all, the approach is systematic, and it gives us a range of reasonable values. It also focuses our attention on which factors are most uncertain and need improvement. For example, we can see that our uncertainty in the rate at which stars form is just a minor fraction of the total uncertainty in the final answer; far more important are the uncertainty in the fraction of life with intelligence and the lifetime of a communicating civilization.

These arguments, and others, have led many astronomers to conclude that complex, technologically advanced, communicating civilizations really are rare in our Milky Way Galaxy, though primitive life such as bacteria might be very common. Other astronomers, however, argue that there are major potential flaws in this reasoning, and such civilizations may be common. It is not clear, for example, whether the apparently "special" conditions of Earth are essential to the development of intelligent life. Moreover, is it necessarily true that if other intelligent creatures evolved, they would choose to colonize space, or have the

Figure 20–16 An artist's conception of a set of radio telescopes on the far side of the Moon, shielded from the Earth's radio signals.

ability to do so? The rather late appearance of technologically advanced life on Earth may also be a statistically unlikely fluke. Finally, some of the arguments relied on the unproven assumption that life elsewhere is quite similar to that on Earth in its properties and evolutionary path.

In any case, unless we make the effort to actually look for signs of intelligent extraterrestrial life, we might never know whether humans are indeed alone in our galaxy, or simply one of many such creatures. Thus, many astronomers support the search for extraterrestrial intelligence, especially using telescopes that are simultaneously doing other types of research projects. Perhaps someday we will be scanning the skies with radio telescopes on the far side of the Moon (Fig. 20–16), shielded by the Moon's bulk from Earth's radio interference. The chances for success might be slim, but all agree that the actual detection of extraterrestrial signals would be one of the most important and mind-blowing discoveries, if not the greatest discovery of all time.

UFOS AND THE SCIENTIFIC METHOD

If some or many astronomers believe that technologically advanced life exists elsewhere in our galaxy, why do they not accept the idea that unidentified flying objects (UFOs) represent visitation from these other civilizations (Fig. 20–17)? The answer to this question leads us not only to explore the nature of UFOs but also to consider the nature of knowledge and truth. The discussion that follows is a personal view of the authors, but one that is shared to a greater or lesser extent by most scientists. (www)

UFOs

The most common UFO is a light that appears in the sky that seems to the observer unlike anything made by humans or any commonly understood natural phenomenon. UFOs may appear as a point or extended, or may seem to vary. But the observations are usually anecdotal, are not controlled as in a scientific experiment, and are not accessible to study by sophisticated instruments.

Most of the sightings of UFOs that are reported can actually be explained in terms of natural phenomena. The Earth's atmosphere can display many strange effects, and these can be used to explain many apparent UFOs. When Venus shines brightly near the horizon, for example, we sometimes get telephone calls from people asking us about the "UFO"—especially if the crescent moon happens to also be in that direction. It is not well-known that a bright star or planet low on the horizon can seem to flash red or green because of atmospheric refraction. Atmospheric effects can affect radar (radio) waves as well as visible light.

Figure 20–17

Sometimes other natural phenomena (flocks of birds, for example) are mistakenly reported as UFOs. Even the more mysterious sightings lack convincing evidence that rules out alternatives. One should not accept explanations that UFOs are flying saucers from other planets before more mundane explanations—including hoaxes, exaggeration, and fraud—are exhausted.

For many of the effects that have been reported, the UFOs would have been defying well-established laws of physics. Where are the sonic booms, for example, from rapidly moving UFOs? Scientists treat challenges to laws of physics very seriously, since our science and technology are based on these laws and they seem to work extremely well.

Most professional astronomers feel that UFOs can be so completely explained by natural phenomena that they are not worthy of more of our time. Although most of us do not categorically deny the possibility that UFOs exist (after all, the Voyager spacecraft might someday pass by a planet orbiting another star), the standard of evidence expected of all claims in science has not yet been met. 🔭

Some individuals may ask why we reject the identification of UFOs with flying saucers, when—they may say—that explanation is "just as good an explanation as any other." Let us go on to discuss what scientists mean by "truth" and how that applies to the above question.

Of Truth and Theories

At every instant, we can explain what is happening in a variety of ways. When we flip a light switch, for example, we assume that the switch closes an electric circuit in the wall and allows the electricity to flow. But it is certainly possible, although not very likely, that the switch activates a relay that turns on a radio that broadcasts a message to the president of the United States. The president then might send back a telepathic message to the electricity to flow, making the light go on. The latter explanation sounds so unlikely that we don't seriously consider it. We would even call the former explanation "true," without qualification.

We usually regard as "true" the simplest explanation that satisfies all the data we have about any given thing. This principle is known as **Occam's Razor;** it is named after a 14th-century British philosopher who originally proposed it. Without this rule, we would always be subject to such complicated doubts that we would accept nothing as known to be true.

Science is based on Occam's Razor, though we don't usually bother to think about it. Sometimes something we call "true" might be more accurately described as a theory (see Chapter 1). An example of a theory is the Newtonian theory of gravitation, which for many years explained almost all of the planetary motions. Albert Einstein's 1916 theory of gravity, known as the general theory of relativity, provided an explanation for a nagging discrepancy in the orbit of Mercury, as we described in Chapter 9. Is Newton's theory "true"? Though we know it is false, it is a good approximation to the truth in most regions of space. Is Einstein's theory "true"? We may say so, although one day a newer theory may come along that is more general than Einstein's in the same way that Einstein's is more general than Newton's. Indeed, as we discussed in Chapter 19, superstring theory is a leading candidate for the unification of general relativity and quantum physics, and hence is a more complete theory than Einstein's.

How does this view of truth tie in with the previous discussion of UFOs? Scientists have assessed the probability of UFOs being flying saucers from other worlds, and most have decided that the probability is so low that we have better things to do with our time and with our national resources. We have so many other, simpler explanations of the phenomena that are reported as UFOs that when we apply Occam's Razor, we call the identifications of UFOs with extraterrestrial visitation "false."

> Occam's Razor, sometimes called the Principle of Simplicity, is a razor in the sense that it is a cutting edge that allows a distinction to be made among theories.

●● CONCLUSION

You have covered a lot of material in this book, and learned much about the Universe. We, the authors, hope that this new knowledge increases your awe and fascination for nature. The understanding of a phenomenon should enhance its beauty, not detract from it. The magnificence of the Universe comes in part from its logical structure—and the foundation is perhaps unexpectedly simple. Indeed, Einstein remarked that "The most incomprehensible thing about the Universe is that it's comprehensible."

The actual consequences of the basic laws can be extraordinarily complicated and varied. The best example is life itself: it is the most complex known structure. Even the simplest cell is more difficult to understand than the formation of galaxies or the structure of stars. What sets humans apart are our highly advanced brains and dexterity. We are able to question, explore, and ultimately understand the inner workings of nature through a process of observation, experimentation, and logical thought. It is almost as if the Universe has found a way to know itself, through us. We are the observers and explorers of the Universe; we are its brains and its conscience. This makes each one of us special to the Universe as a whole. Perhaps it need not have been this way: alter the physical constants ever so slightly, and the Universe may have been stillborn, with no such complexity. But here we are, enjoying life in this beautiful, amazing, mind-blowing Universe.

Concept Review

The study of life elsewhere than on Earth is called **exobiology.** One difficulty in the search for life is that we don't have a clear definition of what life is, and we generally must restrict ourselves to life as we know it, which is based on **organic** compounds—those containing carbon. Specifically, life on Earth is based on chains of carbon called **amino acids,** which link together many at a time to form **proteins,** chief among them DNA. Experiments in which sparks were passed through certain mixtures of gases showed that amino acids easily form, but it is now thought that these mixtures are not representative of Earth's primitive atmosphere. Nevertheless, amino acids seem to form in a variety of environments, including harsh ones.

Primitive life arose very quickly on Earth, suggesting that it is not highly improbable, though the discovery of indigenous life in at least one other location would provide much support for this hypothesis. Within our own Solar System, Mars and Europa are the most likely candidates for life. Extraterrestrial life outside our Solar System, especially intelligent life, is most likely to have formed on moons or planets orbiting Sun-like stars; such stars have a long life and a relatively large **ecosphere,** or range of distances in which conditions suitable for life might be found.

In searching for signs of extraterrestrial life, we can wait for their spacecraft to reach us, but the odds of success are low, even though Einstein's special theory of relativity in principle allows interstellar voyages to be completed in a reasonably short time. A much more likely method of detection is by searching for patterns among electromagnetic signals, especially radio waves. Several such projects are being conducted, mostly notably the "search for extraterrestrial intelligence, at home" (seti@home), in which the general public can participate with personal computers.

Although we do not know how many communicating civilizations exist in the Milky Way Galaxy, an estimate can be attempted by using the **Drake equation.** Among the most uncertain factors are the probability that intelligence arises after the formation of primitive life and the typical lifetime of a communicating civilization. Some scientists think that intelligent life might be common in our galaxy, but many others are concluding that intelligence and technological capability at the level of humans is rather rare. Unless we actually search for signals, however, the answer may always elude us.

Evidence that UFOs visit us is not convincing; most of the reports can be easily explained by atmospheric and other natural phenomena. Although it is possible that some of the sightings are of actual UFOs, **Occam's Razor** (the Principle of Simplicity) suggests otherwise, until the evidence becomes much stronger.

Questions

1. Discuss the most likely places in our Solar System, other than Earth, where there might be primitive (microbial) life.
2. What is the significance of the Miller-Urey experiments?
3. What types of stars are most likely to have planets on which indigenous life developed? Explain.
4. List two means by which we might detect extraterrestrial intelligence.
5. Why have astronomers generally used radio wavelengths in their search for life elsewhere?
6. Describe the radio signals humans have sent to outer space, both intentionally and unintentionally.
†7. On the message we sent to M13, work out the binary value given for the population of Earth, and compare it with the current actual value.
†8. On the message we sent to M13, work out the binary value given for the size of the Arecibo telescope, and compare it with the actual value.
9. Why can't we hope to carry on a conversation at a normal rate with extraterrestrials on a distant star?
10. State the Drake equation and its significance. Discuss the importance of the value of L, the lifetime of the civilization.

†11. In the Drake equation, use your own preferences for each quantity to derive the number of intelligent, communicating civilizations in our galaxy.
†12. If 10% of all stars are of suitable type for life to develop, 30% of all stars have planets, and 20% of planetary systems have a planet or moon at a suitable distance from the star, what fraction of stars have a planet suitable for life? Roughly how many such stars would there be in our Milky Way Galaxy?
†13. Travelling at 100 km/sec, how long would it take a rocket ship to reach a star that is 20 light-years away?
14. Explain how it is possible, in principle, to travel many light-years in a short time interval.
†15. Suppose you are in a rocket ship that is moving at 97% of the speed of light. If your journey to another star appeared to take 30 years from the perspective of an Earth-bound observer, how long did it take in your frame of reference?

†This question requires a numerical solution.

Epilogue

We have completed our grand tour of the Universe. We have seen stars and planets, matter between the stars, giant collections of stars called galaxies and clusters of galaxies, and very distant objects such as quasars. We have witnessed the evolution of stars, in some cases ending with spectacular explosions with compact remnants the size of a city but half a million times more massive than Earth. We have pondered the properties of still more bizarre objects, black holes. We have learned how our Universe began in a hot, compressed state and has been expanding ever since—seemingly faster and faster during the past few billion years, perhaps driven by a cosmic antigravity effect. We have explored the origin of the Universe, galaxies, stars, the chemical elements, planets, and ultimately life itself. If you are thirsty for more information, as the authors would like you to be, you can consult the books listed in the Selected Readings.

Further, we have seen the vitality of contemporary science in general and astronomy in particular. The individual scientists who call themselves astronomers are engaged in fascinating studies, often pushing modern technologies to their limits. New telescopes on the ground and in space, new types of detectors, new computer capabilities for studying data and carrying out calculations, and new theoretical ideas are linked in research about the Universe.

Our views at the cutting edge of astronomy are changing so rapidly that within a few years some of what you read here will have been revised. Science is a dynamic process: New ideas are developed and tested, and modified when necessary. The authors hope that, over the years, you will keep up by following astronomical articles and stories in newspapers, magazines, books, and on television. We would like you to consider the role of scientific research as you vote. And we hope you will remember the methods of science—the mixture of logic and standards of evidence by which scientists operate—that you have seen illustrated in this book.

Artists' conceptions of the Next Generation Space Telescope, an 8-meter infrared telescope to succeed the Hubble Space Telescope around the year 2008. We see the general idea proposed by NASA, *(right)* as well as two proposed specific plans from aerospace companies *(left and center)*. The major ground-based project proposed for the next 10 years, as recommended in 2000 after a study by a high-level committee, is a 30-m segmented telescope, a larger-scale version of the Keck telescopes.

Système International Units

	SI units	SI Abbrev.	Other Abbrev.
length	meter	m	
volume	liter	L	ℓ
mass	kilogram	kg	kgm
time	second	s	sec
temperature	kelvin	K	°K

Other Metric Units

1 micron (μ) = 1 micrometer (μm) = 10^{-6} meter	μm	μ
1 angstrom (Å or A) = 10^{-10} meter = 10^{-8} cm	0.1 nm	Å

Prefixes for Use with Basic Units of Metric System

Prefix	Symbol	Power		Equivalent
yotta	Y	10^{24}	= 1,000,000,000,000,000,000,000,000	
zetta	Z	10^{21}	= 1,000,000,000,000,000,000,000	
exa	E	10^{18}	= 1,000,000,000,000,000,000	
peta	P	10^{15}	= 1,000,000,000,000,000	
tera	T	10^{12}	= 1,000,000,000,000	Trillion
giga	G	10^{9}	= 1,000,000,000	Billion
mega	M	10^{6}	= 1,000,000	Million
kilo	k	10^{3}	= 1,000	Thousand
hecto	h	10^{2}	= 100	Hundred
deca	da	10^{1}	= 10	Ten
—	—	10^{0}	= 1	One
deci	d	10^{-1}	= 0.1	Tenth
centi	c	10^{-2}	= 0.01	Hundredth
milli	m	10^{-3}	= 0.001	Thousandth
micro	μ	10^{-6}	= 0.000001	Millionth
nano	n	10^{-9}	= 0.000000001	Billionth
pico	p	10^{-12}	= 0.000000000001	Trillionth
femto	f	10^{-15}	= 0.000000000000001	
atto	a	10^{-18}	= 0.000000000000000001	
zepto	z	10^{-21}	= 0.000000000000000000001	
yocto	y	10^{-24}	= 0.000000000000000000000001	
Examples:		1000 meters = 1 kilometer = 1 km		
		10^{6} hertz = 1 megahertz = 1 MHz		
		10^{-3} sec = 1 millisecond = 1 msec		

Conversion Factors

1 erg = 10^{-7} joule = 10^{-7} kg $\cdot$ m^2/s^2

1 joule = 6.2419×10^{18} eV

1 in = 25.4 mm = 2.54 cm

1 yd = 0.9144 m

1 mi = 1.6093 km $\approx$ 8/5 km

1 oz = 28.3 g

1 lb = 0.4536 kg

Binary Prefixes

kibi 2^{10} = 1,024

mebi 2^{20} = 1,048,576

gibi 2^{30} = 1.0737×10^9

APPENDIX 2   **Basic Constants**

Physical Constants (1998 CODATA Recommended Values)

Speed of light*	c	= 299 792 458 m/sec (exactly)
Constant of gravitation	G	= $(6.673 \pm 0.010) \times 10^{-11}$ m^3/kg $\cdot$ sec^2
Planck's constant	h	= $(6.626\ 068 \pm 0.000\ 0052) \times 10^{-34}$ J $\cdot$ s
Boltzmann's constant	k	= $(1.380\ 650\ 3 \pm 0.000\ 002\ 4) \times 10^{-23}$ J/K
Stefan-Boltzmann constant	σ	= $(5.670\ 400 \pm 0.000\ 040) \times 10^{-8}$ W/m$^2 \cdot$ K^4
Wien displacement constant	$\lambda_{max}T$	= $(2.897\ 768\ 6 \pm 0.000\ 005) \times 10^7$ A = K
Mass of neutron	m_n	= $(1.674\ 92716 \pm 0.000\ 00013) \times 10^{-27}$ kg
Mass of proton	m_p	= $(1.672\ 621\ 58 \pm 0.000000\ 13) \times 10^{-27}$ kg
Mass of electron	m_e	= $(9.109\ 381\ 88 \pm 0.000000\ 72) \times 10^{-31}$ kg
Rydberg constant	R	= $(10\ 973\ 731.568\ 549 \pm 0.000\ 083)$ m^{-1}

See http://physics.nist.gov/constants

Mathematical Constants (to 100 places just for fun)

π = 3.141 592 653 589 793 238 462 643 383 279 502 884 197 169 399 375 105 820 974 944 592 307 816 406 286 208 998 628 034 825 342 117 067 . . .

e = 2.718 281 828 459 045 235 360 287 471 352 662 497 757 247 093 699 959 574 966 967 627 724 076 630 353 547 594 571 382 178 525 166 427 . . .

Astronomical Constants

Astronomical Unit*	1 A.U.	$= 1.495\ 978\ 70 \times 10^{11}$ m
Solar parallax*	$\pi_\odot$	$= 8.794\ 148$ arcsec
Parsec	1 pc	$= 3.086 \times 10^{16}$ m
		$= 206\ 264.806$ A.U.
		$= 3.261\ 633$ ly
Light-year	1 ly	$= (9.460\ 530) \times 10^{15}$ m
		$= 6.324 \times 10^4$ A.U.
Tropical year (1900)*—(equinox to equinox)		$= 365.242\ 198\ 78$ ephemeris days
Julian century*		$= 36\ 525$ days
Day*		$= 86\ 400$ sec
Sidereal year		$= 365.256\ 366$ ephemeris days
		$= (3.155\ 815) \times 10^7$ sec
Mass of Sun*	$M_\odot$	$= (1.989\ 1) \times 10^{30}$ kg
Radius of Sun*	$R_\odot$	$= 696\ 000$ km
Luminosity of Sun	$L_\odot$	$= 3.827 \times 10^{26}$ J/sec
Mass of Earth*	M_E	$= (5.974\ 2) \times 10^{24}$ kg
Equatorial radius of Earth*	R_E	$= 6\ 378.140$ km
Center of Earth to center of Moon (mean)		$= 384\ 403$ km
Radius of Moon*	R_M	$= 1\ 738$ km
Mass of Moon*	M_M	$= 7.35 \times 10^{22}$ kg
Solar constant	S	$= 1\ 368$ W/m^2
Direction of galactic center (2000.0 precession)	α	$= 17^h 45.6^m$
	δ	$= -28°56'$

*Adopted as "IAU (1976) system of astronomical constants" at the General Assembly of the International Astronomical Union that year. The meter was redefined in 1983 to be the distance travelled by light in a vacuum in 1/299,792,458 second.

APPENDIX 3

APPENDIX 3A   Intrinsic and Rotational Properties

Name	Equatorial Radius km	Equatorial Radius ÷ Earth's	Mass ÷ Earth's	Mean Density (g/cm³)	Oblateness	Surface Gravity (Earth = 1)	Sidereal Rotation Period	Inclination of Equator to Orbit	Apparent Magnitude During 2001
Mercury	2,439.7	0.3824	0.0553	5.43	0	0.378	58.646^d	0.0°	−2.2 to 5.3
Venus	6,051.8	0.9489	0.8150	5.24	0	0.894	243.02^dR	177.3°	−4.6 to −3.9
Earth	6,378.14	1	1	5.515	0.0034	1	$23^h56^m04.1^s$	23.45°	—
Mars	3,397	0.5326	0.1074	3.94	0.006	0.379	$24^h37^m22.662^s$	25.19°	−2.2 to +1.4
Jupiter	71,492	11.194	317.896	1.33	0.065	2.54	9^h50^m to $> 9^h55^m$	3.12°	−2.7 to −1.9
Saturn	60,268	9.41	95.185	0.70	0.098	1.07	$10^h39.9^m$	26.73°	+0.2 to +0.2
Uranus	25,559	4.0	14.537	1.30	0.022	0.8	17^h14^mR	97.86°	+5.7 to +5.9
Neptune	24,764	3.9	17.151	1.76	0.017	1.2	16^h7^m	29.56°	+7.8 to +8.0
Pluto	1,195	0.2	0.0025	2.1	0	0.01	$6^d9^h17^mR$	120°	+13.8 to +13.9

R signifies retrograde rotation.

The masses and radii for Mercury, Venus, Earth, and Mars are the values recommended by the International Astronomical Union in 1976. The radii are from *The Astronomical Almanac 2001*. Surface gravities were calculated from these values. The length of the martian day is from G. de Vaucouleurs (1979). Most densities, oblatenesses, inclinations, and magnitudes are from *The Astronomical Almanac 2001*. Neptune data from *Science*, December 15, 1989 and August 9, 1991. Values for the masses of the giant planets are based on Voyager data for the mass of the Sun divided by the mass of the planet. (E. Myles Standish, Jr., *Astronomical Journal* **105**, 2000, 1992); Jupiter: 1047.3486; Saturn: 3497.898; Uranus: 22902.94; Neptune: 19412.24.

APPENDIX 3B   Orbital Properties

Name	Semimajor Axis A.U.	Semimajor Axis 10^6 km	Sidereal Period Years	Sidereal Period Days	Synodic Period (Days)	Eccentricity	Inclination to Ecliptic
Mercury	0.387 099	57.909	0.240 84	87.96	115.9	0.205 63	7.004 87°
Venus	0.723 332	108.209	0.615 18	224.68	583.9	0.006 77	3.394 71°
Earth	1	149.598	0.999 98	365.25	—	0.016 71	0.000 05°
Mars	1.523 662	227.939	1.880 7	686.95	779.9	0.093 41	1.850 61°
Jupiter	5.203 363	778.298	11.857	4,337	398.9	0.048 39	1.305 30°
Saturn	9.537 070	1429.394	29.424	10,760	378.1	0.054 15	2.484 46°
Uranus	19.191 264	2875.039	83.75	30,700	369.7	0.047 168	0.769 86°
Neptune	30.068 963	4504.450	163.72	60,200	367.5	0.008 59	0.769 17°
Pluto	39.481 687	5915.799	248.02	90,780	366.7	0.248 81	17.141 75°

Mean elements of planetary orbits for 2000, referred to the mean ecliptic and equinox of J2000 (E. M. Standish, X. X. Newhall, J. G. Williams, and D. K. Yeomans, *Explanatory Supplement to the Astronomical Almanac*, P. K. Seidelmann, ed., 1992). Periods are calculated from them.

Planetary Satellites

Satellite	Semimajor Axis of Orbit (km)	Sidereal Revolution Period (d	h	m)	Orbital Eccentricity	Orbital Inclination (°)	Radius (km)	Mass ÷ Mass of Planet	Mean Density (g/cm³)	Discoverer	Visible Magnitude at Mean Opposition Distance
Satellite of the Earth											
The Moon	384,400	27	07	43	0.055	18–29	1738	0.01230002	3.34	—	−12.7
Satellites of Mars											
1 Phobos	9,378	0	07	39	0.015	1.0	13 × 11 × 9	1.5×10^{-8}	1.95	Hall (1877)	11.8
2 Deimos	23,459	1	06	18	0.0005	0.9 – 2.7	8 × 6 × 5	3×10^{-9}	2	Hall (1877)	12.9
Satellites of Jupiter											
XVI Metis	128,000	0	07	04			20	0.5×10^{-10}		Synnott/Voyager 2 (1980)	17.5
XV Adrastea	129,000	0	07	06			13 × 10 × 8	0.1×10^{-10}		Jewett/Danielson (1979)	19.1
V Amalthea	181,000	0	11	57	0.003	0.4	131 × 73 × 67	38×10^{-10}		Barnard (1892)	14.1
XIV Thebe	222,000	0	16	11	0.015	0.8	55 × 45	4×10^{-10}		Synnott/Voyager 1 (1980)	15.6
I Io	422,000	1	18	28	0.004	0.0	1830 × 1819 × 1813	4.68×10^{-5}	3.5	Galileo (1610)	5.0
II Europa	671,000	3	13	14	0.009	0.5	1565	2.52×10^{-5}	3.0	Galileo (1610)	5.3
III Ganymede	1,070,000	7	03	43	0.002	0.2	2634	7.80×10^{-5}	1.9	Galileo (1610)	4.6
IV Callisto	1,883,000	16	16	32	0.007	0.5	2403	5.66×10^{-5}	1.8	Galileo (1610)	5.7
XIII Leda	11,094,000	240			0.148	26.1	5	0.03×10^{-10}		Kowal (1974)	20
VI Himalia	11,480,000	251			0.158	27.6	85	50×10^{-10}		Perrine (1904)	14.8
X Lysithea	11,720,000	260			0.107	29.0	12	0.4×10^{-10}		Nicholson (1938)	18.4
VII Elara	11,737,000	260			0.207	24.8	40	4×10^{-10}		Perrine (1905)	16.8
XII Ananke	21,200,000	671R			0.169	147	10	0.2×10^{-10}		Nicholson (1951)	18.9
XI Carme	22,600,000	692R			0.207	164	15	0.5×10^{-10}		Nicholson (1938)	18.0
VIII Pasiphae	23,500,000	735R			0.378	145	18	1×10^{-10}		Melotte (1908)	17.0
IX Sinope	23,700,000	758R			0.275	153	14	0.4×10^{-10}		Nicholson (1914)	18.3
—										Kowal (1975)	20
Satellites of Saturn											
18 Pan	133,583		13	48			10	8×10^{-12}		Showalter/Voyager 2 (1990)	
15 Atlas	137,670		14	27	0.000	0.3	20 × 10			Terrile/Voyager 1	18
16 Prometheus	139,353		14	43	0.003	0.0	70 × 50 × 40			Collins/Voyager 1	16
17 Pandora	141,700		15	05	0.004	0.0	55 × 45 × 35			Collins/Voyager 1	16
11 Epimetheus	151,422		16	40	0.009	0.3	70 × 60 × 50			Fountain, Larson/ Reitsema/Smith	15
10 Janus	151,472		16	40	0.007	0.1	110 × 100 × 80			Dollfus (1966/80)	14
1 Mimas	185,520		22	37	0.0202	1.5	209 × 196 × 191	8×10^{-8}	1.2	W. Herschel (1789)	12.9
2 Enceladus	238,020	1^h	08^m	32^s	0.00452	0.0	256 × 247 × 245	1.3×10^{-7}	1.1	W. Herschel (1789)	11.7
3 Tethys	294,660	1^h	22^m	15^s	0.00000	1.9	528 × 526	1.3×10^{-6}	1.0	Cassini (1684)	10.2
13 Telesto	294,660	1^h	22^m	15^s			15 × 13 × 8			Smith, Reitsema, Larson, Fountain (1980)	19
14 Calypso	294,660	1^h	22^m	15^s			15 × 8 × 8			Pascu, Seidelmann, Baum, Currie (1980)	19
4 Dione	377,400	2^h	17^m	36^s	0.002230	0.02	560	1.96×10^{-6}	1.4	Cassini (1684)	10.4
12 Helene	377,400	2^h	17^m	45^s	0.005	0.0	18 × 16 × 15			Laques and Lecacheux (1980)	18
5 Rhea	527,040	4^h	12^m	16^s	0.00100	0.4	764	4.4×10^{-6}	1.3	Cassini (1672)	9.7
6 Titan	1,221,830	15^h	21^m	51^s	0.029192	0.3	2575	2.4×10^{-4}	1.9	Huygens (1665)	8.3
7 Hyperion	1,481,100	21^h	06^m	45^s	0.104	0.4	180 × 140 × 113	3×10^{-8}	1.9	Bond, Bond, Lassell (1848)	14.2
8 Iapetus	3,561,300	79^h	03^m	43^s	0.02828	14.7	718	3.3×10^{-6}	1.2	Cassini (1671)	11.1
9 Phoebe	12,952,000	549^h	03^m	33^s	0.16326	177	110	7×10^{-10}		W. Pickering (1898)	16.5
Satellites of Uranus											
6 Cordelia	49,770		08	02	< 0.001	0.1	13			Terrile/Voyager 2 (1986)	
7 Ophelia	53,790		09	02	0.010	0.1	15			Terrile/Voyager 2 (1986)	
8 Bianca	59,170		10	25	< 0.001	0.2	21			Voyager 2 (1986)	
9 Cressida	61,780		11	07	< 0.001	0.0	31			Synnott/Voyager 2 (1986)	
10 Desdemona	62,680		11	22	< 0.001	0.2	27			Synnott/Voyager 2 (1986)	
11 Juliet	64,350		11	50	< 0.001	0.1	42			Synnott/Voyager 2 (1986)	
12 Portia	66,090		12	19	< 0.001	0.1	54			Synnott/Voyager 2 (1986)	
13 Rosalind	69,940		11	54	< 0.001	0.3	27			Synnott/Voyager 2 (1986)	
14 Belinda	75,260		14	57	< 0.001	0.0	33			Synnott/Voyager 2 (1986)	
15 Puck	86,010		18	17	< 0.001	0.3	77			Synnott/Voyager 2 (1985)	
5 Miranda	129,390	1	09	56	0.0027	4.2	240 × 234 × 233	0.2×10^{-5}	1.26	Kuiper (1948)	16.3
1 Ariel	191,020	2	12	29	0.0034	0.3	581 × 578 × 578	1.6×10^{-5}	1.65	Lassell (1851)	14.2
2 Umbriel	266,300	4	03	27	0.0050	0.4	585	1.4×10^{-5}	1.44	Lassell (1851)	14.8
3 Titania	433,910	8	16	56	0.0022	0.1	789	4.1×10^{-5}	1.59	W. Herschel (1787)	13.7
4 Oberon	583,520	13	11	07	0.0008	0.1	761	3.5×10^{-5}	1.50	W. Herschel (1787)	13.9
16 Caliban	7,169,000				0.082	139.7	30			Gladman	22.4
17 Sycorax	12,214,000		1,289		140.509	152.7	60			Nicholson, Burns, Kavelaars (1997)	20.09

Four reported discoveries awaiting confirmation

Satellite	Semimajor Axis of Orbit (km)	Sidereal Revolution Period (d	h	m)	Orbital Eccentricity	Orbital Inclination (°)	Radius (km)	Mass ÷ Mass of Planet	Mean Density (g/cm3)	Discoverer	Visible Magnitude at Mean Opposition Distance
Satellites of Neptune											
3 Naiad	48,230		7^h		< 0.001	4.74	29			Voyager 2 (1989)	24.7
4 Thalassa	50,070		7^h	30^m	< 0.001	0.21	40			Synnott/Voyager 2 (1989)	23.0
5 Despina	52,530		8^h		< 0.001	0.07	74			Synnott/Voyager 2 (1989)	22.6
6 Galatea	61,950		10^h		< 0.001	0.05	79			Synnott/Voyager 2 (1989)	22.3
7 Larissa	73,550		13^h		0.0014	0.20	104 × 89			Reitsema, Hubbard, Lebofsky, Tholen/Voyager 2 (1989)	22.0
8 Proteus	117,650		27^h		0.0004	0.05	218 × 208 × 201			Synnott/Voyager 2 (1989)	20.03
1 Triton	354,760	5^d	21^h	03^mR	0.000016	157,345	1353	2.09×10^{-4}	2.07	Lassell (1846)	13.5
2 Nereid	5,513,400	360^d	5^h		0.75	27.6	170	2×10^{-7}	2.03	Kuiper (1949)	18.7

$+$ *4 reported discoveries awaiting confirmation*

Rings and Ring Arcs of Neptune

Galle	41,900
Leverrier	53,200
Lassell	53,200 – 57,500
Arago	57,500
Adams	62,900
Courage	62,900
Liberté	62,900
Egalité	62,900
Fraternité	62,900

Satellite of Pluto

Satellite	Semimajor Axis of Orbit (km)	Sidereal Revolution Period (d	h	m)	Orbital Eccentricity	Orbital Inclination (°)	Radius (km)	Mass ÷ Mass of Planet	Mean Density	Discoverer	Visible Magnitude
Charon	19,682	6	9	17 R	< 0.001	99	593	0.124		Christy (1978)	16.8

Based on a table by Joseph Veverka in the *Observer's Handbook of the Royal Astronomical Society of Canada* and on the *Astronomical Almanac 2001.* Many of Veverka's values are from J. Burns. Density is average of Pluto and Charon. Inclinations greater than 90° are retrograde; R signifies retrograde revolution. Discoverers from Burns and Matthews, *Satellites* (1986), updated with correspondence with D. Pascu, R. Synnott, H. Reitsema, and R. Terrile (1994).

Figure A4–1   A montage of planetary satellites.

The Brightest Stars

	Star	Name	Position (2000.0) R.A.	Decl.	Apparent Magnitude (V)	Spectral Type	Absolute Magnitude (Mᵥ)	Distance D (ly)	Proper Motion R.A. "/yr	Dec °	Radial Vel. (km/sec)
1.	α CMa A	Sirius	06 45 09	−16 42 58	−1.46	A1 V	+1.5	9	1.324	204	−8
2.	α Car	Canopus	06 23 57	−52 41 44	−0.72	A9 Ib	−5.4	313	0.034	50	+21
3.	α Boo	Arcturus	14 15 31	+19 10 57	−0.04	K2 IIIp	−0.6	37	2.281	209	−5
4.	α Cen A	Rigil Kentaurus	14 39 37	−60 50 02	0.00	G2 V	+4.2	4	3.678	28	−25
5.	α Lyr	Vega	18 36 56	+38 47 01	0.03	A0 V	+0.6	25	0.348	35	−14
6.	α Aur	Capella	05 16 41	+45 59 53	0.08	G6 +G2	−0.8	42	0.430	169	+30
7.	β Ori A	Rigel	05 14 32	−08 12 06	0.12	B8 Ia	−6.6	773	0.004	236	+21
8.	α CMi A	Procyon	07 39 18	+05 13 30	0.38	F2 IV-V	+2.8	11	1.248	214	−3
9.	α Eri	Achernar	01 37 43	−57 14 12	0.46	B3 V	−2.9	144	0.108	105	+19
10.	α Ori	Betelgeuse	05 55 10	+07 24 36	0.50	M2 Iab	−5.0	522	0.028	68	+21
11.	β Cen AB	Hadar	14 03 49	−60 22 22	0.61	B1 III	−5.5	526	0.030	221	−12
12.	α Aql	Altair	19 50 47	+08 52 06	0.77	A7 IV-V	+2.1	17	0.662	54	−26
13.	α Tau A	Aldebaran	04 35 55	+16 30 33	0.85	K5 III	−0.8	65	0.200	161	+54
14.	α Sco A	Antares	16 29 24	−26 25 55	0.96	M1.5 Iab	−5.8	604	0.024	197	−3
15.	α Vir	Spica	13 25 12	−11 09 41	0.98	B1 V	−3.6	262	0.054	232	+1
16.	β Gem	Pollux	07 45 19	+28 01 34	1.14	K0 IIIb	+1.1	34	0.629	265	+3
17.	α Ps A	Formalhaut	22 57 39	−29 37 20	1.16	A3 V	+1.6	25	0.373	116	+7
18.	α Cyg	Deneb	20 41 26	+45 16 49	1.25	A2 Ia	−7.5	1467	0.005	11	−5
19.	β Crucis		12 47 43	−59 41 19	1.25	B0.5 III	−4.0	352	0.042	246	+20
20.	α Leo A	Regulus	10 08 22	+11 58 02	1.35	B7 V	−0.6	77	0.264	271	+4
21.	α Cru A		12 26 35	−63 05 56	1.41	B0.5 IV	−4.0	321	0.030	236	−11
22.	ε CMa A	Adara	06 58 38	−28 59 20	1.50	B2 II	−4.1	431	0.002	27	+27
23.	λ Sco	Shaula	17 33 36	−37 06 14	1.63	B1.5 IV	−3.6	359	0.029	178	0
24.	γ Ori	Bellatrix	05 25 08	+06 20 59	1.64	B2 III	−2.8	243	0.018	221	+18
25.	β Tau	Alnath	05 26 18	+28 36 27	1.65	B7 III	−1.3	131	0.178	172	+8

Based on a table by Robert Garrison in the *Observers Handbook 2001 of the Royal Astronomical Society of Canada*, with the permission of the RASC.

Greek Alphabet

Upper Case	Lower Case		Upper Case	Lower Case	
A	α	alpha	N	ν	nu
B	β	beta	Ξ	ξ	xi
Γ	γ	gamma	O	o	omicron
Δ	δ	delta	Π	π	pi
E	ε	epsilon	P	ρ	rho
Z	ζ	zeta	Σ	σ	sigma
H	η	eta	T	τ	tau
Θ	θ	theta	Υ	υ	upsilon
I	ι	iota	Φ	φ	phi
K	κ	kappa	X	χ	chi
Λ	λ	lambda	Ψ	ψ	psi
M	μ	mu	Ω	ω	omega

APPENDIX 7

Name	R.A. (2000.0) h	m	°	Dec. '	Parallax π "	Distance light-years	Proper Motion μ "/yr	θ °	Radial Velocity km/s	Spectral Type	V	B−V	M_V	Luminosity ($L_\odot = 1$)
1 Sun										G2 V	−26.75	0.65	4.82	1.
2 Proxima Cen	14	29.7	−62	41	0.772	4.21	3.86	282	−22	M5.5 e	11.05	1.90	15.49	0.00005
α Cen A	14	39.6	−60	5	.742	4.40	3.71	278	−22	G2 V	.02	0.65	4.37	1.51
α Cen B							3.69	281	−18	K0 V	1.36	0.85	5.71	0.44
3 Barnard's star	17	57.8	+04	42	.549	5.94	10.37	356	−111	M4 V	9.54	1.74	13.24	0.0004
4 Wolf 359 (CN Leo)	10	56.5	+07	1	.419	7.80	4.69	235	+13	M6 V	13.45	2.0	16.56	0.00002
5 BD +36°2147 HD95735 (Lalande 21185)	11	03.4	+35	58	.392	8.32	4.81	187	−85	M2 V	7.49	1.51	10.46	0.006
6 Sirius A	6	45.1	−16	43	.379	8.61	1.34	204	−8	A1 V	−1.45	0.00	1.44	22.49
Sirius B							1.34	204		DA 2	8.44	−0.03	11.33	0.0025
7 L 726-8, BL Cet = A	1	39.0	−17	57	.374	8.74	3.37	81	+29	M5.5 V	12.41	1.87	15.27	0.00007
UV Cet = B							3.37	81	+28	M6 V	13.25		16.11	0.00003
8 Ross 154 (V1216 Sgr)	18	49.8	−23	50	.337	9.69	0.67	107	−12	M3.5 V	10.45	1.76	13.08	0.0005
9 Ross 248 (HH And)	23	41.9	+44	10	.316	10.31	1.62	177	−78	M5.5 V	12.29	1.91	14.79	0.0001
10 ε Eri	3	32.9	−09	27	.311	10.50	0.98	271	+16	K2 V	3.72	0.88	6.18	0.286
11 CD −36°15693 HD217987 (Lacaille 9352)	23	05.9	−35	51	.304	10.73	6.90	79	+10	M2 V	7.35	1.49	9.76	0.01
12 Ross 128 (FI Vir)	11	47.7	+00	48	.300	10.89	1.36	154	−31	M4 V	11.12	1.76	13.50	0.00034
13 L 789-6 (EZ Aqr) = A	22	38.5	−15	18	.290	11.25	3.25	47	−60	M5 V	12.69	1.99	15.00	0.00008
= B							3.25	47			13.6		15.9	0.00004
14 61 Cyg A	21	06.9	−38	45	.286	11.38	5.28	52	−65	K5 V	5.22	1.17	7.51	0.084
61 Cyg B							5.17	53	−64	K7 V	6.04	1.36	8.32	0.0398
15 Procyon A	7	39.3	+05	14	.286	11.42	1.26	215	−4	F5 IV-V	0.36	0.42	2.64	7.45
Procyon B							1.26	215		DA	10.75		13.03	0.0005
16 BD +43° 44 A (GX And)	0	18.4	+44	1	.280	11.64	2.92	82	+12	M1.5 V	8.08	1.56	10.32	0.006
BD +43° 44 B (GQ And)	0	18.4	+44	2	.280	11.64	2.92	82	+11	M3.5 V	11.05	1.81	13.29	0.0004
17 BD +59 1915 A	18	42.7	+59	38	.280	11.64	2.24	324	−1	M3 V	8.90	1.53	11.14	0.00
BD +59 1915 B	18	42.8	+59	38	.280	11.64	2.27	323	+2	M3.5 V	9.69	1.59	11.93	0.0014
Struve 2398AB = ADS 11632AB														
18 ε Ind	22	03.4	−56	47	.276	11.84	4.71	123	−40	K5 Ve	4.69	1.05	6.89	0.148
19 G 51-15 (DX Cnc)	8	29.8	+26	47	.276	11.84	1.29	243	+25	M6.5 V	14.79	2.07	16.99	0.00001
20 τ Ceti	1	44.1	−15	56	.274	11.90	1.92	296	−17	G8 Vp	3.49	0.72	5.68	0.45
21 L 372-58 = LHS 1565	3	36.0	−44	31	.270	12.07	0.84	119	−20	M5.5 V	13.01	1.90	15.17	0.00007
22 L 725-32 (YZ Cet)	1	12.5	−17	0	.269	12.13	1.37	62	+28	M4.5 V	12.05	1.83	14.20	0.0002
23 BD +5°1668	7	27.4	+05	14	.263	12.39	8.74	171	+18	M3.5 V	9.85	1.57	11.95	0.0014
24 Kapteyn's star	5	11.7	−45	1	.255	12.79	8.66	131	+246	M1 pV	8.85	1.56	10.89	0.004
25 CD −39°14192 (AX Mic) (Lacaille 8760)	21	17.3	−38	52	.253	12.88	3.45	251	+26	M0.5 V	6.68	1.41	8.70	0.028
26 BD +56°2783 A	22	28.0	+57	42	.250	13.08	0.99	242	−33	M3 V	9.79	1.65	11.78	0.002
BD +56°2783 B (DO Cep)							0.99	242	−32	M4 V	11.46	1.8	13.45	0.0004
Krüger 60 AB														
27 Ross 614 A	6	29.4	−02	49	.243	13.44	0.93	132	+17	M4.5 V	11.14	1.72	13.07	0.0005
Ross 614 B (V577 Mon)							0.93	132			14.47		16.40	0.00002
28 BD −12°4523	16	30.3	−12	40	.235	13.89	1.19	185	−22	M3 V	10.08	1.59	11.93	0.0014
29 van Maanen's star	0	49.2	+05	23	.233	14.03	2.98	156	+43	DZ7	12.39	0.55	14.22	0.0002
30 CD −37°15492	0	05.4	−37	21	.229	14.22	6.10	113	+23	M3 V	8.55	1.45	10.35	0.006

Name	R.A. (2000.0) h	m	°	'	Parallax π "	Distance light-years	Proper Motion μ "/yr	θ °	Radial Velocity km/s	Spectral Type	V	B-V	M_V	Luminosity (L$_\odot$ = 1)
31 Wolf 424 A	12	33.3	+09	1	.228		1.81	278	−2	M5.5 V	13.04	1.84	14.83	0.0001
Wolf 424 B (FL Vir)							1.81	278		M7	13.3		15.1	0.00008
32 L 1159-16 (TZ Ari)	2	00.2	+13	3	.225	14.51	2.10	148	−22	M4.5 V	12.27	1.81	14.03	0.0002
33 L 143-23 = LHS 288	10	44.5	−61	12	.223	14.64	1.66	348		M5.5	13.87	1.83	15.61	0.00005
34 BD +68°946	17	36.4	+68	20	.221	14.78	1.31	194	−27	M3 V	9.17	1.49	10.89	0.0037
35 CD −46°11540	17	28.7	−45	54	.220	14.81	1.05	147	−16	M3 V	9.38	1.55	11.10	0.0031
36 LP 731-58 = LHS 292	10	48.2	−11	20	.220	14.81	1.64	159	−2	M6.5 V	15.60	2.10	17.32	0.00001
37 G 208-44 = A	19	53.9	+44	25	.220	14.81	0.74	143	+42	M5.5 V	13.48	1.92	15.19	0.00007
G 208-45 = B							0.74	143	+73	M6 V	14.01	1.97	15.72	0.00004
G 208-44 = C							0.74	143			16.66		18.37	0.000004
38 L 145-141	11	45.7	−64	50	.216	15.07	2.69	97		DQ6	11.50	0.19	13.18	0.00045
39 G 158-27	0	06.7	−07	32	.213	15.30	2.04	204	−28	M5.5 V	13.76	1.97	15.40	0.00006
40 BD −15°6290	22	53.3	−14	16	.213	15.33	1.17	125	−2	M3.5 V	10.16	1.59	11.80	0.0016
41 BD +44°2051 = A	11	05.5	+43	32	.207	15.75	4.51	282	+69	M1 V	8.76	1.55	10.34	0.006
BD +44°2051 = B (WX Uma)	11	05.5	+43	31	.207	15.75	4.51	282	+68	M5.5 V	14.42	2.02	16.00	0.00003
42 BD +50°1725	10	11.4	+49	27	.205	15.88	1.45	250	−26	K7 V	6.59	1.36	8.15	0.0466
43 BD +20°2465 (AD Leo)	10	19.6	+19	52	.205	15.95	0.50	265	+12	M3 V	9.41	1.54	10.96	0.0035
44 CD −49°13515 HD204961	21	33.6	−49	1	.203	16.11	0.82	183	+4	M1 V	8.67	1.48	10.20	0.007
45 LP 944-20	3	39.6	−35	26	.201	16.21	0.44	49		≥ M9 V				
46 CD −44°11909	17	37.0	−44	19	.198	16.44	1.18	217	−41	M4.5 V	10.95	1.66	12.44	0.0009
47 40 Eri A	4	15.3	−07	39	.198	16.49	4.09	213	−43	K1 Ve	4.42	0.82	5.91	0.366
40 Eri B	4	15.4	−7	39	.198	16.47	4.09	213	−17	DA4	9.51	0.04	11.00	0.0034
40 Eri C (DY Eri)							4.09	213	−46	M4.5 V	11.21	1.64	12.70	0.0007
48 BD +43°4305 (EV Lac)	22	46.8	+44	20	.198	16.47	0.84	237	+1	M3.5 V	10.23	1.61	11.71	0.0018
49 BD +02°3482 = A	18	05.5	+02	30	.197	16.60	0.97	173	−7	K0 Ve	4.24	0.78	5.71	0.44
BD +02°3482 = B							0.97	173	−10	K5 Ve	6.01		7.48	0.086
50 Altair	19	50.8	+08	52	.194	16.76	0.66	54	−26	A7 IV-V	0.77	0.2	2.1	12.25

Parallaxes and distances are the new ones from the Hipparcos satellite (1997), courtesy of Hartmut Jahreiss. Parallaxes for the stars not observed by Hipparcos (usually because they are too faint) are from *The General Catalogue of Trigonometric Stellar Parallaxes*, 4th ed., by W. F. van Altena, J. T.-L. Lee, and E. D. Hoffleit (1995). Radial velocities, spectral types, and photometry are from Jahreiss's *Catalogue of Nearby Stars*.

APPENDIX 8

M	NGC	α h	m (2000.0)	δ °	′	m_v	Description	M	NGC	α h	m (2000.0)	δ °	′	m_v	Description
1	1952	5	34.5	+22	01	8.4	Crab Nebula (Tau)	57	6720	18	53.6	+33	02	9.0	Ring N; planetary (Lyr)
2	7089	21	33.5	−00	49	6.5	Globular cluster (Aqr)	58	4579	12	37.7	+11	49	9.8	Spiral galaxy (SBb) (Vir)
3	5272	13	42.2	+28	23	6.4	Glob. cluster (CVn)	59	4621	12	42.0	+11	39	9.8	Elliptical galaxy (Vir)
4	6121	16	23.6	−26	32	5.9	Glob. cluster (Sco)	60	4649	12	43.7	+11	33	8.8	Elliptical galaxy (Vir)
5	5904	15	18.6	+02	05	5.8	Glob. cluster (Ser)	61	4303	12	21.9	+4	28	9.7	Spiral galaxy (Sc) (Vir)
6	6405	17	40.1	−32	13	4.2	Open cluster (Sco)	62	6266	17	01.2	−30	07	6.6	Glob. cluster (Sco)
7	6475	17	53.9	−34	49	3.3	Open cluster (Sco)	63	5055	13	15.8	+42	02	8.6	Spiral galaxy (Sb) (CVn)
8	6523	18	03.8	−24	23	5.8	Lagoon Nebula (Sgr)	64	4826	12	56.7	+21	41	8.5	Spiral galaxy (Sb) (Com)
9	6333	17	19.2	−18	31	7.9	Glob. cluster (Oph)	65	3623	11	18.9	+13	05	9.3	Spiral galaxy (Sa) (Leo)
10	6254	16	57.1	−04	06	6.6	Glob. cluster (Oph)	66	3627	11	20.2	+12	59	9.0	Spiral galaxy (Sb) (Leo)
11	6705	18	51.1	−05	16	5.8	Open cluster (Scu)	67	2682	8	51.4	+11	49	6.9	Open cluster (Can)
12	6218	16	47.2	−01	57	6.6	Glob. cluster (Oph)	68	4590	12	39.5	−26	45	8.2	Glob. cluster (Hyd)
13	6205	16	41.7	+36	28	5.9	Glob. cluster (Her)	69	6637	18	31.4	−32	21	7.7	Glob. cluster (Sgr)
14	6402	17	37.6	−03	15	7.6	Glob. cluster (Oph)	70	6681	18	43.2	−32	18	8.1	Glob. cluster (Sgr)
15	7078	21	30.0	+12	10	6.4	Glob. cluster (Peg)	71	6838	19	53.8	+18	47	8.3	Glob. cluster (Sge)
16	6611	18	18.8	−13	47	6.0	Open cl. & nebula (Ser)	72	6981	20	53.5	−12	32	9.4	Glob. cluster (Aqr)
17	6618	18	20.8	−16	11	7	Omega nebula (Sgr)	73	6994	20	58.9	−12	38		Glob. cluster (Aqr)
18	6613	18	19.9	−17	08	6.9	Open cluster (Sgr)	74	628	1	36.7	+15	47	9.2	Spiral galaxy (Sc) in Pisces
19	6273	17	02.6	−26	16	7.2	Glob. cluster (Oph)	75	6864	20	06.1	−21	55	8.6	Glob. cluster (Sgr)
20	6514	18	02.6	−23	02	8.5	Trifid Nebula (Sgr)	76	650-1	1	42.4	+51	34	11.5	Planetary nebula (Per)
21	6531	18	04.6	−22	30	5.9	Open cluster (Sgr)	77	1068	2	42.7	−0	01	8.8	Spiral galaxy (Sb) (Cet)
22	6656	18	36.4	−23	54	5.1	Glob. cluster (Sgr)	78	2068	5	46.7	+0	03	8	Small emission nebula (Ori)
23	6494	17	56.8	−19	01	5.5	Open cluster (Sgr)	79	1904	5	24.5	−24	33	8.0	Glob. cluster (Lep)
24	6603	18	16.9	−18	29	4.5	Open cluster (Sgr)	80	6093	16	17.0	−22	59	7.2	Glob. cluster (Sco)
25	IC4725	18	31.6	−19	15	4.6	Open cluster (Sgr)	81	3031	9	55.6	+69	04	6.8	Spiral galaxy (Sb) (UMa)
26	6694	18	45.2	−09	24	8.0	Open cluster (Scu)	82	3034	9	55.8	+69	41	8.4	Irregular galaxy (UMa)
27	6853	19	59.6	+22	43	8.1	Dumbbell N., PN (Vul)	83	5236	13	37.0	−29	52	7.6	Spiral galaxy (Sc) (Hyd)
28	6626	18	24.5	−24	52	6.9	Glob. cluster (Sgr)	84	4374	12	25.1	+12	53	9.3	Elliptical galaxy (Vir)
29	6913	20	23.9	+38	32	6.6	Open cluster (Cyg)	85	4382	12	25.4	+18	11	9.2	S0 galaxy (Com)
30	7099	21	40.4	−23	11	7.5	Glob. cluster (Cap)	86	4406	12	26.2	+12	57	9.2	Elliptical galaxy (Vir)
31	224	0	42.7	+41	16	3.4	Andromeda Galaxy (Sb)	87	4486	12	30.8	+12	24	8.6	Elliptical galaxy (Ep) (Vir)
32	221	0	42.7	+40	52	8.2	Elliptical galaxy (And)	88	4501	12	32.0	+14	25	9.5	Spiral galaxy (Sb) (Com)
33	598	1	33.9	+30	39	5.7	Spiral galaxy (Sc) (Tri)	89	4552	12	35.7	+12	33	9.8	Elliptical galaxy (Vir)
34	1039	2	42.0	+42	47	5.2	Open cluster (Per)	90	4569	12	36.8	+13	10	9.5	Spiral galaxy (SBb) (Vir)
35	2168	6	08.9	+24	20	5.1	Open cluster (Gem)	91	4548	12	35.4	+14	30	10.2	M58? (Vir)
36	1960	5	36.1	+34	08	6.0	Open cluster (Aur)	92	6341	17	17.1	+43	08	6.5	Glob. cluster (Her)
37	2099	5	52.4	+32	33	5.6	Open cluster (Aur)	93	2447	7	44.6	−23	52	6.2	Open cluster in (Pup)
38	1912	5	28.7	+35	50	6.4	Open cluster (Aur)	94	4736	12	50.9	+41	07	8.1	Spiral galaxy (Sb) (CVn)
39	7092	21	32.2	+48	26	4.6	Open cluster (Cyg)	95	3351	10	44.0	+11	42	9.7	Barred spiral g. (SBb) (Leo)
40		12	22.4	+58	05	8	Double star (UMa)	96	3368	10	46.8	+11	49	9.2	Spiral galaxy (Sa) (Leo)
41	2287	6	460	−20	44	4.5	Open cluster (CMa)	97	3587	11	14.8	+55	01	11.2	Owl Nebula; planetary (UMa)
42	1976	5	35.4	−5	27	4	Orion Nebula (Ori)								
43	1982	5	35.6	−5	16	9	Orion Nebula; smaller (Ori)	98	4192	12	13.8	+14	54	10.1	Spiral galaxy (Sb) (Com)
44	2632	8	40.1	+19	46	3.1	Praesepe; open cl. (Can)	99	4254	12	18.8	+14	25	9.8	Spiral galaxy (Sc) (Com)
45		3	47.0	+24	07	1.2	Pleiades; open cl. (Tau)	100	4321	12	22.9	+15	49	9.4	Spiral galaxy (Sc) (Com)
46	2437	7	41.8	−14	49	6.1	Open cluster (Pup)	101	5457	14	03.2	+54	21	7.7	Spiral galaxy (Sc) (UMa)
47	2422	7	36.6	−14	30	4.4	Open cluster (Pup)	102							M101; duplication (UMa)
48	2548	8	13.8	−5	48	5.8	Open cluster (Hyd)	103	581	1	33.2	+60	42	7.4	Open cluster (Cas)
49	4472	12	29.8	+8	00	8.4	Elliptical galaxy (Vir)	104	4594	12	40.0	−11	37	8.3	Sombrero N.; spiral (Sa) (Vir)
50	2323	7	02.8	−8	20	5.9	Open cluster (Mon)								
51	5194	13	29.9	+47	12	8.1	Whirlpool Galaxy (Sc) (CVn)	105	3379	10	47.8	+12	35	9.3	Elliptical galaxy (Leo)
52	7654	23	24.2	+61	35	6.9	Open cluster (Cas)	106	4258	12	19.0	+47	18	8.3	Spiral galaxy (Sb) (CVn)
53	5024	13	12.9	+18	10	7.7	Glob. cluster (Com)	107	6171	16	32.5	−13	03	8.1	Glob. cluster (Oph)
54	6715	18	55.1	−30	29	7.7	Glob. cluster (Sgr)	108	3556	11	11.5	+55	40	10.0	Spiral galaxy (Sb) (UMa)
55	6809	19	40.0	−30	58	7.0	Glob. cluster (Sgr)	109	3992	11	57.6	+53	23	9.8	Barred spiral g. (SBc) (UMa)
56	6779	19	16.6	+30	11	8.2	Glob. cluster (Lyr)	110	205	0	40.4	+41	41	8.0	Elliptical galaxy (And)

From Alan Hirshfield and Roger W. Sinnott, *Sky Catalogue 2000.0*, vol. 2, courtesy of Sky Publishing Co. and Cambridge Univ. Press.

Caldwell Catalogue

	NGC/IC	Constellation	Type	R. A. (2000.0)		Dec.		Mag.	Size (')	Notes
1	188	Cepheus	Open cluster	00^h 44.4^m		+85°	20'	8.1	14	
2	40	Cepheus	Planetary nebula	00^h 13.0^m		+72°	32'	12.4	0.8	
3	4236	Draco	Sb galaxy	12^h 16.7^m		+69°	28'	9.7	19 × 7	
4	7023	Cepheus	Nebula	21^h 01.8^m		+68°	12'	—	18 × 18	Bright reflection nebula
5	IC 342	Camelopardalis	SBc galaxy	03^h 46.8^m		+63°	06'	9.2	18 × 17	
6	6543	Draco	Planetary nebula	17^h 58.6^m		+66°	38'	8.1	0.3/5.6	Cat's Eye Nebula
7	2403	Camelopardalis	S0 galaxy	07^h 36.9^m		+65°	36'	8.4	18 × 10	
8	559	Cassiopeia	Open cluster	01^h 29.5^m		+83°	18'	9.5	4	
9	8h2-158	Cepheus	Bright nebula	22^h 56.8^m		+62°	37'	—	50 × 10	Cave Nebula
10	663	Cassiopeia	Open cluster	01^h 46.0^m		+61°	15'	7.1	16	
11	7635	Cassiopeia	Bright nebula	23^h 20.7^m		+61°	12'	—	15 × 8	Bubble Nebula
12	6946	Cepheus	Sc galaxy	20^h 34.8^m		+60°	09'	8.9	11 × 10	
13	457	Cassiopeia	Open cluster	01^h 19.1^m		+58°	20'	6.4	13	Phi Cas Cluster
14	669/884	Perseus	Double cluster	02^h 20.0^m		+57°	08'	4.3	30 and 30	Sword Hand region
15	6826	Cygnus	Planetary nebula	19^h 44.8^m		+50°	31'	8.8	0.5/2.3	Blinking Nebula
16	7243	Lacerta	Open cluster	22^h 15.3^m		+49°	53'	6.4	21	
17	147	Cassiopeia	dE4 galaxy	00^h 33.2^m		+48°	30'	9.3	13 × 8	
18	185	Cassiopeia	dE0 galaxy	00^h 39.0^m		+48°	20'	9.2	12 × 10	
19	IC 6148	Cygnus	Bright nebula	21^h 53.5^m		+47°	16'	—	12 × 12	Cocoon Nebula
20	7000	Cygnus	Bright nebula	20^h 58.8^m		+44°	20'	—	120 × 100	North America Nebula
21	4449	Canes Venatici	Irregular galaxy	12^h 28.2^m		+44°	06'	9.4	5 × 4	
22	7662	Andromeda	Planetary nebula	23^h 25.9^m		+42°	33'	8.3	0.3/2.2	
23	891	Andromeda	Sb galaxy	02^h 22.6^m		+42°	21'	9.9	14 × 3	
24	1275	Perseus	Seyfert galaxy	03^h 19.8^m		+41°	31'	11.6	2.6 × 2	Perseus A radio source
25	2419	Lynx	Globular cluster	07^h 38.1^m		+38°	53'	10.4	4.1	
26	4244	Canes Venatici	S galaxy	12^h 17.5^m		+37°	49'	10.2	16 × 2.5	
27	8888	Cygnus	Bright nebula	20^h 12.0^m		+38°	21'	—	20 × 10	Crescent Nebula
28	752	Andromeda	Open cluster	01^h 57.8^m		+37°	41'	5.7	50	
29	6005	Canes Venatici	Sb galaxy	13^h 10.9^m		+37°	03'	9.8	5.4 × 2	
30	7331	Pegasus	Sb galaxy	22^h 37.1^m		+34°	25'	9.5	11 × 4	
31	IC 405	Auriga	Bright nebula	05^h 16.2^m		+34°	16'	—	30 × 19	Flaming Star Nebula
32	4631	Canes Venatici	S0 galaxy	12^h 42.1^m		+32°	32'	9.3	15 × 3	
33	6992/5	Cygnus	SN remnant	20^h 58.4^m		+31°	43'	—	60 × 6	Eastern Veil Nebula
34	6960	Cygnus	SN remnant	20^h 45.7^m		+30°	43'	—	70 × 6	Western Veil Nebula
35	4889	Coma Berenices	E4 galaxy	13^h 00.1^m		+27°	59'	11.4	3 × 2	Brightest in Coma Cluster
36	4559	Coma Berenices	S0 galaxy	12^h 36.0^m		+27°	58'	9.8	10 × 5	
37	6885	Vulpecula	Open cluster	20^h 12.0^m		+26°	29'	5.9	7	
38	4585	Coma Berenices	Sb galaxy	12^h 36.3^m		+25°	59'	9.6	16 × 3	
39	2392	Gemini	Planetary nebula	07^h 29.2^m		+20°	55'	9.2	0.2/0.7	Eskimo Nebula
40	3626	Leo	Sb galaxy	11^h 20.1^m		+18°	21'	10.9	3 × 2	
41	—	Taurus	Open cluster	04^h 27^m		+16°		0.5	330	Hyades
42	7006	Delphinus	Globular cluster	21^h 01.5^m		+16°	11'	10.6	2.8	Very distant globular
43	7814	Pegasus	Sb galaxy	00^h 03.3^m		+16°	09'	10.5	6 × 2	
44	7479	Pegasus	SBb galaxy	23^h 04.9^m		+12°	19'	11.0	4 × 3	
45	5248	Boötes	Sc galaxy	13^h 37.5^m		+08°	53'	10.2	6 × 4	
46	2261	Monoceros	Bright nebula	08^h 39.2^m		+08°	44'	—	2 × 1	Hubble's Variable Nebula
47	6934	Delphinus	Globular cluster	20^h 34.2^m		+07°	24'	5.9	6.0	
48	2775	Cancer	Sa galaxy	09^h 10.3^m		+07°	02'	10.3	4.5 × 3	
49	2237.9	Monoceros	Bright nebula	06^h 32.3^m		+05°	03'	—	80 × 60	Rosette Nebula
50	2244	Monoceros	Open cluster	05^h 32.4^m		+04°	52'	4.6	24	
51	IO 1613	Cetus	Irregular galaxy	01^h 04.8^m		+02°	07'	9.3	12 × 11	
52	4697	Virgo	E4 galaxy	12^h 48.8^m		−05°	48'	9.3	6 × 4	
53	3115	Sextans	E6 galaxy	10^h 05.2^m		−07°	43'	9.1	6 × 3	Spindle Galaxy
54	2508	Monoceros	Open cluster	08^h 00.2^m		−10°	47'	7.6	7	
55	7009	Aquarius	Planetary nebula	21^h 04.2^m		−11°	22'	8.0	0.4/1.6	Saturn Nebula

	NGC/IC	Constellation	Type	R. A. (2000.0)		Dec.		Mag.	Size (')	Notes
56	248	Cetus	Planetary nebula	00^h	47.0^m	−11°	53′	10.9	3.8	
57	6822	Sagittarius	Irregular galaxy	19^h	44.9^m	−14°	48′	8.8	10 × 9	Barnard's Galaxy
58	2360	Canis Major	Open cluster	07^h	17.8^m	−15°	37′	7.2	13	
59	3242	Hydra	Planetary nebula	10^h	24.8^m	−18°	38′	7.8	0.3/2.1	Ghost of Jupiter
60	4038	Corvus	S0 galaxy	12^h	01.9^m	−18°	52′	10.7	2.8 × 2	
61	4039	Corvus	Sp galaxy	12^h	01.9^m	−18°	53′	10.7	3 × 2	
62	247	Cetus	S galaxy	00^h	47.1^m	−20°	46′	9.1	20 × 7	
63	7283	Aquarius	Planetary nebula	22^h	29.6^m	−20°	48′	7.3	13	Helix Nebula
64	2362	Canis Major	Open cluster	07^h	18.8^m	−24°	57′	4.1	6	Tau CMa Cluster
65	283	Sculptor	Scp galaxy	00^h	47.8^m	−25°	17′	7.1	25 × 7	Sculptor Galaxy
66	5694	Hydra	Globular cluster	14^h	39.6^m	−26°	32′	10.2	3.6	
67	1097	Fornax	SBb galaxy	02^h	46.3^m	−30°	17′	9.2	9 × 7	
68	6729	Corona Australis	Bright nebula	19^h	01.9^m	−36°	57′	—	1.0	R CrA Nebula
69	6302	Scorpius	Planetary nebula	17^h	13.7^m	−37°	06′	9.6	0.8	Bug Nebula
70	300	Sculptor	Sd galaxy	00^h	54.9^m	−37°	41′	8.7	20 × 13	
71	2477	Puppis	Open cluster	07^h	52.3^m	−38°	33′	5.8	27	
72	65	Sculptor	SB galaxy	00^h	14.9^m	−39°	11′	7.9	32 × 6	Brightest in Sculptor
73	1851	Columba	Globular cluster	05^h	14.1^m	−40°	03′	7.3	11	
74	3132	Vela	Planetary nebula	10^h	07.7^m	−40°	26′	9.4	0.8	
75	6124	Scorpius	Open cluster	16^h	25.6^m	−40°	40′	5.8	29	
76	6231	Scorpius	Open cluster	16^h	54.0^m	−41°	48′	2.6	15	
77	6128	Centaurus	Peculiar radio galaxy	13^h	25.5^m	−43°	01′	7.0	18 × 14	Cen A radio source
78	6541	Corona Australis	Globular cluster	18^h	08.0^m	−43°	42′	6.6	13	
79	3201	Vela	Globular cluster	10^h	17.6^m	−46°	25′	6.7	18	
80	5139	Centaurus	Globular cluster	13^h	26.8^m	−47°	29′	3.8	36	Omega Centauri
81	8352	Ara	Globular cluster	17^h	25.5^m	−48°	25′	6.1	7	
82	6193	Ara	Open cluster	16^h	41.3^m	−48°	46′	5.2	15	
83	4945	Centaurus	SBc galaxy	13^h	05.4^m	−49°	28′	8.7	20 × 4	
84	5286	Centaurus	Globular cluster	13^h	46.4^m	−51°	22′	7.6	9	
85	IC 2391	Vela	Open cluster	08^h	40.2^m	−53°	04′	2.5	50	o Vel Cluster
86	8397	Ara	Globular cluster	17^h	40.7^m	−53°	40′	5.6	26	
87	1281	Horologium	Globular cluster	03^h	12.3^m	−55°	13′	8.4	7	
88	5823	Circinus	Open cluster	15^h	05.7^m	−55°	36′	7.9	10	
89	6087	Norma	Open cluster	16^h	18.9^m	−57°	54′	5.4	12	S Nor Cluster
90	2887	Carina	Planetary nebula	09^h	21.4^m	−58°	19′	—	0.2	
91	3632	Carina	Open cluster	11^h	06.4^m	−58°	40′	3.0	55	
92	3372	Carina	Bright nebula	10^h	43.8^m	−59°	52′	—	120 × 120	Eta Carinae Nebula
93	8752	Pavo	Globular cluster	19^h	10.9^m	−59°	59′	5.4	20	
94	4755	Crux	Open cluster	12^h	53.6^m	−60°	20′	4.2	10	Jewel Box Cluster
95	6025	Triangulum Aus.	Open cluster	16^h	03.7^m	−60°	30′	5.1	12	
96	2516	Carina	Open cluster	07^h	58.3^m	−60°	52′	3.8	30	
97	3768	Centaurus	Open cluster	11^h	36.1^m	−61°	37′	5.3	12	
98	4609	Crux	Open cluster	12^h	42.3^m	−62°	58′	8.9	5	
99	—	Crux	Dark nebula	12^h	53^m	−63°		—	400 × 300	Coalsack
100	[IC 2944]	Centaurus	Open cluster	11^h	36.6^m	−63°	02′	4.5	15	Lambda Cen Cluster
101	6744	Pavo	SBb galaxy	19^h	09.8^m	−63°	51′	8.3	16 × 10	
102	IC 2602	Carina	Open cluster	10^h	43.2^m	−64°	24′	1.9	50	Theta Cas Cluster
103	2070	Dorado	Bright nebula	05^h	38.7^m	−69°	06′	—	40 × 25	Tarantula Nebula
104	382	Tucana	Globular cluster	01^h	03.2^m	−70°	51′	6.6	13	
105	4833	Musca	Globular cluster	12^h	59.6^m	−70°	53′	7.3	14	
106	104	Tucana	Globular cluster	00^h	24.1^m	−72°	05′	4.0	31	47 Tucanae
107	6101	Apus	Globular cluster	16^h	25.8^m	−72°	12′	9.3	11	
108	4372	Musca	Globular cluster	12^h	25.8^m	−72°	40′	7.8	19	
109	3195	Chamaeleon	Planetary nebula	10^h	09.5^m	−80°	52′	—	0.6	

The Constellations

Latin Name	Genitive	Abbreviation	Translation	Latin Name	Genitive	Abbreviation	Translation
Andromeda	Andromedae	And	Andromeda*	Lacerta	Lacertae	Lac	Lizard
Antlia	Antliae	Ant	Pump	Leo	Leonis	Leo	Lion
Apus	Apodis	Aps	Bird of Paradise	Leo Minor	Leonis Minoris	LMi	Little Lion
Aquarius	Aquarii	Aqr	Water Bearer	Lepus	Leporis	Lep	Hare
Aquila	Aquilae	Aql	Eagle	Libra	Librae	Lib	Scales
Ara	Arae	Ara	Altar	Lupus	Lupi	Lup	Wolf
Aries	Arietis	Ari	Ram	Lynx	Lyncis	Lyn	Lynx
Auriga	Aurigae	Aur	Charioteer	Lyra	Lyrae	Lyr	Harp
Boötes	Boötis	Boo	Herdsman	Mensa	Mensae	Men	Table (mountain)
Caelum	Caeli	Cae	Chisel	Microscopium	Microscopii	Mic	Microscope
Camelopardalis	Camelopardalis	Cam	Giraffe	Monoceros	Monocerotis	Mon	Unicorn
Cancer	Cancri	Cnc	Crab	Musca	Muscae	Mus	Fly
Canes Venatici	Canum Venaticorum	CVn	Hunting Dogs	Norma	Normae	Nor	Level (square)
Canis Major	Canis Majoris	CMa	Big Dog	Octans	Octantis	Oct	Octant
Canis Minor	Canis Minoris	CMi	Little Dog	Ophiuchus	Ophiuchi	Oph	Ophiuchus* (serpent bearer)
Capricornus	Capricorni	Cap	Goat	Orion	Orionis	Ori	Orion*
Carina	Carinae	Car	Ship's Keel**	Pavo	Pavonis	Pav	Peacock
Cassiopeia	Cassiopeiae	Cas	Cassiopeia*	Pegasus	Pegasi	Peg	Pegasus* (winged horse)
Centaurus	Centauri	Cen	Centaur*	Perseus	Persei	Per	Perseus*
Cepheus	Cephei	Cep	Cepheus*	Phoenix	Phoenicis	Phe	Phoenix
Cetus	Ceti	Cet	Whale	Pictor	Pictoris	Pic	Easel
Chamaeleon	Chamaeleonis	Cha	Chameleon	Pisces	Piscium	Psc	Fish
Circinus	Circini	Cir	Compass	Piscis Austrinus	Piscis Austrini	PsA	Southern Fish
Columba	Columbae	Col	Dove	Puppis	Puppis	Pup	Ship's Stern**
Coma Berenices	Comae Berenices	Com	Berenice's Hair*	Pyxis	Pyxidis	Pyx	Ship's Compass**
Corona Australis	Coronae Australis	CrA	Southern Crown	Reticulum	Reticuli	Ret	Net
Corona Borealis	Coronae Borealis	CrB	Northern Crown	Sagitta	Sagittae	Sge	Arrow
Corvus	Corvi	Crv	Crow	Sagittarius	Sagittarii	Sgr	Archer
Crater	Crateris	Crt	Cup	Scorpius	Scorpii	Sco	Scorpion
Crux	Crucis	Cru	Southern Cross	Sculptor	Sculptoris	Scl	Sculptor
Cygnus	Cygni	Cyg	Swan	Scutum	Scuti	Sct	Shield
Delphinus	Delphini	Del	Dolphin	Serpens	Serpentis	Ser	Serpent
Dorado	Doradus	Dor	Swordfish	Sextans	Sextantis	Sex	Sextant
Draco	Draconis	Dra	Dragon	Taurus	Tauri	Tau	Bull
Equuleus	Equulei	Equ	Little Horse	Telescopium	Telescopii	Tel	Telescope
Eridanus	Eridani	Eri	River Eridanus*	Triangulum	Trianguli	Tri	Triangle
Fornax	Fornacis	For	Furnace	Triangulum Australe	Trianguli Australis	TrA	Southern Triangle
Gemini	Geminorum	Gen	Twins	Tucana	Tucanae	Tuc	Toucan
Grus	Gruis	Gru	Crane	Ursa Major	Ursae Majoris	UMa	Big Bear
Hercules	Herculis	Her	Hercules*	Ursa Minor	Ursae Minoris	UMi	Little Bear
Horologium	Horologii	Hor	Clock	Vela	Velorum	Vel	Ship's Sails**
Hydra	Hydrae	Hya	Hydra* (water monster)	Virgo	Virginis	Vir	Virgin
Hydrus	Hydri	Hyi	Sea serpent	Volans	Volantis	Vol	Flying Fish
Indus	Indi	Ind	Indian	Vulpecula	Vulpeculae	Vul	Little Fox

*Proper names
**Formerly formed the constellation Argo Navis, the Argonauts' ship.

Elements and Solar System Abundances

		Name	Atomic Weight	Abundance			Name	Atomic Weight	Abundance
1	H	hydrogen	1.01	12.00	57	La	lanthanum	138.91	1.20
2	He	helium	4.00	10.99	58	Ce	cerium	140.12	1.61
3	Li	lithium	6.94	1.16	59	Pr	praseodymium	140.91	0.78
4	Be	beryllium	9.01	1.15	60	Nd	neodymium	144.24	1.47
5	B	boron	10.81	2.6	61	Pm	promethium	(145)	
6	C	carbon	12.01	8.55	62	Sm	samarium	150.36	0.97
7	N	nitrogen	14.01	7.97	63	Eu	europium	151.97	0.54
8	O	oxygen	16.00	8.87	64	Gd	gadolinium	157.25	1.07
9	F	fluorine	19.00	4.56	65	Tb	terbium	158.93	0.33
10	Ne	neon	20.18	8.08	66	Dy	dysprosium	162.50	1.15
11	Na	sodium	22.99	6.31	67	Ho	holmium	164.93	0.50
12	Mg	magnesium	24.31	7.58	68	Er	erbium	167.26	0.95
13	Al	aluminum	26.98	6.48	69	Tm	thulium	168.93	0.13
14	Si	silicon	28.09	7.55	70	Yb	ytterbium	173.04	0.95
15	P	phosphorus	30.97	5.57	71	Lu	lutetium	174.97	0.12
16	S	sulfur	32.07	7.27	72	Hf	hafnium	178.49	0.73
17	Cl	chlorine	35.45	5.27	73	Ta	tantalum	180.95	0.13
18	Ar	argon	39.94	—	74	W	tungsten	183.85	0.68
19	K	potassium	39.10	5.13	75	Re	rhenium	186.21	0.27
20	Ca	calcium	40.08	6.34	76	Os	osmium	190.2	1.38
21	Sc	scandium	44.96	3.09	77	Ir	iridium	192.22	1.37
22	Ti	titanium	47.88	4.93	78	Pt	platinum	195.08	1.68
23	V	vanadium	50.94	4.02	79	Au	gold	196.97	0.83
24	Cr	chromium	52.00	5.68	80	Hg	mercury	200.59	1.09
25	Mn	manganese	54.94	5.53	81	Tl	thallium	204.38	0.82
26	Fe	iron	55.85	7.51	82	Pb	lead	207.2	2.05
27	Co	cobalt	58.93	4.91	83	Bi	bismuth	208.98	0.71
28	Ni	nickel	58.69	6.25	84	Po	polonium	(209)	
29	Cu	copper	63.55	4.27	85	At	astatine	(210)	
30	Zn	zinc	65.39	4.65	86	Rn	radon	(222)	
31	Ga	gallium	69.72	3.13	87	Fr	fancium	(223)	
32	Ge	germanium	72.61	3.63	88	Ra	radium	(226)	
33	As	arsenic	74.92	2.37	89	Ac	actinium	(227)	
34	Se	selenium	78.96	3.35	90	Th	thorium	232.04	0.08
35	Br	bromine	79.90	2.63	91	Pa	protractinium	231.04	
36	Kr	krypton	83.80	3.23	92	U	uranium	238.03	−0.49
37	Rb	rubidium	85.47	2.40	93	Np	neptunium	(237)	
38	Sr	strontium	87.62	2.93	94	Pu	plutonium	(244)	
39	Y	yttrium	88.91	2.22	95	Am	americium	(243)	
40	Zr	zirconium	91.22	2.61	96	Cm	curium	(247)	
41	Nb	niobium	92.91	1.40	97	Bk	berkelium	(247)	
42	Mo	molybdenum	95.94	1.96	98	Cf	californium	(251)	
43	Tc	technetium	(98)		99	Es	einsteinium	(252)	
44	Ru	ruthenium	101.07	1.82	100	Fm	fermium	(257)	
45	Rh	rhodium	102.91	1.09	101	Md	mendelevium	(260)	
46	Pd	palladium	106.42	1.70	102	No	nobelium	(259)	
47	Ag	silver	107.87	1.24	103	Lr	lawrencium	(262)	
48	Cd	cadmium	112.41	1.76	104	Rf	rutherfordium	(261)	
49	In	indium	114.82	0.82	105	Db	dubnium	(262)	
50	Sn	tin	118.71	2.14	106	Sg	seaborgium	(266)	
51	Sb	antimony	121.75	1.04	107	Bh	bohrium	(264)	
52	Te	tellurium	127.60	2.24	108	Hs	hassium	(267)	
53	I	iodine	126.90	1.51	109	Mt	meitnerium	(268)	
54	Xe	xenon	131.29	2.23	110			(269)	
55	Cs	cesium	132.91	1.12	111			(272)	
56	B	barium	137.33	2.21	112			(277)	
					114				
					116				
					118				

Abundances given are logarithmic, based on hydrogen = 12.00.

The abundances given are solar photospheric for elements 3–10 and meteoritic for other elements; when both are available, differences are almost always slight. The abundance data were supplied by Nicolas Grevesse in 1994, and represent an update from Edward Anders and Nicolas Grevesse, *Geochimica et Cosmochima Acta*, **53**, 197–214 (1989).

Atomic weights are from the 1988 IUPAC (International Union of Pure and Applied Chemistry) report, plus the longest halflife for the heaviest elements.

In 1997, a committee of the International Union of Pure and Applied Chemistry recommended the names given for elements 101 through 109.

I thank Peter Armbruster of Gesellschaft für Schwerionen forschung mbH Darmstad for information about the heaviest elements, many of which he discovered.

Extra-Solar Planets

	Star Name	Upper Limit (Jupiter masses)	Period (d)	Semi-Major Axis (A.U.)	Eccen-tricity
1	HD 187123	0.52	3.097	0.042	0.00
2	τ Bootis	3.64	3.3126	0.042	0.00
3	HD 75289	0.42	73.508	0.047	0.00
4	51 Pegasi	0.44	4.2308	0.051	0.01
5	υ Andromedae (b)	0.69	4.617	0.059	0.04
6	υ Andromedae (c)	2.0	241.3	0.82	0.23
7	υ Andromedae (d)	4.1	1280.6	2.4	0.31
8	HD 217107	1.28	7.11	0.07	0.14
9	ρ_1 55 Cancri	0.85	14.656	0.12	0.03
10	Gliese 86	3.6	15.8	0.11	0.04
11	HD 195019	3.43	18.3	0.14	0.05
12	ρ Corona Borealis	1.1	39.6	0.23	0.1
13	HD 168443	5.04	58	0.28	0.54
14	Gliese 876	2.1	60.9	0.21	0.27
15	HD114762	11.0	84	0.41	0.33
16	70 Virginis	7.4	116.7	0.47	0.40
17	HD 210277	1.36	437	1.15	0.45
18	16 Cygni B	1.74	802.8	1.70	0.68
19	47 Ursae Majoris	2.42	1093	2.08	0.10
20	14 Herculis	4	2000	3	0.35
21	ι Horologii	2.26	320	0.91	0.16
22	HD130322	1.08	10.7	0.103	0.03
23	HD192263	0.76	24.1	0.18	0.12
24	HD10697	6.59	1083	2.0	0.12
25	HD37124	1.04	155	0.585	0.19
26	HD134987	1.58	260	0.78	0.24
27	HD177830	1.28	391	1.00	0.43
28	HD222582	5.4	576	1.35	0.71
29	HD209458	0.63	3.5238	0.045	0.00

Pulsar Planets around PSR 1257+12

A: mass >0.015 Earth masses, semimajor axis 0.19 AU, period 25.34 days
B: mass >3.4 Earth masses, semimajor axis 0.36 AU, period 66.54 days
C: mass >2.8 Earth masses, semimajor axis 0.47 AU, period 98.22 days

Selected Readings

Monthly Non-Technical Magazines in Astronomy

Sky and Telescope, P.O. Box 9111, Belmont, MA 02138, 800 253 0245; www.skypub.com.

Astronomy, 21027 Crossroads Circle, P.O. Box 1612, Waukesha, WI 53187, 800 533 6644; astronomy.com.

Mercury, Astronomical Society of the Pacific, 390 Ashton Ave., San Francisco, CA 94112, 800 962 3412; www.aspsky.org.

StarDate, 2609 University, Rm. 3.118, University of Texas, Austin, TX 78712, 800 STARDATE; stardate.utexas.edu.

The Griffith Observer, 2800 East Observatory Road, Los Angeles, CA 90027; www.griffithobs.org.

Magazines and Annuals Carrying Articles on Astronomy

Science News, P.O. Box 1925, Marion, OH 43305, 800 552 4412; www.sciserv.org. Published weekly.

Scientific American, P.O. Box 3186, Harlan, IA 51593-2377, 800 333 1199.

National Geographic, P.O. Box 96583, Washington, D.C. 20078-9973, 800 NGC LINE.

Natural History, P.O. Box 5000, Harlan, IA 51593-5000, 800 234 5252.

New Scientist, 151 Wardour St., London W1V 4BN, UK, 888 822 4342, rbi.subscriptions@rbi.co.uk; www.newscientist.com.

Physics Today, American Institute of Physics, 2 Huntingdon Quadrangle, Melville, NY 11747, 800 344 6902; www.aip.org/pt; www.physicstoday.org.

Science Year, World Book Encyclopedia, Inc., Post Office Box 11207, Des Moines, IA 50340-1207, 800 504 4425. The World Book Science Annual.

Smithsonian, P.O. Box 420311, Palm Coast, FL 32142-0311, Washington, D.C. 20560, 800 766 2149; www.smithsonianmag.si.edu.

Discover, P.O. Box 42105, Palm Coast, FL 34142-0105, 800 829 9132.

The Planetary Report, The Planetary Society, 65 North Catalina Avenue, Pasadena, CA 91106-2301, 818 793 5100; planetary.org/tps.

Observing Reference Books

Pasachoff, J. M., *A Field Guide to the Stars and Planets*, 4th ed. (Boston: Houghton Mifflin Co., 2000). All kinds of observing information, including monthly maps and the 2000.0 sky atlas by Wil Tirion, and Graphic Timetables to locate planets and special objects like clusters and galaxies.

Pasachoff, J. M., *Peterson's First Guide to Astronomy*. (Boston: Houghton Mifflin Co., 1997). A brief, beautifully illustrated introduction to observing the sky. Tirion monthly maps.

Pasachoff, J. M., *Peterson's First Guide to the Solar System*. (Boston: Houghton Mifflin Co., 1997). Color illustrations and simple descriptions mark this elementary introduction. Tirion maps of Mars, Jupiter, and Saturn's positions through 2010.

Arnold, H. J. P., P. Doherty, and P. Moore, *The Photographic Atlas of the Stars*. (Bristol, UK: Institute of Physics, 1997).

The Astronomical Almanac (yearly). U.S. Government Printing Office, Washington, D.C. 20402.

Dickinson, T., *Nightwatch: A Practical Guide to Viewing the Universe*, 3rd ed. (Willowdale, Ontario: Firefly Books, 1998).

Harrington, P. S., *The Deep Sky: An Introduction*. (Cambridge, MA: Sky Publishing Corp., 1997).

Hirshfeld, A., and R. W. Sinnott, and F. Ochsenbein, *Sky Catalogue 2000.0*, 2nd ed., 2 vols. (Cambridge, MA: Sky Publishing Corp. 1985–91). Vol. 1 is stars and Vol. 2 is full of tables of all other objects.

Kitchen, C., and R. W. Forrest, *Seeing Stars: The Night Sky Through Small Telescopes*. (New York: Springer-Verlag, 1998).

Levy, D. H., *Observing Variable Stars: A Guide for the Beginner*. (New York: Cambridge University Press).

Mullaney, J., *Celestial Harvest: 3000-Plus Showpieces of the Heavens for Telescope Viewing & Contemplation*. (self published, P.O. Box 1146, Exton, PA 19341, 1998); jimullaneysm@email.msn.com.

Observer's Handbook (yearly), Royal Astronomical Society of Canada, 136 Dupont Street, Toronto, Ontario M5R 1V2 Canada.

O'Meara, S. J., and D. H. Levy, *The Messier Objects*. (New York: Cambridge University Press, 1998).

Ottewell, G., *Astronomical Calendar* (yearly) and *The Astronomical Companion*. Department of Physics, Furman University, Greenville, SC 29613; 864 294 2208; guyverno@aol.com; www.kalend.com.

Ridpath, I., ed., *Norton's Star Atlas and Reference Handbook (Epoch 2000.0)*, 19th ed. (Longman, 1998). The old standard, updated.

Ridpath, I., *Eyewitness Handbooks: Stars and Planets*. (DK Publishing, 1998).

Sinnott, R. W., ed., *NGC 2000.0: The Complete New General Catalogue and Index Catalogues of Nebulae and Star Clusters*. (Cambridge, MA: Sky Publishing Corp. and New York: Cambridge University Press, 1998). A centennial reissue of Dreyer's work with updated data.

Sinnott, R. W., and M. A. C. Perryman, *Millennium Star Atlas.* (Cambridge, MA: Sky Publishing Corp., 1997). Uses the Hipparcos data.

Tirion, W., *Cambridge Star Atlas*, 2nd ed. (Cambridge: Cambridge University Press, 1996). A naked-eye star atlas in full color. A Moon map, 24 monthly sky maps, 20 detailed star charts, and 6 all-sky maps.

Tirion, W., and R. W. Sinnott, *Sky Atlas 2000.0*, 2nd ed., (Cambridge, MA: Sky Publishing Corp. and New York: Cambridge University Press, 1998). Large-scale star charts.

Tirion, W., B. Rappaport, and G. Lovi, *Uranometria 2000.0*, 2 vols. (Richmond, VA: Willmann-Bell, 1986). Vol. 1 covers +90° to −6°. Vol. 2 covers +6° to −90°. Star maps to magnitude 9.5 and an essay on historical atlases.

Willmann-Bell catalogue, *Astronomy Books*, P.O. Box 35025, Richmond, VA 23235, 800 825 7827; www.willbell.com. They publish many observing guides and distribute all kinds of astronomy books.

For Information About Amateur Societies

American Association of Variable Star Observers (AAVSO), 25 Birch St., Cambridge, MA 02138; www.aavso.org.

American Meteor Society, Dept. of Physics and Astronomy, SUNY, Geneseo, NY 14454; www.amsmeteors.org.

Astronomical League, the umbrella group of amateur societies. For their newsletter, *The Reflector*, write The Astronomical League, Executive Secretary, c/o Science Service Building, 1719 N St., N.W., Washington, D.C. 20030; www.astroleague.org.

Astronomical Society of the Pacific, 390 Ashton Ave., San Francisco, CA 94112; www.aspsky.org.

International Dark Sky Association, c/o David Crawford, 3545 N. Stewart, Tucson, AZ 85716, 877 600 5888; www.darksky.org.

The Planetary Society, 65 North Catalina Ave., Pasadena, CA 981106-2301, 818 793 5100; planetary.org/tps.

British Astronomical Association, Burlington House, Piccadilly, London W1V 0NL, England; www.ast.cam.ac.uk/~baa.

Royal Astronomical Society of Canada, 124 Merton St., Toronto, Ontario M4S 2Z2; Canada, www.rasc.ca.

Careers in Astronomy

AAS Education Office, Adler Planetarium & Astronomy Museum, 1300 S. Lake Shore Dr., Chicago IL 60605; aased@aas.org; www.aas.org. A free booklet, *A Career in Astronomy*, is available on request.

Teaching

Pasachoff, J. M., and J. R. Percy, *The Teaching of Astronomy.* (Cambridge University Press, 1990; paperbound, 1993). The proceedings of International Astronomical Union Colloquium #105, held in Williamstown in 1988.

Percy, J. R., *Astronomy Education: Current Developments, Future Coordination.* (San Francisco: Astronomical Society of the Pacific, 1996). An ASP conference, held in College Park, MD, in 1995.

Gouguenheim, L., D. McNally, and J. R. Percy, *New Trends in Astronomy Teaching.* (Cambridge: Cambridge University Press, 1997). The proceedings of International Astronomical Union Colloquium #162, held in London in 1996.

General Reading and Reference

Parker, S. P., and J. M. Pasachoff, *McGraw-Hill Encyclopedia of Astronomy*, (New York: McGraw-Hill, 1993).

Audouze, J., and G. Israel, eds. *The Cambridge Atlas of Astronomy*, 3rd ed. (Cambridge: Cambridge University Press, 1994). Coffee-table size with fantastic photos and authoritative text.

Greeley, R., and R. Batson, *The NASA Atlas of the Solar System.* (Cambridge: Cambridge University Press, 1997). Coffee-table size with fantastic photos and authoritative text.

Lausten, S., C. Madsen, and R. M. West, *Exploring the Southern Sky: A Pictorial Atlas from the European Southern Observatory.* (New York: Springer-Verlag, 1987). Beautiful and impressive.

Malin, D., *The Invisible Universe.* (Boston: Bulfinch Press, 1999). Fantastic color photographs and interesting historical and scientific.

Maran, S. P., ed., *The Astronomy and Astrophysics Encyclopedia.* (New York: Van Nostrand, 1992). Essays on all aspects of astronomy.

Mitton, J., *The Penguin Dictionary of Astronomy.* (New York: Penguin, 1993). Short entries on all aspects of astronomy.

Morrison, P., and P. Morrison, *Powers of 10.* (New York: Scientific American Library, 1982). A classic progression through different size scales in the Universe. A video version, a CD, and a DVD are available.

Trefil, J., *Other Worlds: Images of the Cosmos from Earth and Space.* (National Geographic Society, 1999).

Tyson, N. G., C. Liu, and R. Irion, *The Universe: At Home in the Cosmos.* (Washington D.C., Joseph Henry Press, 2000). Clearly written and lavishly illustrated.

History

Hoskin, M., ed., *Cambridge Illustrated History: Astronomy.* (Cambridge: Cambridge University Press, 1997).

Cohen, I. B., *The Newtonian Revolution.* (Cambridge: Cambridge University Press, 1981).

Gingerich, O., *The Eye of Heaven: Ptolemy, Copernicus and Kepler.* (New York: American Institute of Physics, 1993).

Kolb, R., *Blind Watchers of the Sky.* (Reading, MA: Addison-Wesley, 1996). The evolution of world views, beginning with Tycho Brahe.

Leverington, D., *A History of Astronomy: From 1890 to the Present.* (New York: Springer-Verlag, 1995).

Reston, J., *Galileo: A Life.* (New York: HarperCollins, 1994).

Voelkel, J., *Kepler.* (New York: Oxford University Press, 1999). A short biography in the Oxford Portraits series.

Art and Astronomy

Olson, R. J. M., and J. M. Pasachoff, *Fire in the Sky: Comets and Meteors, the Decisive Centuries, in British Art and Science.* (Cambridge: Cambridge University Press, 1998, 1999). From the time of Newton and Halley to the present.

Clair, J., ed., *The Cosmos: From Romanticism to the Avant-Garde.* (Prestel, 1999). A mixture of objects, paintings, photographs, and so on, from a museum exhibition, with accompanying essays.

Solar System

Alvarez, W., *T.-Rex and the Crater of Doom.* (Princeton: Princeton University Press, 1997). A personal account of the quest to understand the extinction of the dinosaurs.

Beatty, J. K., C. C. Petersen, and A. Chaikin, *The New Solar System*, 4th ed. (Cambridge, MA: Sky Publishing Corp. and Cambridge University Press, 1999). Each chapter written by a different expert.

Beebe, R., *Jupiter: The Giant Planet*, 2nd ed. (Washington, D.C.: Smithsonian Institution Press, 1996).

Cernan, E., and D. David, *The Last Man on the Moon.* (New York: St. Martin's Press, 1999). A first-person account.

Chaikin, A., *A Man on the Moon: The Voyages of the Apollo Astronauts.* (New York: Viking, 1994). The story of the missions and the people on them.

Chapman, C., and D. Morrison, *Cosmic Catastrophes.* (New York: Plenum, 1989). An interesting account of the many ways in which life on Earth is threatened.

Croswell, K., *Planet Quest: The Epic Discovery of Alien Solar Systems.* (New York: Free Press, 1997). Excellent early account of the search for extrasolar planets.

Elliot, J., and R. Kerr, *Rings.* (Cambridge, MA: MIT Press, 1984). Includes first-person accounts and other stories of the discoveries.

Espenak, F., *Fifty Year Canon of Lunar Eclipses: 1968–2035.* (NASA Ref. Pub. 1216).

Goldsmith, D., and T. Owen, *The Search for Life in the Universe.* (New York: Addison-Wesley, 1992). Describes the conditions thought to be necessary for life as we know it.

Goldsmith, D., *The Hunt for Life on Mars.* (New York: Dutton, 1997).

Goldsmith, D. W., *Worlds Unnumbered: The Quest to Discover Other Solar Systems.* (Sausalito, CA: University Science Books, 1997).

Harland, D. M., *Exploring the Moon: The Apollo Expeditions.* (New York: Springer-Verlag, 1999).

Hunt, G. E., and P. Moore, *Atlas of Neptune.* (Cambridge: Cambridge University Press, 1994, 1989; order from Sky Publishing Corp.). Maps and tables.

Lemonick, M., *Other Worlds: The Search for Life in the Universe.* (New York: Simon & Schuster, 1998).

Maor, E., *June 8, 2004–Venus in Transit.* (Princeton, NJ: Princeton University Press, 2000).

McNab, D., and J. Younger, *The Planets.* (New Haven, CT: Yale University Press, 1999). To accompany a TV series on planetary exploration.

Morrison, D., *Exploring Planetary Worlds.* (New York: Scientific American Library, 1993).

Morrison, D., and T. Owen, *The Planetary System*, 2nd ed. (Reading, MA: Addison-Wesley, 1996). All about the solar system.

Rogers, J. H., *The Giant Planet Jupiter.* (Cambridge: Cambridge University Press, 1995).

Sagan, C., *Pale Blue Dot.* (New York: Random House, 1994). One of astronomy's most eloquent spokesperson's last works.

Stern, S. Alan, *Our Worlds: The Magnetism and Thrill of Planetary Exploration: As Described by Leading Planetary Scientists.* (Cambridge: Cambridge University Press, 1999).

Stern, S. A., and J. Mitton, *Pluto & Charon: Ice Worlds on the Ragged Edge of the Solar System.* (New York: Wiley, 1997).

Ward, P. D., and D. Brownlee. *Rare Earth: Why Complex Life is Uncommon in the Universe.* (New York: Copernicus Books, 2000). Argues that intelligent, technologically advanced life is very rare in the cosmos.

Wilhelms, D. E., *To a Rocky Moon: A Geologist's History of Lunar Exploration.* (Tucson: University of Arizona Press, 1993).

Zuckerman, B., and M. Hart, eds., *Extraterrestrials—Where Are They?* 2nd ed. (Cambridge: Cambridge University Press, 1995).

The Sun

Espenak, F., *Fifty Year Canon of Solar Eclipses.* (NASA Ref. Pub. 1178, Rev. 1987). Maps and tables.

Golub, L., and J. M. Pasachoff, *The Sun.* (Cambridge: Harvard University Press, 2001). A non-technical, trade book.

Golub, L., and J. M. Pasachoff, *The Solar Corona.* (Cambridge: Cambridge University Press, 1997). An advanced text.

Lang, K. R., *Sun, Earth, and Sky.* (New York: Springer-Verlag, 1995).

Phillips, K. J. H., *Guide to the Sun.* (New York: Cambridge University Press, 1992).

Taylor, P. O., and N. L. Hendrickson, *Beginner's Guide to the Sun.* (Waukesha, WI: Kalmach Books, 1995).

Zirker, J., *Journey From the Center of the Sun.* (Princeton: Princeton University Press, 2001). A non-technical, trade book.

Stars and the Milky Way Galaxy

Aller, L. H., *Atoms, Stars, and Nebulae*, 3rd ed. (Cambridge: Cambridge University Press, 1991). A classic, recently revised.

Croswell, K., *The Alchemy of the Heavens.* (New York: Anchor, 1995). An excellent account of our Milky Way Galaxy.

Henbest, N., and H. Couper, *The Guide to the Galaxy.* (Cambridge: Cambridge University Press, 1994). Exciting, contemporary results; profusely illustrated.

Kaler, J. B., *Cosmic Clouds: Birth, Death, and Recycling in the Galaxy.* (W. H. Freeman, 1997).

Kaler, J. B., *Stars and Their Spectra.* (Cambridge: Cambridge University Press, 1989). OBAFGKM.

Mann, A., *Shadow of a Star.* (W. H. Freeman, 1997). The story of the neutrinos from SN 1987A.

Marschall, L., *The Supernova Story*. (Princeton: Princeton University Press, 1994.) Supernovae in general plus SN 1987A.

Mook, D., and T. Vargish, *Inside Relativity*. (Princeton: Princeton University Press, 1987). An excellent introduction to special relativity.

Thorne, K., *Black Holes and Time Warps: Einstein's Outrageous Legacy*. (Norton, 1994). An excellent account of general relativity.

Tucker, W. H., *On X-Ray Astronomy*. (Cambridge, MA: Harvard University Press, 2001).

Galaxies and Cosmology

Adams, F., and G. Laughlin, *The Five Ages of the Universe*. (New York: Free Press, 1999). Tells the story of what will happen as the Universe expands forever.

Aschenbach, B., H. M. Hahn, and J. Truemper, *The Invisible Sky: ROSAT and the Age of X-Ray Astronomy*. (New York: Copernicus, 1998).

Begelman, M., and M. Rees, *Gravity's Fatal Attraction: Black Holes in the Universe*. (New York: Scientific American Library, 1996).

Berendzen, R., R. Hart, and D. Seeley, *Man Discovers the Galaxies*. (New York: Neale Watson Academic Publications, 1976). A historical review.

Dressler, A., *Voyage to the Great Attractor*. (New York: Knopf, 1994). Personal and scientific stories.

Ferris, T., *The Whole Shebang: A State of the Universe(s) Report*. (New York: Simon & Schuster, 1997, 1998). Contemporary cosmology, from a gifted journalist.

Goldsmith, D., *The Astronomers*. (New York: St. Martin's Press, 1991). The companion to a PBS television series.

Goldsmith, D., *The Runaway Universe*. (Cambridge, MA: Perseus Books, 2000). Describes the discovery of a nonzero cosmological constant.

Greene, B., *The Elegant Universe*. (New York: Norton, 1999). Superstrings and the Universe; Phi Beta Kappa book award winner.

Guth, A. H., *The Inflationary Universe: The Quest for a New Theory of Cosmic Origins*. (New York: Helix Books/Addison Wesley, 1997). By the originator of the inflationary theory.

Hawking, S. W., *A Brief History of Time, updated and expanded*. (New York: Bantam Books, 1998). A best-selling discussion of fundamental topics. *The Illustrated Brief History of Time* (1997) is also available.

Hodge, P. W., *Galaxies*. (Cambridge, MA: Harvard University Press, 1986). A non-technical study of galaxies.

Kaku, M., *Hyperspace*. (New York: Doubleday, 1994). A very readable introduction to string theory and other dimensions.

Lederman, L. M., and D. N. Schramm, *From Quarks to the Cosmos*, 2nd ed. (New York: Scientific American Library, 1995).

Lemonick, M., *The Edge of the Universe*. (New York: Villard, 1993). The story of the Universe and the people who study it.

Lightman, A., and R. Brawer, *Origins: The Lives and Worlds of Modern Cosmologists*. (Cambridge, MA: Harvard University Press, 1990). Interviews with 27 cosmologists.

Malin, D., *The Invisible Universe*. (Boston: Bulfinch Press, 1999). By the master of ground-based astronomical color photography.

Overbye, D., *Lonely Hearts of the Cosmos: The Scientific Quest for the Secrets of the Universe*. (New York: HarperCollins, 1991). Humanizing the cosmologists.

Pasachoff, J. M., H. Spinrad, P. Osmer, and E. S. Cheng, *The Farthest Things in the Universe*. (Cambridge, MA: Cambridge University Press, 1995). The cosmic background radiation, quasars, and distant galaxies.

Rees, M., *Before the Beginning: Our Universe and Others*. (New York: Helix Books/Addison Wesley, 1997). By the Astronomer Royal, a major researcher in the field.

Rowan-Robinson, M., *The Nine Numbers of the Cosmos*. (Oxford: Oxford University Press, 1999).

Rubin, V., *Bright Galaxies, Dark Matters*. (Woodbury, New York: American Institute of Physics, 1997). Science and biography.

Sandage, A., and J. Bedke, *The Carnegie Atlas of Galaxies*. (Washington, D.C.: Carnegie Institution of Washington, 1994). Images to stare at and pore over.

Silk, J., *A Short History of the Universe*. (New York: Scientific American Library, 1994).

Smolin, L., *The Life of the Cosmos*. (Oxford: Oxford University Press, 1997). Suggests that the properties of our Universe resulted from natural selection.

Smoot, G., and K. Davidson, *Wrinkles in Time*. (New York: Morrow, 1993). A personal account of the discovery of ripples in the cosmic background radiation.

Weinberg, S., *The First Three Minutes*, 2nd ed. (New York: Basic Books, 1993). A readable discussion of the first minutes after the Big Bang, including a discussion of the cosmic background radiation.

Zuckerman, B., and M. A. Malkan, *The Origin and Evolution of the Universe*. (Sudbury, MA: Jones and Bartlett, 1996).

Glossary

absolute magnitude The magnitude that a star would appear to have if it were at a distance of ten parsecs from us.

absorption line Wavelengths at which the intensity of radiation is less than it is at neighboring wavelengths.

absorption nebula Gas and dust seen in silhouette.

accelerating universe The model for the Universe based on observations late in the 1990s that the expansion of the Universe is speeding up over time, rather than slowing down in the way that gravity alone would modify its expansion.

accretion disk Matter that an object has taken up and that has formed a disk around the object.

active galactic nucleus (AGN) A galaxy with an exceptionally bright nucleus in some part of the spectrum; includes radio galaxies, Seyfert galaxies, quasars, and QSOs.

active galaxy A galaxy radiating much more than average in some part of the spectrum, revealing high-energy processes; radio galaxies and Seyfert galaxies are examples.

active region Regions on the Sun where sunspots, plages, flares, etc., are found.

active sun The group of solar phenomena that vary with time, such as active regions and their phenomena.

AGN Active galactic nucleus.

albedo The fraction of light reflected by a body.

allowed states The energy values that atoms can have by laws of quantum theory.

alpha particles A helium nucleus; consists of two protons and two neutrons.

alt-azimuth A two-axis telescope mounting in which motion around one of the axes, which is vertical, provides motion in azimuth, and motion around the perpendicular axis provides up-and-down (altitude) motion.

altitude (a) Height above the surface of a planet, or (b) for a telescope mounting, elevation in angular measure above the horizon.

amino acid A type of molecule containing a chain of carbon atoms and the group NH_2 (the amino group). Amino acids are fundamental building blocks of life.

Amor asteroids A group of asteroids with semimajor axes greater than Earth's and between 1.017 A.U. and 1.3 A.U.; about half cross the Earth's orbit.

angstrom A unit of length equal to 10^{-8} cm.

angular momentum An intrinsic property of a system corresponding to the amount of its revolution or spin. The amount of angular momentum of a body orbiting around a point is the mass of the orbiting body times its (linear) velocity of revolution times its distance from the point. The amount of angular momentum of a spinning sphere is the amount of inertia, an intrinsic property of the distribution of mass, times the angular velocity of spin. The conservation of angular momentum is a law that states that the total amount of angular momentum remains constant in a system that is undisturbed from outside itself.

angular resolution See *resolution*.

angular velocity The rate at which a body rotates or revolves expressed as the angle covered in a given time (for example, in degrees per hour).

anisotropy Deviation from isotropy; changing with direction.

annular eclipse A type of solar eclipse in which a ring (annulus) of solar photosphere remains visible.

anthropic principle The idea that since we exist, the Universe must have certain properties or it would not have evolved so that life would have formed and humans would have evolved.

antimatter A type of matter in which each particle (antiproton, antineutron, etc.) is opposite in charge and certain other properties to a corresponding particle (proton, neutron, etc.) of the same mass of the ordinary type of matter from which the Solar System is made.

antiparticle The antimatter corresponding to a given particle.

aperture The diameter of the lens or mirror that defines the amount of light focused by an optical system.

aperture synthesis The use of several smaller telescopes together to give some of the properties, such as resolution, of a single larger aperture.

aphelion For an orbit around the Sun, the farthest point from the Sun.

Apollo asteroids A group of asteroids, with semimajor axes greater than Earth's and less than 1.017 A.U., whose orbits overlap the Earth's.

apparent magnitude The brightness of a star as seen by an observer, given in a specific system in which a difference of five magnitudes corresponds to a brightness ratio of one hundred times; the scale is fixed by correspondence with a historical background.

archaebacteria A primitive type of organism, different from either plants or animals, perhaps surviving from billions of years ago.

association A physical grouping of stars; in particular, we talk of O and B associations.

asterism A special apparent grouping of stars, part of a constellation.

asteroid A "minor planet," a non-luminous chunk of rock smaller than planet-size but larger than a meteoroid, in orbit around a star.

asteroid belt A region of the Solar System, between the orbits of Mars and Jupiter, in which most of the asteroids orbit.

astrometric binary   A system of two stars in which the existence of one star can be deduced by study of its gravitational effect on the proper motion of the other star.

astrometry   The branch of astronomy that involves the detailed measurement of the positions and motions of stars and other celestial bodies.

Astronomical Unit   The average distance from the Earth to the Sun.

astrophysics   The science, now essentially identical with astronomy, applying the laws of physics to the Universe.

Aten asteroids   A group of asteroids with semimajor axes smaller than 1 A.U.

atom   The smallest possible unit of a chemical element. When an atom is subdivided, the parts no longer have properties of any chemical element.

atomic clock   A system that uses atomic properties to provide a measure of time.

atomic number   The number of protons in an atom.

atomic weight   The number of protons and neutrons in an atom, averaged over the abundances of the different isotopes.

aurora   Glowing lights visible in the sky, resulting from processes in the Earth's upper atmosphere and linked with the Earth's magnitude field.

aurora australis   The southern aurora.

aurora borealis   The northern aurora.

autumnal equinox   Of the two locations in the sky where the ecliptic crosses the celestial equator, the one that the Sun passes each year when moving from northern to southern declinations.

A.U.   Astronomical Unit.

azimuth   The angular distance around the horizon from the northern direction, usually expressed in angular measure from 0° for an object in the northern direction, to 180° for an object in the southern direction, around to 360°.

background radiation   See *primordial background radiation*.

Baily's beads   Beads of light visible around the rim of the Moon at the beginning and end of a total solar eclipse. They result from the solar photosphere shining through valleys at the edge of the Moon.

Balmer series   The set of spectral absorption or emission lines resulting from a transition down to or up from the second energy level (first excited level) of hydrogen.

bar   The straight structure across the center of some spiral galaxies, from which the arms unwind.

baryons   Nuclear particles (protons, neutrons, etc.) subject to the strong nuclear force; made of quarks.

basalt   A type of rock resulting from the cooling of lava.

baseline   The distance between points of observation when it determines the accuracy of some measurement.

beam   The cone within which a radio telescope is sensitive to radiation.

Becklin-Neugebauer object   An object visible only in the infrared in the Orion Molecular Cloud, apparently a very young star.

belts   Dark bands around certain planets, notably Jupiter.

Big Bang theory   A cosmological model, based on Einstein's general theory of relativity, in which the Universe was once compressed to infinite density and has been expanding ever since.

binary pulsar   A pulsar in a binary system.

binary star   Two stars revolving around each other.

bipolar flow   A phenomenon in young or forming stars in which streams of matter are ejected from the poles.

black body   An object that absorbs all radiation that hits it and emits radiation that exactly follows Planck's law.

black-body curve   A graph of brightness vs. wavelength that follows Planck's law; each such curve corresponds to a given temperature. Also called a Planck curve.

black-body radiation   Radiation whose distribution in wavelength follows Planck's law, the black-body curve.

black dwarf   A non-radiating ball of gas that results when a white dwarf radiates all its energy.

black hole   A region of space from which, according to the general theory of relativity, neither radiation nor matter can escape.

black hole era   The future era, following the degenerate era, when the only objects (besides photons and subatomic particles) the Universe will contain will be black holes.

blueshifted   A shift of optical wavelengths toward the blue or in all cases towards shorter wavelengths; when the shift is caused by motion, from a velocity of approach.

B-N object   See *Becklin-Neugebauer object*.

Bohr atom   Niels Bohr's model of the hydrogen atom, in which the energy levels are depicted as concentric circles of radii that increase as (level number)^2.

bound-free transition   An atomic transition in which an electron starts bound to the atom and winds up free from it.

brown dwarf   A self-gravitating, self-luminous object, not a satellite and insufficiently massive for nuclear fusion to being (less than 0.08 solar mass).

carbonaceous   Containing a lot of carbon.

carbon cycle   A chain of nuclear reactions, involving carbon as a catalyst at some of its intermediate stages, that transforms four hydrogen atoms into one helium atom with a resulting release in energy. The carbon cycle is important only in stars hotter than the Sun.

carbon-nitrogen cycle   The carbon cycle, acknowledging that nitrogen also plays an intermediary role.

carbon-nitrogen-oxygen cycle   A set of variations of the carbon cycle including nitrogen and oxygen isotopes as intermediaries.

Cassegrain   (telescope) A type of reflecting telescope in which the light focused by the primary mirror is intercepted short of its focal point and refocused and reflected by a secondary mirror through a hole in the center of the primary mirror.

Cassini   (a) Jean Dominique Cassini (1625–1712); (b) the NASA mission to orbit Saturn and send a probe into Titan upon its arrival there in 2004.

Cassini's division   The major division in the rings of Saturn.

catalyst   A substance that participates in a reaction but that is left over in its original form at the end.

CCD   Charge-coupled device, a solid-state imaging device.

celestial equator   The intersection of the celestial sphere with the plane that passes through the Earth's equator.

celestial poles   The intersection of the celestial sphere with the axis of rotation of the Earth.

celestial sphere   The hypothetical sphere centered at the center of the Earth to which it appears that the stars are affixed.

centaur   A Solar-System object, with 2060 Chiron as the first example, intermediate between comets and icy planets or satellites and orbiting the Sun between the orbits of Jupiter and Neptune.

center of mass   The "average" location of mass; the point in a body or system of bodies at which we may consider all the mass to be located for the purpose of calculating the gravitational effect of that mass or its mean motion when a force is applied.

central star   The hot object at the center of a planetary nebula, which is the remaining core of the original star.

Cepheid variable   A type of supergiant star that oscillates in brightness in a manner similar to the star δ Cephei. The periods of Cepheid variables, which are between 1 and 100 days, are linked to the average luminosity of the stars by known relationships; this allows the distances to Cepheids to be found.

Chandra X-ray Observatory   The major NASA x-ray observatory, the former Advanced X-ray Astrophysics Facility, launched in 1999 to make high-resolution observations in the x-ray part of the spectrum.

Chandrasekhar Limit   The limit in mass, about 1.4 solar masses, above which electron degeneracy cannot support a star, and so the limit above which white dwarfs cannot exist.

chaos   The condition in which very slight differences in initial conditions typically lead to very different results. Also, specifically, examples of orbits or rotation directions that change after apparently random intervals of time because of such sensitivity to initial conditions.

charm   An arbitrary name that corresponds to a property that distinguishes certain elementary particles, including types of quarks, from each other.

chromatic aberration   A defect of lens systems in which different colors are focused at different points.

chromosphere   The part of the atmosphere of the Sun (or another star) between the photosphere and the corona. It is probably entirely composed of spicules and probably roughly corresponds to the region in which mechanical energy is deposited.

circle   A conic section formed by cutting a cone perpendicularly to its axis.

circumpolar stars   For a given observing location, stars that are close enough to the celestial pole that they never set.

classical   When discussing atoms, not taking account of quantum mechanical effects.

closed universe   A big-bang universe with positive curvature; it has finite volume and will eventually contract.

cluster   (a) Of stars, a physical grouping of many stars; (b) of galaxies, a physical grouping of at least a few galaxies.

CNO tri-cycle   See *carbon-nitrogen-oxygen tri-cycle*.

COBE (Cosmic Background Explorer)   A spacecraft launched in 1989 to study the cosmic background radiation.

coherent radiation   Radiation in which the phases of waves at different locations in a cross-section of radiation have a definite relation to each other; in noncoherent radiation, the phases are random. Only coherent radiation shows interference.

cold dark matter   Non-luminous matter that moves slowly through the expanding Universe, such as mini black holes and exotic nuclear particles.

color   (a) Of an object, a visual property that depends on wavelength; (b) an arbitrary name assigned to a property that distinguishes three kinds of quarks.

coma   (a) Of a comet, the region surrounding the head; (b) of an optical system, an off-axis aberration in which the images of points appear with comet-like asymmetries.

comet   A type of object orbiting the Sun, often in a very elongated orbit, that when relatively near to the Sun shows a coma and may show a tail.

comparative planetology   Studying the properties of Solar-System bodies by comparing them.

comparison spectrum   A spectrum of known elements on Earth photographed in order to provide a known set of wavelengths for zero Doppler shift.

composite spectrum   The spectrum that reveals a star is a binary system, since spectra of more than one object are apparent.

condensation   A region of unusually high mass or brightness.

conic sections   Geometric shapes obtained by slicing a cone.

conjunction   Where two celestial objects reach the same celestial longitude; approximately corresponds to their closest apparent approach in the sky. When only one body is named, it is understood that the second body is the Sun.

conservation law   A statement that the total amount of some property (angular momentum, energy, etc.) of a body or set of bodies does not change.

constellation   One of 88 areas into which the sky has been divided for convenience in referring to the stars or other objects therein.

continental drift   The slow motion of the continents across the Earth's surface, explained in the theory of plate tectonics as a set of shifting regions called plates.

continuous spectrum   A spectrum with radiation at all wavelengths but with neither absorption nor emission lines.

continuum (plural: **continua**)   The continuous spectrum that we would measure from a body if no spectral lines were present.

convection   The method of energy transport in which the rising motion of masses from below carries energy upward in a gravitational field. Boiling is an example.

convection zone   The subsurface zone in certain types of stars in which convection dominates energy transfer.

coordinate systems   Methods of assigning positions with respect to suitable axes.

core   The central region of a star or planet.

corona, solar or stellar   The outermost region of the Sun (or of other stars), characterized by temperatures of millions of kelvins.

coronagraph   A type of telescope with which the corona (and, sometimes, merely the chromosphere) can be seen in visible light at times other than that of a total solar eclipse by occulting (hiding) the bright photosphere.

Also, any telescope in which a bright central object is occulted to reveal fainter things.

coronal holes Relatively dark regions of the corona having low density; they result from open field lines.

correcting plate A thin lens of complicated shape at the front of a Schmidt camera that compensates for spherical aberration.

cosmic abundances The overall abundances of elements in the Universe.

cosmic background radiation The isotropic glow of 3K radiation.

cosmic rays Nuclear particles or nuclei travelling through space at high velocity.

cosmogony The study of the origin of the Universe, usually applied in particular to the origin of the Solar System.

cosmological constant A constant arbitrarily added by Einstein to an equation in his general theory of relatively in order to provide a solution in which the Universe did not expand or contract. It was only subsequently discovered that the Universe does expand after all.

cosmological principle The principle that on the whole the Universe looks the same in all directions and in all regions.

cosmology The study of the Universe as a whole.

coudé focus A focal point of large telescopes in which the light is reflected by a series of mirrors so that it comes to a point at the end of a polar axis. The image does not move even when the telescope moves, which permits the mounting of heavy equipment.

critical density The density of matter that would allow the Universe to expand forever, but at a rate that would decrease to 0 at infinite time. The density of the Universe with respect to the critical density defines whether the Universe will expand forever or eventually contract.

crust The outermost solid layer of some objects, including neutron stars and some planets.

dark era The final era of the Universe when only low-energy photons, neutrinos, and some elementary particles (the ones that did not find a partner to annihilate) remain.

dark matter Non-luminous matter. See *hot dark matter* and *cold dark matter.*

dark nebula Dust and gas seen in silhouette.

daughter molecules Relatively simple molecules in comets resulting from the breakup of more complex molecules.

declination Celestial latitude, measured in degrees north or south of the celestial equator.

deferent In the Ptolemaic system of the Universe, the larger circle, centered at the Earth, on which the centers of the epicycles revolve.

degenerate era The future era when the Universe will be filled with cold brown dwarfs, white dwarfs, and neutron stars.

density Mass divided by volume.

density wave A circulating region of relatively high density, important, for example, in models of spiral arms.

density-wave theory The explanation of spiral structure of galaxies as the effect of a wave of compression that rotates around the center of the galaxy and causes the formation of stars in the compressed region.

deuterium An isotope of hydrogen that contains 1 proton and 1 neutron.

deuteron A deuterium nucleus, containing 1 proton and 1 neutron.

diamond-ring effect The last Baily's bead glowing brightly at the beginning of the total phase of a solar eclipse, or its counterpart at the end of totality.

differential forces A net force resulting from the difference of two other forces; a tidal force.

differential rotation Rotation of a body in which different parts have different angular velocities (and thus different periods of rotation).

differentiation For a planet, the existence of layers of different structure or composition.

diffraction A phenomenon affecting light as it passes any obstacle, spreading it out in a complicated fashion.

diffraction grating A very closely ruled series of lines that, through diffraction of light, provides a spectrum of radiation that falls on it.

dirty snowball A theory explaining comets as amalgams of ices, dust, and rocks.

discrete Separated; isolated.

disk (a) Of a galaxy, the disk-like flat portion, as opposed to the nucleus or the halo; (b) of a star or planet, the two-dimensional projection of its surface.

dispersion (a) Of light, the effect that different colors are bent by different amounts when passing from one substance to another; (b) of the pulses of a pulsar, the effect that a given pulse, which leaves the pulsar at one instant, arrives at the Earth at different times depending on the different wavelength or frequency at which it is observed. Both of these effects arise because light of different wavelengths travels at different speeds except in a vacuum.

D lines A pair of lines from sodium that appear in the yellow part of the spectrum.

DNA Deoxyribonucleic acid, a long chain of molecules that contains the genetic information of life.

Dobsonian An inexpensive type of large-aperture amateur telescope characterized by a thin mirror, composition tube, and Teflon bearings on an alt-azimuth mount.

Doppler effect A change in wavelength that results when a source of waves and the observer are moving relative to each other.

Doppler shift The change in wavelength that arises from the Doppler effect, caused by relative motion toward or away from the observer.

double star A binary star; two or more stars orbiting each other.

double-lobed structure An object in which radio emission comes from a pair of regions on opposite sides.

Drake Equation An equation advanced by Frank Drake and popularized also by Carl Sagan that attempts to calculate the number of civilizations by breaking the calculation down into a series of steps that can be assessed individually, such as rate of star formation, the fraction of stars with planets, and the average lifetime of a civilization.

dust tail The dust left behind a comet, reflecting sunlight.

dwarf ellipticals Small, low-mass elliptical galaxies.

dwarf stars Main-sequence stars.

dwarfs Dwarf stars.

dynamo   A device that generates electricity through the effect of motion in the presence of a magnetic field.

dynamo theories   Explaining sunspots and the solar activity cycle through an interaction of motion and magnetic fields.

$E = mc^2$ Einstein's formula (special theory of relativity) for the equivalence of mass and energy.

early universe   The Universe during its first minutes.

earthshine   Sunlight illuminating the Moon after having been reflected by the Earth.

eccentric   Deviating from a circle.

eccentricity   A measure of the flatness of an ellipse, defined as the distance between the foci divided by the major axis.

eclipse   The passage of all or part of one astronomical body into the shadow of another.

eclipsing binary   A binary star in which one member periodically hides the other.

ecliptic   The path followed by the Sun across the celestial sphere in the course of a year.

ecliptic plane   The plane of the Earth's orbit around the Sun.

ecosphere   The region around a star in which conditions are suitable for life, normally a spherical shell.

electric field   A force set up by an electric charge.

electromagnetic force   One of the four fundamental forces of nature, giving rise to electromagnetic radiation.

electromagnetic radiation   Radiation resulting from changing electric and magnetic fields.

electromagnetic spectrum   Energy in the form of electromagnetic waves, in order of wavelength.

electromagnetic waves   Waves of changing electric and magnetic fields, travelling through space at the speed of light.

electromagnetism   The combined force of electricity and magnetism, which follows the formulae unified by Maxwell.

electron   A particle of one negative charge, 1/1830 the mass of a proton, that is not affected by the strong force. It is a lepton.

electron volt (eV)   The energy necessary to raise an electron through a potential of one volt.

electroweak force   The unified electromagnetic and weak forces, according to a recent theory.

element   A kind of atom, characterized by a certain number of protons in its nucleus. All atoms of a given element have similar chemical properties.

elementary particle   One of the constituents of an atom.

ellipse   A curve with the property that the sum of the distances from any point on the curve to two given points, called the foci, is constant.

elliptical galaxy   A type of galaxy characterized by elliptical appearance.

emission line   Wavelengths (or frequencies) at which the brightness of radiation is greater than it is at neighboring wavelengths (or frequencies).

emission nebula   A glowing cloud of interstellar gas.

emulsion   A coating whose sensitivity to light allows photographic recording of incident radiation.

energy   A fundamental quantity usually defined in terms of the ability of a system to do something that is technically called "work," that is, the ability to move an object by application of force, where the work is the force times the displacement.

energy, law of conservation of   Energy is neither created nor destroyed, but may be changed in form.

energy level   A state corresponding to an amount of energy that an atom is allowed to have by the laws of quantum mechanics.

energy problem   For quasars, how to produce so much energy in such a small emitting volume.

ephemeris   A listing of astronomical positions and other data that change with time. From the same root as *ephemeral.*

epicycle   In the Ptolemaic theory, a small circle, riding on a larger circle called the deferent, on which a planet moves. The epicycle is used to account for retrograde motion.

equal areas, law of   Kepler's second law.

equant   In Ptolemaic theory, the point equally distant from the center of the deferent as the Earth but on the opposite side, around which the epicycle moves at a uniform angular rate.

equator   (a) Of the Earth, a great circle on the Earth, midway between the poles; (b) celestial, the projection of the Earth's equator onto the celestial spheres; (c) galactic, the plane of the disk as projected onto a map.

equatorial mount   A type of telescope mounting in which one axis, called the polar axis, points toward the celestial pole and the other axis is perpendicular. Motion around only the polar axis is sufficient to completely counterbalance the effect of the Earth's rotation.

equinox   An intersection of an ecliptic and the celestial equator. The center of the Sun is geometrically above and below the horizon for equal lengths of time on the two days of the year when the Sun passes the equinoxes; if the Sun were a point and atmospheric refraction were absent, then day and night would be of equal length on those days.

erg   A unit of energy in the metric system, corresponding to the work done by a force of one dyne (the force that is required to accelerate one gram by one cm/sec^2) producing a displacement of one centimeter.

ergosphere   A region surrounding a rotating black hole (or other system satisfying Kerr's solution) from which work can be extracted.

escape velocity   The initial velocity that an object must have to escape the gravitational pull of a mass.

event horizon   The sphere around a black hole from within which nothing can escape; the place at which the exit cones close.

evolutionary track   The set of points on a temperature–luminosity diagram showing the changes of a star's temperature and luminosity with time.

excitation   The raising of atoms to higher energy states than the lowest possible.

excited level   An energy level of an atom above the ground level.

exit cone   The cone that, for each point within the photon sphere of a black hole, defines the directions of rays of radiation that escape.

exobiology   The study of life located elsewhere than Earth.

exponent   The "power" representing the number of times a number is multiplied by itself.

exponential notation   The writing of numbers as a power of 10 times a number with one digit before the decimal point.

extended objects   Objects with detectable angular size.

extinction   The dimming of starlight by scattering and absorption as the light traverses interstellar space.

extragalactic   Exterior to the Milky Way Galaxy.

extragalactic radio sources   Radio sources outside our galaxy.

eyepiece   The small combination of lenses at the eye end of a telescope, used to examine the image formed by the objective.

field of view   The angular expanse viewable.

filament   A feature of the solar surface seen in Hα as a dark wavy line; a prominence projected on the solar disk.

fireball   An exceptionally bright meteor.

fission, nuclear   The splitting of an atomic nucleus.

flare   An extremely rapid brightening of a small area of the surface of the Sun, usually observed in Hα and other strong spectral lines and accompanied by x-ray and radio emission.

flash spectrum   The solar chromospheric spectrum seen in the few seconds before or after totality at a solar eclipse.

flat (critical) Universe   A Universe in which Euclid's parallel postulate holds, one that is barely expanding forever. It has infinite volume and its age is 2/3 the Hubble time.

flatness problem   One of the problems solved by the inflationary theory, that the Universe is exceedingly close to being flat for no obvious reason.

flavors   A way of distinguishing quark; up, down, strange, charmed, truth (top), beauty (bottom).

fluorescence   The transformation of photons of relatively high energy to photons of lower energy through interactions with atoms, and the resulting radiation.

flux   The amount of something (such as energy) passing through a surface per unit time.

flux tube   A torus through which charged particles circulate.

focal length   The distance from a lens or mirror to the point to which rays from an object at infinity are focused.

focus (plural: **foci**)   (a) A point to which radiation is made to converge; (b) of an ellipse, one of the two points the sum of the distances to which remains constant.

force   In physics, something that can or does cause change of momentum, measured by the rate of change of momentum with time.

Fraunhofer lines   The absorption lines of a solar or other stellar spectrum.

frequency   The rate at which waves pass a given point.

full moon   The phase of the Moon when the side facing the Earth is fully illuminated by sunlight.

fusion   The amalgamation of nuclei into heavier nuclei.

fuzz   The faint light detectable around nearby quasars.

galactic cannibalism   The incorporation of one galaxy into another.

galactic year   The length of time the Sun takes to complete an orbit of our galactic center.

Galilean satellites   Io, Europa, Ganymede, and Callisto: the four major satellites of Jupiter, discovered by Galileo in 1610.

Galileo   (a) The Italian scientist Galileo Galilei (1564–1642); (b) the NASA spacecraft in orbit around the planet Jupiter that sent a probe into Jupiter's clouds on December 7, 1995.

gamma rays   Electromagnetic radiation with wavelengths shorter than approximately 0.1 Å.

gamma-ray burst   A brief burst of gamma rays, studied especially by the Compton Gamma-ray Observatory during 1991–2000, which found thousands of such bursts uniformly distributed in the sky; discovered to come, at least in many cases, from exceedingly distant and thus extremely powerful sources of energy.

gas tail   The puffs of ionized gas trailing a comet.

general theory of relativity   Einstein's 1916 theory of gravity.

geocentric   Earth-centered.

geology   The study of the Earth, or of other solid bodies.

geothermal energy   Energy from the Earth's surface.

giant   A star that is larger and brighter than main-sequence stars of the same color.

giant ellipticals   Elliptical galaxies that are very large.

giant molecular cloud   A basic building block of our galaxy, containing dust, which shields the molecules present.

giant planets   Jupiter, Saturn, Uranus, and Neptune; or even larger extrasolar planets.

giant star   A star more luminous than a main-sequence star of its spectral type; a late stage in stellar evolution.

gibbous moon   The phases between half moon and full moon.

globular cluster   A spherically symmetric type of collection of stars that shared a common origin.

gluon   The particle that carries the color force (and thus the strong nuclear force).

grand unified theories (GUTs)   Theories unifying the electroweak force and the strong force.

granulation   Convection cells on the Sun about 1 arc sec across.

grating   A surface ruled with closely spaced lines that, through diffraction, breaks up light into its spectrum.

gravitational force   One of the four fundamental forces of nature, the force by which two masses attract each other.

gravitational instability   A situation that tends to break up under the force of gravity.

gravitational lens   In the gravitational lens phenomenon, a massive body changes the path of electromagnetic radiation passing near it so as to make more than one image (or a brightening) of an object. A double quasar was the first example to be discovered.

gravitationally   Controlled by the force of gravity.

gravitational redshift   A redshift of light caused by the presence of mass, according to the general theory of relativity.

gravitational waves   Waves that many scientists consider to be a consequence, according to the general theory of relativity, of changing distributions of mass.

gravity   The tendency for all masses to attract each other; described in a formula by Newton and more recently described by Einstein as a result of a warping of space and time by the presence of a mass.

gravity assist   Using the gravity of one celestial body to change a spacecraft's energy.

grazing incidence Striking at a low angle.

great circle The intersection of a plane that passes through the center of a sphere with the surface of that sphere; the largest possible circle that can be drawn on the surface of a sphere.

Great Dark Spot A giant circulating region on Neptune seen by Voyager 2 in 1989; it has since disappeared.

Great Red Spot A giant circulating region on Jupiter.

greenhouse effect The effect by which the atmosphere of a planet heats up above its equilibrium temperature because it is transparent to incoming visible radiation but opaque to the infrared radiation that is emitted by the surface of the planet.

Gregorian calendar The calendar in current use, with normal years that are 365 days long, with leap years every fourth year except for years that are divisible by 100 but not by 400.

ground level An atom's lowest possible energy level.

ground state See *ground level.*

GUTs See *grand unified theories.*

H₀ The Hubble constant.

H I region An interstellar region of neutral hydrogen.

H II region An interstellar region of ionized hydrogen.

H line The spectral line of ionized calcium at 3968 Å.

Hα The first line of the Balmer series of hydrogen, at 6563 Å.

half-life The length of time for half a set of particles to decay through radioactivity or instability.

halo Of a galaxy, the region of the galaxy that extends far above and below the plane of the galaxy, containing globular clusters.

head Of a comet, the nucleus and coma together.

heat flow The flow of energy from one location to another.

heavyweight stars Stars of more than about 8 solar masses.

heliocentric Sun-centered; using the Sun rather than the Earth as the point to which we refer. A heliocentric measurement, for example, omits the effect of the Doppler shift caused by the Earth's orbital motion.

helium flash The rapid onset of fusion of helium into carbon through the triple-alpha process that takes place in most red-giant stars.

Herbig-Haro objects Blobs of gas ejected in star formation.

hertz The measure of frequency, with units of /sec (per second); formerly called cycles per second.

Hertzsprung-Russell diagram A graph of temperature (or equivalent) vs. luminosity (or equivalent) for a group of stars.

high-energy astrophysics The study of x-rays, gamma rays, and cosmic rays, and of the processes that make them.

highlands Regions on the Moon or elsewhere that are above the level that may have been smoothed by flowing lava.

homogenous Uniform throughout.

horizon problem One of the problems of cosmology solved by the inflationary theory: why the Universe is identical in all directions, even though widely separated regions could never have been in thermal equilibrium with each other since they are beyond each other's horizon.

hot dark matter Non-luminous matter with large speeds, like neutrinos.

hour circle The great circles passing through the celestial poles.

H-R diagram Hertzsprung-Russell diagram.

Hubble constant (H₀) The constant of proportionality in Hubble's law linking the speed of recession of a distant object and its distance from us.

Hubble Deep Field A small part of the sky in Ursa Major extensively studied by the Hubble Space Telescope in December 1995 and since also studied by a wide variety of telescopes on the ground and in space; the long exposures allowed astronomers to see "deep" into space, that is, to great distances and therefore far back in time. See http://www.stsci.edu/ftp/science/hdf.html. A Hubble Deep Field—South was observed subsequently (http://www.stsci.edu/ftp/science/hdfsouth/hdfs.html).

Hubble flow The assumed uniform expansion of the Universe, on which any peculiar motions of galaxies or clusters of galaxies is superimposed.

Hubble time The amount of time since the big bang, assuming a constant speed for any given galaxy; the Hubble time is calculated by tracing the Universe backward in time using the current Hubble constant.

Hubble type Hubble's galaxy classification scheme; E0, E7, Sa, SBa, etc.

Hubble's law The linear relation between the speed of recession of a distant object and its distance from us, $v = H_0 d$.

hyperfine level A subdivision of an energy level caused by such relatively minor effects as changes resulting from the interactions among spinning particles in an atom or molecule.

hypothesis The first step in the traditional formulation of the scientific method; a tentative explanation of a set of facts that is to be tested experimentally or observationally.

IC *Index Catalogue,* one of the supplements to Dreyer's *New General Catalogue.*

igneous Rock cooled from lava.

inclination Of an orbit, the angle of the plane of the orbit with respect to the ecliptic plane.

inclined Tilted with respect to some other body, usually describing the axis of rotation or the plane of an orbit.

Index Catalogue See *IC.*

inferior conjunction An inferior planet's reaching the same celestial longitude as the Sun's.

inferior planet A planet whose orbit around the Sun is within the Earth's, namely, Mercury and Venus.

inflation The theory that the Universe expanded extremely fast, by perhaps 10 to the 100th power, in the first fraction of a second after the big bang. The concept of inflation solves several problems in cosmology, such as the horizon problem.

inflationary universe A model of the expanding Universe involving a brief period of extremely rapid expansion.

infrared Radiation beyond the red, about 7000 Å to 1 mm.

interference The property of radiation, explainable by the wave theory, in which waves in phase can add (constructive interference) and waves out of phase can subtract (destructive interference); for light, this gives alternate light and dark bands.

interferometer   A device that uses the property of interference to measure such properties of objects as their positions or structure.

interferometry   Observations using an interferometer.

intergalactic medium   Material between galaxies in a cluster.

interior   The inside of an object.

international date line   A crooked imaginary line on the Earth's surface, roughly corresponding to 180° longitude, at which, when crossed from east to west, the date jumps forward by one day.

interplanetary medium   Gas and dust between the planets.

interstellar medium   Gas and dust between the stars.

interstellar reddening   The relatively greater extinction of blue light by interstellar matter than of red light.

inverse-square law   Decreasing with the square of increasing distance.

ion   An atom that has lost one or more electrons.

ionized   Having lost one or more electrons.

ionosphere   The highest region of the Earth's atmosphere.

ion tail   See *gas tail*.

IRAS   The Infrared Astronomical Satellite (1983).

iron meteorites   Meteorites with a high iron content (about 90%); most of the rest is nickel.

irons   Iron meteorites.

irregular galaxy   A type of galaxy showing no regular shape or symmetry.

ISO (Infrared Space Observatory)   A European Space Agency project, 1995–1998.

isotope   A form of chemical element with a specific number of neutrons.

isotopic   Being the same in all directions.

joule   The SI unit of energy, $1 \text{ kg} \cdot \text{m}^2/\text{s}^2$.

Jovian planet   Same as giant planet.

JPL   The Jet Propulsion Laboratory in Pasadena, California, funded by NASA and administered by Caltech; a major space contractor.

Julian calendar   The calendar with 365-day years and leap years every fourth year without exception; the predecessor to the Gregorian calendar.

Julian day   The number of days since noon on January 1, 713 B.C. Used for keeping track of variable stars or other astronomical events. January 1, 2000, noon, begins Julian day 2,451,545.

K line   The spectral line of ionized calcium at 3933 Å.

Keplerian   Following Kepler's law.

Kuiper belt   A reservoir of perhaps hundreds of thousands of Solar-System objects, each hundreds of kilometers in diameter, orbiting the Sun outside the orbit of Neptune. It is the source of short-period comets.

Kuiper belt object   An object in the Kuiper belt; many comets and perhaps even the planet Pluto are examples.

large-scale structure   The network of filaments and voids or other shapes distinguished when studying the Universe on the largest scales of distance.

laser   An acronym for "**l**ight **a**mplification by **s**timulated **e**mission of **r**adiation," a device by which certain energy levels are populated by more electrons than normal, resulting in an especially intense emission of light at a certain frequency when the electrons drop to a lower energy level.

latitude   Number of degrees north or south of the equator measured from the center of a coordinate system.

law of equal areas   Kepler's second law.

leap year   A year in which a 366th day is added.

lens   A device that focuses waves by refraction.

lenticular   A galaxy of type S0.

light   Electromagnetic radiation between about 4000 and 7000 Å.

light curve   The graph of the brightness of an object vs. time.

light pollution   Excess light in the sky.

light-year   The distance that light travels in a year.

lighthouse model   The explanation of a pulsar as a spinning neutron star whose beam we see as it comes around.

lightweight stars   Stars between about 0.07 and 4 solar masses.

limb   The edge of a star or planet.

line profile   The graph of the intensity of radiation vs. wavelength for a spectral line.

lithosphere   The crust and upper mantel of a planet.

lobes   Of a radio source, the regions to the sides of the center from which high-energy particles are radiating.

local   In our region of the Universe.

Local Group   The two dozen or so galaxies, including the Milky Way Galaxy, that form a small cluster.

Local Supercluster   The supercluster of galaxies in which the Virgo Cluster, the Local Group, and other clusters reside.

logarithmic   A scale in which equal intervals stand for multiplying by ten or some other base, as opposed to linear, in which increases are additive.

longitude   The angular distance around a body measured along the equator from some particular point; for a point not on the equator, it is the angular distance along the equator to a great circle that passes through the poles and through the point.

longperiod variables   Mira variables.

look-back time   The duration over which light from an object has been travelling to reach us.

luminosity   The total amount of energy given off by an object per unit time; its power.

luminosity class   Different regions of the H-R diagram separating objects of the same spectral type: supergiants (I), bright giants (II), giants (III), subgiants (IV), dwarfs (V).

lunar eclipse   The passage of the Moon into the Earth's shadow.

lunar occultation   An occultation by the Moon.

lunar soils.   Dust and other small fragments on the lunar surface.

Lyman alpha   The spectral line (1216 Å) that corresponds to a transition between the two lowest major energy levels of a hydrogen atom.

Lyman-alpha forest   The many Lyman-alpha lines, each differently Doppler-shifted, visible in the spectra of some quasars.

Lyman lines   The spectral lines that correspond to transitions to or from the lowest major energy level of a hydrogen atom.

⊙   A unit equal to one solar mass, that is, the mass of the Sun.

M⊙   Solar mass; the mass of the Sun, used as a unit of measurement. For example, a star with $5M_\odot$ has 5 times the mass of the Sun.

MACHOs   **M**assive **C**ompact **H**alo **O**bjects—like dim stars, brown dwarfs, or black holes—that could account for some of the dark matter. The name was chosen to contrast with WIMPs.

Magellanic Clouds   Two small irregular galaxies, satellites of the Milky Way Galaxy, visible in the southern sky.

magnetic-field lines   Directions mapping out the direction of the force between magnetic poles; the packing of the lines shows the strength of the force.

magnetic lines of force   See *magnetic field lines.*

magnetic monopole   A single magnetic charge of only one polarity; may or may not exist.

magnetosphere   A region of magnetic field around a planet.

magnification   An apparent increase in angular size.

magnitude   A factor of $\sqrt[5]{100} = 2.511886\ldots$ in brightness. See *absolute magnitude* and *apparent magnitude.* An *order of magnitude* is a power of ten.

magnitude scale   The scale of apparent magnitudes and absolute magnitudes used by astronomers, in which each factor of 100 in brightness corresponds to a difference of 5 magnitudes.

main sequence   A band on a Hertzsprung-Russell diagram in which stars fall during the main, hydrogen-burning phase of their lifetimes.

major axis   The longest diameter of an ellipse; the line from one side of an ellipse to the other that passes through the foci. Also, the length of that line.

mantle   The shell of rock separating the cores of a differentiated planet from its thin surface crust.

mare (plural: **maria**)   One of the smooth areas of the Moon or on some of the other planets.

mascon   A concentration of mass under the surface of the Moon, discovered from its gravitational effect on spacecraft orbiting the Moon.

maser   An acronym for "**m**icrowave **a**mplification by **s**timulated **e**mission of **r**adiation," a device by which certain energy levels are more populated than normal, resulting in an especially intense emission of radio radiation at a certain frequency when the system drops to a lower energy level.

mass   A measure of the inherent amount of matter in a body.

mass-luminosity relation   A well-defined relation between the mass and luminosity for main-sequence stars.

mass number   The total number of protons and neutrons in a nucleus; also known as *atomic weight.*

Maunder minimum   The period 1645–1715, when there were very few sunspots, and no periodicity, visible.

mean solar day   A solar day for the "mean Sun," which moves at a constant rate during the year.

merging   The interaction of two galaxies in space with a single galaxy as the result; the merging of two spiral galaxies to form an elliptical galaxy, or of two galaxies interacting with a ring galaxy resulting, as examples.

meridian   The great circle on the celestial sphere that passes through the celestial poles and the observer's zenith.

mesosphere   A middle layer on the Earth's atmosphere, where the temperature again rises above the stratosphere's decline.

Messier numbers   Numbers of fuzzy-looking objects in the 18th-century list of Charles Messier.

metal   (a) For stellar abundances, any element higher in atomic number than 2, that is, more massive than helium. (b) In general, neutral matter that is a good conductor of electricity.

meteor   A track of light in the sky from rock or dust burning up as it falls through the Earth's atmosphere.

meteor shower   The occurrence at yearly intervals of meteors at a higher-than-average rate as the Earth goes through a comet's orbit and the dust left behind by that comet.

meteorite   An interplanetary chunk of rock after it impacts on a planet or Moon, especially on the Earth.

meteoroid   An interplanetary chunk of rock smaller than an asteroid.

micrometeorite   A tiny meteorite. The micrometeorites that hit the Earth's surface are sufficiently slowed down that they can reach the ground without being vaporized.

mid-Atlantic ridge   The spreading of the sea floor in the middle of the Atlantic Ocean as upwelling material forces the plates to move apart.

midnight sun   The Sun seen around the clock from locations sufficiently far north or south at the suitable season.

Milky Way   The band of light across the sky from the stars and gas in the plane of the Milky Way Galaxy.

Milky Way Galaxy   The collection of gas, dust, and perhaps 400 billion stars in which we live; the Milky Way is our view of the plane of the Milky Way Galaxy.

mini black hole   A black hole the size of a pinhead (and thus the mass of an asteroid) or less, thought to be left over from the big bang; Stephen Hawking deduced that such mini black holes should seem to emit radiation at a rate that allows a temperature to be assigned to it.

minor axis   The shortest diameter of an ellipse; the line from one side of an ellipse to the other that passes midway between the foci and is perpendicular to the major axis. Also, the length of that line.

minor planets   Asteroids.

Mira variable   A long-period variable star similar to Mira (omicron Ceti).

missing-mass problem   The discrepancy between the mass visible and the mass derived from calculating the gravity acting on members of clusters of galaxies.

molecule   Bound atoms that make the smallest collection that exhibits a certain set of chemical properties.

momentum   A measure of the tendency that a moving body has to keep moving. The momentum in a given direction (the "linear momentum") is equal to the mass of the body times its speed in that direction. See also *angular momentum.*

mountain ranges   Sets of mountains on the Earth, Moon, etc.

multiverse   The set of parallel universes that may exist, with our observable universe as only one part.

naked singularity   A singularity that is not surrounded by an event horizon and therefore kept from our view.

neap tides   The tides when the gravitational pulls of the Sun and Moon are perpendicular, making them relatively low.

Near-Earth object   A comet or asteroid whose orbit is sufficiently close to Earth's that a collision is possible or even likely in the long term. See *Amor asteroids, Apollo asteroids.*

nebula (plural: **nebulae**)   Interstellar regions of dust or gas.

nebular hypothesis   The particular nebular theory for the formation of the Solar System advanced by Laplace.

nebular theories   The theories that the Sun and the planets formed out of a cloud of gas and dust.

neutrino   A spinning, neutral elementary particle with little rest mass, formed in certain radioactive decays.

neutron   A massive, neutral elementary particle, one of the fundamental constituents of an atom.

neutron star   A star that has collapsed to the point where it is supported against gravity by neutron degeneracy.

New General Catalogue   *A New General Catalogue of Nebulae and Clusters of Stars* by J. L. E. Dreyer, 1888.

new moon   The phase when the side of the Moon facing the Earth is the side that is not illuminated by sunlight.

Newtonian   (telescope) A reflecting telescope where the beam from the primary mirror is reflected by a flat secondary mirror to the side.

NGC   *New General Catalogue.*

non-thermal radiation   Radiation that cannot be characterized by a single number (the temperature). Normally, we derive this number from Planck's law, so that radiation that does not follow Planck's law is called non-thermal.

nova (plural: **novae**)   A star that suddenly increases in brightness; an event in a binary system when matter from the giant component falls on the white dwarf component.

nuclear bulge   The central region of spiral galaxies.

nuclear burning   Nuclear fusion.

nuclear force   The strong force, one of the fundamental forces.

nuclear fusion   The amalgamation of lighter nuclei into heavier ones.

nucleosynthesis   The formation of the elements.

nucleus (plural: **nuclei**)   (a) Of an atom, the core, which has a positive charge, contains most of the mass, and takes up only a small part of the volume; (b) of a comet, the chunks of matter, no more than a few km across, at the center of the head; (c) of a galaxy, the innermost region.

O and B association   A group of O and B stars close together.

objective   The principal lens or mirror of an optical system.

oblate   With equatorial greater than polar diameter.

occultation   The hiding of one astronomical body by another.

Occam's razor   The principle of simplicity, from the medieval philosopher William of Occam (approximately 1285–1349): the simplest explanation for all the facts will be accepted.

Olbers's paradox   The observation that the sky is dark at night contrasted to a simple argument that shows that the sky should be uniformly bright.

one atmosphere   The air pressure at the Earth's surface.

one year   The length of time the Earth takes to orbit the Sun.

Oort comet cloud   The trillions of incipient comets surrounding the Solar System in a 50,000 A.U. sphere.

open cluster   A galactic cluster, a type of star cluster.

open universe   A big-bang cosmology in which the Universe has infinite volume and will expand forever.

opposition   An object's having a celestial longitude 180° from that of the Sun.

optical   In the visible part of the spectrum, 4000–7000 Å, or having to do with reflecting or refracting that radiation.

optical double   A pair of stars that appear extremely close together in the sky even though they are at different distances from us and are not physically linked.

organic   Containing carbon in its molecular structure.

Orion Molecular Cloud   The giant molecular cloud in Orion behind the Orion Nebula, containing many young objects.

oscillating universe   The version of a closed universe in which our cycle is but one of many.

Ozma   One or two projects that searched nearby stars for radio signals from extraterrestrial civilizations.

ozone layer   A region in the Earth's upper stratosphere and lower mesosphere where O_3 absorbs solar ultraviolet radiation.

paraboloid   A 3-dimensional surface formed by revolving a parabola around its axis.

parallax   (a) Trigonometric parallax, half the angle through which a star appears to be displaced when the Earth moves from one side of the Sun to the other (2 A.U.); it is inversely proportional to the distance. (b) Other ways of measuring distance, as in spectroscopic parallax.

parallel light   Light that is neither converting nor diverging.

parent molecules   Molecules in a comet that break up into daughter molecules.

parsec   The distance from which 1 A.U. subtends one second of arc. (Approximately 3.26 ly.).

particle physics   The study of elementary nuclear particles.

peculiar motion   The motion of a galaxy with respect to the Hubble flow.

penumbra   (a) For an eclipse, the part of the shadow from which the Sun is only partially occulted; (b) of a sunspot, the outer region, not as dark as the umbra.

perfect cosmological principle   The assumption that on a large scale the Universe is homogenous and isotropic in space and unchanging in time.

perihelion   The near point to the Sun of the orbit of a body orbiting the Sun.

period   The interval over which something repeats.

phase   (a) Of a planet, the varying shape of the lighted part of a planet or moon as soon from some vantage point; (b) the relation of the variations of a set of waves.

phase transitions   Changes from one state of matter to another, as from solid to liquid or liquid to gas; phase transitions in the early Universe marked the separation of the fundamental forces.

photometry The electronic measurement of the amount of light.

photomultiplier An electronic device that through a series of internal stages multiplies a small current that is given off when light is incident on it; a large current results.

photon A packet of energy that can be thought of as a particle travelling at the speed of light.

photon sphere The sphere around a black hole, 3/2 the size of the event horizon, within which exit cones open and in which light can orbit.

photosphere The region of a star from which most of its light is radiated.

plage The part of a solar active region that appears bright when viewed in Hα.

Planck's constant The constant of a proportionality, h, between the frequency of an electromagnetic wave and the energy of an equivalent photon. $E = hv = hc/\lambda$.

Planck's law The formula that predicts, for an opaque object at a certain temperature, how much radiation there is at every wavelength.

Planck time The time very close to the big bang, 10^{-43} seconds, before which a quantum theory of gravity would be necessary to explain the Universe and which is therefore thus currently inaccessible to our computations.

planet A celestial body of substantial size (more than about 1000 km across), basically non-radiating and of insufficient mass for nuclear reactions ever to begin, ordinarily in orbit around a star.

planetary nebulae Shells of matter ejected by low-mass stars after their main-sequence lifetime, ionized by ultraviolet radiation from the star's remaining core.

planetesimal One of the small bodies into which the primeval solar nebula condensed and from which the planets formed.

plasma An electrically neutral gas composed of exact equal numbers of ions and electrons.

plates Large flat structures making up a planet's crust.

plate tectonics The theory of the Earth's crust, explaining it as plates moving because of processes beneath.

plumes Thin structures in the solar corona near the poles.

point objects Objects in which no size is distinguishable.

polar axis The axis of an equatorial telescope mounting that is parallel to the Earth's axis of rotation.

pole star A star approximately at a celestial pole; Polaris is now the pole star; there is no south pole star.

poor cluster A cluster of stars or galaxies with few members.

positive ion An atom that has lost one or more electrons.

positron An electron's antiparticle (charge of $+e$), where e is the unit of electric charge.

precession The very slowly changing position of stars in the sky resulting from variations in the orientation of the Earth's axis; a complete cycle takes 26,000 years.

precession of the equinoxes The slow variation of the position of the equinoxes (intersections of the ecliptic and celestial equator) resulting from variations in the orientation of the Earth's axis.

pre-main-sequence star A ball of gas in the process of slowly contracting to become a star as it heats up before beginning nuclear fusion.

pressure Force per unit area.

primary cosmic rays The cosmic rays arriving at the top of the Earth's atmosphere.

primary distance indicators Ways of measuring distance directly, as in trigonometric parallax.

prime focus The location at which the main lens or mirror of a telescope focuses an image without being reflected or refocused by another mirror or other optical element.

primeval solar nebula The early stage of the Solar System in nebular theories.

primordial background radiation Isotropic millimeter and submillimeter radiation following a black-body curve for about 3 K; interpreted as a remnant of the big bang.

primordial nucleosynthesis The formation of the nuclei of isotopes of hydrogen (such as deuterium), helium, and lithium in the first 10 minutes of the Universe.

primum mobile In Ptolemaic and Aristotelian theory, the outermost sphere around the Earth, which gave its natural motion to inner spheres.

principal quantum number The integer n that determines the main energy levels in an atom.

prograde motion The apparent motion of the planets when they appear to move forward (from west to east) with respect to the stars; see also *retrograde motion*.

prolate Having the diameter along the axis of rotation longer than the equatorial diameter.

prominence Solar gas protruding over the limb, visible to the naked eye only at eclipses but also observed outside the eclipses by its emission-line spectrum.

proper motion Angular motion across the sky with respect to a framework of galaxies or fixed stars.

proteins Long chains of amino acids; fundamental components of all cells of life as we know it.

proton Elementary particle with positive charge $+e$, where e is the unit of electric charge, one of the fundamental constituents of an atom.

proton-proton chain A set of nuclear reactions by which four hydrogen nuclei combine one after the other to form one helium nucleus, with a resulting release of energy.

protoplanets The loose collections of particles from which the planets formed.

protosun The Sun in formation.

pulsar A celestial object that gives off pulses of radio waves; thought to be a rotating neutron star with a very strong magnetic field.

QSO Quasi-stellar objects, formally a radio-quiet version of a quasar, though now radio-loud quasars are also sometimes included.

quantized Divided into discrete parts.

quantum A bundle of energy.

quantum fluctuation The spontaneous formation and disappearance of virtual particles; it does not violate the law of conservation of energy because of Heisenberg's Uncertainty Principle.

quantum mechanics The branch of 20th-century physics that describes atoms and radiation.

quantum theory The set of theories that evolved in the early 20th century to explain atoms and radiation as limited to discrete energy levels or quanta of energy, with quantum mechanics emerging as a particular version.

quark   One of the subatomic particles of which modern theoreticians believe such elementary particles as protons and neutrons are composed. The various kinds of quarks have positive or negative charges of $1/3e$ or $2/3e$, where e is the unit of electric charge.

quasar   A very-large-redshift object that is almost stellar (point-like) in appearance, but has a very non-stellar spectrum consisting of broad emission lines thought to be the nucleus of a galaxy with an accreting supermassive black hole.

quiescent prominence   A long-lived and relatively stationary prominence.

quiet Sun   The collection of solar phenomena that do not vary with the solar activity cycle.

radar   The acronym for **ra**dio **d**etection **an**d **r**anging, an active rather than passive radio technique in which radio signals are transmitted and their reflections received and studied.

radial velocity   The velocity of an object along a line (the radius) joining the object and the observer; the component of velocity toward or away from the observer.

radiant   The point in the sky from which all meteors in a meteor shower appear to be coming.

radiation   Electromagnetic radiation. Sometimes also particles such as protons, electrons, and helium nuclei.

radiation belts   Belts of charged particles surrounding planets.

radioactive   Having the property of spontaneously changing into another isotope or element.

radio galaxy   A galaxy that emits radio radiation orders of magnitude stronger than those from normal galaxies.

radio telescope   An antenna or set of antennas, often together with a focusing reflecting dish, that is used to detect radio radiation from space.

radio waves   Electromagnetic radiation with wavelengths longer than about one millimeter.

red giant   A post-main-sequence stage of the lifetime of a star; the star becomes relatively luminous and cool.

reddened   See *reddening*.

reddening   The phenomenon by which the extinction of blue light by interstellar matter is greater than the extinction of red light so that the redder part of the continuous spectrum is relatively enhanced.

redshifted   A shift of optical wave lengths towards the red, or in all cases toward longer wavelengths.

red supergiant   Extremely luminous, cool, and large stars; a post-main-sequence phase of evolution of stars of more than about 8 solar masses.

reflecting telescope   A type of telescope that uses a mirror or mirrors to form the primary image.

reflection nebula   Interstellar gas and dust that we see because it is reflecting light from a nearby star.

refracting telescope   A type of telescope in which the primary image is formed by a lens or lenses.

refraction   The bending of electromagnetic radiation as it passes from one medium to another or between parts of a medium that has varying properties.

refractory   Having a high melting point.

relativistic   Having a speed that is such a large fraction of the speed of light that the special theory of relativity must be applied.

resolution   The ability of an optical system to distinguish detail; also called angular resolution.

red mass   The mass an object would have if it were not moving with respect to the observer.

rest wavelength   The wavelength radiation would have if its emitter were not moving with respect to the observer.

retrograde motion   The apparent motion of the planets when they appear to move backwards (from east to west) with respect to the stars from the direction that they move ordinarily.

retrograde rotation   The rotation of a moon or planet opposite to the dominant direction in which the Sun rotates and the planets orbit and rotate.

revolution   The orbiting of one body around another.

revolve   To move in an orbit around another body.

rich cluster   A cluster of many galaxies.

ridges   Raised surface features on the Moon, apparently volcanic, perhaps having developed on the crust of lava lakes, as volcanic vents, or from faulting; Mercury also has ridges.

right ascension   Celestial longitude, measured eastward along the celestial equator in hours of time from the vernal equinox.

rims   The raised edges of craters.

Roche limit   The sphere for each mass inside of which blobs of gas cannot agglomerate by gravitational interaction without being torn apart by tidal forces; normally about $2\frac{1}{2}$ times the radius of a planet.

rotate   To spin on one's own axis.

rotation   Spin on an axis.

rotational curve   A graph of the speed of rotation vs. distance from the center of a rotating object like a galaxy.

RR Lyrae variable   A short-period variable star. All RR Lyrae stars have approximately equal luminosity and so are used to determine distances.

S0   A transition type of galaxy between ellipticals and spirals; has a disk but no arms.

scarps   Lines of cliffs; found on Mercury, Earth, the Moon, and Mars.

scattered   Light absorbed and then reemitted in all directions.

Schmidt telescope (Schmidt camera)   A telescope that uses a spherical mirror and a thin lens to provide photographs of a wide field.

Schwarzschild radius   The radius that, according to Schwarzschild's solutions to Einstein's equations of the general theory of relativity, corresponds to the event horizon of a black hole.

scientific method   No easy definition is possible, but it has to do with a way of testing and verifying hypotheses.

scientific notation   Exponential notation.

secondary cosmic rays   High-energy particles generated in the Earth's atmosphere by primary cosmic rays.

secondary distance indicators   Ways of measuring distances that are calibrated by primary distance indicators.

sedimentary   Formed from settling in a liquid or from deposited material, as for a type of rock.

seeing The steadiness of the Earth's atmosphere as if it affects the resolution that can be obtained in astronomical observations.

seismic waves Waves travelling through a solid planetary body from an earthquake or impact.

seismology The study of waves propagating through a body and the resulting deduction of the internal properties of the body. "Seismo-" comes from the Greek for earthquake.

semimajor axis Half the major axis, that is, for an ellipse, half the longest diameter.

semiminor axis Half the minor axis, that is, for an ellipse, half the shortest diameter.

Seyfert galaxy A type of galaxy that has an unusually bright nucleus and whose spectrum shows broad emission lines that cover a wide range of ionization stages.

Shapley-Curtis debate The 1920 debate (and its written version) on a scale of our galaxy and of "spiral nebulae."

shear wave A type of twisting seismic wave.

shock wave A front marked by an abrupt change in pressure caused by an object moving faster than the speed of sound in the medium through which the object is traveling.

shooting stars Meteors.

showers A time of many meteors from a common cause.

sidereal With respect to the stars.

sidereal day A day with respect to the stars.

sidereal rotation period A rotation with respect to the stars.

sidereal time The hour angle of the vernal equinox; equal to the right ascension of objects on your meridian.

sidereal year A circuit of the Sun with respect to the stars.

significant figure A digit in a number that is meaningful (within the accuracy of the data).

singularity A point in space where quantities become exactly zero or infinitely large; one is present in a black hole.

slit A long, thin gap through which light is allowed to pass.

SOHO The Solar and Heliospheric Observatory, a joint NASA and European Space Agency mission to study the Sun. It is 1 million kilometers toward the Sun from Earth, a location from which it can view the Sun constantly with its dozen instruments, which include coronagraphs, cameras with filters that show the corona, and particle detectors.

solar activity cycle The 11-year cycle with which solar activity like sunspots, flares, and prominences varies.

solar atmosphere The photosphere, chromosphere, and corona of the Sun.

solar constant The total amount of energy that would hit each square centimeter of the top of the Earth's atmosphere at the Earth's average distance from the Sun.

solar day A full rotation with respect to the Sun.

solar dynamo The generation of sunspots by the interaction of convection, turbulence, differential rotation, and magnetic field.

solar flares An explosive release of energy on the Sun.

solar mass The mass of the Sun, 1.99×10^{33} grams, about 330,000 tons the Earth's mass.

solar rotational period The time for a complete rotation with respect to the Sun.

solar time A system of timekeeping with respect to the Sun such that the Sun is overhead of a given location at noon.

solar wind An outflow of particles from the Sun representing the expansion of the corona.

solar year (tropical year) An object's complete circuit of the Sun; a tropical year in between vernal equinoxes.

solstice The point on the celestial sphere of northernmost or southernmost declination of the Sun in the course of a year; colloquially, the time when the Sun reaches that point.

space velocity The velocity of a star with respect to the Sun.

spallation The break-up of heavy nuclei that undergo nuclear collisions.

special theory of relativity Einstein's 1905 theory of relative motion.

speckle interferometry A method obtaining higher resolution of an image by analysis of a rapid series of exposures that freeze atmospheric blurring.

spectra classes See *spectrum*.

spectral lines Wavelengths at which the brightness is abruptly different from the brightness at neighboring wavelengths.

spectral type One of the categories O, B, A, F, G, K, M, L, into which stars can be classified from study of their spectral lines, or extensions of this system. The above decreasing sequence of spectral types corresponds to a sequence of surface temperature.

spectrograph A device to make and record a spectrum.

spectrometer A device to make and electronically measure a spectrum.

spectroscopic binary A type of binary star that is known to have more than one component because of the changing Doppler shifts of the spectral lines that are observed.

spectroscopic parallax The distance to a star derived by comparing its apparent brightness with its luminosity magnitude deduced from study of its position on a temperature-luminosity diagram (determined by observing its spectrum—spectral type and luminosity class).

spectroscopy The use of spectrum analysis.

spectrum (plural: **spectra**) A display of electromagnetic radiation spread out by wavelength or frequency.

speed of light By Einstein's special theory of relativity, the speed at which all electromagnetic radiation travels, and the largest possible speed of an object moving through space.

spherical aberration For an optical system, a deviation from perfect focusing by having a shape that is too close to that of a sphere.

spicule A small jet of gas at the edge of the quiet Sun, approximately 1000 km in diameter and 10,000 km high, with a lifetime of about 15 minutes.

spin-flip Transition in the relative orientation of the spins of an electron and the nucleus it is orbiting.

spiral arms Bright regions looking like a pinwheel.

spiral density wave A wave that travels around a galaxy in the form of a spiral, compressing gas and dust to relatively high density and thus beginning star formation at those compressions.

spiral galaxy A class of galaxy characterized by arms that appear as though they are unwinding like a pinwheel.

spiral nebula   The old name for a shape in the sky, seen with telescopes, with arms spiraling outward from the center; proved in the 1920s actually to be a galaxy.

sporadic   Not regularly.

sporadic meteor   A meteor not associated with a shower.

spring tides   The tides at their highest, when the Earth, Moon, and Sun are in a line (from "to spring up").

stable   Tending to remain in the same condition.

star   A self-luminous ball of gas that shines or has shone because of nuclear reaction in its interior.

star clusters   Groupings of stars of common origin.

stationary limit   In a rotating black hole, the radius within which it is impossible for an object to remain stationary, no matter what it does.

steady-state theory   The cosmological theory based on the perfect cosmological principle, in which the Universe is unchanging over time.

Stefan-Boltzmann law   The radiation law that states that the energy emitted by a black body varies with the fourth power of its surface temperature.

stellar atmosphere   The outer layers of stars not completely hidden from our view.

stellar chromosphere   The region above a photosphere that shows an increase in temperature.

stellar corona   The outermost region of a star characterized by temperatures of 10^6 K and high ionization.

stellar evolution   The changes of a star's properties with time.

stelliferous era   The current era, with lots of stars in the Universe.

stones   A stony type of meteorite.

stratosphere   An upper layer of a planet's atmosphere, above the weather, where the temperature begins to increase. The Earth's stratosphere is at 20–50 km.

streamers   Coronal structures at low solar latitudes.

string theories   See *superstring theories*.

strong force   The nuclear force, the strongest of the four fundamental forces of nature.

strong nuclear force   The strong force.

subtend   The angle that an object appears to take up in your field of view; for example, the full Moon subtends $\frac{1}{2}°$.

sunspot   A region of the solar surface that is dark and relatively cool; it has an extremely high magnetic field.

sunspot cycle   The 11-year cycle of variation of the number of sunspots visible on the Sun.

superbolt   Giant lightning.

supercluster   A cluster of clusters of galaxies.

supercooled   The condition in which a substance is cooled below the point at which it would normally make a phase change; for the Universe, the point in the early Universe at which it cooled below a certain temperature without breaking its symmetry; the strong and electroweak forces remained unified.

supergiant   A post-main-sequence phase of evolution of stars of more than about 8 solar masses. They fall in the upper right of the temperature–luminosity diagram; luminosity class I.

supergranulation   Convection cells on the solar surface about 20,000 km across and vaguely polygonal in shape.

supergravity   A theory unifying the four fundamental forces.

superluminal speed   An apparent speed greater than that of light.

supermassive black hole   A black hole of millions or billions of times the Sun's mass, as is found in the centers of galaxies and quasars.

supermassive star   A stellar body of more than about $100\ M_\odot$.

supernova (plural: **supernovae**)   The explosion of a star with the resulting release of tremendous amounts of radiation.

supernova remnants   The gaseous remainder of the star destroyed in a supernova.

superstring (string) theories   A possible unification of quantum theory and general relativity in which fundamental particles are really different vibrating forms of a tiny, one-dimensional "string" instead of being localized at single points.

symmetric   A correspondence of shape so that rotating or reflecting an object gives you back an identical form; symmetry of forces in the early Universe corresponds to forces acting identically that now act differently as the four fundamental forces: gravity, electromagnetism, weak nuclear force, and strong nuclear force.

synchronous orbit   An orbit of the same period; a satellite in geostationary orbit has the same period as the Earth's rotation and so appears to hover.

synchronous rotation   A rotation of the same period as an orbiting body.

synchronous radiation   Nonthermal radiation emitted by electrons spiralling at relativistic velocities in a magnetic field.

synodic   Measured with respect to an alignment of astronomical bodies other than or in addition to the Sun or the stars.

syzygy   An alignment of three celestial bodies.

tail   Gas and dust left behind as a comet orbits sufficiently close to the Sun, illuminated by sunlight.

temperature–luminosity diagram   A diagram of a group of stars with temperatures on the horizontal axis and intrinsic brightness on the vertical axis; also called a temperature–magnitude diagram or a Hertzsprung–Russell diagram.

terminator   The line between night and day on a moon or planet; the edge of the part that is lighted by the Sun.

terrestrial planets   Mercury, Venus, Earth, and Mars.

theory   A later stage of the traditional form of the scientific method, in which a hypothesis has passed enough of its tests that it is generally accepted.

thermal pressure   Pressure generated by the motion of particles that can be characterized by temperature.

thermal radiation   Radiation whose distribution of intensity over wavelength can be characterized by a single number (the temperature). Black-body radiation, which follows Planck's law, is thermal radiation.

thermosphere   The uppermost layer of the atmosphere of the Earth and some other planets, the ionosphere, where absorption of high-energy radiation heats the gas.

3° background radiation   The isotropic black-body radiation at 3 K, thought to be a remnant of the big bang.

tide A periodic variation of the force of gravity on a body, based on the difference in the strength of gravitational pull from one place on the body to another (the tidal force); on Earth, the tide shows most obviously where the ocean meets the shore as a period variation of sea level.

time dilation According to relativity theory, the slowing of time perceived by an observer watching another object moving rapidly or located in a strong gravitational field.

TRACE The **T**ransition **R**egion **a**nd **C**oronal **E**xplorer, a solar satellite to study the solar corona and the transition region between the chromosphere and corona at high spatial resolution.

transit The passage of one celestial body in front of another celestial body. When a planet is *in transit*, we understand that it is passing in front of the Sun. Also, *transit* is the moment when a celestial body crosses an observer's meridian, or the special type of telescope used to study such events.

transition zone The thin region between a chromosphere and a corona.

Trans-Neptunian Object (TNO) One of the sub-planetary objects in the Kuiper belt; it was proposed but not accepted that Pluto should be considered as one.

transparency Clarity of the sky.

transverse velocity Velocity along the plane of the sky.

trigonometric parallax See *parallax*.

triple-alpha process A chain of fusion processes by which three helium nuclei (alpha particles) combine to form a carbon nucleus.

Trojan asteroids A group of asteroids that precede or follow Jupiter in its orbit by 60°.

tropical year The length of time between two successive vernal equinoxes.

troposphere The lowest level of the atmosphere of the Earth and some other planets, in which all weather takes place.

T Tauri star A type of irregularly varying star, like T Tauri, whose spectrum shows broad and very intense emission lines. T Tauri stars have presumably not yet reached the main sequence and are thus very young.

tuning-fork diagram Hubble's arrangement of types of elliptical, spiral, and barred spiral galaxies.

21-cm line The 1420-MHz line from neutral hydrogen's spin-flip.

twinkle A scintillation—rapid changing in brightness—and slight changing in position of stars as their light passes through the Earth's atmosphere.

Type Ia supernova A supernova whose distribution in all types of galaxies, and the lack of hydrogen in its spectrum, make us think that it is an event in low-mass stars, probably resulting from the incineration of a white dwarf in a binary system.

Type II supernova A supernova associated with spiral arms, and that has hydrogen in its spectrum, making us think that it is the explosion of a massive star.

ultraviolet The region of the spectrum 100–4000 Å, also used in the restricted sense of ultraviolet radiation that reaches the ground, namely, 3000–4000 Å.

umbra (plural: **umbrae**) (a) Of a sunspot, the dark central region; (b) of an eclipse shadow, the part from which the Sun cannot be seen at all.

uncertainty principle Heisenberg's statement that the product of uncertainties of position and momentum is greater than or equal to Planck's constant. Consequently, both position and momentum cannot be known to infinite accuracy.

universal gravitation constant The constant G of Newton's law of gravity: force = Gm_1m_2/r^2.

valleys Depressions in the landscapes of solid objects.

Van Allen belts Regions of high-energy particles trapped by the magnetic field of the Earth.

variable star A star whose brightness changes over time.

vernal equinox The equinox crossed by the Sun as it moves to northern declinations.

Very Large Array The National Radio Astronomy Observatory's set of radio telescopes in New Mexico, used together for interferometry.

Very Large Telescope The set of instruments, including four 8.2-m reflectors and several smaller telescopes, erected by the European Southern Observatory on a mountaintop in Chile through 2001.

Very-Long-Baseline Array The National Radio Astronomy Observatory's set of radio telescopes dedicated to very-long-baseline interferometry and spread over an 8000-km baseline across the United States.

very-long-baseline interferometry The technique using simultaneous measurements made with radio telescopes at widely separated locations to obtain extremely high resolution.

visible light Light to which the eye is sensitive, 4000–7000 Å.

visual binary A binary star that can be seen through a telescope to be double.

VLA See *Very Large Array*.

VLBA See *Very-Long-Baseline Array*.

VLBI See *very-long-baseline interferometry*.

VLT See *Very Large Telescope*.

void A giant region of the Universe in which few galaxies are found.

volatile Evaporating (changing to a gas) readily.

wave front A plane in which parallel waves are in step.

wavelength The distance over which a wave goes through a complete oscillation.

weak nuclear force One of the four fundamental forces of nature, weaker than the strong force and the electromagnetic force. It is important only in the decay of certain elementary particles, such as neutrons.

weight The force of the gravitational pull on a mass.

white dwarf The final stage of the evolution of a star initially between 0.08 and about 8 solar masses but with a final mass no larger than 1.4 M. It is supported by electron degeneracy. White dwarfs are found to the lower left of the main sequence of the temperature–luminosity diagram.

white light All the light of the visible spectrum together.

Wien's displacement law The expression of the inverse relationship of the temperature of a black body and the wavelength of the peak of its emission.

WIMPs **W**eakly **I**nteracting **M**assive **P**articles, an as yet undiscovered massive particle, interacting with other elementary particles only through the weak nuclear force, that may be a major type of cold dark matter.

winter solstice For northern-hemisphere observers, the southernmost declination of the Sun, and its date.

Wolf-Rayet star A type of O star whose spectrum shows very broad emission lines.

wormhole The hypothetical connection or bridge between two black holes; they probably do not exist except in idealized black holes.

x-rays Electromagnetic radiation between 0.1 and 100 Å.

year The period of revolution of a planet around its central star; more particularly, the Earth's period of revolution around the Sun.

Zeeman effect The splitting of certain spectral lines in the presence of a magnetic field.

zenith The point in the sky directly overhead an observer.

zero-age main sequence The curve of a temperature–luminosity diagram determined by the locations of stars at the time they begin nuclear fusion.

zodiac The band of constellations through which the Sun, Moon, and planets move in the course of a year.

zodiacal light A glow in the nighttime sky near the ecliptic from sunlight reflected by interplanetary dust.

Illustration Acknowledgments

The many photographs from the Hubble Space Telescope were created with support to Space Telescope Science Institute, operated by the Association of Universities for Research in Astronomy, Inc., from NASA contract NAS5-26555, and are reproduced with permission from AURA/STScI. The Hubble Space Telescope is a project of international cooperation between NASA and the European Space Agency (ESA). The Solar and Heliospheric Observatory (SOHO) is a project of international cooperation between ESA and NASA.

Front cover: NASA and the Early Release Observations Team: Andrew Fruchter, Sylvia Baggett, and Zoltan Levay (Space Telescope Science Institute) and Richard Hook (Space Telescope—European Coordinating Facility).

Back cover: European Southern Observatory with the Very Large Telescope.

Frontispiece: NASA and the Early Release Observations Team: Andrew Fruchter, Sylvia Baggett, and Zoltan Levay (Space Telescope Science Institute) and Richard Hook (Space Telescope—European Coordinating Facility).

Contents Overview: Spiral Galaxy NGC4414, Hubble Heritage Team (AURA/STScI/NASA); Mars, Steve Lee (U. Colorado), Jim Bell (Cornell U.), Mike Wolff (STScI), and NASA.

Table of Contents: Hickson Compact Group 87: Hubble Heritage Team (STScI/AURA/NASA); Keck Telescopes: © Andrew Perala; Space Telescope Science Institute/NASA and the Hubble Heritage Team (STScI/AURA); Jupiter and Comet Shoemaker-Levy 9: STScI, NASA; solar eclipse: © Jay M. Pasachoff and Wendy Carlos; M4 star cluster: European Southern Observatory; planetary nebula NGC 6751: Hubble Heritage Team (STScI/AURA/NASA); RCW38 star-forming region: European Southern Observatory; gravitational lensing: Kavan Ratnatunga of Johns Hopkins U and colleagues, NASA/STScI; Hubble Deep Field–South: R. Williams (STScI), the HDF-South Team, and NASA.

Preface: Trapezium, J. Bally, D. Devine, R. Sutherland/ NASA/Space Telescope Science Institute: Next Generation Space Telescope: NASA/Goddard Space Flight Center; Mars, Dan Britt, U. Tennessee, and Caltech/JPL/NASA; Veil Nebula: Nigel Sharp, National Optical Astronomy Observatories; Copernicus's Notebook: Jay M. Pasachoff; Hubble Space Telescope: Johnson/Space Center/NASA; Jupiter: Reta Beebe (New Mexico State University), and NASA; Star cluster NGC1850, European Southern Obervatory.

Chapter 1—Opener NASA/Johnson Space Center; **p. 1** NGC 4013, Blair Savage, Chris Howk (U. Wisconsin)/N.A. Sharp (NOAO)/WIYN/NSF; © WIYN Consortium, Inc. All rights reserved; **Fig. 1–2** National Optical Astronomy Observatories/Kitt Peak; **Fig. 1–3A** European Southern Observatory; **Fig. 1–3B** NASA/Chandra X-ray Observatory Center/SAO; **Fig. 1–4** Photo courtesy of Gemini Observatory; **Fig. 1–5** European Southern Observatory; **Fig. 1–6** Don Figer (Space Telescope Science Institute) and NASA; **Fig. 1–7** © 1985 Royal Observatory Edinburgh/Anglo-Australian Observatory, photograph by David F. Malin from original U.K. Schmidt Plates; **p. 6 top left** © 1995 David Scharf; **p. 6 top middle** Jay M. Pasachoff; **p. 6 top right** Neil Leifer for Sports Illustrated; © Time Warner, Inc.; **p. 6 bottom left** Joseph Distefano, City of Boston, and Abrams Aerial Survey Corp.; **p. 6 bottom middle and bottom right** NASA; **p. 7 top left and right** © NASA/JPL/ Caltech; **p. 7 middle** with NASA/JPL/Caltech insets; **p. 7 bottom right** Axel Mellinger, Universität Potsdam; **p. 8 top** Hubble Heritage Team (AURA/STScI/NASA); **p. 8 middle right** Hubble Heritage Team (AURA/STScI/NASA); **p. 8 middle left** Margaret Geller and John Huchra, Harvard-Smithsonian Center for Astrophysics; **p. 8 bottom** NASA COBE Science Team; **Fig. 1–8** © 1919 by The New York Times Co. Reprinted by permission; **Fig. 1–9** Drawing by Handelsman; © 1978 The New Yorker Magazine, Inc.; **Fig. 1–10** Jay M. Pasachoff.

Chapter 2—Opener European Southern Observatory; **Fig. 2–4** Jay M. Pasachoff; **Fig. 2–6** American Institute of Physics, Niels Bohr Library, Uhlenbeck Collection; **Fig. 2–9** © 1987 Royal Observatory Edinburgh/Anglo-Australian Observatory, photography by David F. Malin from U.K. Schmidt plates.

Chapter 3—Opener NASA/Chandra X-ray Observatory Center/SAO, courtesy of Martin Zombeck; **Fig. 3–1** Jay M. Pasachoff, after Ewen Whitaker, Lunar and Planetary Laboratory, U. Arizona; **Fig. 3–2A** Williams College–Hopkins Observatory, photo by Stephan Martin; **Fig. 3–2B** Jay M. Pasachoff; **Fig. 3–3** Akira Fujii; **Fig. 3–4** drawn from a plot made with *Visible Universe*; **Fig. 3–8** © 1986 Columbia Pictures Industries, Inc. All rights reserved; **Fig. 3–12** Yerkes Observatory; **Fig. 3–13 background** © 1999 Subaru Telescope, National Astronomical Observatory of Japan; **Fig. 3–14** photo © Andrew Perala; **Fig. 3–15** Jay M. Pasachoff; **Fig. 3–16** and **3–17** European Southern Observatory; **Fig. 3–19** © 1987 and 1979, ROE/AAO. **Fig. 3–20** Jay M. Pasachoff; **Fig. 3–22** NASA/Johnson Space Center; **Fig. 3–23** Raytheon Optical Systems; **Fig. 3–24** Robert Williams and the Hubble Deep Field Team (STScI) and NASA; **Fig. 3–25** Alan Title, Lockheed Martin Research Laboratories; and Leon Golub, Harvard-Smithsonian Center for Astrophysics; **Fig. 3–28** Eastman Kodak; **Fig. 3–29** Raytheon Optical Systems; **Fig. 3–31** ESA/ISO and LWS Consortium; **Fig. 3–32** NASA COBE Science Team; **Fig. 3–33A** Debra Elmegreen, Vassar College; **Fig. 3–33A (inset)** David B. Jansky, Courtesy of Woodruff Sullivan and the Astronomical Society of the Pacific; **Fig. 3–33B** Robert W. Wilson; **Fig. 3–34** ABC National Radio Astronomy Observatories; **Fig. 3–35** Jay M. Pasachoff; **Fig. 3–37** Courtesy of Leo Blitz, U. California at Berkeley; **Fig. 3–38** Riccardo Giovanelli and Martha Haynes, Cornell U.; **Fig. 3–39B** Jay M. Pasachoff; p. 43 Chris Jones, Union College.

Chapter 4—Opener Raymond Pojman, Carbondale, CO; **Fig. 4–1** Pekka Parviainen; **Fig. 4–3** Williams College—Hopkins Observatory, by Kevin Reardon, now at Osservatorio di Capodimonte, Naples; **Fig. 4–5** NASA/JPL/Caltech; **Fig. 4–6AB** Cartoons by Charles Schulz. © 1970 United Feature Syndicate, Inc.; **Fig. 4–9** © 1997 Richard Wainscoat; **Fig. 4–10** © 1982 Richard E. Hill; **Fig. 4–15A** Clifford Holmes; **Fig. 4–16** Jay M. Pasachoff; **Fig. 4–17 to 4–20** Akira Fujii; **Fig. 4–22** Emil Schulthess, Black Star; **Fig. 4–23** Map Creation Ltd.; **Fig. 4–24** Jay M. Pasachoff; **Fig. 4–26** Mt. Vernon Ladies Association of the Union.

Chapter 5—Opener Jay M. Pasachoff; **Fig. 5–1** NASA; **Fig. 5–2** Naval Research Laboratory; **Fig. 5–3** NASA; **Fig. 5–4** NASA/JPL/Caltech; **Fig. 5–5** NASA/AURA/STScI; **Fig. 5–6** NASA/JPL/Caltech; **Fig. 5–7** Geoff Chester, Einstein Planetarium, Smithsonian Institution; **Fig. 5–8** Photo Vatican Museums; **Fig. 5–9** Courtesy of the Houghton Library, Harvard University; **Fig. 5–10** Biblioteca Nazionale Marciana, Venice; **Fig. 5–12** Photograph by Charles Eames, reproduced courtesy of Eames Office and Owen Gingerich. © 1990, 1998 Lucia Eames dba Eames Office; **Fig. 5–13** Jay M. Pasachoff; **Fig. 5–14** Huntington Library, San Marino, CA; **p. 75** Torun Museum, courtesy of Marek Demianski; **p. 76** Jay M. Pasachoff; **Fig. 5–16,** Photograph Courtesy of the Royal Ontario Museum © ROM; **Fig. 5–17** Jay M. Pasachoff; **Fig. 5–18B** Jay M. Pasachoff; p. 78 Jay M. Pasachoff; **Fig. 5–21** William C. Livingston, National Solar Observatory, AURA; **Fig. 5–22** Jay M. Pasachoff; **Fig. 5–23** U. Michigan Library, Dept. of Rare Books and Special Collections; translation by Stillman Drake, reprinted courtesy of *Scientific American*; **p. 81** National Maritime Museum; **Fig. 5–26** Jay M. Pasachoff; **Fig. 5–27** Jay M. Pasachoff; p. 84 National Portrait Gallery, London; **Fig. 5–28** Jay M. Pasachoff; **Fig. 5–29** Jay M. Pasachoff; **Fig. 5–32** Alan P. Boss, Carnegie Institute of Washington; **Fig. 5–33** Al Schultz (Computer Sciences Corp. and STScI), Sally Heap (NASA/Goddard Space Flight Center), and NASA; **Fig. 5–34** Jane Greaves, J. S. Greaves; W. S. Holland; G. Moriarty-Schieven; T. Jenness; W. R. F. Dent; B. Zuckerman; C.

McCarthy; R. A. Webb; H. M. Butner; W. K. Gear; and H. J. Walker, *Astrophys. J.* 506, L133–137, 1998 October 20.

Chapter 6—Opener NASA/JPL/Caltech; **Fig. 6–1** NASA/JPL/Caltech; **Fig. 6–2** NASA/Ames Research Center; **Fig. 6–3** Cassini Imaging Team/U. Arizona and NASA/JPL/Caltech; **Fig. 6–4** U.S. Geological Survey and NASA; **Fig. 6–5** Jay M. Pasachoff; **Fig. 6–6** Italian Space Agency; **Fig. 6–7** NASA/JPL/Caltech and U. Arizona; **Fig. 6–8** M. F. Corrin, L. M. Gahagan, and L. A. Lawver, PLATES Project, Univesity of Texas Institute for Geophysics; **Fig. 6–9** Jay M. Pasachoff; **Fig. 6–11** Pieter Tans, Climate Monitoring and Diagnostics Laboratory, National Oceanic and Atmospheric Administration, Carbon Cycle Group, Boulder, CO; and Charles D. Keeling, Scripps Institute of Oceanography, La Jolla, CA; **Fig. 6–12** TOMS, NASA/Goddard Space Flight Center; **Fig. 6–13** NASA; **Fig. 6–14** Pekka Parviainen; **Fig. 6–15** Jay M. Pasachoff; **Fig. 6–17ABC** NASA/Johnson Space Center; **Fig. 6–18** Gerald Wasserburg, Caltech; **Fig. 6–19** Drawings by Donald E. Davis under the guidance of Don E. Wilhelms of the U.S. Geological Survey; reproduced with the assistance of the Hansen Planetarium; **Fig. 6–20** U.S. Naval Observatory; **Fig. 6–21** NASA; **Fig. 6–22** Naval Research Laboratory; **Fig. 6–23** NASA; **Fig. 6–25** NASA; **Fig. 6–26** A. G. W. Cameron, Harvard-Smithsonian Center for Astrophysics; **Fig. 6–27** NASA/Johnson Space Center; **Fig. 6–31** courtesy of Robert G. Strom from his *Mercury: The Elusive Planet* (Smithsonian Institution Press), artwork by Karen Denomy; **Fig. 6–32** Robert G. Strom, Lunar and Planetary Laboratory, U. Arizona; **Fig. 6–33** NASA; **Fig. 6–34** NASA; **Fig. 6–35** NASA; **Fig. 6–36** David Mitchell, U. California, Berkeley; **Fig. 6–37** Jay M. Pasachoff; **Fig. 6–40** NASA/JPL/Caltech; **Fig. 6–41** Courtesy of Donald B. Campbell, Arecibo Observatory; **Fig. 6–42** NASA/Ames Research Center; **Fig. 6–43** NASA/JPL/Caltech; **Fig. 6–44** Courtesy of Valeriy Barsukov and Yuri Surkov; **Fig. 6–45** Donald Parker; **Fig. 6–46** NASA/JPL/Caltech; **Fig. 6–47** Dan Britt, U. Tennessee, and NASA; **Fig. 6–48** NASA/JPL/Caltech and Malin Space Science Systems; **Fig. 6–49** Mosaic by Jody Swann, USGS, Branch of Astrogeology, Flagstaff; **Fig. 6–50** NASA/JPL/Caltech and Malin Space Science Systems; **Fig. 6–51** NASA/JPL/Caltech and Malin Space Science Systems; **Fig. 6–52** NASA/JPL/Caltech and Malin Space Science Systems; **Fig. 6–53** Steve Lee (U. Colorado), Jim Bell (Cornell Univ.), Mike Wolff (STScI), and NASA; **Fig. 6–54** NASA/JPL/Caltech; **Fig. 6–55** NASA/JPL/Caltech; **Fig. 6–56** NASA.

Chapter 7—Opener Prepared for NASA by Stephen P. Meszaros; **Fig. 7–2** NASA/JPL/Caltech; **Fig. 7–3** NASA/JPL/Caltech; **Fig. 7–4** NASA/JPL/Caltech; **Fig. 7–5A** NASA/JPL/Caltech; **Fig. 7–5B** J. Clarke (U. Michigan) and NASA; **Fig. 7–7** Mark Showalter, NASA/Ames; **Fig. 7–8** NASA/JPL/Caltech; **Fig. 7–9** NASA/JPL/Caltech; **Fig. 7–10** U.S. Geological Survey, processing by A. McEwen, T. Rock, and L. Soderblom; **Fig. 7–11** NASA/JPL/Caltech; **Fig. 7–12** NASA/JPL/Caltech; **Fig. 7–13** NASA/JPL/Caltech; **Fig. 7–14** NASA/JPL/Caltech; **Fig. 7–15** Reta

Beebe (New Mexico State U.), D. Gilmore, L. Gergeron (STScI), and NASA; **Fig. 7–17** Erich Karkoschka, Lunar and Planetary Laboratory, U. Arizona, and NASA; **Fig. 7–18** NASA/JPL/Caltech; **Fig. 7–19** NASA/JPL/Caltech; **Fig. 7–20** NASA/JPL/Caltech; **Fig. 7–21** redrawn from NASA/JPL/Caltech; **Fig. 7–22** redrawn from Augustin Sanchez-Lavega, J. F. Rojas, and P. V. Sada; **Fig. 7–23** J. P. Trauger (JPL) and NASA; **Fig. 7–24** NASA/JPL/Caltech; crescent Saturn: NASA/JPL/Caltech; **Fig. 7–25A** Peter H. Smith and Mark T. Lemmon (U. Arizona) and NASA; **Fig. 7–25B** Claire Max, Bruce Macintosh, and Seran Gibbard, Lawrence Livermore National Laboratory, and W. M. Keck Observatory; **Fig. 7–26** NASA/JPL/Caltech; **Fig. 7–28** Erich Karkoschka (U. Arizona) and STScI; **Fig. 7–29** NASA/JPL/Caltech; **Fig. 7–30** James L. Elliot, MIT; **Fig. 7–31** NASA/JPL/Caltech; **Fig. 7–32** Master and Fellows of St. Johns College of Cambridge; **Fig. 7–33** Charles Kowal, Space Telescope Science Institute; **Fig. 7–34** Tobias Owen, Institute for Astronomy, U. Hawaii, after R. Danhy; **Fig. 7–35** Heidi Hammel, Space Science Institute, Boulder; **Fig. 7–36** NASA/JPL/Caltech; **Fig. 7–37** NASA/JPL/Caltech; **Fig. 7–38** NASA/JPL/Caltech; **Fig. 7–39A** Richard Dekany (JPL), Thomas Hayward, Bernhard Brandl, and Don Banfield (Cornell); the AO system was developed by a JPL team led by Richard Dekany: imaged at the Palomar Observatory; **Fig. 7–39B** W. M. Keck Observatory; **Fig. 7–40** Information from Norman Ness; style from *Sky & Telescope*; **Fig. 7–41** NASA/JPL/Caltech, reprocessed by Caroline Porco, U. Arizona; **Fig. 7–42** Christophe Dumas and Richard J. Terrile (JPL), and the NICMOS team; **Fig. 7–43** NASA/JPL/Caltech, reprocessed by USGS; **Fig. 7–44** NASA/JPL/Caltech; **Figs. 7–45** and **7–46** Geoff Marcy, San Francisco State U. and U. California, Berkeley.

Chapter 8—Opener © Tony and Daphne Hallas/Astro Photo; **Fig. 8–1** Jay M. Pasachoff; **Fig. 8–2** Lowell Observatory photographs; **Fig. 8–3** U.S. Naval Observatory/James W. Christy, U.S. Navy photograph; **Fig. 8–4** John R. Spencer, Institute for Astronomy, U. Hawaii; **Fig. 8–5A** Keith S. Noll, STScI, and NASA; **Fig. 8–5B** Photo courtesy of Gemini Observatory; **Fig. 8–6** S. Alan Stern, Southwest Research Institute, and Marc Buie, Lowell Observatory, and NASA; **Fig. 8–7** James L. Elliot, MIT; **Fig. 8–8** David Jewitt, Institute for Astronomy, U. Hawaii; **Fig. 8–9** Raymond Pojman; **Fig. 8–10** Akira Fujii; **Fig. 8–11** Akira Fujii; **Fig. 8–12** Naval Research Laboratory, LASCO on the Solar and Heliospheric Observatory: SOHO is a NASA and ESA collaboration; **Fig. 8–13** National Portrait Gallery, London; **Fig. 8–14** © 1986 Royal Observatory Edinburgh/Anglo-Australian Observatory, photography by David Malin from U.K. Schmidt plates; **Fig. 8–15** © Max-Planck-Institut für Aeronomie, Lindau/Hartz, FRG; photographed with the Halley Multicolour Camera aboard the European Space Agency's Giotto spacecraft, courtesy of H. U. Keller, additional processing by Harold Reitsema, Ball Aerospace; **Fig. 8–16** Jay M. Pasachoff and Steven Souza; **Fig. 8–17AB** Soviet VEGA team; **Fig. 8–18** Susan Wyckoff, Arizona State U., Tempe; **Fig. 8–19** Jay M. Pasachoff; **Fig. 8–20** H. A. Weaver (Applied Research Corp.), P. D. Feldman

(The Johns Hopkins U.), and NASA; **Fig. 8–21** NASA/STScI; **Fig. 8–22AB** Heidi Hammel, Space Science Institute, Boulder, and NASA/STScI; **Fig. 8–22C** W. M. Keck Observatory/Imke de Pater, James Graham, Garrett Jernigan, U. California, Berkeley; **Fig. 8–23** Jay M. Pasachoff; **Fig. 8–24** © 1998 Tony and Daphne Hallas, Astro Photo; **Fig. 8–25** Jay M. Pasachoff; **Fig. 8–26** Jay M. Pasachoff; **Fig. 8–27** Allan E. Morton, courtesy of Meteor Crater, Northern Arizona, USA; **Fig. 8–28A** John Bortle; **Fig. 8–28B** Bill Menke, Lamont Doherty Earth Observatory; **Fig. 8–29** Jay M. Pasachoff; **Fig. 8–30ABCDE** Thierry Pauwels, Koninklijke Sterrenwacht van Belgie (Observatoire Royal de Belgique–Royal Observatory of Belgium); **Fig. 8–31** Andrew Chaikin; **p. 190 left** © 1992 Universal City Studios and Amblin; **p. 190 right** Virgil Sharpton, Lunar and Planetary Institute; **Fig. 8–32** U.S. Geological Survey; **Fig. 8–33** NASA/JPL/Caltech; **Fig. 8–34** Johns Hopkins University Applied Physics Laboratory; **Fig. 8–35A** NASA/JPL/Caltech; **Fig. 8–35B** Johns Hopkins University Applied Physics Laboratory; **Fig. 8–36** Illustration from *The Little Prince* by Antoine de Saint-Exupéry, copyright 1943 and renewed 1971 by Harcourt Brace & Company, reproduced by permission of the publisher.

Chapter 9—Opener Alan Title, Lockheed Martin Advanced Technology Center, and Leon Golub, Harvard-Smithsonian Center for Astrophysics; **Fig. 9–1** Jay M. Pasachoff, Williams College—Hopkins Observatory, with Dan Seaton; **Fig. 9–2** Stanford Lockheed Institute for Space Research from SOHO SOI/MDI; **Fig. 9–4** © 1998 Jay M. Pasachoff and Wendy Carlos; **Fig. 9–5** William C. Livingston, National Solar Observatory/AURA; **Fig. 9–8** Kanzelhöhe Solar Observatory of the Karl Frazens-University Graz; **Fig. 9–9** NASA/Goddard Space Flight Center, courtesy of Joe Gurman; SOHO is a joint project of NASA and the European Space Agency; **Fig. 9–10** Eclipse image © 1999 Jay M. Pasachoff and Wendy Carlos, merged with a SOHO/EIT image from NASA/Goddard Space Flight Center, courtesy of Joe Gurman; **Fig. 9–11** Alan Title, Lockheed Martin Advanced Technology Center, and Leon Golub, Harvard-Smithsonian Center for Astrophysics; **Fig. 9–12** Sacramento Peak Observatory, National Solar Observatory/AURA; **Fig. 9–13** Robert A. E. Fosbury, European Southern Observatory; **Fig. 9–14** Yohkoh image courtesy of Loren Acton, U. Montana; **Fig. 9–15** Alan Title, Lockheed Martin Advanced Technology Center, and Leon Golub, Harvard-Smithsonian Center for Astrophysics; **Fig. 9–16** 7" Starfire refractor, Wolfgang Lille/Baade Planetarium, Germany; **Fig. 9–17** Jay M. Pasachoff; **Fig. 9–18** Data from Sunspot Index Data Center, Royal Observatory of Belgium; **Fig. 9–19** National Solar Observatory/NOAO; **Fig. 9–20** NASA and ESA; **Fig. 9–21** and **9–25** to **9–27** Jay M. Pasachoff; **Fig. 9–28** Courtesy of the Archives, California Institute of Technology; **Fig. 9–29B** AP/Wide World Photos, Eric Risberg; **Fig. 9–31** Huntington Library, San Marino, CA, and the Albert Einstein Archives, courtesy of the Hebrew University of Jerusalem, Jewish National University and Library; **Fig. 9–32** © 1919 The New York Times Co. Reprinted by permission.

Chapter 10—Opener Hubble Heritage Team; AURA/STScI/NASA; **Fig. 10–4** Harvard College Observatory; **Figs. 10–5** and **10–6** Roger Bell, University of Maryland, and Michael Briley, U. Wisconsin at Oshkosh; **Fig. 10–7** Yves Grosdidier, Université de Montreal and Observatoire de Strasbourg; Antony Moffat, Université de Montreal; Gilles Joncas, Université Laval; Agnes Acker, Observatoire de Strasbourg; and NASA; **Fig. 10–8** Hubble Heritage Team (AURA/STScI/NASA); **Fig. 10–12** American Institute of Physics, Niels Bohr Library; Margaret Russell Edmondson Collection; **Fig. 10–13** Dorrit Hoffleit–Yale U. Observatory; courtesy of American Institute of Physics, Niels Bohr Library.

Chapter 11—Opener Image created from 2 separate pointings of HST (WFPC2): (1) Francesco R. Ferraro (ESO, Bologna Obs.), Barbara Paltrinieri (U. La Sapienza), Robert T. Rood (U. Virginia), and Ben Dorman (Raytheon/STX); (2) Michael Shara (STScI, American Museum of Natural History), David Zurek (STScI), and Laurent Drissen (U. Laval); **Fig. 11–1** Dan Overcash; **Figs. 11–2** and **11–3** Copyright U.C. Regents, UCO/Lick Observatory image; **Fig. 11–7** Jay M. Pasachoff; **Fig. 11–8** AAVSO; **Fig. 11–9** observations obtained for Jay M. Pasachoff with the Automatic Photoelectric Telescope; **Fig. 11–10** J. D. Fernie and R. McGonegal, reprinted from *Astrophys. J.* 275, p. 735, with permission of the U. Chicago Press; **Fig. 11–11** Akira Fujii; **Fig. 11–12** Wendy L. Freedman, Observatories of the Carnegie Institution of Washington, and NASA; **Fig. 11–13** © 1977 Anglo-Australian Observatory, photograph by David Malin; **Fig. 11–14** © 1984 Anglo-Australian Observatory, photograph by David Malin; **Fig. 11–15** Canada-France-Hawaii Telescope Corp., © Regents of the University of Hawaii, photo by Laird Thompson, then Institute for Astronomy, U. Hawaii, now U. Illinois; **Fig. 11–17** Shigeto Hirabayashi.

Chapter 12—Opener Y. Totsuka, U. Tokyo, Inst. for Cosmic Ray Research; **Fig. 12–1** © 1979 Royal Observatory Edinburgh/Anglo-Australian Observatory, photography by David Malin from original U.K. Schmidt plates; **Fig. 12–2** Alain Abergel, Institut d'Astrophysique Spatiale, Orsay, with the Infrared Space Observatory; **Fig. 12–3** C. R. O'Dell and Z. Wen (Rice U.)/NASA/STScI; **Fig. 12–4** (*left*) Jeff Hester and Paul Scowen (Arizona State U.), and NASA/STScI/ESA; (*right*) Mark McCaughrean with the Calar Alto 3.5-m telescope; **Fig. 12–5** Jeff Hester (Arizona State U.), the WFPC2 Investigation Definition Team, and NASA; **Fig. 12–9** Don F. Figer, Mark Morris, Ian F. McLean (UCLA), Gene Serabyn (Caltech) and R. Michael Rich (Columbia); **Fig. 12–11** (*left*) Palomar Observatory, Caltech; (*right*) T. Nakajima and S. Kulkarni (Caltech), S. Durrance and F. Golimowski (Johns Hopkins U.), and NASA; **Fig. 12–12** Brookhaven National Laboratory; **Fig. 12–13** Y. Totsuka, U. Tokyo, Inst. for Cosmic Ray Research; **Fig. 12–14** Lawrence Berkeley National Laboratory and Sudbury Neutrino Observatory.

Chapter 13—Opener Hubble Heritage Team (AURA/ STScI/NASA); **Fig. 13–1A** © 1979 Royal Observatory Edinburgh/Anglo-Australian Observatory, photography by David Malin from original U. K. Schmidt plates; **Fig. 13–1B** Robert O'Dell and Kerry P. Handron (Rice University), and NASA; **Fig. 13–2** U. Washington—U.S. Naval Observatory—Cornell U.–Arcetri Obs. (Florence, Italy) collaboration, courtesy of Bruce Balick, U. Washington; **Fig. 13–3** Canada-France-Hawaii Telescope Corp.; **Fig. 13–4A** Raghvendra Sahai and John Trauger (JPL), the WFPC2 Science Team, and NASA; **Fig. 13–4B** Rodger Thompson, Marcia Rieke, Glenn Schneider, Dean Hines (University of Arizona); Raghvendra Sahai (JPL); NICMOS Instrument Definition Team; and NASA; **Fig. 13–6** Harvey Richer (U. British Columbia), and NASA; **Fig. 13–7** McDonald Observatory, U. Texas; **Fig. 13–8** G. Dana Berry, then at Space Telescope Science Institute; **Fig. 13–8** Andrea Dupree (Harvard-Smithsonian Center for Astrophysics), Ronald Gilliland (STScI), NASA and ESA; **Fig. 13–10** Courtesy Encyclopaedia Britannica, Inc.; from the 1989 *Britannica Yearbook of Science and the Future*; illustration by Jane Meredith; **Fig. 13–11** Weidong Li and Alex Filippenko, UC Berkeley; **Fig. 13–12A** Jeff Hester (Arizona State U.), and NASA; **Fig. 13–12B** Joachim Trümper, Max-Planck-Institut für Extraterrestrische Physik; **Fig. 13–13A** David F. Malin (Anglo-Australian Observatory) and Jay M. Pasachoff, from Palomar Observatory plates made available by the California Institute of Technology, courtesy of Robert Brucato; **Fig. 13–13B** John P. Hughes (Rutgers U.) and NASA/ Chandra X-ray Observatory Center/SAO; **Fig. 13–14** © 1987 Anglo-Australian Observatory, photograph by David Malin; **Fig. 13–15A** Hubble Heritage Team; **Fig. 13–15B** and **13–16AB** NASA, Peter Challis and Robert Kirshner (Harvard-Smithsonian Center for Astrophysics), Peter Garnavich (University of Notre Dame), and the SINS Collaboration; The Supernova Intensive Survey Team includes Dick McCray (University of Colorado, Boulder); Nino Panagia (Space Telescope Science Institute, Baltimore, MD); Nick Suntzeff (Cerro Tololo Inter-American Obervatory, Chile); George Sonneborn and Jason Pun (NASA Goddard Space Flight Center, Greenbelt, MD); **Fig. 13–16c** NASA/CXC/SAO/ Pennsylvania State U/D Burrows et al.; **Fig. 13–17** © 1988 Anglo-Australian Observatory, photograph by David Malin; **Fig. 13–18** Chien Peng (U. Arizona) and Alex Filippenko (UC Berkeley); **Fig. 13–19** background: NASA-Corbis; **Fig. 13–20** © Royal Observatory Edinburgh, Brian Hadley; **Fig. 13–21AB** Master and Fellows of Churchill College, Cambridge U.; **Fig. 13–22** Joseph H. Taylor, Jr., at the Princeton Pulsar Physics Laboratory; **Fig. 13–24** Joachim Trümper, Max-Planck-Institut für Extraterrestrische Physik; **Fig. 13–25A** Jeff Hester and Paul Scowen (Arizona State U.), and NASA; **Fig. 13–25B** NASA/Chandra X-ray ׀Observatory Center/SAO; **Fig. 13–26** Walter Brisken, at the Princeton Pulsar Physics Laboratory; **Fig. 13–28** From a model by Mordechai Milgrom, Bruce Margon, Jonathan Katz, and George Abell; **Fig. 13–29** Courtesy of M. Watson, R. Willingale, Jonathan E. Grindlay, and Frederick D. Seward; see *Astrophysical J.* 273, p. 688 (1983), courtesy of U. Chicago Press; **Fig. 13–30** after a Cornell University visualization, imaging by Chris Hildreth and Wayne Lytle; courtesy of Alex Wolzczsan, Pennsylvania State U.

Chapter 14—Opener VLA of NRAO; **Fig. 14–1** Roy Bishop, Acadia University; **Fig. 14–6** Drawing by Charles Addams, © 1974 The New Yorker Magazine, Inc.; **Fig. 14–8** Chapin Library, Williams College; **Fig. 14–9** NASA/Goddard Space Flight Center; **Fig. 14–10** Palomar Observatory/California Institute of Technology photograph by Jerome Kristian; **p. 259** Image Alex Filippenko (UC Berkeley) and Chien Peng (U. Arizona); radial velocity curve: Alex Filippenko and Tom Matheson (UC Berkeley); **Fig. 14–11** © Julian Calder/Woodfin Camp; **Fig. 14–12** L. Ferrarese (Johns Hopkins U.) and NASA; **Fig. 14–13** NASA/Chandra X-ray Observatory Center/SAO; **Fig. 14–14** Dave Finley (National Radio Astronomy Observatory), Bill Junor (University of New Mexico), John Biretta and Mario Livio (Space Telescope Science Institute), and NASA.

Chapter 15—Opener Wolfgang Brandner (JPL/IPAC), Eva K. Grebel (U. Washington), You-Hua Chu (U. Illinois at Urbana-Champaign), and NASA; **Fig. 15–1** Jayanne English (CGPS/STScI) using data acquired by the Canadian Galactic Plane Survey (NRC/NSERC) and produced with the support of Russ Taylor (U. Calgary); **Fig. 15–2** Axel Mellinger, Universität Potsdam; **Fig. 15–4** C. R. O'Dell (Rice U.) and NASA; **Fig. 15–5** A. Caulet (ST-ECF, ESA) and NASA; **Fig. 15–6** © 1979 Royal Observatory Edinburgh/Anglo-Australian Observatory, photograph from original U.K. Schmidt Plates by David Malin; **Fig 15–7A** © 1985 Royal Observatory Edinburgh/Anglo-Australian Observatory, photograph from original U.K. Schmidt Plates by David Malin; **Fig. 15–7B** EXOSAT Observatory; **Fig. 15–7C** Richard E. White, Smith College; **Fig. 15–8** European Southern Observatory; **Fig. 15–9** European Southern Observatory; **Fig. 15–11** NASA/Goddard Space Flight Center Data Facility; **Fig. 15–12** Don Figer (STScI) and NASA; **Fig. 15–13** Image processing at the Naval Research Laboratory using DoD High Performance Computing Resources; produced by N. E. Kassim, D. S. Briggs, T. J. W. Lazio, T. N. LaRosa, J. Imamura, and S. D. Hyman; original data from the NRAO Very Large Array, courtesy of A. Pedlar, K. Anantharamiah, M. Goss, and R. Ekers; **Fig. 15–14** Neil E. Killeen (CSIRO, ATNF, Australia) and K. Y. Lo, U. Illinois at Urbana-Champaign, with the VLA of NRAO; **Fig. 15–15** Mark Morris (UCLA), Farhad Yusef-Zadeh (Northwestern U.), and Don Chance (STScI), with the VLA of NRAO; **Fig. 15–16** Lund Observatory, Sweden; **Fig. 15–17** M. Hauser (STScI) and NASA; **Fig. 15–18** EGRET Science Team/NASA, courtesy of Carl Fichtel, NASA/Goddard Space Flight Center; **Fig. 15–19** NASA and the BATSE Science Team, courtesy of Gerald G. Fishman, NASA/Marshall Space Flight Center; **Fig. 15–20AB** Jay M. Pasachoff; **Fig. 15–21** © 1981 NOAO/CTIO, photo by Gabriel Martin; **Fig. 15–22** after Agris Kalnajs; **Fig. 15–24A** © 1977 Anglo-Australian Observatory, photograph by David Malin; **Fig. 15–24B** Jeff Hester (Arizona State U.) and NASA; **Fig. 15–27** Dap Hartmann, Harvard-Smithsonian Center for Astrophysics, and W. B. Burton at Leiden Obs.; see their *Atlas of Galactic Neutral Hydrogen*, Cambridge U. Press, 1997; **Fig. 15–29** Dan P. Clemens (Boston U.), D. Sanders (U.

Hawaii), and N. Scoville (Caltech); **Fig. 15–30AB** Ian Gatley (NOAO); **Fig. 15–31** C. R. O'Dell (Rice U.) and NASA; **Fig. 15–32** after Ben Zuckerman (UCLA); **Fig. 15–33** WFPC2 image: C. Robert O'Dell and Shui Kwan Wong (Rice U.) and NASA; NICMOS image: Rodger Thompson, Susan Stolovy, and Marcia Rieke (U. Arizona), Glenn Schneider (Steward Obs., U. Arizona), Edwin Erickson (SETI Institute/NASA Ames Research Center), and David Axon (STScI); and NASA; **Fig. 15–34** Jay M. Pasachoff; **Fig. 15–35** Jay M. Pasachoff.

Chapter 16—Opener European Southern Observatory; **Fig. 16–1A** Jack Newton; **Fig. 16–1B** © National Astronomical Observatory of Japan, Subaru Telescope. All rights reserved; **Fig. 16–2** U. of California Observatories/Lick Observatory; **Fig. 16–3** Tim Puckett; **Fig. 16–4** © Tony and Daphne Hallas, Astro Photo; **Fig. 16–5** From the historical collection of the Observatories of the Carnegie Institution of Washington; **Fig. 16–6B** Ẑeljko Ivezić and Robert Lupton for the Sloan Digital Sky Survey Collaboration; **Fig. 16–7A** © 1991 Richard J. Wainscoat/Institute for Astronomy, U. Hawaii; **Fig. 16–7B** European Southern Observatory; **Fig. 16–8** European Southern Observatory; **Fig. 16–9** ESA/ISO/ISOPHOT and M. Haas *et al.*; **Fig. 16–10A** © 1987 Anglo-Australian Observatory, photograph by David Malin; **Fig. 16–10B** Tod R. Lauer (NOAO), and NASA; **Fig. 16–11AB** © 1984 Royal Observatory Edinburgh/Anglo-Australian Observatory, photograph from original U.K. Schmidt plates by David Malin; **Fig. 16–12A** John Hibbard (NRAO) at the VLA of NRAO; **Fig. 16–12B** Brad Whitmore (STScI) and NASA; **Fig. 16–13** Curt Struck and Philip Appleton (Iowa State U.), Kirk Borne (Hughes STX Corp.), and Ray Lucas (STScI), and NASA; (NRAO) at the VLA of NRAO; **Fig. 16–14** European Southern Observatory; **Fig. 16–15** black-and-white image from Palomar Observatory, Caltech; color HST image from Hui Yang (University of Illinois) and NASA; **Fig. 16–16** © 1987 Anglo-Australian Observatory, photograph by David Malin; **Fig. 16–17** © 1981 Royal Observatory Edinburgh/Anglo-Australian Observatory, photograph from original U.K. Schmidt plates by David Malin; **Fig. 16–18** NOAO/Nigel Sharp; **Fig. 16–19** Rudolph Schild, Smithsonian Astrophysical Observatory; **Fig. 16–20** NASA/Chandra X-ray Observatory Center/SAO; **Fig. 16–21A** John Huchra, Margaret Geller, and V. de Lapparent, © 1989 Smithsonian Astrophysical Observatory; **Fig. 16–21B** John Huchra and Margaret Geller, © Smithsonian Astrophysical Observatory; **Fig. 16–23** Vera Rubin and Mark Godfrey; **Fig. 16–24AB** Dana Berry, STScI; **Fig. 16–25A** NASA and the Early Release Observations Team: Andrew Fruchter, Sylvia Baggett, and Zoltan Levay (Space Telescope Science Institute) and Richard Hook (Space Telescope—European Coordinating Facility); **Fig. 16–25B** W. N. Colley and E. Turner (Princeton U.) and J. A. Tyson (Bell Labs, Lucent Technologies, and NASA; **Fig 16–27A** National Academy of Sciences; **Fig. 16–27B** From E. Hubble and M. L. Humason, *Astrophysical Journal* 74, p. 77, 1931, courtesy of U. Chicago Press; **Fig. 16–28** Robert Williams and the Hubble Deep Field Team (STScI), and NASA; **Fig. 16–29** Robert Williams and the

Hubble Deep Field–South Team (STScI), and NASA; **Fig. 16–30** Robert Williams and the Hubble Deep Field Team (STScI), and NASA; with redshifts measured at the Keck Telescope, compiled by M. Dickinson and Z. Levay; **Fig. 16–31AB** Arjun Dey (NOAO) and Hyron Spinrad (U. California Berkeley); **Fig. 16–32** NASA and the Astro-1 (UIT) mission; **Fig. 16–33** Mark Dickinson (STScI) and NASA; **Fig. 16–34** Rogier Windhorst and Sam Pascarelle (Arizona State U.), and NASA; **Fig. 16–35** Joshua Barnes, Institute for Astronomy, U. Hawaii, and Lars Hernquist, Harvard U.; both then at Institute for Advanced Study; **Fig. 16–36** Virgo Consortium, U. of Durham.

Chapter 17—Opener John N. Bahcall and Sofia Kirhakos (Institute for Advanced Study), and Donald P. Schneider (Pennsylvania State U.), and NASA; **Fig. 17–1** Chien Peng (U. Arizona), Alex Filippenko (U. California Berkeley), and NASA; **Fig. 17–2** Credit: Radio sky: NRAO/AUI, observers James J. Condon, John J. Proderick, and Goerge A. Seielstad; optical image from Laird Thompson, U. Illinois; *Astrophys J.* 279, L47 (1984), courtesy of U. Chicago Press; (*lower inset*) Alan Stockton, U. Hawaii; (*upper inset*) Fernando Cabrera Guerra and NASA/ESA; **Fig. 17–3** William C. Keel (U. Alabama), Karl-Heinz Mack, Dayton Jones (JPL), and Rick Perley (NRAO); **17–4A:** © 1980 Anglo-Australian Observatory; **Fig. 17–4B:** NASA/Chandra X-ray Observatory Center/Smithsonian Astrophysical Observatory; **Fig. 17–5** Alex Filippenko and Thomas Matheson, U. California, Berkeley; **Fig. 17–6** Palomar Observatory, Caltech; **Fig. 17–7** Herman-Josef Roeser (ESO NTT), Herman-Josef Roeser and NASA (HST), and Palomar Observatory, Caltech (POSS); **Fig. 17–8** Stephen P. Unwin, Caltech; **Fig. 17–9** Maarten Schmidt/Palomar Observatory/Caltech photograph; **Fig. 17–11** Dan Stern (JPL), Hyron Spinrad (U. California Berkeley), *et al.*; **Fig. 17–12** T. J. Balonek *et al.*, Colgate U., Foggy Bottom Observatory; **Fig. 17–15** NASA/Chandra X-ray Observatory Center/SAO; **Fig. 17–16** Richard Pogge, Ohio State U.; **Fig. 17–17AB** John N. Bahcall and Sofia Kirhakos (Institute for Advanced Study), and Donald P. Schneider (Pennsylvania State U.), and NASA; **Fig. 17–18** Tony Tyson (Bell Labs, Lucent Technologies), W. A. Baum (Lowell Obs.) and Tobias J. N. Kreidl (Northern Arizona U. and Lowell Obs.); **Fig. 17–19** Joe Miller and Andy Sheinis, U. California, Santa Cruz; **Fig. 17–20** Based on work of Maarten Schmidt, Caltech; **Fig. 17–21** John N. Bahcall and Sofia Kirhakos (Institute for Advanced Study), Donald P. Schneider (Pennsylvania State U.), Mike Disney (U. of Wales), and NASA; **Fig. 17–22** European Southern Observatory; from an ASP slide set by William C. Keel, U. Alabama; **Fig. 17–23** Lincoln J. Greenhill, Harvard U., and Holland Ford, STScI/Johns Hopkins U.; **Figs 17–24A** and **17–25** Holland Ford, STScI/Johns Hopkins U.; Richard Harms, Applied Research Corp.; Zlatan Tsvetanov, Arthur Davidsen, and Gerard Kriss, Johns Hopkins U.; Ralph Bohlin and George Hartig, STScI; Linda Dressel and Ajay K. Kochar, Applied Research Corp.; and Bruce Margon, U. Washington; NASA/ESA/STScI; **Fig. 17–24B** Duccio Macchetto, NASA/ESA; **Fig. 17–26** Andrea Ghez, *et al.*, U. California, Los Angeles; **Fig. 17–27A** Stephen C. Unwin,

JPL/Caltech, and Ann Wehrle, IPAC/JPL. **Fig. 17–27B** VLA of NRAO; observers; Shudong Zhou (U. Illinois), John A. Biretta (STScI), and Frazer N. Owen (NRAO); **Fig. 17–28** Richard Pogge, Ohio State U., and Alex Filippenko, U. California at Berkeley; **Fig. 17–29** Zlatan Tsvetanov (Johns Hopkins U.), and Clive Tadhunter (U. Sheffield); **Fig. 17–30** William C. Keel (U. Alabama), Michael Rauch (Caltech), and NASA; **Fig. 17–31A** William C. Keel (U. Alabama) and NASA; **Fig. 17–31B,C:** Alan Stockton, U. Hawaii; **Fig. 17–33A** Jacqueline N. Hewitt (MIT), *et al.*; **Fig. 17–33B** Pierre Magain, *et al.*, observations at ESO; **Fig. 17–33C** NASA and ESA; **Fig. 17–33C** Kavan Ratnatunga (Johns Hopkins University), and NASA.

Chapter 18—Opener © Andrew Perala; **Fig. 18–2** Jay M. Pasachoff; **Fig. 18–4** images from N. A. Sharp (NOAO); computer spectra models from Ned Wright (UCLA), from B. Zuckerman and M. Malkan, eds., *The Origin and Evolution of the Universe*, © 1996: Jones & Bartlett Publishers, Sudbury, MA, www.jbpub/com. Reprinted with permission; **Fig. 18–9** based on a figure by Ned Wright (UCLA), from B. Zuckerman and M. Malkan, *The Origin and Evolution of the Universe*, © 1996: Jones & Bartlett Publishers, Sudbury, MA www.jbpub.com. Reprinted with permission; **Fig. 18–10** Michael Rich, Kenneth Mighell, James D. Neill (Columbia U.), Wendy Freedman (Carnegie Observatories), and NASA; **Fig. 18–12** © 1986 Anglo-Australian Observatory, photograph by David Malin; **Fig. 18–13** Andrew Fraknoi, Astronomical Society of the Pacific; **Fig. 18–14** Wendy L. Freedman, Observatories of the Carnegie Institution of Washington, and NASA; **Fig. 18–15** J. Trauger (JPL) and NASA; **Fig. 18–16** © 1995. Reprinted with permission of *Discover Magazine*; **Fig. 18–17** © 1987 Royal Observatory Edinburgh/Anglo-Australian Observatory, Johns Hopkins University, photography by David Malin from original U.K. Schmidt plates; **Fig. 18–18** Adam Riess (STScI); **Fig. 18–25** Richard Griffiths (Johns Hopkins U.); and the Medium Deep Survey Team, and NASA; **Fig. 18–27** Brian P. Schmidt (Australian National U.) and the High-z Supernova Search Team, and NASA; **Fig. 18–28** Peter Garnavich, Notre Dame (then Harvard-Smithsonian Center for Astrophysics), the High-z Supernova Search Team, and NASA; **Fig. 18–30** Alex Filippenko (U. California Berkeley), Adam Riess (STScI), and the High-z Supernova Search Team.

Chapter 19—Opener Robert Williams and the Hubble Deep Field Team (STScI), and NASA; **Fig. 19–1** Hubble Space Telescope WFPC Team and NASA, courtesy of William A. Baum, U. Washington; **Fig. 19–2** Clemson U. and Donald D. Clayton (Clemson U.); **Fig. 19–3** Bell Laboratories, Lucent Technologies; **Fig. 19–4** NASA COBE Science Team, courtesy of Nancy Boggess; **Fig. 19–5** NASA COBE Science Team, special graphing courtesy of E. S. Cheng; **Fig. 19–7** NASA COBE Science Team; **Fig. 19–8** Max Tegmark, Angelica de Oliveira-Costa, Marc Devlin, Barth Netterfield, Lyman Page, and Ed Wollack, in *Astrophys. J.* 474, L77–80, 1996; **Fig. 19–9A** Andrew Lange (Caltech), Paolo de Bernardis (U. Rome), and the BOOMERANG (Balloon Observations of Millimetric Extra-

References to illustrations, either photographs or drawings, are in italics. References to tables are followed by t. References to Appendices are prefaced by A.

Significant initial numbers followed by letters are alphabetical under their spellings, for example, 21 cm is alphabetized as *twenty-one*. Less important initial numbers are ignored in alphabetizing. For example, 3C 273 appears at the beginning of the Cs. M1 appears at the beginning of the Ms. Greek letters are alphabetized under their English equivalents.

SPRING SKY

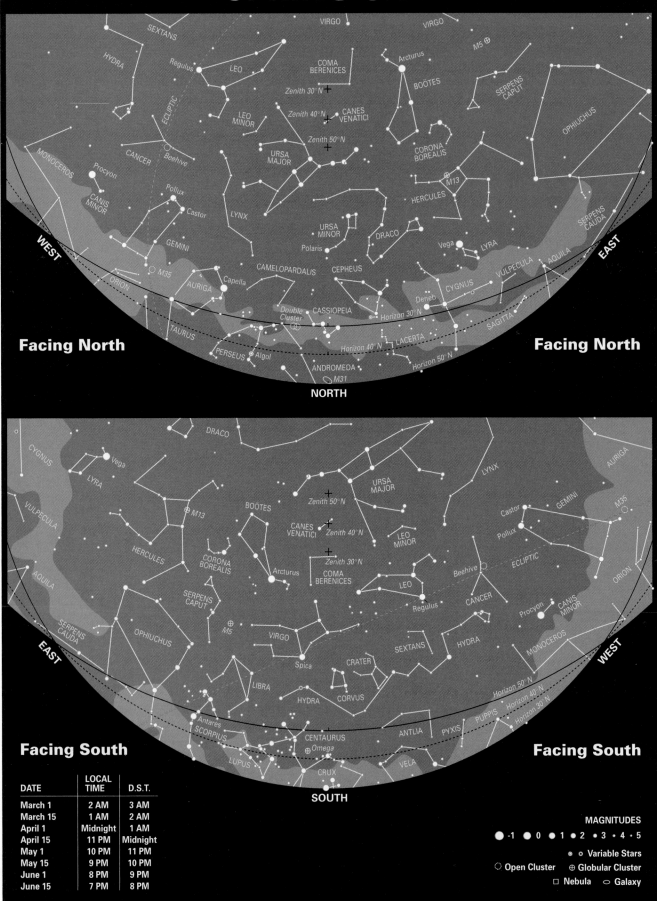

Facing North

NORTH

Facing South

SOUTH

DATE	LOCAL TIME	D.S.T.
March 1	2 AM	3 AM
March 15	1 AM	2 AM
April 1	Midnight	1 AM
April 15	11 PM	Midnight
May 1	10 PM	11 PM
May 15	9 PM	10 PM
June 1	8 PM	9 PM
June 15	7 PM	8 PM

MAGNITUDES

-1 0 1 2 3 4 5

Variable Stars
Open Cluster   Globular Cluster
Nebula   Galaxy

MAP BY WIL TIRION; FOR JAY M. PASACHOFF